Physiological Imaging
of the Brain
with PET

Physiological Imaging of the Brain with PET

Edited by

Albert Gjedde

PET Center
Aarhus University Hospital
Aarhus, Denmark

Søren B. Hansen

PET Center
Aarhus University Hospital
Aarhus, Denmark

Gitte M. Knudsen

University Hospital Rigshospitalet
Neurobiology Research Unit
Copenhagen, Denmark

Olaf B. Paulson

University Hospital Rigshospitalet
Neurobiology Research Unit
Copenhagen, Denmark

ACADEMIC PRESS

A Harcourt Science and Technology Company

San Diego   San Francisco   New York   Boston   London   Sydney   Tokyo

Front cover photograph: The common pattern of most impaired cerebral metabolism characterizing vegetative state patients. (For more details, see Chapter 48, Figure 1.)

This book is printed on acid-free paper. ∞

Academic Press
A Harcourt Science and Technology Company
525 B Street, Suite 1900, San Diego, California 92101-4495, USA
http://www.academicpress.com

Academic Press
Harcourt Place, 32 Jamestown Road, London NW1 7BY, UK
http://www.academicpress.com

Library of Congress Catalog Card Number: 00-104310

International Standard Book Number: 0-12-285751-8

PRINTED IN THE UNITED STATES OF AMERICA
00   01   02   03   04   05   EB   9   8   7   6   5   4   3   2   1

Contents

SECTION

III

SPECIFIC COMPARTMENT KINETICS

SECTION

IV

PARAMETRIC IMAGING

SECTION

X

PHARMACOLOGY AND EXPERIMENTAL TOMOGRAPHY: THE FUTURE

Contributors

The numbers in parentheses indicate the pages on which the authors' contributions begin.

Junko Abe (353) Department of Rehabilitation, Nagoya City Rehabilitation Center, 1-2, Mikanyama, Mizuho-ku, Nagoya, 467-8622 Japan

Anissa Abi-Dargham (219, 249) Departments of Psychiatry and Radiology, Columbia University College of Physicians and Surgeons, and Division of Brain Imaging, New York, New York 10032; Department of Neuroscience, New York State Psychiatric Institute, New York, New York 10032

A. J. Abrunhosa (51) Department of Biophysics, IBILI, Faculty of Medicine, Az. de Santa Comba, Celas, 3000-354 Coimbra, Portugal

Joel Aerts (319, 329) Cyclotron Research Center, University of Liège–Sart Tilman, 4000 Liège, Belgium

Jon R. Anderson (305) PET Imaging Service, Minneapolis Veterans Affairs Medical Center, Minneapolis, Minnesota 55417

Itzhak Angel (347) D-Pharm Ltd., Kiryat Weizmann Science Park, Rehovot 76123, Israel

John Ashburner (45) Functional Imaging Laboratory, Queen Square, London, United Kingdom

M.-C. Asselin (73) Department of Physics and Astronomy, McMaster University, Hamilton, Ontario, Canada L8S 4K1

John Aston (45) MRC Cyclotron Unit, Hammersmith Hospital, London W12 0HS, United Kingdom

A. M. Aupée (369) INSERM U320, Cyceron and Department of Neurology, University of Caen, Caen 14074, France

Richard B. Banati (361) MRC Cyclotron Unit, Imperial College School of Medicine, Hammersmith Hospital, London W12 0NN, United Kingdom

J. C. Baron (369) INSERM U320, Cyceron and Department of Neurology, University of Caen, Caen 14074, France

Jorge R. Barrio (193) Departments of Medical and Molecular Pharmacology, UCLA School of Medicine, Los Angeles, California 90095

Dirk Bender (237) PET Center, Aarhus University Hospital, 8000 Aarhus C, Denmark

J. R. Bergman (179) Radiochemistry Laboratory, Turku PET Centre, University of Turku, FIN-20521 Turku, Finland

Marvin Bergsneider (159) Division of Neurosurgery, Department of Surgery, UCLA School of Medicine, Los Angeles, California 90095

A. Bertoldo (165) Department of Electronics and Informatics, University of Padova, 35131 Padova, Italy

Gunnar Blomqvist (273) Uppsala University PET Center, UAS, 75185 Uppsala, Sweden; INSERM U334, Service Hospitalier Frédéric Joliot, CEA/DSV, 91401 Orsay cedex, France

P. M. Bloomfield (73) MRC Cyclotron Unit, Hammersmith Hospital, London W12 0HS, United Kingdom

F. Brady (51) MRC Cyclotron Unit, Hammersmith Hospital, London W12 0NN, United Kingdom

Matthew Brett (29) MRC Cyclotron Unit, Imperial College School of Medicine, Hammersmith Hospital, London W12 0NN, United Kingdom

David J. Brooks (361) MRC Cyclotron Unit, Imperial College School of Medicine, Hammersmith Hospital, London W12 0NN, United Kingdom

Theodore Bryan (105) Division of Nuclear Medicine, Department of Internal Medicine, University of Michigan School of Medicine, Ann Arbor, Michigan 48109

Elizabeth R. Butch (105) Department of Chemistry, Eastern Michigan University, Ypsilanti, Michigan 48197

Annachiara Cagnin (361) MRC Cyclotron Unit, Imperial College School of Medicine, Hammersmith Hospital, London W12 0NN, United Kingdom

Weidong Cai (35, 147) Biomedical and Multimedia Information Technology (BMIT) Group, Basser Department of Computer Science, The University of Sydney, Sydney, New South Wales 2006, Australia

Richard E. Carson (205, 229) PET Department, National Institutes of Health, Bethesda, Maryland 20892

Grace Chan (127, 131) UBC/TRIUMF PET Group, Vancouver, British Columbia, V6T 2A3 Canada

Michael A. Channing (205) PET Department, National Institutes of Health, Bethesda, Maryland 20892

Diane C. Chugani (381) Departments of Radiology and Pediatrics, Children's Hospital of Michigan PET Center, Detroit, Michigan 48201

Harry T. Chugani (381) Departments of Radiology and Pediatrics, and Neurology, Children's Hospital of Michigan PET Center, Detroit, Michigan 48201

C. Cobelli (165) Department of Electronics and Informatics, University of Padova, 35131 Padova, Italy

C. Crouzel (273) INSERM U334, Service Hospitalier Frédéric Joliot, CEA/DSV, 91401 Orsay cedex, France

Paul Cumming (257) PET Center, Aarhus University Hospital, 8000 Aarhus C, Denmark

Vincent J. Cunningham (29, 45, 73, 361) MRC Cyclotron Unit, Imperial College School of Medicine, Hammersmith Hospital, London W12 0NN, United Kingdom

Christian Degueldre (319, 329) Cyclotron Research Center, University of Liège–Sart Tilman, 4000 Liège, Belgium

V. de la Sayette (369) INSERM U320, Cyceron and Department of Neurology, University of Caen, Caen 14074, France

Guy Del Fiore (319) Cyclotron Research Center, University of Liège–Sart Tilman, 4000 Liège, Belgium

J. J. de Lima (51) Department of Biophysics, IBILI, Faculty of Medicine, Az. de Santa Comba, Celas, 3000-354 Coimbra, Portugal

B. Desgranges (369) INSERM U320, Cyceron and Department of Neurology, University of Caen, Caen 14074, France

Vijay Dhawan (3) Department of Neurology, North Shore University Hospital, New York University School of Medicine, New York, New York 10003

L. Di Giamberardino (273) INSERM U334, Service Hospitalier Frédéric Joliot, CEA/DSV, 91401 Orsay cedex, France

I. Doignon (273) INSERM U334, Service Hospitalier Frédéric Joliot, CEA/DSV, 91401 Orsay cedex, France

D. J. Doudet (127) Neurodegenerative Disorders Centre, University of British Columbia, Vancouver, British Columbia V6T 2B5 Canada

Wayne C. Drevets (211) Department of Psychiatry and Radiology, University of Pittsburgh School of Medicine, Pittsburgh, Pennsylvania 15213

William C. Eckelman (205, 229) PET Department, National Institutes of Health, Bethesda, Maryland 20892

David Eidelberg (3) Department of Neurology, North Shore University Hospital, New York University School of Medicine, New York, New York 10003

F. Eustache (369) INSERM U320, Cyceron and Department of Neurology, University of Caen, Caen 14074, France

Alan Evans (3) McConnell Brain Imaging Center, Montreal Neurological Institute, McGill University, Montreal, Quebec, H3A 2B4 Canada

Marie-Elisabeth Faymonville (319, 329) Department of Anesthesiology and Intensive Care Medicine, Centre Hospitalier Universitaire–Sart Tilman, 4000 Liège, Belgium

Dagan Feng (35, 121, 147) Biomedical and Multimedia Information Technology (BMIT) Group, Basser Department of Computer Science, The University of Sydney, Sydney, New South Wales 2006, Australia

Georges Franck (329) Department of Neurology, Centre Hospitalier Universitaire–Sart Tilman, 4000 Liège, Belgium

K. A. Frey (197) Division of Nuclear Medicine, University of Michigan, Ann Arbor, Michigan 48109

Kiyoshi Fukushi (139) Division of Advanced Technology for Medical Imaging, National Institute of Radiological Sciences, Chiba-shi, Chiba 263, Japan

Michael J. Fulham (121) Department of PET and Nuclear Medicine, Royal Prince Alfred Hospital, Camperdown, Sydney, New South Wales 2050, Australia

Roger Fulton (35, 147) Department of PET and Nuclear Medicine, Royal Prince Alfred Hospital, Camperdown, Sydney, New South Wales 2050, Australia

Masami Futatsubashi (377) Central Research Laboratory, Hamamatsu Photonics K.K., Hamakita, Japan

Antony Gee (237) PET Center, Aarhus University Hospital, 8000 Aarhus C, Denmark

Mehran Ghaemi (335, 341) Max-Planck-Institut für neurologische Forschung, D-50931 Cologne, Germany

Morad Ghaemi (335) Klinik für Neurologie der Universitat zu Köln, D-50924 Cologne, Germany

Albert Gjedde (237, 257) PET Center, Aarhus University Hospital, 8000 Aarhus C, Denmark

Serge Goldman (329) PET/Biomedical Cyclotron Unit, ULB Erasme, Brussels, Belgium

Michael M. Graham (115) Department of Nuclear Medicine, University of Iowa, Iowa City, Iowa 52242

Martin Grond (335) Klinik für Neurologie der Universitat zu Köln, D-50924 Cologne, Germany

Roger N. Gunn (45, 73, 171, 361) MRC Cyclotron Unit, Hammersmith Hospital, London W12 0HS, United Kingdom

Neeraja Gunupudi (105) Department of Chemistry, Eastern Michigan University, Ypsilanti, Michigan 48197

Ningning Guo (249) Department of Psychiatry, Columbia University College of Physicians and Surgeons, and Division of Brain Imaging, New York, New York 10032; Department of Neuroscience, New York State Psychiatric Institute, New York, New York 10032

M. T. Haaparanta (179) Radiochemistry Laboratory, Turku PET Centre, University of Turku, FIN-20521 Turku, Finland

Lars Kai Hansen (9) Section for Digital Signal Processing, Department of Mathematical Modelling, Technical University of Denmark, DK-2800 Lyngby, Denmark

Søren Baarsgard Hansen (23, 237) PET Center, Aarhus University Hospital, 8000 Aarhus C, Denmark

Kazuo Hashikawa (309) Department of Nuclear Medicine, Osaka University Graduate School of Medicine, 2-2 Yamada-oku, Suita, 565-0871 Osaka, Japan

Walter F. Haupt (335) Klinik für Neurologie der Universitat zu Köln, D-50924 Cologne, Germany

Wolf-Dieter Heiss (281, 335, 341) Max-Planck-Institut für neurologische Forschung, D-50931 Cologne, Germany

Karl Herholz (281) Max-Planck-Institut für neurologische Forschung, D-50931 Cologne, Germany

Peter Herscovitch (205, 229) PET Department, National Institutes of Health, Bethesda, Maryland 20892

J. Dee Higley (229) Laboratory of Clinical Studies, DICBR, National Institute on Alcohol Abuse and Alcoholism, Bethesda, Maryland 20892

Ella Hirani (171) MRC Cyclotron Unit, Hammersmith Hospital, London W12 0HS, United Kingdom

James E. Holden (127, 131) Department of Medical Physics, Medical Sciences Center, University of Wisconsin, Madison, Wisconsin 53706

Daniel P. Holt (265) Pet Facility, Department of Radiology, University of Pittsburgh School of Medicine, Pittsburgh, Pennsylvania 15213

Daniel Hommer (229) Laboratory of Clinical Studies, DICBR, National Institute on Alcohol Abuse and Alcoholism, Bethesda, Maryland 20892

Masatsugu Hori (309) Department of Medicine and Therapeutics, Osaka University Graduate School of Medicine, 2-2 Yamada-oku, Suita, 565-0871 Osaka, Japan

Sung-Cheng Huang (159, 193) Division of Nuclear Medicine and Biophysics, Department of Molecular and Medical Pharmacology, UCLA School of Medicine, Los Angeles, California 90095

Susan P. Hume (171) MRC Cyclotron Unit, Hammersmith Hospital, London W12 0HS, United Kingdom

John L. Humm (15) Medical Physics Department, Memorial Sloan Kettering Cancer Center, New York, New York 10021

Dah-Ren Hwang (249) Department of Psychiatry, Columbia University College of Physicians and Surgeons, and Division of Brain Imaging, New York, New York 10032; Department of Neuroscience, New York State Psychiatric Institute, New York, New York 10032

Akihiko Iida (353) Department of Radiology, Nagoya City Rehabilitation Center, 1-2, Mikanyama, Mizuho-ku, Nagoya 467-8622, Japan

Yoko Ikoma (59, 91, 109) Uchiyama Laboratory, Department of Electronics, Information and Communication Engineering, School of Science and Engineering, Waseda University, 55N-06-10, 3-4-1, Okubo, Shinjuku-ku, Tokyo 169-8555, Japan

Toshiaki Irie (139) Division of Advanced Technology for Medical Imaging, National Institute of Radiological Sciences, Chiba-shi, Chiba 263, Japan

Kenji Ishii (109) Positron Medical Center, Tokyo Metropolitan Institute of Gerontology, 35-2 Sakaecho, Itabashi-ku, Tokyo 173-0015, Japan

Koichi Ishizu (237) PET Center, Aarhus University Hospital, 8000 Aarhus C, Denmark

Kiyoshi Iwabuchi (389) Department of Neurology and Psychiatry, Kanagawa Rehabilitation Center, 516 Narusawa, Atsugi-City, Kanagawa 243-0121, Japan

Nathalie Janssens (319) Department of Anesthesiology and Intensive Care Medicine, Centre Hospitalier Universitaire–Sart Tilman, 4000 Liège, Belgium

A. Jobert (273) INSERM U334, Service Hospitalier Frédéric Joliot, CEA/DSV, 91401 Orsay cedex, France

Terry Jones (51, 361) MRC Cyclotron Unit, Hammersmith Hospital, London W12 0NN, United Kingdom

Csaba Juhasz (381) Department of Radiology, Children's Hospital of Michigan PET Center, Detroit, Michigan 48201

Hidehiro Kabasawa (353) Department of Rehabilitation, Nagoya City Rehabilitation Center, 1-2, Mikanyama, Mizuho-ku, Nagoya 467-8622, Japan

Tatsushi Kamiya (389) Division of Neurology, Second Department of Internal Medicine, Nippon Medical School, 1-1-5 Sendagi, Bunkyo-ku, Tokyo 113-8603, Japan

Toshihiko Kanno (377) Positron Medical Center, Hamamatsu Medical Center, 5000 Hirakuchi, Hamakita 434-0041, Japan

H. Karbe (341) Max-Planck-Institut für neurologische Forschung, D-50931 Cologne, Germany; Neurobologische Universitätsklinik Köln, 50931 Cologne, Germany.

Yasuo Katayama (389) Division of Neurology, Second Department of Internal Medicine, Nippon Medical School, 1-1-5 Sendagi, Bunkyo-ku, Tokyo 113-8603, Japan

Lawrence S. Kegeles (219) Departments of Psychiatry and Radiology, Columbia University College of Physicians and Surgeons, and Division of Brain Imaging, New York, New York 10032; Department of Neuroscience, New York State Psychiatric Institute, New York, New York 10032

J. Kessler (341) Max-Planck-Institut für neurologische Forschung, D-50931 Cologne, Germany; Neurobologische Universitätsklinik Köln, 50931 Cologne, Germany.

Imtiaz Khan (171) MRC Cyclotron Unit, Hammersmith Hospital, London W12 0HS, United Kingdom

Michael R. Kilbourn (105, 197) Division of Nuclear Medicine, Department of Internal Medicine, University of Michigan School of Medicine, Ann Arbor, Michigan 48109

Yuichi Kimura (91, 109) Positron Medical Center, Tokyo Metropolitan Institute of Gerontology, 35-2 Sakaecho, Itabashi-ku, Tokyo 173-0015, Japan

Paul E. Kinahan (211) PET Facility, Department of Radiology, University of Pittsburgh School of Medicine, Pittsburgh, Pennsylvania 15213

R. A. Koeppe (197) Division of Nuclear Medicine, University of Michigan, Ann Arbor, Michigan 48109

Alex Kozak (347) D-Pharm Ltd., Kiryat Weizmann Science Park, Rehovot 76123, Israel

Michael Krakovsky (347) D-Pharm Ltd., Kiryat Weizmann Science Park, Rehovot 76123, Israel

Kenneth A. Krohn (115) Division of Nuclear Medicine, Department of Radiology and Department of Neurology, University of Washington, Seattle, Washington 98195

Martin Krzywinski (131) UBC/TRIUMF PET Group, Vancouver, British Columbia, V6T 2A3 Canada

D. E. Kuhl (197) Division of Nuclear Medicine, University of Michigan, Ann Arbor, Michigan 48109

C. Lalevée (369) INSERM U320, Cyceron and Department of Neurology, University of Caen, Caen 14074, France

Bernard Lambermont (329) Department of Internal Medicine, Centre Hospitalier Universitaire–Sart Tilman, 4000 Liège, Belgium

Maurice Lamy (319, 329) Department of Anesthesiology and Intensive Care Medicine, Centre Hospitalier Universitaire–Sart Tilman, 4000 Liège, Belgium

Steven M. Larson (15) Nuclear Medicine Service, Memorial Sloan Kettering Cancer Center, New York, New York 10021

Marc Laruelle (65, 219, 249) Departments of Psychiatry and Radiology, Columbia University College of Physicians and Surgeons, and Division of Brain Imaging, New York, New York 10032; Department of Neuroscience, New York State Psychiatric Institute, New York, New York 10032

Steven Laureys (319, 329) Cyclotron Research Center, University of Liège–Sart Tilman, 4000 Liège, Belgium

K. L. Leenders (153, 295) Groningen NeuroImaging Program, University and University Hospital Groningen, 9700 RB Groningen, The Netherlands; Department of Physics, University of Surrey, Guildford, United Kingdom

Edward F. Leonard (15) Center for Biomedical Engineering, Columbia University, New York, New York 10027

Thomas K. Lewellen (115) Division of Nuclear Medicine, Department of Radiology and Department of Neurology, University of Washington, Seattle, Washington 98195

Jeanne M. Link (115) Division of Nuclear Medicine, Department of Radiology and Department of Neurology, University of Washington, Seattle, Washington 98195

Markku Linnoila (deceased) (229) Laboratory of Clinical Studies, DICBR, National Institute on Alcohol Abuse and Alcoholism, Bethesda, Maryland 20892

J.-S. Liow (289) Department of Radiology, University of Minnesota; PET Imaging Service, Veterans Affairs Medical Center, Minneapolis, Minnesota 55417

Brian J. Lopresti (211, 265) PET Facility, Department of Radiology, University of Pittsburgh School of Medicine, Pittsburgh, Pennsylvania 15213

S. Luthra (51) MRC Cyclotron Unit, Hammersmith Hospital, London W12 0NN, United Kingdom

André Luxen (319, 329) Cyclotron Research Center, University of Liège–Sart Tilman, 4000 Liège, Belgium

Yilong Ma (3) McConnell Brain Imaging Center, Montreal Neurological Institute, McGill University, Montreal, Quebec, H3A-2B4 Canada; Department of Neurology, North Shore University Hospital, New York University School of Medicine, New York, New York 10003

R. P. Maguire (153, 295) Groningen NeuroImaging Program, University and University Hospital Groningen, 9700 RB Groningen, The Netherlands; Department of Physics, University of Surrey, Guildford, United Kingdom

J. John Mann (219, 249) Departments of Psychiatry and Radiology, Columbia University College of Physicians and Surgeons, and Division of Brain Imaging, New York, New York 10032; Department of Neuroscience, New York State Psychiatric Institute, New York, New York 10032

Pierre Maquet (319, 329) Cyclotron Research Center, University of Liège–Sart Tilman, 4000 Liège, Belgium

Mitsuhito Mase (353) Department of Neurosurgery, Nagoya City University Medical School, 1-Kawasumi, Mizuho-ku, Nagoya 467-8602, Japan

Chester A. Mathis (211, 265) PET Facility, Department of Radiology and Pharmaceutical Sciences, University of Pittsburgh School of Medicine, Pittsburgh, Pennsylvania 15213

Masayasu Matsumoto (309) Department of Medicine and Therapeutics, Department of Neurology, Osaka University Graduate School of Medicine, 2-2 Yamada-oku, Suita, Osaka 565-0871, Japan

Takashi Matsumoto (353) Department of Neurosurgery, Nagoya City University Medical School, 1-Kawasumi, Mizuho-ku, Nagoya 467-8602, Japan

Osama R. Mawlawi (15, 65, 249) Brain Imaging Division, Departments of Psychiatry and Radiology, Columbia University, New York, New York 10032; Nuclear Medicine Service, Memorial Sloan Kettering Cancer Center, New York, New York 10021; Center for Biomedical Engineering, Columbia University, New York, New York 10027

J. M. Meador-Woodruff (197) Division of Nuclear Medicine, University of Michigan, Ann Arbor, Michigan 48109

Steven R. Meikle (121) Department of PET and Nuclear Medicine, Royal Prince Alfred Hospital, Camperdown, Sydney, New South Wales 2050, Australia

Carolyn Cidis Meltzer (265) Pet Facility, Department of Radiology, University of Pittsburgh School of Medicine, Pittsburgh, Pennsylvania 15213

Ernst Meyer (97) McConnell Brain Imaging Centre, Montreal Neurological Institute, McGill University, Montreal, Quebec, H3A-2B4 Canada

Masahiro Mishina (389) Division of Neurology, Second Department of Internal Medicine, Nippon Medical School, 1-1-5 Sendagi, Bunkyo-ku, Tokyo 113-8603, Japan

Peter Moldt (237) NeuroSearch, Pederstrupvej 93, 2750 Ballerup, Denmark

Gustave Moonen (319) Department of Neurology, Centre Hospitalier Universitaire–Sart Tilman, 4000 Liège, Belgium

Robert Y. Moore (265) Department of Neurology, University of Pittsburgh Medical Center, Pittsburgh, Pennsylvania 15213

J. J. Moreno-Cantú (289) Department of Radiology, University of Minnesota; PET Imaging Service, Veterans Affairs Medical Center, Minneapolis, Minnesota 55417

Ole Lajord Munk (23) PET Center, Aarhus University Hospital, 8000 Aarhus C, Denmark

Mark Muzi (115) Division of Nuclear Medicine, Department of Radiology and Department of Neurology, University of Washington, Seattle, Washington 98195

Otto Muzik (381) Department of Radiology, Children's Hospital of Michigan PET Center, Detroit, Michigan 48201

Ralph Myers (361) MRC Cyclotron Unit, Imperial College School of Medicine, Hammersmith Hospital, London W12 0NN, United Kingdom

Jouni Mykkänen (39) Department of Computer and Information Sciences, University of Tampere, 33014 Tampere, Finland

Yuri Nagano (353) Department of Rehabilitation, Nagoya City Rehabilitation Center, 1-2, Mikanyama, Mizuho-ku, Nagoya 467-8622, Japan

Hideki Nagatomo (389) Department of Neurology and Psychiatry, Kanagawa Rehabilitation Center, 516 Narusawa, Atsugi-City, Kanagawa 243-0121, Japan

Shin-Ichiro Nagatsuka (139) ADME/TOX Research Institute, Daiichi Pure Chemicals, Co., Ltd., 2117 Muramatsu, Tokai, Ibaraki 319-1198, Japan

C. Nahmias (73) Department of Nuclear Medicine, McMaster University Medical Centre, Hamilton, Ontario, Canada L8N 3Z5

Mohammad Namavari (193) Laboratory of Structural Biology and Molecular Medicine (DOE), UCLA School of Medicine, Los Angeles, California 90095

Hiroki Namba (139) Division of Advanced Technology for Medical Imaging, National Institute of Radiological Sciences, Chiba-shi, Chiba 263, Japan

Tadashi Nariai (301) Department of Neurosurgery, Tokyo Medical and Dental University, 1-5-45 Yushima, Bunkyouku, Tokyo 113, Japan

Peter Neelin (3) McConnell Brain Imaging Center, Montreal Neurological Institute, McGill University, Montreal, Quebec, H3A-2B4 Canada

Gerald Nestadt (257) Department of Psychiatry, Johns Hopkins University, Baltimore, Maryland 21205

Elsebet Østergaard Nielsen (237) NeuroSearch, Pederstrupvej 93, 2750 Ballerup, Denmark

Hiroshi Nishimura (309) Department of Tracer Kinetics, Osaka University Graduate School of Medicine, 2-2 Yamada-oku, Suita, Osaka 565-0871, Japan

Tsunehiko Nishimura (309) Department of Tracer Kinetics, Osaka University Graduate School of Medicine, 2-2 Yamada-oku, Suita, Osaka 565-0871, Japan

Shuji Nobezawa (377) Positron Medical Center, Hamamatsu Medical Center, 5000 Hirakuchi, Hamakita 434-0041, Japan

Masaru Nukata (309) Department of Medicine and Therapeutics, Osaka University Graduate School of Medicine, 2-2 Yamada-oku, Suita, Osaka 565-0871, Japan

Keiichi Oda (59) Positron Medical Center, Tokyo Metropolitan Institute of Gerontology, 1-1 Naka-cho Itabashi-ku, Tokyo 173-0022, Japan

Tetsuo Ogawa (353) Department of Rehabilitation, Nagoya City Rehabilitation Center, 1-2, Mikanyama, Mizuho-ku, Nagoya 467-8622, Japan

Vesa Juhani Oikonen (179, 187) Turku PET Centre, University of Turku, FIN-20521 Turku, Finland

Hiroyuki Okada (377) Central Research Laboratory, Hamamatsu Photonics K.K., Hamakita, Japan

Naohiko Oku (309) Department of Nuclear Medicine, Osaka University Graduate School of Medicine, 2-2 Yamada-oku, Suita, Osaka 565-0871, Japan

Jolanta Opacka-Juffry (171) MRC Cyclotron Unit, Hammersmith Hospital, London W12 0HS, United Kingdom

Leif Østergaard (237) Department of Neuroradiology, Aarhus University Hospital, 8000 Aarhus C, Denmark

Finbarr O'Sullivan (115) Department of Statistics, University College, Cork, Ireland

Yasuomi Ouchi (377) Positron Medical Center, Hamamatsu Medical Center, 5000 Hirakuchi, Hamakita 434-0041, Japan

S. Pappata (273) INSERM U334, Service Hospitalier Frédéric Joliot, CEA/DSV, 91401 Orsay cedex, France

Ramin Parsey (249) Department of Psychiatry, Columbia University College of Physicians and Surgeons, and Division of Brain Imaging, New York, New York 10032; Department of Neuroscience, New York State Psychiatric Institute, New York, New York 10032

Gunter Pawlik (281) Max-Planck-Institut für neurologische Forschung, D-50931 Cologne, Germany

Peter Alshede Philipsen (9) Section for Digital Signal Processing, Department of Mathematical Modelling, Technical University of Denmark, DK-2800 Lyngby, Denmark

Christophe Phillips (329) Cyclotron Research Center, University of Liège–Sart Tilman, 4000 Liège, Belgium; The

Wellcome Department of Cognitive Neurology, Institute of Neurology, WC1N3BG London, United Kingdom

Uwe Pietrzyk (281) Max-Planck-Institut für neurologische Forschung, D-50931 Cologne, Germany

Michael Polyak (347) D-Pharm Ltd., Kiryat Weizmann Science Park, Rehovot 76123, Israel

Peter Høst Poulsen (237) PET Center, Aarhus University Hospital, 8000 Aarhus C, Denmark

Julie C. Price (211, 265) PET Facility, Department of Radiology, University of Pittsburgh School of Medicine, Pittsburgh, Pennsylvania 15213

J. O. Rinne (179) Turku PET Centre, University of Turku, FIN-20521 Turku, Finland

David A. Rottenberg (289, 305) PET Imaging Service, Minneapolis Veterans Affairs Medical Center, Minneapolis, Minnesota 55417; Departments of Radiology and Neurology, University of Minnesota Medical School, Minneapolis, Minnesota 55455

Olivier Rousset (3) McConnell Brain Imaging Center, Montreal Neurological Institute, McGill University, Montreal, Quebec, H3A-2B4 Canada

Jobst Rudolf (335) Klinik für Neurologie der Universitat zu Köln, D-50924 Cologne, Germany; Max-Planck-Institut für neurologische Forschung, D-50931 Cologne, Germany

Ulla Ruotsalainen (39, 187) DMI/Signal Processing Laboratory, Tampere University of Technology, 33101 Tampere, Finland

Hanna Maria Ruottinen (179, 187) Department of Neurology, University of Turku, 20520 Turku, Finland

Thomas J. Ruth (127, 131) UBC/TRIUMF PET Group, Vancouver, British Columbia, V6T 2A3 Canada

M. Sajjad (289) PET Imaging Service, Veterans Affairs Medical Center, Minneapolis, Minnesota 55417

Masanobu Sakamoto (377) Department of Neurology, Hamamatsu Medical Center, Hamamatsu, Japan

Masaharu Sakoh (237) PET Center, Aarhus University Hospital, 8000 Aarhus C, Denmark

N. Satyamurthy (193) Laboratory of Structural Biology and Molecular Medicine (DOE), UCLA School of Medicine, Los Angeles, California 90095

Jørgen Scheel-Krüger (237) NeuroSearch, Pederstrupvej 93, 2750 Ballerup, Denmark

Bernard Schmall (229) PET Department, National Insitutes of Health Clinical Center, Bethesda, Maryland 20892

K. C. Schmidt (83) Laboratory of Cerebral Metabolism, National Institute of Mental Health, Bethesda, Maryland 20892

Alexander Schuster (281) Max-Planck-Institut für neurologische Forschung, D-50931 Cologne, Germany

Yujiro Seike (309) Department of Nuclear Medicine, Osaka University Graduate School of Medicine, 2-2 Yamadaoku, Suita, Osaka 565-0871, Japan

Michio Senda (59, 91, 109, 301, 389) Positron Medical Center, Tokyo Metropolitan Institute of Gerontology, 35-2 Sakaecho, Itabashi-ku, Tokyo, 173-0015, Japan

Chenggang Shen (381) Department of Radiology, Children's Hospital of Michigan PET Center, Detroit, Michigan 48201

Ann Shinn (249) Department of Psychiatry, Columbia University College of Physicians and Surgeons, and Division of Brain Imaging, New York, New York 10032; Department of Neuroscience, New York State Psychiatric Institute, New York, New York, 10032

Hitoshi Shinotoh (139) Division of Advanced Technology for Medical Imaging, National Institute of Radiological Sciences, Chiba-shi, Chiba 263, Japan

Susan E. Shoaf (229) Otsuka America Pharmaceutical, Inc., Rockville, Maryland 20850; Laboratory of Clinical Studies, DICBR, National Institute on Alcohol Abuse and Alcoholism, Bethesda, Maryland 20892

Kooresh Shoghi-Jadid (193) Departments of Medical and Molecular Pharmacology, UCLA School of Medicine, Los Angeles, California 90095

John J. Sidtis (305) Department of Neurology, University of Minnesota Medical School, Minneapolis, Minnesota 55455

Norman Simpson (249) Department of Psychiatry, Columbia University College of Physicians and Surgeons, and Division of Brain Imaging, New York, New York 10032; Department of Neuroscience, New York State Psychiatric Institute, New York, New York 10032

Mark Slifstein (65, 249) Departments of Psychiatry and Radiology, Columbia University College of Physicians and Surgeons, New York, New York 10032

Donald F. Smith (237) Institute for Basic Research in Psychiatry, Department of Biological Psychiatry, Psychiatric Hospital, Skovagervej 2, 8240 Risskov, Denmark

Gwenn S. Smith (265) Department of Radiology and Psychiatry, Hillside Hospital, Division of LIJ, Glen Oaks, New York 11004

Scott E. Snyder (105) Division of Nuclear Medicine, Department of Internal Medicine, University of Michigan School of Medicine, Ann Arbor, Michigan 48109

Jan Sobesky (335) Max-Planck-Institut für neurologische Forschung, D-50931 Cologne, Germany

O. H. Solin (179) Radiochemistry Laboratory, Turku PET Centre, University of Turku, FIN-20521 Turku, Finland

Vesna Sossi (127, 131) UBC/TRIUMF PET Group, Vancouver, British Columbia, V6T 2A3 Canada

Alexander M. Spence (115) Department of Neurology, University of Washington, Seattle, Washington 98195

N. M. Spyrou (153, 295) Groningen NeuroImaging Program, University and University Hospital Groningen, 9700 RB Groningen, The Netherlands; Department of Physics, University of Surrey, Guildford, United Kingdom

A. Jonathan Stoessl (127, 131) Neurodegenerative Disorders Centre, University of British Columbia, Vancouver, British Columbia V6T 2B5 Canada

David B. Stout (193) Departments of Medical and Molecular Pharmacology, UCLA School of Medicine, Los Angeles, California 90095

Stephen C. Strother (289, 305) PET Imaging Service, Minneapolis Veterans Affairs Medical Center, Minneapolis, Minnesota 55417; Departments of Radiology and Neurology, University of Minnesota Medical School, Minneapolis, Minnesota 55455

Brigitte Szelies (335) Klinik für Neurologie der Universitat zu Köln, D-50924 Cologne, Germany

Kazuhiro Takazawa (301) Department of Electronics, Information and Communication Engineering, School of Science and Engineering, Waseda University, 55N-06-10, Uchiyama Laboratory, 3-4-1, Okubo, Shinjuku-ku, Tokyo 169-8555, Japan

Shuji Tanada (139) Division of Advanced Technology for Medical Imaging, National Institute of Radiological Sciences, Chiba-shi, Chiba 263, Japan

Noriko Tanaka (139) Division of Advanced Technology for Medical Imaging, National Institute of Radiological Sciences, Chiba-shi, Chiba 263, Japan

B. Tavitian (273) INSERM U334, Service Hospitalier Frédéric Joliot, CEA/DSV, 91401 Orsay cedex, France

S. F. Taylor (197) Division of Nuclear Medicine, University of Michigan, Ann Arbor, Michigan 48109

Akiro Terashi (389) Division of Neurology, Second Department of Internal Medicine, Nippon Medical School, 1-1-5 Sendagi, Bunkyo-ku, Tokyo 1138603, Japan

Tadanori Teratani (309) Department of Tracer Kinetics, Osaka University Graduate School of Medicine, 2-2 Yamada-oku, Suita, Osaka 565-0871, Japan

Alexander Thiel (281, 341) Max-Planck-Institut für neurologische Forschung, D-50931 Cologne, Germany

Jussi Tohka (39) DMI/Signal Processing Laboratory, Tampere University of Technology, 33101 Tampere, Finland

Tatsuo Torizuka (377) Positron Medical Center, Hamamatsu Medical Center, 5000 Hirakuchi, Hamakita 434-0041, Japan

Paule-Joanne Toussaint (97) McConnell Brain Imaging Centre, Montreal Neurological Institute, McGill University, Montreal, Quebec, H3A-2B4 Canada

Hinako Toyama (59, 91, 109, 301) Medical Information Processing Office, Research Center of Charged Particle Therapy, National Institute of Radiological Sciences, Anagawa 4-9-1, Inago-ku, Chiba 263-8555, Japan

N. Turjanski (73) MRC Cyclotron Unit, Hammersmith Hospital, London W12 0HS, United Kingdom

Federico E. Turkheimer (29, 361) MRC Cyclotron Unit, Imperial College School of Medicine, Hammersmith Hospital, London W12 0NN, United Kingdom

Akihiko Uchiyama (59, 91) Uchiyama Laboratory, Department of Electronics, Information and Communication Engineering, School of Science and Engineering, Waseda University, 55N-06-10, 3-4-1, Okubo, Shinjuku-ku, Tokyo 169-8555, Japan

Masayuki Ueda (389) Department of Neurology, University of Tennessee College of Medicine, Memphis, Tennessee 38163; Division of Neurology, Second Department of Internal Medicine, Nippon Medical School, 1-1-5 Sendagi, Bunkyo-ku, Tokyo 113-8603, Japan

Koji Uemura (59, 91, 109, 301) Department of Electronics, Information and Communication Engineering, School of Science and Engineering, Waseda University, 55N-06-10, 3-4-1 Okubo, Shinjuku-ku, Tokyo, 169–8555, Japan

Ronald L. Van Heertum (219, 249) Departments of Psychiatry and Radiology, Columbia University College of Physicians and Surgeons, and Division of Brain Imaging, New York, New York 10032; Department of Neuroscience, New York State Psychiatric Institute, New York, New York 10032

F. Viader (369) INSERM U320, Cyceron and Department of Neurology, University of Caen, Caen 14074, France

Victor Villemagne (265) PET Facility, Department of Radiology, University of Pittsburgh School of Medicine, Pittsburgh, Pennsylvania 15213

Dimitris Visvikis (29) Institute of Nuclear Medicine, University College London Medical School, Middlesex Hospital, London, United Kingdom

Hans-Martin von Stockhausen (381) Max-Planck Institute for Neurological Research, D-50931 Cologne, Germany

Bik-kee Vuong (205) PET Department, National Institutes of Health, Bethesda, Maryland 20892

L. M. Wahl (73) Theoretical Biology, Institute for Advanced Study, Princeton, New Jersey 08540

Hiroshi Watabe (205) Department of Investigative Radiology, National Cardiovascular Center Research Institute, 5-7-1 Fujishiro-Dai, Suita 565-8565, Japan

Klaus Wienhard (281) Max-Planck-Institut für neurologische Forschung, D-50931 Cologne, Germany

Wendol A. Williams (229) Laboratory of Clinical Studies, DICBR, National Institute on Alcohol Abuse and Alcoholism, Bethesda, Maryland 20892

Dean F. Wong (257) Department of Radiology, Division of Nuclear Medicine, Johns Hopkins University, Baltimore, Maryland 21205

Koon-Pong Wong (121) Biomedical and Multimedia Information Technology (BMIT) Group, Basser Department of Computer Science, The University of Sydney, Sydney, New South Wales 2006, Australia

Kazuo Yamada (353) Department of Neurosurgery, Nagoya City University Medical School, 1-Kawasumi, Mizuho-ku, Nagoya 467-8602, Japan

Haruko Yamamoto (309) Department of Nuclear Medicine, Osaka University Graduate School of Medicine, 2-2 Yamada-oku, Suita, Osaka 565-0871, Japan

Randa E. Yee (193) Departments of Medical and Molecular Pharmacology, UCLA School of Medicine, Los Angeles, California 90095

Fuji Yokoi (257) Department of Radiology, Division of Nuclear Medicine, Johns Hopkins University, Baltimore, Maryland 21205

Etsuji Yoshikawa (377) Central Research Laboratory, Hamamatsu Photonics K.K., Hamakita, Japan

Yolanda Zea-Ponce (219) Departments of Psychiatry and Radiology, Columbia University College of Physicians and Surgeons, and Division of Brain Imaging, New York, New York 10032; Department of Neuroscience, New York State Psychiatric Institute, New York, New York 10032

Yun Zhou (159) Division of Nuclear Medicine and Biophysics, Department of Molecular and Medical Pharmacology, UCLA School of Medicine, Los Angeles, California 90095

Foreword

Positron emission tomography is undergoing an important transition. In some respects, the transition represents a return to the origins of PET. The original promise of PET was the opportunity to measure biochemical and biological variables in discrete anatomical locations of living organisms. Applied to the brain, PET promised anatomically accurate correlations among brain circulation and energy metabolism in functionally discrete conditions. As such, PET of the brain was an extension of the approaches introduced by Seymour Kety, Louis Sokoloff, and Niels A. Lassen. Borrowing from one element of the method of functional brain imaging developed by Lassen and his co-workers (the step-wise subtraction approach), PET went through a phase of statistics parameter mapping in which the maps did not reflect a well-defined biological entity but rather a meta-analytical macro-parameter of uncertain physiological or pathophysiological significance. To an increasing extent, these maps are now prepared by other techniques, freeing PET for the kinds of experiments for which it was originally developed, namely, biochemical studies involving measurements of picomolar concentration sensitivity and millimeter anatomical resolution. As a biological tool, PET now probes deeply into second messenger and transcription events.

This volume attests to the continuing refinement of the technological efforts to improve the picomolar sensitivity and the millimeter resolution of PET. The contributions of radiochemists and physicists contained in these pages are as essential to this improvement as the insights of the computational scientists, the statisticians, and the mathematicians. These contributors are joined by pharmacologists and kineticists who endeavor to make sense of the signals recorded by PET. This collaboration with biologists is impressively interdisciplinary.

No single scientific aim informs the interdisciplinary enterprise, of course, but a limited number of themes nonetheless underpin the work. The themes revolve around the attempt to understand the missing links between neurotransmission and brain energy metabolism on one hand and brain function on the other. If the BrainPET99 meeting and this volume of the transactions contribute to reaching this worthy goal, it is in no small part due to the quality of the interdisciplinary collaborations exemplified in these pages.

Albert Gjedde
Søren B. Hansen
Gitte M. Knudsen
Olaf B. Paulson

Acknowledgments

Local Editorial Assistance

Flemming Andersen
Dirk Bender
Paul Cumming
Erik H. Danielsen

Peter Johannsen
Susanne Keiding
Ron Kupers
Ole Lajord Munk

Anders Bertil Rodell
Donald F. Smith
Manouchehr Vafaee

Further Acknowledgments

We thank the BrainPET constituency, as well as the Society for Cerebral Blood Flow and Metabolism and its Ad Hoc BrainPET Committee: Richard Carson, Peter Herscovitch, Terry Jones, Iwao Kanno, Gitte Moos Knudsen, Adriaan Lammertsma, Steven Strother.
We also thank: Pia Farup, Dorthe Givard, Palle Monefeldt, and Karin Suffolk for excellent secretarial assistance.

International Scientific Advisory Board for Brain '99 and BrainPET99

N. J. Abbott
K. Abe
B. A. Ardekani
T. Asano
L. M. Auer
R. N. Auer
A. Baethmann
J.-C. Baron
M. F. Beal
A. L. Betz
J. Biller
R. G. Blasberg
J. Bogousslavsky
D. Brooks
R. M. Bryan, Jr.
A. Buchan
R. Carson
S. R. Cherry
V. Cunningham
G. J. del Zoppo

N. H. Diemer
W. D. Dietrich
U. Dirnagl
S. Dohi
L. Edvinsson
D. Eidelberg
E. F. Ellis
A. C. Evans
G. Z. Feuerstein
C. Fieschi
S. Finklestein
K. Friston
Y. Fukuuchi
M. D. Ginsberg
A. Gjedde
D. I. Graham
P. Grasby
J. H. Greenberg
R. Gunn
A. M. Hakim

E. D. Hall
C. Halldin
J. M. Hallenbeck
E. Hamel
S. B. Hansen
D. R. Harder
W.-D. Heiss
D. D. Heistad
M. G. Hennerici
K. Herholz
P. Herscovitch
S. Holm
K-A. Hossmann
C. Y. Hsu
C. Iadecola
H. Iida
M. Jensen
T. Jones
I. Kanno
O. Kempski

C. Kennedy
M. Kiessling
H. K. Kimelberg
T. Kirino
I. Klatzo
K. Kogure
E. Kozniewska
J. Krieglstein
D. E. Kuhl
W. Kuschinsky
J. C. LaManna
M. Lauritzen
T. J-F. Lee
K. L. Leenders
G. L. Lenzi
L. Litt
A. H. Lockwood
W. D. Lust
E. MacKenzie
F. Marcoux

M. P. Mattson
J. C. Mazziotta
J. McCulloch
T. McIntosh
B. S. Meldrum
J. S. Meyer
M. E. Moseley
M. A. Moskowitz
H. Naritomi
M. Nedergaard
W. D. Obrist
J. Olesen

D. A. Pelligrino
J. D. Pickard
E. Pinard
D. J. Reis
M. Reivich
J. Risberg
D. A. Rottenberg
E. Ryding
T. Sawada
K. C. Schmidt
A. Schousboe
N. Secher

M. Senda
J. Seylaz
F. R. Sharp
K. Shimazu
E. Shohami
R. P. Simon
L. Sokoloff
G. K. Steinberg
A. J. Strong
S. Strother
A. Tamura
M. M. Todd

M. Tomita
R. J. Traystman
C. G. Wasterlain
K. M. A. Welch
F. A. Welsh
D. Wilson
K. Yamada
T. Yanagihara
W. L. Young

SECTION

I

INSTRUMENTATION AND IMAGE ANALYSIS

Validation of a Dynamic PET Simulator in Clinical Brain Studies

YILONG MA,[*,†] **OLIVIER ROUSSET,**[*] **PETER NEELIN,**[*] **ALAN EVANS,**[*] **VIJAY DHAWAN,**[†]
and DAVID EIDELBERG[†]

*McConnell Brain Imaging Center,
Montreal Neurological Institute,
McGill University, Montreal, Canada
†Department of Neurology,
North Shore University Hospital,
New York University School of Medicine,
New York, New York 10003

This work has been undertaken to evaluate the accuracy of 3D dynamic simulations in neurological imaging protocols with positron emission tomography (PET). We used [^{18}F]fluorodopa PET images from a normal brain and a Parkinsonian brain. Spatially correlated MR images were segmented into several tissue types and anatomical structures. Voxels within every structure were assigned with the time–activity curves (TACs) derived from clinical studies after correcting for partial volume effects. Both noise-free and noisy projection data of this brain model were created and reconstructed as in the real scans. TACs were then generated from the dynamic images and compared with the measured data. The results show good agreement between the simulated and observed TACs in the normal brain. However, the match is poorer in the Parkinson's brain, particularly in striatal structures. This suggests a possible mismatch between the simulated true activity distribution and that in the diseased brain. Both normal and patient data have root-mean-square errors of 2% in cortical gray matter and <10% in striatum without and with noise. This tool can be used to optimize temporal sampling strategy and parameter estimation algorithms.

I. INTRODUCTION

Positron emission tomography (PET) permits quantitative analysis of a wide array of biological markers in the living human brain. This has become an important methodology for probing metabolic and pharmacokinetic processes in both healthy brains and neurological diseases. However, clinical images are heavily smoothed to suppress statistical noise, resulting in relatively poor image resolution and non-linear partial volume distortions in many functional brain structures.

Estimation of physiological parameters with PET poses serious problems because of unknown bias and variance in recorded time–activity curves (TACs). The problems are normally evaluated by performing simple simulations on a set of tissue TACs which ignore object- and camera-dependent distortions (Blomqvist *et al.*, 1995; Feng *et al.*, 1998). While this method is useful for relative comparisons of tracer kinetic models, it generally gives very limited information. Any practical solution to this type of problem must consider contributions from data acquisition and reconstruction algorithms specific to each PET imaging experiment.

Simulation tools based on projection data have been used in several imaging centers to compare image processing or restoration algorithms (Huang *et al.*, 1998; Sastry and Carson, 1997). By modeling image acquisition from raw data

Data flow in the validation experiment

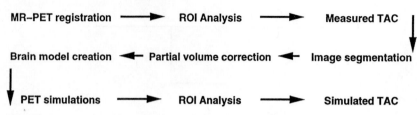

FIGURE 1.   A block diagram showing the generation of simulated time–activity curves from the measured data.

this approach allows correct simulation of the magnitude and distribution of noise in the image space. However, none of the previous methods have been validated. We have implemented a comprehensive PET simulator by combining tomograph physical characteristics with realistic tracer biodistribution data and segmented MR images. It models 3D resolution, attenuation, scatter, randoms, and Poisson statistics as described previously (Ma and Evans, 1997). This chapter describes a preliminary study to validate its performance in clinical brain scans. We want to demonstrate the usage of computer simulations in both normal brain function and pathological cases.

II. MATERIALS AND METHODS

The chief goal of this work is to generate simulated regional TACs and compare the results with the clinical measurements from the scanner. This is based on several experimental procedures illustrated schematically in Fig. 1 and computation steps are summarized below.

A. Clinical Image Data

We used typical MR data and PET images of [^{18}F]fluorodopa (FDOPA) acquired from a normal subject and a patient with Parkinson's disease. Each subject had a MRI scan as part of the PET imaging protocol. This was done on a Philips 1.5 T scanner employing a 3D gradient echo sequence (TE = 10.3 ms, TR = 18 ms, flip angle = 30°, NEX = 1). Images were saved as 128 256 × 256 slices with a voxel size of 1 × 1 × 1 mm^3.

PET images were obtained on a Scanditronix PC2048 brain scanner following a bolus injection of 185 MBq (5 mCi), FDOPA. The total scan duration was 90 min with a time sequence of 6 × 0.5 min, 7 × 1 min, 5 × 2 min, 4 × 5 min, 5 × 10 min. Image volumes consisted of 27 frames and were reconstructed with a 12-mm Hanning filter. Images were stored on a 15 × 128 × 128 matrix with a voxel size of 6 × 2 × 2 mm^3.

B. Individual Brain Model

In each case, MRI data were spatially registered to PET transmission images automatically. This method provided better MR-PET registration in neuroreceptor PET imaging studies as compared with conventional approaches based on emission images. TACs were then extracted in gray matter (GM), white matter (WM), caudate nucleus (CN), and putamen (PU) using regional masks drawn from MRI data. A 3D brain phantom was created by segmenting the MRI volume into a set of functional and anatomical structures. Tissue maps and small structures were labeled using an automated tissue classifier and semiautomatic tools, respectively. Figure 2 shows typical MR images and distinct tissue types used in this study.

In order to obtain true regional values of FDOPA uptake we performed a fully 3D partial volume correction to the observed TACs (Rousset *et al.*, 1998). The accuracy of this method has been validated using a 3D brain phantom of human basal ganglia. To remove local variability and statistical uncertainty inherent in the corrected data we computed an average,

$$A_i = \sum_j^N T_j a_j / \sum_j^N a_j \qquad (1)$$

where T_j and a_j are the corrected TAC and geometric area of each region of interest (ROI) and N is the total number of ROIs over several slices and both hemispheres. In order to further reduce potential errors from discrete time sampling, the resulting values A_i were fitted by linear regression (Fig. 3).

In this preliminary analysis we assume that the tracer is uniformly distributed over the volume of each tissue. Thus we assigned the corrected TAC to each tissue in the segmented brain phantom to represent true activity distribution of FDOPA accumulation. Soft tissue and skull bone were also assigned appropriate photon attenuation values.

C. Dynamic Simulations

A series of projection data were simulated with all the physical effects in PET scans as described elsewhere (Ma

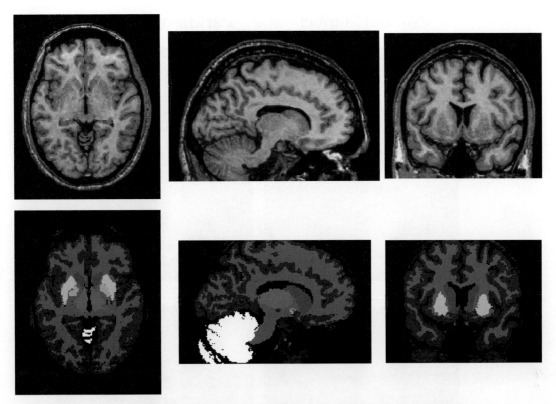

FIGURE 2. Volumetric MR images and segmented tissue maps in the 3D brain phantom. Note the clear definition of striatum, thalamus, and cerebellum. Both imaging volumes are created after registration with the PET data.

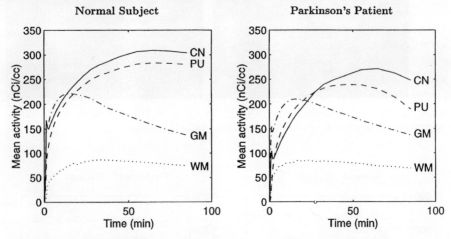

FIGURE 3. True tissue time–activity curves in gray matter (GM), white matter (WM), caudate nucleus (CN), and putamen (PU).

and Evans, 1997). This included photon attenuation and scatter modeling in addition to other camera- and study-specific factors. The resulting data were reconstructed into tomographic images using the scanner reconstruction program. Both procedures were done using acquisition and reconstruction parameters that were identical to the real scan data. TACs were then generated from the dynamic images with and without counting noise. Comparison with the measured scan data was performed by computing the root-mean-square (rms) distances of the relative discrepancy,

$$D_i = \frac{(A_s - A_r)}{A_r}, \tag{2}$$

averaged over frames. A_s and A_r represented the TACs from the simulated and real data, respectively. The first time point was excluded from the calculation since it approaches zero.

Normal Subject   Parkinson's patient

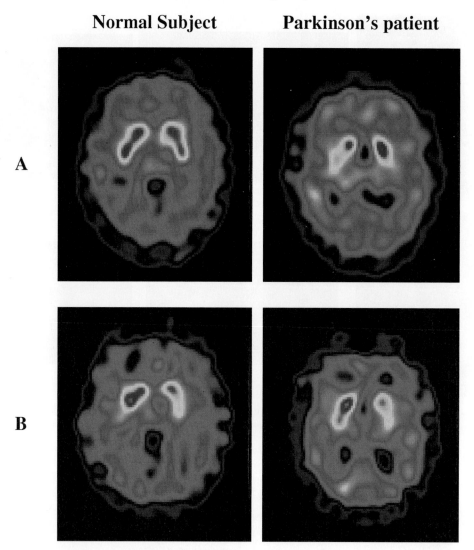

FIGURE 4.   Real (A) and simulated (B) FDOPA images with the highest tracer uptake in the striatum. Images contain about 300 K slice counts and reconstructed with a 12 mm Hanning filter.

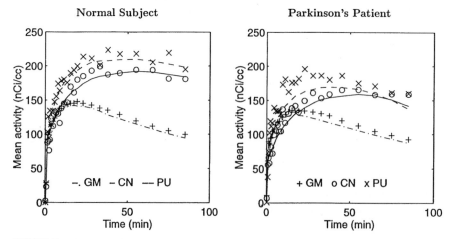

FIGURE 5.   Simulated (lines) and real (symbols) time–activity curves in gray matter (GM) and striatal structures (CN, PU). Data are mean values over basal ganglia slices on both sides of the brain.

III. RESULTS AND DISCUSSION

Simulation results reproduce general patterns of FDOPA PET images in both subjects (Fig. 4). Figure 5 presents TACs in gray matter, caudate nucleus, and putamen. It shows good agreement between the simulated and observed TACs in the normal brain. We see similar results after adding Poisson noise in the projection data. This implies that anatomical structures and activity concentrations in the 3D brain model correlate with the uptake regions of FDOPA in the normal brain.

However, we observe a somewhat poor match in the Parkinson's brain, particularly in the basal ganglia (Fig. 5). This may come from several sources. The first one is inhomogeneity of radioactivity distribution in the diseased brain. Second, striatal structures derived from MR images may not correspond to uptake areas of FDOPA visualized on PET scans. Table 1 lists the rms distances. Both normal and patient data have rms errors of 2% in gray matter, 3% in white matter, and <10% in striatum without and with noise. The residual discrepancy is likely to stem from small PET-MRI misregistration and possible errors in MR image segmentation.

The large differences between the observed and the true TACs reflect partial volume problems, as we have documented before. Note that the true kinetic curves in Fig. 3 show higher caudate uptake in both normal and Parkinson's diseases about 20 min after the injection. However, the observed TACs in caudate are lower than those in putamen. This is expected since caudate is generally smaller than putamen across the brain volume and the data are averaged over basal ganglia slices and left–right sides of the brain. More importantly the data shown here demonstrate that partial volume effects can drastically alter the kinetic relationship between normal and Parkinson's brains. It suggests that the use of uncorrected data may impair accurate diagnosis of disease severity.

A much simpler simulation is widely used to compare kinetic data analysis algorithms. It usually works by solving compartmental equations while adding varying amounts of random noise to theoretical TACs (Blomqvist et al., 1995; Feng et al., 1998). This type of work has limited value since it ignores image bias and variance introduced by the projection and backprojection process. To avoid such problems several groups have developed projection-based simulation tools for PET brain images. We have expanded the previous methods and validated the simulated data and images with brain phantoms (Ma and Evans, 1997). In particular the technique allows us to simulate both emission and transmission scans rapidly.

The work reported here represents the first attempt to compare simulations with clinical data. This could not be achieved in the past because the measured data are always biased by partial volume effects. By correcting the bias explicitly we have now overcome this obstacle. Since the true activity distribution in the brain is known this permits more realistic evaluation of many neuroimaging methods operating at both pixel and regional levels. One useful example is the automatic tissue segmentation algorithms based on unique features of TACs in different cerebral structures (Gunn et al., 1998). Recently there has been growing interest in probing ligand and receptor interactions with multiple radiotracers. Simulations using a Hoffman brain phantom have demonstrated the feasibility of performing a dual-tracer paradigm in a single session (Koeppe et al., 1998). The methods validated here rely on both structural and functional data of the individual brain. With proper use, this powerful tool offers an objective basis for designing optimal imaging protocols in a large number of clinical applications.

IV. SUMMARY

In this chapter we have simulated and compared dynamic data from FDOPA PET imaging in the human brain. Preliminary data indicate that the simulated and measured TACs match with reasonably good accuracy. It is still necessary to model spatially variant tracer distribution from MR images and clinical uptake information. Nevertheless, the results presented here confirm the validity of the PET simulator in predicting the shape and magnitude of TACs in normal and abnormal brains. This would allow researchers to evaluate data acquisition and analysis methods in clinical imaging studies. One immediate use is in optimizing temporal sampling strategy and parameter estimation algorithms.

Acknowledgments

This work was supported by the Medical Research Council of Canada and the International Consortium on Brain Mapping. The authors gratefully acknowledge the excellent technical assistance from the staff at the Brain Imaging Center for carrying out MRI and PET studies.

TABLE 1.   Percentage Root-Mean-Square Distances between Regional Data

Noise	GM	WM	CN	PU
Normal subject				
No	1.31	2.73	8.28	4.51
Yes	1.37	2.40	8.70	8.63
Parkinson's patient				
No	2.11	2.24	5.41	7.14
Yes	2.05	2.52	5.61	6.51

Note. Values are calculated from (simulation − real)/real averaged over the dynamic frames. Notice the similarity between the noiseless and the noisy data.

References

Blomqvist, G., Lammertsma, A. A., Mazoyer, B., and Wienhard, K. (1995). Effect of tissue heterogeneity on quantification in positron emission tomography. *Eur. J. Nucl. Med.* **22**: 652–663.

Feng, D., Wong, K. P., Wu, C. M., and Siu, W. C. (1998). Simultaneous extraction of physiological and input function parameters from PET measurements. *In "Quantification of Brain Function Using PET."* (R. E. Carson, M. E. Daube-Witherspoon, and P. Herscovitch, Eds.), pp. 321–327. Academic Press, San Diego.

Gunn, R. N., Lammertsma, A., and Cunningham, V. (1998). Parametric imaging of ligand–receptor interactions using a reference tissue model and cluster analysis. *In "Quantification of Brain Function Using PET"* (R. E. Carson, M. E. Daube-Witherspoon, and P. Herscovitch, Eds.), pp. 401–406. Academic Press, San Diego.

Huang, S. C., Yang, J., Yu, C. L., and Lin, K. P. (1998). Performance characteristics of a feature-matching axial smoothing method for brain PET images. *In "Quantification of Brain Function Using PET"* (R. E. Carson, M. E. Daube-Witherspoon, and P. Herscovitch, Eds.), pp. 85–90. Academic Press, San Diego.

Koeppe, R. A., Ficaro, E. P., Raffel, D. M., Minoshima, S., and Kilbourn, M. R. (1998). Temporally overlapping dual tracer PET studies. *In "Quantification of Brain Function Using PET"* (R. E. Carson, M. E. Daube-Witherspoon, and P. Herscovitch, Eds.), pp. 359–366. Academic Press, San Diego.

Ma, Y., and Evans, A. C. (1997). Analytical modeling of PET imaging with correlated functional and structural images. *IEEE Trans. Nucl. Sci.* **44**: 2439–2444.

Rousset, O., Ma, Y., and Evans, A. C. (1998). Correction for partial volume effects in PET: Principle and validation. *J. Nucl. Med.* **39**(5): 904–911.

Sastry, S., and Carson, R. E. (1997). Multimodality Bayesian algorithm for image reconstruction in PET: A tissue composition model. *IEEE Trans. Med. Imaging* **16**: 750–761.

2

PET Reconstruction with a Markov Random Field Prior

PETER ALSHEDE PHILIPSEN and LARS KAI HANSEN

Section for Digital Signal Processing, Department of Mathematical Modelling, Building 321, Technical University of Denmark, DK-2800 Lyngby, Denmark

In the field of reconstruction of positron emission tomography (PET) images a host of approaches including filtered backprojection (FB), algebraic reconstruction technique, and expectation maximization (EM) have been proposed [Jain, 1989]. When reconstructing sampled noisy projections we face an ill-posed problem, and the basic difference among the approaches is how they constrain the solution. Therefore, the interest in using statistical image reconstruction methods is high, also in a Bayesian framework. Here we present a Bayesian method for reconstructing PET data with additional prior information on the reconstructed image. The prior used is a Markov random field model which have the advantages of integrating smoothing and preservation of sharp edges/discontinuities. In the solution found by our Bayesian method we clearly see that the edge structure induced by the discontinuous smoothness prior, provides an effective kind of activity segmentation. The proposed method improves on both EM and FB and is tested on both phantom and real data.

I. INTRODUCTION

Reconstruction of positron emission tomography (PET) images is a lively research field with a host of approaches including filtered backprojection (FB), algebraic reconstruction technique (ART), and expectation maximization (EM) (Jain, 1989). Reconstruction of sampled noisy projections is an ill-posed problem; hence, the basic difference among the models is how they regularize the solution. Research in using statistical methods in image reconstruction and image restoration in a Bayesian framework is extensive (Li,

1995; Mumcuoglu *et al.*, 1996; Chen *et al.*, 1990; Zerubia and Chellappa, 1993; Johnson *et al.*, 1991; Geman and Geman, 1984).

We present a Bayesian method for reconstructing PET data using additional prior information on the reconstructed image.

II. MATERIALS AND METHODS

A. Introduction to Bayesian Signal Processing

The basic idea in Bayesian modeling is to consider all the processes of the imaging system as stochastic processes. This means that both the source signal and the processes of the imaging system are stochastic processes. A useful estimate of the reconstructed signal is given by the *maximum a posterior* (MAP) estimate. The observation model is derived from the Radon transform, in the sense that the observed signal is the sinogram of the PET image.

The Bayes formula can then be used to obtain the distribution $P(\mathbf{u}|\mathbf{d})$ of the reconstructed signal $\mathbf{u}$, conditioned on the observed degraded signal $\mathbf{d}$:

$$P(\mathbf{u}|\mathbf{d}) = \frac{P(\mathbf{d}|\mathbf{u})P(\mathbf{u})}{P(\mathbf{d})} \propto P(\mathbf{d}|\mathbf{u})P(\mathbf{u}). \quad (1)$$

This conditional distribution is the product of the distribution of the imaging system process, also called the observation model: $P(\mathbf{d}|\mathbf{u}) \equiv P(\mathbf{u} \rightarrow \mathbf{d})$, and the *prior* distribution of the reconstructed signal $P(\mathbf{u})$. $P(\mathbf{u}|\mathbf{d})$ of Eq. (1) is referred to as the *posterior* distribution. Under a given observation, $\mathbf{d}$, $P(\mathbf{d})$ is constant.

B. The Observation Model

The observation model ($P(\mathbf{d} \to \mathbf{u})$) includes all external influence on the signal, e.g., physical incorrectness in the recording system. This could be several effects like the intrinsic blurring of PET and the Radon transform, in the sense that the observed signal is the sinogram of the PET image.

We consider the case with a linear transformation $\varphi(\cdot)$ with additive spatially independent, white, Gaussian noise. This resolves in an observation model like

$$P(\mathbf{d}|\mathbf{u}) = Z_2 \exp\left(-[\mathbf{d} - \varphi(\mathbf{u})]^T \Sigma_{Gauss}^{-1}[\mathbf{d} - \varphi(\mathbf{u})]\right), \quad (2)$$

where $Z_2 = \prod_i \frac{1}{\sqrt{2\pi\sigma_i^2}}$ and Σ_{Gauss}^{-1}

$$= diag\left(\frac{1}{2\sigma_1^2}, \cdots, \frac{1}{2\sigma_i^2}, \cdots, \frac{1}{2\sigma_I^2}\right).$$

The linear transformation can be the forward Radon transformation combined with blurring used to model the forward PET model for reconstruction,

$$\varphi(\mathbf{u}) = \mathbf{RBu}, \quad (3)$$

where $\mathbf{B}$ is the blurring and $\mathbf{R}$ the Radon transform. The blurring is set to match the intrinsic in PET and is assumed to be Gaussian.

C. Prior Model

The prior used is a Markov random field model including both neighbor connectivity, which introduces smoothness in the image, and discontinuities between the image elements. In this model it is possible to incorporate extra information about the desired signal from additional sources. While we have used only simple smoothness regularization in this work, we have previously used similar methods to include anatomical data, e.g., tissue boundaries extracted from MR images (Philipsen *et al.*, 1997). Similar MRF priors have been used, e.g., by Johnson *et al.* (1991).

D. Markov Random Field Model

The basic Markov random field model used is a model with edge elements (also called line-elements or line-processes) (e.g., in Geman and Geman, 1984), and the edge elements have influence on the individual cliques[1] in a four-neighborhood system in two dimensions (Fig. 1, left) and in the six-neighborhood system in 3D (Fig. 1, right).

The edge elements are labeled binary as {edge, nonedge}, implemented as {+1, −1}. In 3D the edges are introduced as p_j, q_j, and r_j connected to edges in the x, y, and z directions. The edge element p_j has the label {edge} if the signal elements u_j and u_{j+dx} are independent. This corresponds to eliminating the horizontal clique between u_j and u_{j+dx} and gives the possibility of incorporating discontinuities in the image.

Similar considerations are to be formed for the edge elements and an energy function for the combined the Markov random field can be expressed. The coupling between Markov random field and Gibbs random field gives the prior distribution a Gibbs form,[2]

$$P(\mathbf{u}, \mathbf{p}, \mathbf{q}, \mathbf{r}|\theta) = Z_1^{-1} \exp\left(-E_{\text{prior}}(\mathbf{u}, \mathbf{p}, \mathbf{q}, \mathbf{r}|\theta)\right), \quad (4)$$

where $E_{\text{prior}}(\mathbf{u}, \mathbf{p}, \mathbf{q}, \mathbf{r}|\theta)$ is the energy or cost function for the MRF model, θ are the model parameters, and Z_1^{-1} is a normalization constant.

---

[1] A clique is the simplest connectivity between image elements in the neighborhood system.

[2] A distribution in the form $P(f) = Z^{-1} \exp\left(-E(f)/T\right)$, where $E(f)$ is a cost function (bounded from below), T is a parameter, and Z a normalization constant, is in Gibbs form.

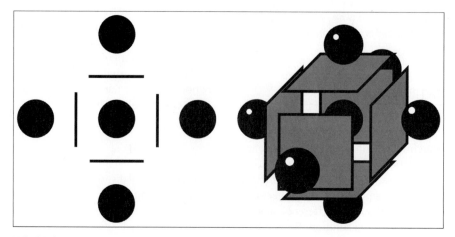

FIGURE 1. (Left) 2D four-neighborhood system with edge elements. (Right) 3D six-neighborhood system with edge elements.

E. 2D Model with Connected Edge Elements

The prior function can be more or less complicated; therefore a simple 2D model is presented and examined before this model is expanded. The 2D model can have the an outline for a chosen part of the image (Fig. 2).

The smoothness prior in the image is normally implemented as a penalty on the gradients within the defined cliques. The gradient at pixel i along the x axis is normally defined as $u'_j = (u_{j+dx} - u_j)$, but higher order approximations can be used, and similarly for the y direction. If a constant gradient in the image is wanted the constraints are to be applied on the second-order derivative.

To obtain closed edge curves several possible solutions have been proposed (see Zerubia and Chellappa, 1993; Li, 1995; Hertz *et al.*, 1991; Nadabar and Jain, 1996; Geman and Geman, 1984). All of those methods use local connections between the edge elements. Most of the methods prefer long straight lines connecting each dimension independently (Zerubia and Chellappa, 1993). In this context only a simple connection term is presented which allows turning edge curves and coupling between the dimensions.

For the simple four-neighborhood system the smoothness prior is implemented as a cost based on the absolute value of the gradient, normally squared here with local edge connectivity. The energy function has the following form:

$$E_{\text{prior}}(\mathbf{u}, \mathbf{p}, \mathbf{q} \mid \theta) = \frac{\nu}{2} \sum_j^J (1 - p_j)(u_j - u_{j+dx})^2$$

$$+ \frac{\nu}{2} \sum_j^J (1 - q_j)(u_j - u_{j+dy})^2$$

$$+ \mu^p \sum_j^J p_j + \mu^q \sum_j^J q_j$$

$$- \lambda \sum_j p_j p_{j+dy} q_j q_{j+dx}.$$

The last three terms are added to control the number of active edges and local connection between the edges. For a description of parameter choice, see Philipsen (1998).

$u_{j-dx+dy}$	$p_{j-dx+dy}$	u_{j+dy}	p_{j+dy}	$u_{j+dx+dy}$
q_{j-dx}		q_j		q_{j+dx}
u_{j-dx}	p_{j-dx}	u_j	p_j	u_{j+dx}
$q_{j-dx-dy}$		q_{j-dy}		$q_{j+dx-dy}$
$u_{j-dx-dy}$	$p_{j-dx-dy}$	u_{j-dy}	p_{j-dy}	$u_{j+dx-dy}$

FIGURE 2. The 2D model can have the following outline for a chosen part of the image.

F. Maximum a Posterior Estimation

We estimate the reconstructed image as the image with highest posterior probability, maximum a posterior (MAP):

$$\mathbf{u}_{MAP} = \arg\max_{\mathbf{u}} P(\mathbf{u}|\mathbf{d}) = \arg\max_{\mathbf{u}} P(\mathbf{d}|\mathbf{u}) P(\mathbf{u}). \quad (5)$$

The model of the total system can now be expressed as

$$P(\mathbf{u}|\mathbf{d}) \propto \exp\left[-(E_{observation} + E_{prior})\right], \quad (6)$$

where the normalization constant can be ignored and the MAP estimate of Eq. (6) corresponds to finding the minimum of the energy function.

The energy function usually has many local minima, and methods based only on gradient decent do only find the optimal solution if the search starts close to the optimal solution.

To find the MAP estimate we use *mean field annealing* (MFA). For details concerning annealing schedule, parameter choice, etc., and for generalizations of the approach to 3D reconstruction, see Philipsen (1998).

III. RESULTS AND DISCUSSION

The Markov random field model using mean field annealing (MRF-MFA) algorithm is tested against FB and maximum likelihood expectation maximation (ML-EM) on:

- Phantom studies
- PET data from the GE-PET scanner at National University Hospital in Copenhagen.

A. Phantom Studies

The Phantom includes multiple objects both with and without sharp edges and is a modified version of the Shepp–Logan phantom (Jain, 1989). The sinograms are generated using simulated Poisson processes with an increasing number of counts (Fig. 3). A pixel-based second-order error measure, called L_2, is used (Philipsen, 1998).

The results using a simulated Poisson-based sinogram with 8×10^6 counts can be seen in Table 1.

TABLE 1. L_2 Errors Using Simulated Poisson-Based Sinogram with 8×10^6 Counts

Algorithm	L_2 error
Filtered backprojection	0.520
Expectation maximization	0.314
Mean field annealing	0.309

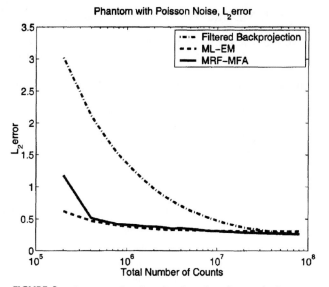

FIGURE 3. L_2 error as function of total number of counts in sinogram.

B. Results of Real Data

The three algorithms are also tested on real data from the GE-PET scanner at National University Hospital in Copenhagen. The sinogram is intensity corrected using transmission and blank scans (Fig. 4a). The results are presented below. The FB solution is very noisy (Fig. 4b), while traces of the EM instability artifacts are seen as radial stripes in center panel of Fig. 4c. In the solution found by the Bayes method (Fig. 4d), we clearly see the edge structure induced by the discontinuous smoothness prior, providing a kind of activity segmentation.

C. Conclusion

In the solution found by the Bayes method we clearly see the edge structure induced by the discontinuous smoothness

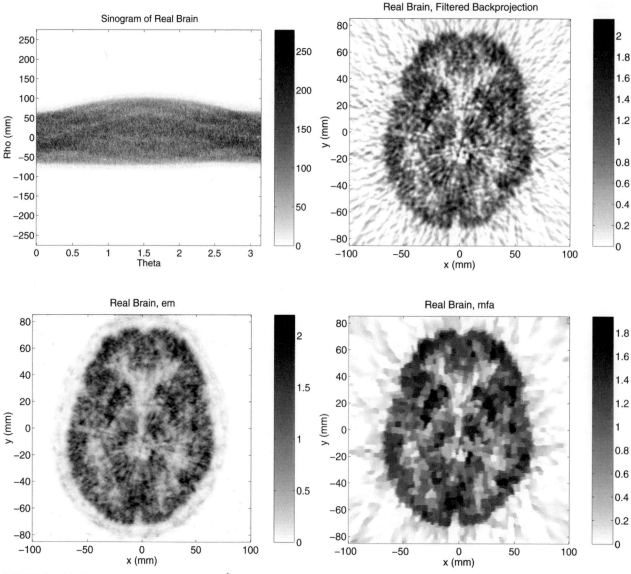

FIGURE 4.   (a) Sinogram corresponding to 3×10^6 counts. (b) Reconstruction with filtered backprojection. (c) Reconstruction with EM. (d) Reconstruction with MRF-MFA.

prior, providing a kind of activity segmentation. The proposed method improves on both EM and FB. FB suffers from noise contamination, while the EM result shows instability artifacts.

Acknowledgements

We acknowledge Søren Holm at the PET center at the National University Hospital in Copenhagen for providing the PET data and Peter Toft for discussing PET reconstruction. This research was supported by the Danish Research Councils through the Computational Neural Network Center (Connect).

References

Chen, C.-T., Johnson, V. E., Wong, W. H., Hu, X., and Metz, C. E. (1990). Bayesian image reconstruction in poistron emission tomography. *IEEE Trans. Nuclear Sci.* 37(2).

Geman, D., and Geman, S. (1984). Stochastic relaxation, Gibbs distributions and the Bayesian restoration of images. *IEEE Trans. Pattern Anal. Mach. Intelligence* 6: 721–741.

Hertz, J., Krogh, A., and Palmer, R. (1991). *"Introduction to the Theory of Neural Computation."* Addison Wesley, New York.

Jain, A. K. (1989). *"Fundamentals of Digital Image Processing."* Prentice Hall, New York.

Johnson, V. E., Wong, W. H., Hu, X., and Chen, C.-T. (1991). Image restoration using gibbs priors: Boundary modeling, treatment of blurring, and selection of hyperparameter. *IEEE Trans. Pattern Anal. Mach. Intelligence* 13(5): 413–425.

Li, S. Z. (1995). *"Markov Random Fields in Computer Vision."* Springer-Verlag, Tokyo.

Mumcuoglu, E. U., Leahy, R. M., and Cherry, S. R. (1996). Bayesian reconstruction of pet images: Methodology and performance analysis. *Phys. Med. Biol.* 41:1777–1807.

Nadabar, S. G., and Jain, A. K. (1996). Parameter estimation in Markov random field contextual models using geometric models of objects. *IEEE Trans. Pattern Anal. Mach. Intelligence* 18(3).

Philipsen, P. A. (1998). *"Reconstruction and Restoration of PET Images."* Ph.D. thesis, Department of Mathematical Modelling, Technical University of Denmark.

Philipsen, P. A., Kjems, U., Toft, P., and Hansen, L. K. (1997). Restoring functional pet images using anatomical mr images. In *"Porceedings of the Interdisciplinary Inversion Workshop 5,"* ISBN 87-90400-12-7. Department of Earth Sciences, Aarhus University.

Zerubia, J., and Chellappa, R. (1993). Mean field annealing using compound Gauss–Markov random fields for edge detection and image estimation. *IEEE Trans. Neural Networks* 4(4): 703–709.

A New Method for Correcting Head Movement Artifact in Positron Emission Tomography

OSAMA R. MAWLAWI,*,†,‡ EDWARD F. LEONARD,‡ JOHN L. HUMM,§ and STEVEN M. LARSON†

*Division of Neuroscience, Departments of Psychiatry and Radiology, Columbia University, New York, New York 10032
†Nuclear Medicine Service, Memorial Sloan Kettering Cancer Center, New York, New York 10021
‡Center for Biomedical Engineering, Columbia University, New York, New York 10027
§Medical Physics Department, Memorial Sloan Kettering Cancer Center, New York, New York 10021

We have developed a new method for correcting image blurring due to patient head motion during positron emission tomography (PET). The method is based on acquiring a temporal 3D record of the head position simultaneously with the head's emission data. The head's position record is used to realign all the acquired lines of response (LORs) back to their original orientation before the onset of motion. The correction method is designed for 3D PET studies of rigid bodies only. It is performed during the image reconstruction process and is applied to the PET data while in sinogram space. We tested this method on a GE ADVANCE PET scanner while using a new device for measuring head position. The new measuring device consisted of two charge coupled device cameras that captured images of the head from two different viewpoints. These images were then combined to generate a 3D representation of the head location throughout the imaging session. Two tests were conducted to evaluate the movement correction method. (a) The first method was a "stationary" test where a micropositioner moved a radioactive source to different locations within the field of view of the PET scanner. At each new location, PET data were acquired and then corrected for motion as measured by the movement measuring device. (b) The second method was a "movement simulation" test where a programmable micropositoner continuously moved the source while the PET scanner acquired consecutive frames of data. The results of these tests showed that the correction method is capable of producing images with an average movement error of 0.42 mm.

I. INTRODUCTION

Positron emission tomography (PET) imaging has long suffered from movement artifacts. These artifacts are the result of involuntary subject movements that occur due to long imaging intervals or to illnesses that prevent patients from maintaining a constant position during an imaging session. Several techniques have been proposed to correct movement artifacts in PET images (Kearfott *et al.*, 1984; Bergstrom *et al.*, 1981; Thornton *et al.*, 1991; Bettinardi *et al.*, 1991; Dubey *et al.*, 1993; Daube-Witherspoon *et al.*, Picard and Thompson, 1995a,b; Green *et al.*, 1994). The majority of these techniques however, have been only partially successful at minimizing this effect (Bettinardi *et al.*, 1991; Dubey *et al.*, 1993; Picard and Thompson, 1995a; Green *et al.*, 1994). Moreover this partial success came either at the expense of patient discomfort or at high cost. In addition, the current advances in PET camera hardware and software have led to PET images that are characterized by high resolution and high signal-to-noise ratio. This means that slight movements during the data acquisition period are reflected as large degradations in the refined PET image resolution and hence are viewed as a potential source of patient misdiagnosis.

In this chapter we introduce a new method for correcting image blurring due to patient motion. The method is designed for rigid bodies only and is used to correct image artifacts pertaining to the movement of the head only. The method is optimized for the GE ADVANCE PET scanner and is applied to emission data that have been acquired in the three-dimensional (3D) mode.

The movement correction method is based on the concurrent acquisition of a 3D head position record along with the PET emission data of the patient's head. The head position information is used to correct any displaced lines of response (LORs) back to their original location. This method is similar to the technique of "rebinning the mispositioned lines of response" that was initially presented by Daube-Witherspoon *et al.*, 1990 and later by Menke *et al.*, 1996. In this work, however, a complete implementation from detecting and generating the head position record to correcting the mispositioned LORs and generating a blur-free image is presented.

The ability of the proposed correction method to produce a blur-free image is dependent on the spatial accuracy and frequency of the 3D head position record. Daube-Witherspoon *et al.*, 1990, 1997 and Mawlawi *et al.*, 1998 have previously shown that the spatial head position should be recorded at a frequency and resolution of at least 1 Hz and ± 1 mm, respectively. In this paper, a charge coupled device (CCD)-based video imaging system that is capable of these characteristics is presented, evaluated, and used in the implementation of the movement correction method (see Experiments and Results, below).

In the next section (Methodology), a description of the image deblurring method followed by a description of the mathematical manipulations required to transform the LORs back to their original locations is presented. The results of experiments that were conducted to test and evaluate the video imaging system and the efficacy of the correction process are given in the last two sections (Experiments and Discussion).

II. METHODOLOGY

The image deblurring method requires two steps. The first is the determination of a position and orientation record of a patient's head over time. The second is the use of this record to transform the acquired LORs back to their original location.

The acquisition of the 3D head position record is performed by monitoring the patient's head position with a video imaging system and employing the principles of photogrammetry (Abdel-Aziz and Karara, 1971). Two CCD cameras, placed behind the PET scanner, were positioned at a collimation angle Φ that allowed the field of view (FOV) of both cameras to capture the same area on the surface of a patient's head (Fig. 1a). The CCD cameras viewed the patient's head via an angled mirror (Fig. 1b). At least four noncollinear markers placed on the patient's forehead were used as reference points to represent the spatial lo-

cation of the head. Pictures were then captured simultaneously from both cameras via a frame grabber; the reference markers were identified in both pictures; and their two-dimensional (2D) locations were determined. Photogrammetric techniques were then used to generate the 3D location of each marker from its respective pair of 2D images. These 3D locations were then used to determine the spatial location of the patient's head. This process was repeated throughout the PET imaging session and a corresponding 3D head position record was determined.

Prior to the generation of the 3D position record, CCD cameras were calibrated to fix the position of the 2D CCD images with respect to the PET scanner's coordinate system.

The second step in the image deblurring method is to use the head position record to correct the acquired PET emission data (LORs). Each entry in the head position record $(t_1, t_2, \ldots)$ is compared with the original location of the patient's head position (t_0). From these two locations, a transformation (translation and rotation) matrix (T_i), which quantifies the head movement from its original location to its current one, is calculated (Abdel-Aziz and Karara, 1971). These transformation matrices are used to transform the acquired LORs back to their original locations (Fig. 1a).

The LOR transformation process presumes that the head position record and the PET emission data (LORs) are synchronized. That is, for each 3D head position (t_i) a corresponding set of emission data (LORs) is expected (LORs acquired during $\Delta t = t_i. - t_{i-1}$). Several synchronization techniques are possible (see Discussion). In this paper, the emphasis is placed on showing, that the correction method produces blur-free images irrespective of how the images are synchronized. To this end, the feasibility of the correction method was tested in two ways: (1) acquiring PET emission data of a radioactive source before and after moving from one position to the next (motion between frames); (2) acquiring consecutive frames of PET data along with a corresponding set of 3D position information of a radioactive source as it was moved within the FOV of the scanner (motion simulation test).

III. SOFTWARE

Several software programs were written to carry out the image deblurring method. These programs were designed to perform the following algorithms:

(A) Calculate the CCD camera parameters in order to determine the 2D CCD images in the PET scanner's reference frame (CCD camera calibration).
(B) Calculate the 3D marker positions from their 2D camera pictures.
(C) Calculate the transformation matrix that defines the patient's head motion between time points t_0 and t_i.

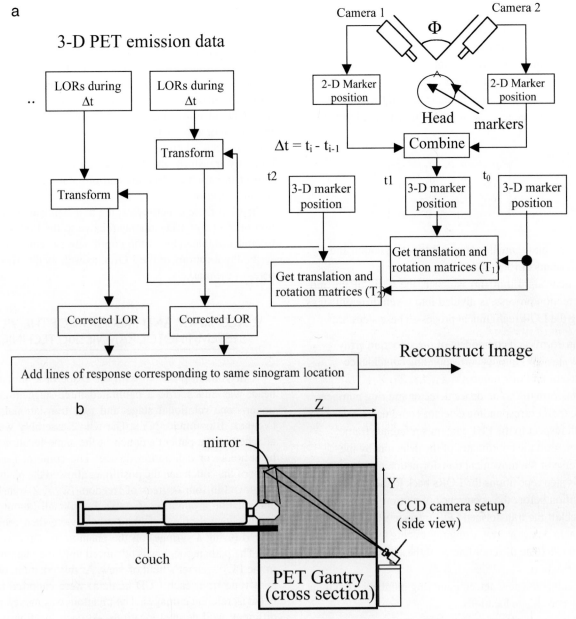

FIGURE 1. (a) Block diagram of the movement correction method. (b) CCD cameras and mirror setup.

(D) Transform the PET raw data according to the calculated transformation matrix in order to correct the mispositioned lines of response.

Algorithm A is run only once during the prescanning procedure. Algorithms B and C are executed "on the fly" each time a pair of camera pictures is acquired. Algorithm D is employed as a postprocessing task. Ideally this step should be carried out whole the pictures are acquired. However, due to the large number of LORs that can occur during a scanning interval (25.2 million in 3D mode), on-the-fly processing cannot be achieved. Algorithms A and B were writ-

ten following the principles of camera calibration and photogrammetry (Abdel-Aziz and Karara, 1971; Tsai, 1987). Algorithm C is based on the following derivation:

Let A and P be the matrices corresponding to the 3D marker locations at times t_0 and t_i, then

$$A = [a_1, a_2, a_3, a_4] \quad \text{and} \quad P = [p_1, p_2, p_3, p_4],$$

where

$$a_i = [x_i, y_i, z_i]^T \quad \text{and} \quad p_i = [x_i, y_i, z_i]^T.$$

Let A^* be the matrix formed by adding a row of ones to A. The columns of A^* are the homogeneous coordinates of the columns of A. Similarly for P^*.

Then, assuming the four points $\{a_1, a_2, a_3, a_4\}$ are not coplanar, A^* is invertable and $T = P^* A^{*^{-1}}$ is the (4×4) transformation matrix which translates and rotates at the same time. The general form of this transformation matrix is

$$T = \begin{bmatrix} R_{11} & R_{12} & R_{13} & t_x \\ R_{21} & R_{22} & R_{23} & t_y \\ R_{31} & R_{32} & R_{33} & t_x \\ 0 & 0 & 0 & 0 \end{bmatrix},$$

where t_x, t_y, and t_z are its components of translation and $R_{i,j}$ its components of rotation.

The final algorithm (D) which is the basis of the motion correction process, is divided into seven different steps showing the LOR realignment process. These steps are:

(1) Transform each component of the sinogram array $A(u, v, \theta, \Phi)$ into its corresponding coincidence detector and ring numbers (X_1, X_2, Z_1, Z_2).

(2) Transform the coincidence detector and ring numbers into their corresponding detector coordinates measured with respect to the PET camera's reference frame.

(3) Transform the coordinates of the detectors by the inverse of the movement transformation matrix (T). This step repositions the LORs back to their initial position before the occurrence of motion.

(4) Check if the transformed LOR intersects the PET camera detector array using ray-tracing techniques and then calculate the coordinates of these points of intersection.

(5) Calculate the new detector and ring numbers from their new coordinate locations.

(6) Determine the location in the sinogram array $A(u, v, \theta, \Phi)$ that corresponds to the new detector and ring numbers.

(7) Increment the value of the new location to reflect the detection of a new event.

The mathematical formulation for these different steps is shown in reference Mawlawi (1998).

To carry out algorithm D, the 3D PET raw data (sinograms) must be accessible. For this reason, the breakpointing utility of the GE ADVANCE PET scanner was used. Care was taken to breakpoint at the appropriate location in the reconstruction path to ensure correct data preprocessing such as normalization, geometric, scatter, and deadtime corrections.

IV. EXPERIMENTS AND RESULTS

Two sets of experiments were conducted to test the ability of the correction method to generate blur-free images. These experiments were divided into two categories. The first category focused on characterizing the accuracy and precision of the CCD video imaging system/photogrammetric technique in determining the 3D position of the markers. The second category focused on the ability of the correction method to correctly transform the LORs of an object that has been moved in various directions and orientations back to their original location.

In all of these experiments, the first step was the calibration of the two CCD cameras relative to the PET scanner's reference frame. This is the coordinate system used to describe the locations of the LORs, as well as the 3D coordinates of the markers.

V. ACCURACY AND PRECISION OF THE VIDEO SYSTEM/PHOTOGRAMMETRIC TECHNIQUE

A Styrofoam phantom with the shape of a human forehead, was attached to a calibrated three-stage micropositioner—one rotational stage and two translational stages (Velmex, Bloomfield, NY). The whole assembly was then attached to the patient's couch in the same location as the head-holder or calibration device. The combined arrangement of the couch and the positioner allowed the phantom to be moved in four degrees of freedom (X, Y, Z translations and rotation around the Z axis). Four small (about 1 mm in diameter) droplets of hand cream were then arbitrarily placed (using a syringe) on the phantom to serve as markers. The patient couch was advanced until the phantom was in the PET scanner's field of view. At this position, two images (one from each CCD camera) were captured and labeled as reference images. The phantom was moved to nine different, well-defined locations. At each location two images were captured and saved to disk. The 3D positions of the markers were then determined and compared to their true values. Table 1 shows the results of these measurements, reported as the average and maximum error of the measured values along each of the three axes (X, Y, Z).

VI. MOVEMENT CORRECTION

A. Motion between Frames

A phantom made of five lucite cylinders of different diameters (3.5, 5.6, 8, 11.3, and 16 mm) each filled with 0.04 MBg/ml of F-18 were placed in a Styrofoam block. The block was attached to a three-stage micropositioner and

TABLE 1.

Position No.	Actual position coordinates in mm (X, Y, Z)	Average error along X (mm)	Average arror along Y (mm)	Average error along Z (mm)	Maximum error along X (mm)	Maximum error along Y (mm)	Maximum error along Z (mm)
1	Baseline						
2	Baseline +(0, 0 − 10)	−0.0074	−0.0200		−0.0346	−0.7325	
3	Baseline +(0, 0 − 20)	0.0180	−0.5446		0.0286	−0.9951	
4	Baseline +(0, 0 − 30)	−0.0538	0.1207		−0.2007	0.5544	
5	Baseline +(0, −10, −30)	0.0513		0.2504	0.0843		0.3117
6	Baseline +(0, −20, −30)	0.0612		−0.0657	0.1293		−0.1742
7	Baseline +(0, −30, −30)	−0.05037		−0.4521	−0.0948		−0.5363
8	Baseline +(6.35, −30, −30)		0.6256	0.2020		1.0678	0.28169
9	Baseline +(12.7, −30, −30)		0.0401	0.1787		−0.5867	0.25764
10	Baseline +(19.0, −30, −30)		0.4275	0.0938		0.8915	0.2231

Note. The X location was determined by the micropositioner whereas the Y and Z locations were determined by the positional encoder of the patient couch.

moved into the FOV of the PET scanner. Emission data were collected in 3D mode for 3 min (scan 1). The phantom was then translated along the X, Y, and Z axes by 13.5, 42.5, −30 mm, respectively, and rotated around the Z axis by 5°. At this position, emission data were collected in 3D mode for 4 min (scan2). The choice of rotation around the Z axis was made in order to simulate as closely as possible a typical left–right head movement of a patient while inside a PET scanner.

The motion correction algorithm was then applied to realign the LORs of scan 2 back to those of scan 1. The sinogram and image results are shown in Figs. 2A (before motion), 2B (after motion), and 2C (after motion correction).

The transformation matrix used to perform the LOR realignment is

$$T = \begin{bmatrix} 0.99619 & 0 & -0.08715 & 13.5 \\ 0 & 1 & 0 & 42.5 \\ 0.08715 & 0 & 0.99619 & -30 \\ 0 & 0 & 0 & 1 \end{bmatrix}.$$

B. Motion Simulation

A radioactive rod source (Ge-68) was placed in a Styrofoam block that was marked with four droplets of hand cream. The rod source/Styrofoam block was then placed on a programmable micropositioner and the whole assembly was positioned in the FOV of the scanner. The micropositioner was programmed to simulate motion by laterally advancing the Styrofoam block along the X axis at a speed of 1 mm/s. At the same time, the PET scanner was set to acquire consecutive frames of 1 s duration. The total travel distance of the micropositoner was set to 2 cm, which resulted in 20 s of PET scanning time.

A baseline PET frame was initially acquired before the micropositioner was set in motion to act as a reference PET frame. The CCD video imaging system was also set to acquire a pair of pictures at a frequency of 1 s with the first set of pictures being acquired 0.5 s after the onset of motion (i.e., after 0.5 mm of motion). This arrangement ensured that the PET data were acquired during the movement of the rod source and that the position of the source was captured while it was half-way through its travel distance per frame.

All emission data were acquired in 3D mode and analyzed as described above. Figure 3 shows the sinogram and image data with 1, 2, 5, and 10 mm of motion. The 2-, 5-, and 10-mm data were generated by consecutively adding the PET frames as the source was moving. The effect of motion is shown in the sinogram data as a progressive broadening in the detector response while the corresponding images show this effect as a blurring artifact. Also shown in the figure is the sinogram and image data of the source after motion correction. These last two pictures show no broadening and hence no blurring artifact.

VII. DISCUSSION

The movement correction method was developed for rigid bodies (e.g., head) only. Movements of other parts of the body such as the torso cannot be corrected using this method since the external surface of the torso is deformable. Furthermore, the internal organs of the torso are not rigid with respect to the external body contour. To correct for the movement of nonrigid bodies, other methods such as gating must be used.

Our experiments showed that the connection method was capable of producing blur-free images with an accuracy that

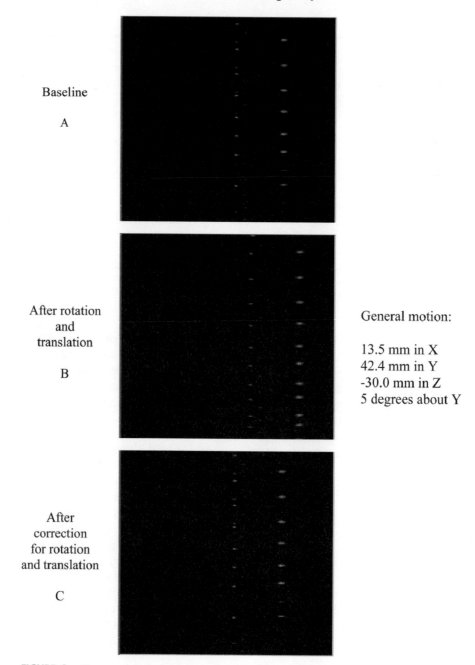

Baseline

A

After rotation
and
translation

B

General motion:

13.5 mm in X
42.4 mm in Y
-30.0 mm in Z
5 degrees about Y

After
correction
for rotation
and translation

C

FIGURE 2.   Sinogram data for the "motion between frames" study before and after motion correction (sinogram 120).

depended on how precisely the 3D head position data were determined. The accuracy depends also on the ability to synchronize the acquisition of the PET data to the head movement data. A mismatch in synchronization would lead to correcting LORs by the wrong amount of motion. In general the process of data synchronization may be performed by: (1) synchrozing the PET emission data and the head position record at a preset time, (2) acquiring the PET emission data into a new frame whenever the head position exceeds a certain predefined displacement threshold, and (3) acquiring the PET emission data in LIST mode.

All of these methods are possible with commercially available PET scanners but they require large amounts of disk space for their implementation. In the first method, the storage requirement depends upon the duration of the preset time interval. Shortening the duration of the acquisition time frames increases the accuracy of motion correction, but at the expense of larger memory requirements. In the second

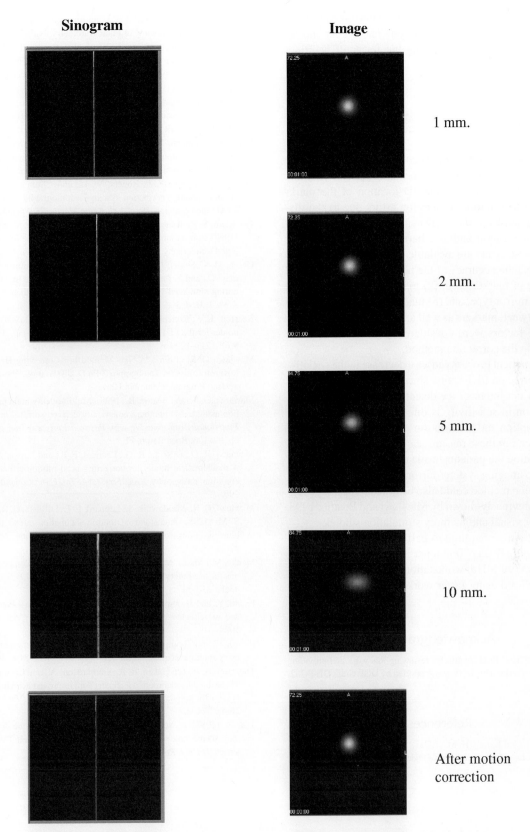

FIGURE 3. Sinogram and image data of rod source with 1, 2, 5, and 10 mm of motion. Also shown is the sinogram and image data after motion correction.

method, the storage space depends on the movement threshold and hence the patient's condition. The bigger threshold results in more image blurring but fewer PET frames and smaller storage requirements. Finally, acquiring PET data using LIST mode also requires a large amount of storage space. However, with the current trend toward acquiring PET data in 3D mode and the increase in detector number per scanner, this method might require the least amount of memory.

Several factors affect the accuracy of the correction method. The direct linear transformation (DLT) method, which is used to determine the 3D location of the markers, does not take into consideration errors due to the distortion in the lenses of the CCD cameras. These errors may result in a pincushion and/or a barrel effect. Methods that correct for these errors are available (Tsai, 1987). Other factors that affect the accuracy of the photogrammetric method are: (1) type of marker used, (2) marker location, (3) room lighting, (4) mirror type, and (5) marker identification (identification between markers as well as within a CCD picture). All of these factors were considered and optimized while implementing the correction method.

The realignment process causes a decrease in the sensitivity of the corrected PET frame since some of the realigned LORs might not intersect the detector array of the scanner. This reduction in sensitivity is directly proportional to the amount of motion and the location of the object in the FOV of the scanner. For these reasons, care must be taken to comfortably position the patients inside the scanner.

The video imaging system/mirror setup used in the movement correction method could also be utilized for image registration as previously shown by Mawlawi and Beattie, 1998, as well as for visual and memory stimulation studies.

Finally, the time required to perform the correction process (algorithm D) was 20 min per 3D frame while running on an HP 735 (99 MHz) workstation. This time can be substantially reduced using faster workstations utilizing multiple processors.

Acknowledgments

The authors thank Brad Beattie for reviewing this work and providing some valuable remarks. This work was supported by DOE Grant DE-FG02-86ER60407.

References

Abdel-Aziz, Y. I., and Karara, H. M. (1971). "Direct Linear Transformation from Comparator Coordinates into Object Space Coordinates in Close Range Photogrammetry," *ASP Symposium on Close-Range Photogrammetry*. American Society of Photogrammetry, Falls Church, VA.

Bergstrom, M., Boethius, J., Eriksson, L., Greitz, T., Ribbe, T., and Widen, L. (1981). Head fixation device for reproducible position alignment in transmission CT and positron emission tomography. *J. Comput. Assist Tomogr.* **5**(1): 136–141.

Bettinardi, V., Scardaoni, R., Gilardi, M. C., Rizzo, G., Perani, D., Paulesu, E., Striano, G., Triulzi, F., and Fazio, F. (1991). Head holder for PET, CT, and MR studies. *J. Comput. Assist. Tomogr.* **15**(5): 886–892.

Daube-Witherspoon, M. E., Yan, Y. C., Green, M. V., and Carson, K. M. (1990). Correction for motion distortion in PET by dynamic monitoring of patient position. *Proc. 37th Annu. Meeting J. Nucl. Med.* **31**: 861.

Dubey, A., Edgeworth, J., Rogers, W. L., and Clinthorne, N. H. (1993). A patient motion correction system for tomographic imaging based on CCD video cameras. *Proc. 40th Annu. Meeting J. Nucl. Med.* **34**.

Goldstein, S. R., Daube-Witherspoon, M. E., Green, M. V., and Eidsath, A. (1997). A head motion measurement system suitable for emission computed tomography. *IEEE Trans. Med. Imaging* **16**(1): 17–27.

Green, M. V., Seidel, J., Stein, S. D., Tedder, T. E., Kempner, K. M., Kertzman, C., and Zeffiro, T. A. (1994). Head movement in normal subjects during simulated PET brain imaging with and without head restraint. *J. Nucl. Med.* **35**(9): 1538–1546.

Kearfott, K. J., Rottenberg, D. A., and Knowles, R. J. (1984). A new head holder for the PET, CT, and NMR imaging. *J. Comput. Assist. Tomogr.* **8**(6): 1217–1220.

Mawlawi, O. R. (1998). "A New Method For Correcting Head Motion in Positron Emission Tomography." Ph.D. dissertation, Biomedical Engineering Program, Columbia Univ.

Mawlawi, O. R., and Beattie, B. (1998). Multimodality brain image registration using a 3-D photogrammetrically derived surface. *In Quantitative Functional Brain Imaging with Positron Emission Tomography*, CRC, pp. 99–106. Boca Raton, FL.

Mawlawi, R. O., DiResta, R. G., Leonard, F. E., and Larson, M. S. (1995). A mathematical model for correcting head motion artifact in positron emission tomography. *Proc. First LAAS Int. Conf. Comput. Simul.* 283–290.

Mawlawi, O. R., Miodownik, S., Leonard, E. F., DiResta, G. R., and Larson, S. M. (1998). A customized motion acquisition circuit for image deblurring in positron emission tomography. *IEEE Trans. Instrum. Meas.* **47**(2).

Menke, M., Atkins, M. S., and Buckley, K. R. (1996). Compensation methods for head motion detected during PET scans. *IEEE Trans. Nucl. Sci.* **43**(1): 310–317.

Picard, Y., and Thompson, C. J. (1995a). Digitized video subject positioning and surveillance system for PET. *IEEE Trans. Nucl. Sci.* **42**(4): 1024–1029.

Picard, Y., and Thompson, C. J. (1995b). Motion correction of PET images using multiple acquisition frames. *Conf. Rec. IEEE Med. Imaging Conf.*

Thornton, A. F., Ten Haken, R. K., Gerhardson, A., and Correll, M. (1991). Three dimensional motion analysis of an improved head immobilization system for simulation, CT, MRI, and PET imaging. *Radiother. Oncol.* **20**(4): 224–228.

Tsai R. (1987). A versitile camera calibration technique for high accuracy 3D machine vision meterology using off-the-shelf TV cameras and lenses. *IEEE J. Rob. Autom.* **RA-3**(4): 323–44.

4

Automated Registration of PET Brain Scans Using Neural Networks

OLE LAJORD MUNK and SØREN BAARSGAARD HANSEN

PET Center, Aarhus University Hospital, DK-8000, Denmark

Image registration is often necessary to align images where patients move during quantitative positron emission tomography (PET) examinations. We propose an automated method for registration of dynamic PET brain scans using artificial neural networks. Neural networks are fast and robust algorithms capable of learning a desired input–output mapping without explicitly having to be told the rules—instead they are adjusted according to presented examples. A striatal phantom was scanned in order to provide at set of completely aligned reference images. Misalignments were produced by applying a random set of three translations in the range ±10 mm and three rotations in the range ±5°. Features extracted from sets of images were used to create examples, and a neural network was trained to predict the transformation necessary to align the images. We used feedforward neural networks with a single layer of hidden nodes and a variant of backpropagation as a learning algorithm. The neural network method was compared to a standard automated image registration algorithm. Results showed that neural networks were more accurate. This new registration method is automated, provides accurate registrations, and can be applied retrospectively. The method is suitable for PET images since the image features presented to the networks are insensitive to noise and changes in activity distribution.

I. INTRODUCTION

Patient movement during dynamic positron emission tomography (PET) examinations is a well-known problem, and image registration is often necessary to allow kinetic analysis of small anatomical regions. Two registration strategies are commonly used; manual registration by matching known structural landmarks or automated image registration (AIR) where the image is iteratively resampled into new spatial orientations until a cost function is optimized (Woods *et al.*, 1992). So far, no optimal registration procedure has been found for registration of PET images. Manual registration is subjective, needs operator intervention, and does not work well on PET images where anatomical structures are difficult to identify. Automated registration algorithms usually require a high degree of similarity between the two images and are therefore not suitable for dynamic PET studies where images have different tracer distributions and signal-to-noise ratios. Furthermore, automated algorithms can be slow using many iterations to optimize the cost function.

We have applied a new automated method for registration of PET brain scans using feed-forward neural networks where the computational nodes are arranged in layers, and each node has internal connections, i.e., weights, to all nodes in the next layer. Neural networks are trainable algorithms that have been successfully applied to a wide range of pattern recognition problems. During training, the weights are adjusted automatically according to the presented examples, and the network obtains an improved input–output mapping. This learning process is time consuming, but once trained, neural networks have the advantage of being fast and noise tolerant due to the parallel structure of the nodes.

II. MATERIALS AND METHODS

A. Dynamic FDOPA Phantom Study

A brain phantom with four small anatomical structures (striatum) was filled with ^{18}F using higher concentration in the striatum than in the surrounding space. The phantom was placed in a headholder and scanned using a Siemens ECAT EXACT HR-47 camera and our [^{18}F]fluorodopa (FDOPA) protocol in order to provide a set of 28 completely aligned reference images simulating a dynamic FDOPA study. Data were acquired for 120 min using the following frame structure: 6×30 s, 7×1 min, 5×2 min, 4×5 min, 5×10 min, 1×30 min. A transmission scan with external rod sources was performed and used for photon attenuation correction, and images were reconstructed with filtered backprojection using a Hanning filter with a cutoff at 0.5 Nyquist frequency. The resulting 3D images contained $128 \times 128 \times 47$ voxels with a size of $2.0 \times 2.0 \times 3.1$ mm and central spatial resolution of 4.6 mm full-width half-maximum.

B. Generation of Examples for Neural Network

Neural networks learn from examples that consist of an input vector describing the problem of interest and an output vector containing the solution we want the network to predict. The potential network performance, i.e., the ability to predict the output vector, is highly dependent on the quality of the features we present to the network as an input vector. It is important to use features with a high signal-to-noise ratio with respect to the problem and thereby permit the network to achieve the best possible performance.

Using the simulated FDOPA study, a large number of examples were produced that were appropriate for training a neural network to predict the transformation between two misaligned images. Different feature extraction techniques were tried and it was decided to use the following procedure. Misaligned images were produced by applying three translations (x, y, and z; range ± 10 mm) and three rotations (transaxial, coronal, and sagittal; range $\pm 5°$) to images from the phantom study. Each example was generated using two images: one reference and one misaligned with a known transformation. Image processing was used to extract a small number of robust features suitable as input for the network. The two images were filtered using a median filter, which does not blur the edges. The filtered images were thresholded, and the two binary images were subtracted to obtain an image highly sensitive to misalignments. This image processing made the subtracted image insensitive to differences in activity distribution and noise. Finally, the subtracted image was approximated using 3D moments m_{rst} of order $r + s + t$,

$$m_{rst} = \sum_x \sum_y \sum_z x^r y^s z^t V(x, y, z), \qquad (1)$$

where $V(x, y, z)$ is the activity concentration in the voxel with coordinates (x, y, z). Twenty low-order moments were calculated using Eq. (1) and used as an input vector. The six parameters describing the transformation necessary to align the images were used as an output vector. All registration steps are summarized in Fig. 1. The neural network was presented to a wide range of registration problems by making examples using two random images and applying transformation parameters that were uniformly randomized in the

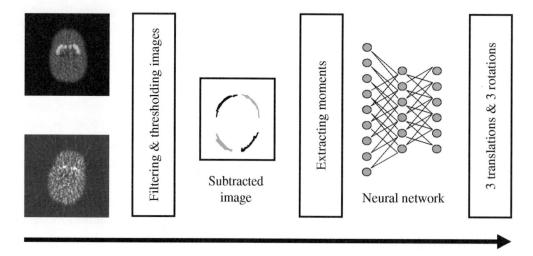

FIGURE 1. The registration method using neural networks. Two misaligned PET images are median filtered, thresholded, and subtracted. The resulting image is approximated using 20 moments, which are used as input to a neural network trained to predict the transformation needed to align the images.

defined range. This was done in order to cover the entire parameter space since networks can be expected to accurately predict only transformations similar to what they have been presented to during training.

In all data sets, individual inputs were normalized to mean value 0.0 with standard deviation 1.0, and outputs were rescaled to ±1.0, corresponding to the output range from the activation function used in the networks.

C. Network Architecture and Training

The network architecture used to solve the registration problem was a two-layer feed-forward network with an input layer consisting of 20 nodes, a hidden layer, and an output layer with 6 nodes. Successive layers were fully connected by weighted connections w_{ij} between all nodes, and a threshold node clamped to minus one had weighted connections w_{0j} to nodes in the hidden layer and output layer (Hertz *et al.*, 1991). The output O_i from node i is then calculated using the activation function g on the sum of weighted input,

$$O_i = g\left(\sum_{k=0}^{N} w_{ik}\xi_k\right), \qquad (2)$$

where ξ_k is the output from node k in the previous layer and w_{ik} is the weight from node k in the previous layer to node i. The networks used $g = \tanh$ as the activation function. Outputs were produced by applying the moments to the input layer and calculating the transformation by forward propagation through the network using Eq. (2).

Networks are generally trained using the backpropagation learning algorithm based on gradient descent (Rumelhart *et al.*, 1986). Weights are iteratively changed proportional to the gradient of an error function with global minimum corresponding to perfect prediction of the presented examples. The error of the individual parameters is calculated by comparing the expected value from the output vector to the value predicted by the network. The network performance on a data set is calculated as the mean square error (MSE), which is the sum of squared errors divided by the number of outputs and the number of examples in the data set.

Before training, all network weights were assigned randomized values in the range ±0.1. We used a variant of backpropagation with adaptive learning rate, which was automatically adjusted according to the change in MSE on the training set. If MSE was reduced during the epoch (i.e., one pass through the training set), the learning rate was increased, otherwise it was decreased. The ability of modifying the learning parameters during training was found to be useful since the best initial values were not optimal after many epochs. Compared to traditional backpropagation, this decreased total training time and risk of the algorithm being caught close to a local minimum of the error func-

tion. The adaptive algorithm was also found to be faster than backpropagation with momentum (Hertz *et al.*, 1991) for the present problem.

To obtain networks with high generalization ability, each data set of examples was divided into three separate sets used for training and evaluation: training (50%), validation (25%), and test (25%). Only examples in the training set were presented to the network during training and used to adjust the weights using the learning algorithm. Too much training led to overfitting, reducing the ability to generalize in response to new inputs, but this was avoided by calculating the network error on the validation set after each epoch and stopping the training process at minimum validation error. After training, the independent test set was used to assess the network error when performing new registrations.

III. RESULTS AND DISCUSSION

A. Optimal Network Architecture

The optimal network architecture depends on the problem that the network is trained to solve. No rules exist that describe which architecture to choose for the registration problem. Therefore, this was investigated experimentally. The best two-layer 20–X–6 architecture for the registration problem was found in an experiment where the registration error was measured as function of the number of hidden nodes X (Fig. 2).

It cannot be expected to achieve good performance from very simple networks, and Fig. 2 shows that at least 6 hid-

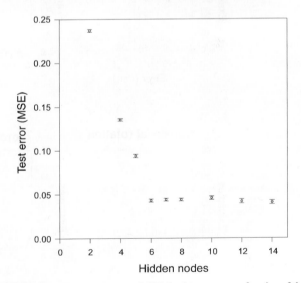

FIGURE 2. Mean square error (MSE) on the test set as a function of the number of hidden nodes. Each architecture was trained 10 times using a large data set ($N = 2300$). Slightly different test errors were found depending on the initial weights and the figure shows the mean error with error bars.

den nodes are required. However, as a rule of thumb small networks with fewer degrees of freedom will usually have a better generalization ability than very complex networks. This would cause a slightly increasing MSE as a function of the number of hidden nodes but the effect was not significant in our results using up to 14 hidden nodes. Possibly because our training method uses a validation set, serious overfitting was prevented. Instead, MSE reaches a plateau after 6 hidden nodes. It is, however, reasonable to use the small network with 6 hidden nodes, since smaller networks are faster and usually require less learning time. We conclude that the optimum number of hidden nodes is 6 using a two-layer architecture. Some three-layer architectures were tried but adding a hidden layer did not improve the prediction accuracy. The one-layer network (i.e., no hidden layer) turned out to be too simple to describe the data.

Independent experiments showed that the number of examples in all data sets used in later experiments ($N = 784$) was sufficient to describe the complexity of the registration problem using the 20–6–6 architecture. Fewer examples were found to increase the test error and more examples did not improve network performance.

We examined the importance of the individual input by pruning of input nodes but removal of the least important moments from the input vector did not improve the test er-

ror. A slightly decreased test error was observed when m_{000}, m_{100}, m_{010}, and m_{001} were replaced by higher-order moments, but no notable improvement was found by using more than 20 moments. Pruning of internal connections (Reed, 1993) in 20–6–6 networks was tried in order to decrease the number of free parameters, but removing the least important weights and retraining the network did not improve performance. Extensive pruning was found to worsen network performance considerably.

In conclusion, we found the optimal network architecture to be a fully connected two-layer 20–6–6 network with 168 internal weights including threshold nodes.

B. Neural Network Registration Error

The results show that a single network trained to perform registration in the range ± 10 mm (all translations) and $\pm 5°$ (all rotations) does not predict all transformation parameters equally well. The error distributions for a 20–6–6 network are shown in Fig. 3. The relatively poor estimates of the transaxial and coronal rotation can be attributed to the circular shape of the brain around these axes. Rotations have smaller effect than the translations on the moments, thus making it difficult for the network to learn to correct for them.

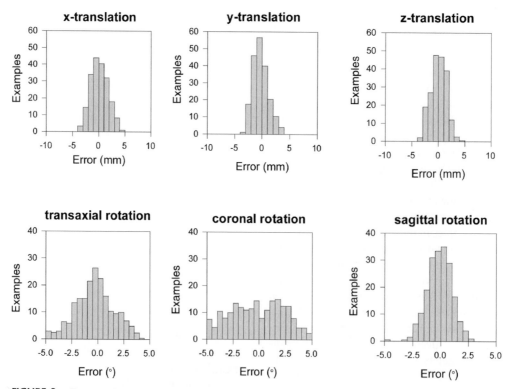

FIGURE 3.   Test error distribution for each transformation parameter using a 20–6–6 neural network. The examples in the test set ($N = 196$) had transformations randomized in the ranges ± 10 mm (all translations) and $\pm 5°$ (all rotations). The transaxial and coronal rotations are poorly predicted compared to the translations.

The registration accuracy was improved by using two networks where the second network was trained to correct registration error from the first network. Feature extraction, network architecture, and training procedure were exactly as for the first network, but the second network was trained using examples generated from a smaller parameter space. It is clearly seen from Figs. 3 and 4 that the use of two networks provides more accurate transformations and that the prediction of the rotations is improved.

Registration errors from the neural network method, using two 20–6–6 networks, were compared to data from a standard AIR algorithm using the sum of absolute differences (Eberl *et al.*, 1996) as the cost function. We also tried the standard deviation of pixel-by-pixel ratios (Woods *et al.*, 1992) as cost functions and found similar results.

The registration method using two networks provides better registration accuracy than the AIR algorithm. The neural network results presented in Table 1 could be further improved by using more than two networks at the expense of producing more examples, training more networks, and increased time for the network method to perform the individual registrations.

In conclusion, our results show that the neural network method can perform accurate registration of images with different noise level. Registration accuracy can be improved by using more networks and we found that two networks produced more accurate registrations than a standard automated algorithm. Furthermore, neural networks are an order of magnitude faster (once trained) than traditional algorithms since no iterations are needed. Image preprocessing and the structure of neural networks make the method robust to noise and differences in activity distribution. The limitation of the current method is that an automated segmentation algorithm is needed to obtain a general and tracer-independent registration method for human studies.

TABLE 1.   Registration Comparison

	Neural Network		AIR algorithm	
	Error	Max. error	Error	Max. error
Translations (mm)	0.32 ± 0.05	1.1	0.9 ± 0.1	5.2
Rotations (°)	0.9 ± 0.1	5.3	1.8 ± 0.2	7.4

Note. Fifty image registrations were performed using two neural networks and a standard automated algorithm. The results are presented as the mean of the absolute values of the deviation from the correct values $\pm$ S.D. The maximum error is also shown.

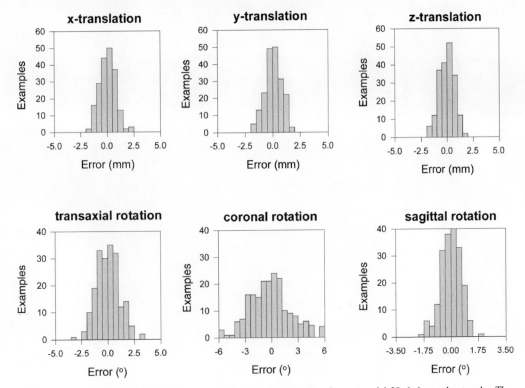

FIGURE 4.   Test error distribution for each transformation parameter using two serial 20–6–6 neural networks. The examples in the test set ($N = 196$) had transformations randomized in the ranges ± 10 mm (all translations) and $\pm 5°$ (all rotations). The use of a second network clearly improves the registration accuracy.

References

Eberl, S., Kanno, I., Fulton, R. R., Ryan, A., Hutton, B. F., and Fulham, M. J. (1996). Automated Interstudy Image Registration Technique for SPECT and PET. *J. Nucl. Med.* **37**, 137–145.

Hertz, J., Krogh, A., and Palmer, R. G. (1991). *"Introduction to the Theory of Neural Computation."* Adisson-Wesley, Redwood City, CA.

Reed, R. (1993). Pruning algorithms—A survey (1993). *IEEE Trans. Neural Networks* **4**(5), 740–747.

Rumelhart, D. E., Hinton, G. E., and Williams R. J. (1986). Learning internal representation by error propagation. *In "Parallel Distributed Processing: Explorations in the Microstructure of Cognition"* (D. E. Rumelhart and J. L. McClelland, Eds.), pp. 318–362. MIT Press, Cambridge, MA.

Woods, R. P., Cherry, S. R., and Mazziotta J. C. (1992). Rapid automated algorithm for aligning and reslicing PET images. *J. Comput. Assist. Tomogr* **16**(4), 620–633.

5

Statistical Estimation of PET Images in the Wavelet Domain

FEDERICO E. TURKHEIMER, MATTHEW BRETT,* DIMITRIS VISVIKIS, and
VINCENT J. CUNNINGHAM

PET Methodology Group, MRC Cyclotron Unit, Hammersmith Hospital, London, UK
*Institute of Nuclear Medicine, University College London Medical School, Middlesex Hospital, London, UK

The theoretical framework of this chapter addresses the problem of the use of the wavelet transform in the estimation of PET-SPECT images. The solution of the problem of "estimation" allows the solution of the equivalent problems of optimal filtering, maximum compression, and statistical testing. In particular, theory and algorithms are presented that allow current wavelet methodology to deal with the two main characteristics of nuclear medicine images: low signal to noise ratios and correlated noise. The technique is applied to a number of clinical images. Results show the ability of wavelets to model images and to estimate the signal generated by cameras of different resolutions in a wide variety of noise conditions. Moreover, the same methodology can be used for the multiscale analysis of statistical maps. The wavelet transform is shown to be a valuable tool for the numerical treatment of images in nuclear medicine, and the methods described here may be a starting point for further developments for the processing of PET images.

I. INTRODUCTION

The wavelet transform (WT) is a mathematical tool that has been extensively used in many areas of signal processing and has been shown to be particularly suitable for application in the biomedical field (Unser and Aldroubi, 1996). The present work develops a theoretical framework for the use of WT in the estimation of PET images. The issue of estimation is not usually addressed in PET image analysis.

Current methods rely mainly on monoresolution approaches that smooth the image with filters of fixed FWHM.

In the case of statistical maps, the features of interest of an image are searched by hypothesis testing approaches that model the noise, but not the signal. WT methods efficiently address the problem of image estimation and so, at the same time, the equivalent problems of optimal filtering, maximum compression, and statistical testing.

Two main difficulties hamper the use of WT in PET. The first is due to the assumption of "white noise" in WT methods. This assumption is not tenable given the correlation among pixels introduced by the PET camera. The second problem is that usually, in PET images, the signal is quite sparse and noise levels are high. Such images are known to be unsuitable for WT (Coifman and Donoho, 1995).

In the following section the techniques and algorithms developed will be briefly described. For a complete treatment of the topic and background material on wavelets, the interested reader is referred to Turkheimer *et al.* (1999).

II. METHODS

The first methodological issue stands in the selection of the appropriate wavelet base for the application. The choice of Battle–Lemarie wavelets is motivated by previous results that showed their optimality for PET because they are orthogonal and achieve maximum decorrelation within and between wavelet channels (Ruttiman *et al.*, 1996). In statistical terms this offers the great advantage of dealing with

Level j

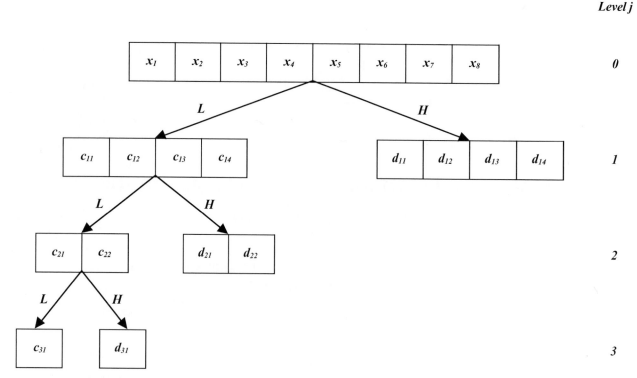

FIGURE 1. The discrete wavelet transform is obtained through application of complementary low-pass and high-pass filters and subsequent decimation (here described as H and L). Both H and L are applied to data vector $x_1, x_2, \ldots x_8$. The output of h are the four wavelet coefficients for the first resolution. The output of L are the four coefficients of the scaling function. The wavelets coefficients of the other resolution levels are obtained by iterating the low- and high-pass filtering steps on the coefficients of the scaling function.

coefficients in the new functional base that are normally distributed and independent.

Unfortunately, the correlation of the noise on the original images causes the noise in the wavelet domain to vary among resolution channels (Johnstone and Silverman, 1997). Therefore, a "colored" noise model for wavelets was developed; the variance of wavelet coefficients is computed from the correlation function of images and the correlation function of the filters used to compute the WT (Turkheimer *et al.*, 1999). The correlation function of images was modeled either with a Gaussian point-spread function (clinical images) or computed directly on residual images in the case of statistical maps (Turkheimer *et al.*, 1999).

Third, the use of traditional discrete wavelet transform (DWT) (see Figs. 1 and 2) does not produce satisfactory results with images with low signal-to-noise ratios (Coifman and Donoho, 1995; Turkheimer *et al.*, 1999). The problem stems from the fact that the DWT does not enjoy the translation invariant property. In other words, if the signal is not aligned with the wavelet base, its coefficients will not appear on the correct resolution channel but will be spread through all resolutions with values below the noise level (Simoncelli *et al.*, 1992). Coifman and Donoho (1995) introduced the translation-invariant DWT (DWT-TI) that pro-

duces a translation-invariant representation of the signal with minimal increase of complexity. This approach was further extended to 2 and n dimensions (Turkheimer *et al.*, 1999). The resulting set of the DWT-TI wavelet coefficients is over-complete ($N \log(N)$) and therefore does not imply a unique inverse transform. The approach of Coifman and Donoho (1995) and Turkheimer *et al.* (1999) constructs a unique inverse by averaging all reconstructions obtained by all shifted DWTs.

III. RESULTS

Results on the application of wavelet filters are shown in Figs. 3–5. Figure 3 shows the denoised output of an image of [^{11}C]raclopride. Note how the well known pattern of distribution on the striatum is carefully recovered.

Figure 4 shows the signal recovery for an image of a [^{11}C]PK11195 study of peripheral benzodiazepine receptors in the brain. The removal of the noise preserved most of the high-resolution signal.

Finally, Fig. 5 shows a typical application of wavelets to statistical maps. In this instance, the figure shows the recovered pattern of changes in blood flow for a visuo-motor

Level j   *0*   *1*   *2*

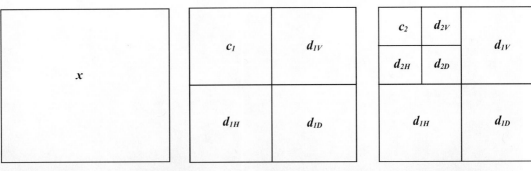

FIGURE 2. Illustration of the two-dimensional wavelet transform. The transform is performed on images by applying the filters L and H to rows and columns. This operation produces a quadrant c containing the coefficients of the scaling functions, and three quadrants of wavelet coefficients d usually labeled as horizontal, vertical, and diagonal (H, V, D, respectively). Wavelet coefficients of higher resolutions are obtained by applying the above filtering steps to quadrants c.

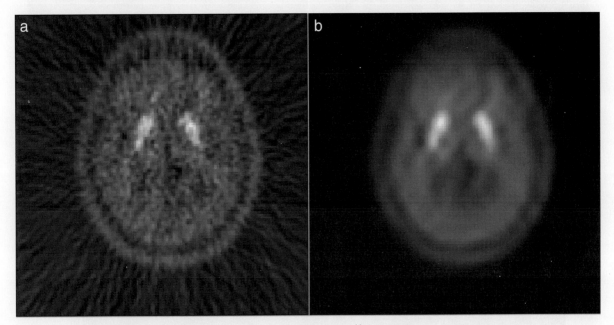

FIGURE 3. Results on the application of the wavelet filters on a [^{11}C]raclopride emission image scanned with an ECAT 953B PET camera (CTI/Siemens, Knoxville, TN) and reconstructed with a ramp filter. The noise image (a) is filtered so that the pattern of distribution is efficiently recovered (b).

study (Brett *et al.*, 1998). It is important to note that while the current approach for the analysis of statistical fields is based on simple thresholding, wavelets allow the estimation of the patterns of signal in the noise field (Turkheimer *et al.*, 1999).

IV. DISCUSSION

The current work aims to lay down a general framework for wavelet-based methods in PET. The new developments allow the use of a rigorous statistical framework of WT; estimation can be carried out according to a number of different criteria (minimax, least squares, control of false positives) (Turkheimer *et al.*, 1999). In the case of images, WT filters allow the efficient modeling of the signal as they adapt to the content of signal at different resolution levels in a theoretically optimal fashion. WT is also an alternative to traditional hypothesis testing methods (Worsley *et al.*, 1996) for the multiscale analysis of statistical maps as it *estimates* the significant profiles of SPMs of whatever shape and resolution, with strict control of the error rate.

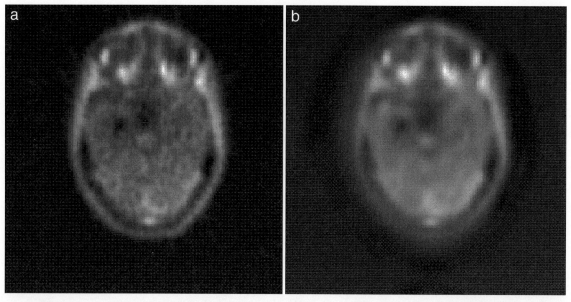

FIGURE 4.   The original image (a) is taken from a PK 11195 study of the distribution of peripheral benzodiazepine receptors in the brain (images kindly provided by R. Banati). The camera used was the EXACT 3D PET scanner (CTI/Siemens). The image is reconstructed with a ramp filter at the highest resolution (~4.5 mm). The WT reconstruction (b) efficiently denoises the image while preserving edges and details.

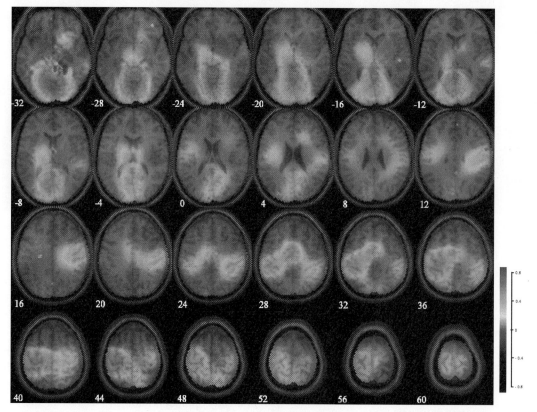

FIGURE 5.   The figure shows the results of the wavelet procedure when applied to the analysis of a water activation study that activated the motor and visual areas. The filtered activation image is displayed overlaid on a gray scale MRI of the MNI standard brain. Values represent the pattern of z-scores estimated with an error rate $\alpha = 0.05$ over all 24 slices. The left side of the brain is displayed at the left of the slices. Slices are labeled with their distance in millimeters from the transverse plane containing the anterior and posterior commisures.

References

Brett, M., Stein, J. F., and Brooks, D. J. (1998). The role of the premotor cortex in imitated and conditional praxis. *Neuroimage* **7**(4): S978.

Coifman, R. R., and Donoho, D. L. (1995). Translation-invariant denoising. *Lect. Notes Statist.* **103**: 125–150.

Johnstone, I. M., and Silverman, B. W. (1997). Wavelet threshold estimators for data with correlated noise. *J. R. Statist. Soc.* B **59**: 319–351.

Ruttiman, U. E., Unser, M., Thevenaz, P., Lee, C., Rio, D., and Hommer, D. W. (1996). Statistical analysis of image differences by wavelet decomposition. *In* "*Wavelets in Medicine and Biology*" (A. Aldroubi and M. Unser, Eds.), pp. 115–144. CRC Press, Boca Raton, FL.

Simoncelli, E. P., Freeman, W. T., Adelson, E. H., and Heeger, D. J. (1992). Shiftable multi-scale transforms. *IEEE Trans. Inf. Theory* **38**: 587–607.

Turkheimer, F. E., Brett, M., Visvikis, D., and Cunningham, V. J. (1999). Multiresolution analysis of emission tomography images in the wavelet domain. *J. Cereb. Blood Flow Metab.*, **19**(11): 1189–1208.

Unser, M., and Aldroubi, A. (1996). A review of wavelets in biomedical applications. *IEEE Proc.* **84**: 626–638.

Worsley, K. J., Marret, S., Neelin, P., and Evans, A. C. (1996). Searching scale space for activation in PET images. *Hum. Brain Mapping* **4**: 74–90.

6

FIPS: A Functional Image Processing System for PET Dynamic Studies

DAGAN FENG,[*,†] **WEIDONG CAI,**[*,†] **and ROGER FULTON**[†,‡]

[*]*Center for Multimedia Signal Processing, Department of Electronic and Information Engineering, Hong Kong Polytechnic University, Hong Kong*
[†]*Biomedical and Multimedia Information Technology (BMIT) Group, Basser Department of Computer Science, The University of Sydney, Australia*
[‡]*Department of PET and Nuclear Medicine, Royal Prince Alfred Hospital, Australia*

Positron emission tomography (PET) imaging can provide quantitative physiological information, in addition to qualitative information. However, most of PET image handling software packages have not included modeling and quantitative analysis functions. In this paper, based on some of our systematic studies on biomedical functional image data processing and modeling, we developed a functional image processing system, which is a completely interactive window-based processing system that includes features for brain PET dynamic studies. The system contains several special modules: (1) basic image statistical analysis; (2) optimal PET image sampling schedule design; (3) physiological knowledge-based clustering analysis; (4) dynamic image data compression; (5) most of the major fast algorithms for generation of functional images; etc. This system would therefore be very useful for the design of proper data acquisition protocol for brain PET dynamic studies, functional imaging data analysis, compression, modeling, and simulation.

I. INTRODUCTION

Functional imaging techniques such as positron emission tomography (PET) have matured into a crucial and powerful tool for modern biomedical research and clinical diagnosis (Feng *et al.*, 1997). One of the major advantages of PET imaging is that it can provide quantitative physiological information, in addition to qualitative information. However, most of the PET image handling software packages have not included modeling and quantitative analysis functions yet. In this paper, based on our systematic studies on biomedical functional image data processing and modeling, we developed a functional image processing system, called FIPS.

II. SYSTEM OVERVIEW

FIPS is an interactive window-based processing system that includes features for brain PET dynamic studies. The software was implemented in C programming language and IDL on a SUN Ultra-2 workstation running Solaris 2.5, as well as on an Intel Pentium-based platform using RedHat Linux 5.1. The system contains several special modules: (1) basic image statistical analysis, such as tissue time–activity curve (TTAC) and plasma time–activity curve (PTAC) viewer, dynamic image sequence player, profile, histogram, and surface, etc; (2) optimal PET image sampling schedule (OISS) design (Li *et al.*, 1996; Ho-Shon *et al.*, 1996); (3) physiological knowledge-based clustering analysis (KCA); (4) dynamic image data compression (Ho *et al.*, 1997); (5) most of the major fast algorithms for generation of functional images, such as the Patlak graphic approach (PGA) (Patlak *et al.*, 1983), the linear least squares method (LLS) (Feng *et al.*, 1995), and the generalized linear least squares algorithm (GLLS) (Feng *et al.*, 1996). Figure 1 illustrates the system structure of the FIPS. Some of the functional modules and related research are described and summarized in the following sections.

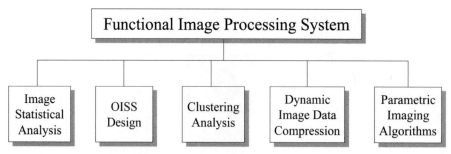

FIGURE 1.   The system structure of the functional image processing system (FIPS).

A. Optimal PET Image Sampling Schedule

An OISS for dynamic PET has recently been derived theoretically and investigated by computer simulation and clinical studies (Li *et al.*, 1996; Ho-Shon *et al.*, 1996). It has been demonstrated that the use of OISS is an effective way of reducing image storage requirements while providing comparable parameter estimates.

Finding the optimal image sampling schedule involves minimizing the determinant of the convariance matrix of the estimated parameters or, conversely, maximizing the determinant of the Fisher information matrix by rearranging the sample intervals, using the minimum number of required samples. It has been shown (Li *et al.*, 1996) that the minimum number of temporal frames required is equal to the number of model parameters to be estimated. For example, for the five-parameter [^{18}F]-2-*fluoro*-2-*deoxy*-D-*glucose* (FDG) model, only 5 temporal frames are sufficient to obtain parameter estimates of equivalent statistical accuracy and reliability to the conventional technique which typically requires the acquisition of more than 20 temporal frames (Feng *et al.*, 1998). Since fewer temporal image frames need to be reconstructed, the computational burden is substantially reduced. Figure 2 is a graphic interface of the FIPS system. The image on the right-bottom corner illustrates an example of the OISS for the five-parameter FDG model.

B. Knowledge-Based Clustering Analysis (KCA)

In general, a time–activity curve (TAC) can be obtained from each pixel in dynamic PET image. However, many TACs may have similar kinetics. Based on domain specific physiological kinetic knowledge related to dynamic PET images and physiological tracer kinetic modeling, a physiological KCA algorithm has been developed (Ho *et al.*, 1997) to classify imagewide TACs, $C_i(t)$ (where $i = 1, 2, \ldots, R$, and R is the total number of image pixels), into S cluster groups C_j (where $j = 1, 2, \ldots, S$, and $S \ll R$) by measurement of the magnitude of natural association (similarity characteristics). Each pixel TAC is classified as belonging to one of the cluster groups. An index table indexed by cluster group contains the mean TAC for each cluster group. This

KCA module can be used as a processing tool to extract physiological information from dynamic image data.

C. Dynamic Image Data Compression

We have recently proposed a three-stage technique for dynamic image data compression (Ho *et al.*, 1997). First, we apply the OISS design to reduce the number of temporal frames. In order to obtain more accurate and precise results from clinical dynamic PET data, the OISS with five-parameter FDG model (OISS-5) was used to correct for the cerebral blood volume (CBV) and partial volume (PV) effects (Cai *et al.*, 1998). Second, the physiological knowledge-based clustering analysis algorithm can be used to further compress the reduced set of temporal PET image frames into a single indexed image. In this stage, noisy background pixels were removed prior to cluster analysis to improve reliability. Finally, we compress and store the indexed image using the portable network graphics (PNG) format. Details of this three-stage compression technique can be found in (Ho *et al.*, 1997). This compression algorithm can reduce image storage space by more than 95% without sacrificing image quality. It could benefit the current expansion in medical imaging and image data management.

D. Fast Algorithms for Generation of Parametric Images

With the recent development of high spatial and temporal resolution PET, a number of parametric imaging algorithms have been developed (Feng *et al.*, 1995), most of which involve certain strong assumptions or require considerable computational time. The GLLS algorithm has been shown to offer some advantages over other methods for parameter estimation in nonuniformly sampled biomedical systems (Feng *et al.*, 1996). We found that, compared with existing algorithms (Feng *et al.*, 1995), the GLLS algorithm: (1) can directly estimate continuous model parameters; (2) does not require initial parameter values; (3) is generally applicable to a variety of models with different structures; (4) can estimate individual model parameters as well as physiological parameters; (5) requires very little computing time; and (6) can

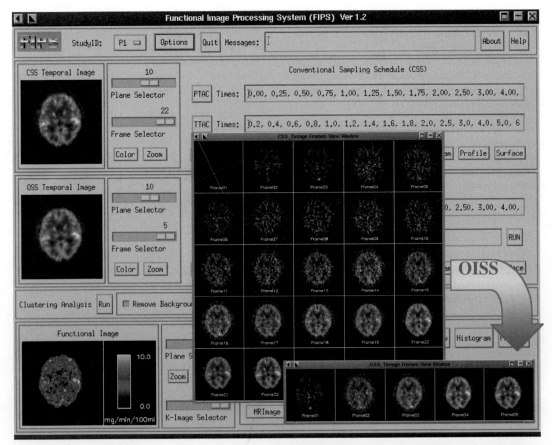

FIGURE 2. Illustration of an interface of the FIPS system. The image on the bottom right shows an example of compressing 22 brain PET FDG image frames obtained from the traditional sampling schedule into 5 frames according to the OISS design. The image on the bottom left illustrates a parametric image of the local cerebral metabolic rates of glucose (LCMRGlc) generated using the built-in fast GLLS algorithm.

produce unbiased estimates (Feng *et al.*, 1996). Details of the GLLS algorithm can be found in Feng *et al.* (1996). The GLLS algorithm has been integrated into the FIPS system, as part of the fast-parametric-image-algorithm module. We also integrate the famous PGA (Patlak *et al.*, 1983) and LLS algorithm (Feng *et al.*, 1995), etc., into the FIPS system. On bottom left of Fig. 2, we can see the parametric image of the local cerebral metabolic rates of glucose (LCMRGlc) generated by using the GLLS algorithm.

III. CONCLUSIONS

Based on some of our systematic studies on biomedical functional image data processing and modeling, we developed a FIPS, which integrates several special modules for brain PET dynamic studies, such as the OISS design, the KCA, the GLLS fast algorithm, the three-stage algorithm for dynamic image data compression and some image statistical analysis functions. In addition, the development of some

special modules, including physiological knowledge-based image smoothing, image segmentation, extracting PTAC from dynamic PET, and data visualization, is actively conducted in our group. The FIPS is expected to be very useful in brain PET functional imaging data acquisition, analysis, compression, modeling, and simulation.

Acknowledgments

This research is supported by the RGC and ARC Grants. The authors are grateful to the National PET/Cyclotron Center, Taipei Veterans General Hospital Taiwan, for their kind assistance.

References

Cai, W., Feng, D., and Fulton, R. (1998). "Clinical Investigation of a Knowledge-Based Data Compression Algorithm for Dynamic Neurologic FDG-PET Images," *The 20th Annual International Conference of IEEE Engineering in Medicine and Biology Society*, Oct. 29–Nov. 1, 1998, Hong Kong, pp. 1270–1273.

Feng, D., Cai, W., and Fulton, R. (1998). "An Optimal Image Sampling Schedule Design for Cerebral Blood Volume and Partial Volume Correction in Neurologic FDG-PET Studies," *The 29th Annual Scientific Meeting of A&NZ Society of Nuclear Medicine*, April 4–8, 1998, Melbourne, Australia, *Aust. NZ J. Med.* **28**(3), 361.

Feng, D., Ho, D., Chen, K., Wu, L. C., Wang, J. K., Liu, R. S., and Yeh, S. H. (1995). An evaluation of the algorithms for determining local cerebral metabolic rates of glucose using positron emission tomography dynamic data. *IEEE Trans. Med. Imag.* **14**, 697–710.

Feng, D., Ho, D., Iida, H., and Chen, K. (1997). Techniques for functional imaging. *In "Medical Imaging Systems Techniques and Applications: General Anatomy"* (C. T. Leondes, Ed.), pp. 85–145. Gordon and Breach Science, Amsterdam.

Feng, D., Ho, D., Lau, K. K., and Siu, W. C. (1998). GLLS for optimally sampled continuous dynamic system modeling: Theory and algorithm. *In "Quantitative Functional Brain Imaging with Positron Emission Tomography,"* (R. E. Carson *et al.*, Eds.), pp. 339–345. Academic Press, San Diego.

Feng, D., Huang, S. C., Wang, Z., and Ho, D. (1996). An unbiased parametric imaging algorithm for non-uniformly sampled biomedical system parameter estimation. *IEEE Trans. Med. Imag.* **15**, 512–518.

Ho, D., Feng, D., and Chen, K. (1997). Dynamic image data compression in spatial and temporal domains: Theory and algorithm. *IEEE Trans. Info. Tech. Biomed.* **1**(4), 219–228.

Ho-Shon, K., Feng, D., Hawkins, R. A., Meikle, S., Fulham, M., and Li, X. (1996). Optimised sampling and parameter estimation for quantification in whole body PET. *IEEE Trans. Biomed. Eng.* **43**, 1021–1028.

Li, X., Feng, D., and Chen, K. (1996). Optimal image sampling schedule: A new effective way to reduce dynamic image storage space and functional image processing time. *IEEE Trans. Med. Imag.* **15**, 710–719.

Patlak, C. S., Blasberg, R. G., and Fenstermacher, J. (1983). Graphical evaluation of blood to brain transfer constants from multiple-time uptake data. *J. Cereb. Blood Flow Metab.* **3**, 1–7.

Automated Delineation of Brain Structures with Snakes in PET

JOUNI M. MYKKÄNEN,* **JUSSI TOHKA,**† **and ULLA RUOTSALAINEN**†

**Department of Computer and Information Sciences, P.O. Box 607, FIN-33014, University of Tampere, Finland*
†*Tampere University of Technology, DMI/Signal Processing Laboratory, Finland*

Delineation of target objects from positron emission tomography (PET) images is difficult to automate because of relatively low resolution, noise, and highly variable functionality between individuals. Deformable models, called snakes, have been studied intensively with high-resolution magnetic images for segmentation. We applied the snake method to PET images and found it to be a useful tool for the automated delineation. A snake can delineate functional structures, improve registration alignment, and correct initialization errors.

I. INTRODUCTION

In brain positron emission tomography (PET), delineation of target objects is an important method for analyzing a reconstructed image. Because of relatively low resolution, noise, and highly variable functional structures, automated methods may encounter difficulties and delineation is usually done by using manual methods; for example, see Huesman *et al.* (1998). New PET scanners produce more and more data, and a fully manual delineation is a laborious and time-consuming task. This puts more pressure on finding automated solutions for PET delineation.

With magnetic resonance (MR) images, there are studies (Kelemen *et al.*, 1998; McInerney and Terzopoulos, 1995; Davatzikos and Bryan, 1996) for the automated delineation with deformable models, snakes (Kass *et al.*, 1988). Snakes can overcome many limitations of traditional segmentation methods such as an easy use of a priori knowledge and a

capability for handling missing data. A survey of deformable models can be found in McInerney and Terzopoulos (1996). In general, a semiautomatic method seems to produce better results than a fully automated method.

Snakes can be used for the delineation of targets in fuzzy biological contexts such as PET images where traditional edge detection methods may fail. A snake can be initialized by a priori information, such as a segmentation obtained from an MR image, and it can be adjusted to the automated delineation of region of interest (ROI) from a PET image.

The aims of this study are to apply the snake method, to find out how to automate it, and learn how to apply a priori information for PET delineation.

II. THEORETICAL ASPECTS OF SNAKE

We applied Lai's (1994) generalized snakes. A snake is a mathematical model for representing shapes and its aim is to minimize its energy (Lai, 1994),

$$E(V) = (1 - \lambda)E_{\text{ext}}(V) + \lambda E_{\text{int}}(V), \qquad (1)$$

where V is a snake, $\lambda \in [0, 1]$ is a regularization parameter, E_{ext} is external image energy, and E_{int} is the internal energy of the snake. The two-dimensional snake V is a set of discrete points; i.e., $V = [v_1, v_2, \ldots, v_n]$, where $v_i = (x, y)$ and $x, y \in \{1, 2, \ldots, N\}$. The internal energy is denoted

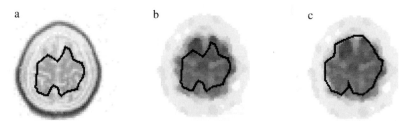

FIGURE 1. Delineation of subject 1, plane 6/35: (a) MRI segmentation, (b) the template snake, and (c) delineation result with the snake from the corresponding FDG-PET image.

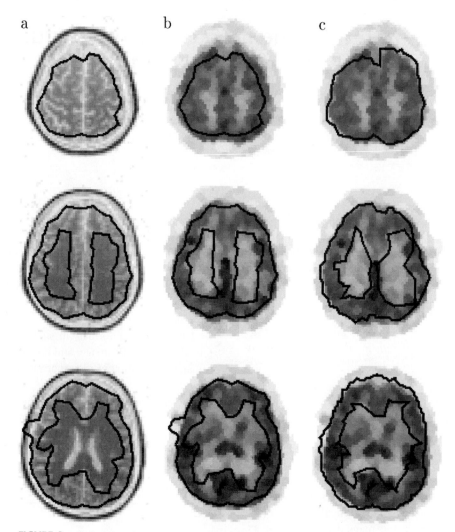

FIGURE 2. Delineation of subject 2, planes 7, 11, and 14/35: (a) MRI segmentation, (b) the template snakes, and (c) delineation results with snakes from the corresponding FDG-PET image.

(Lai, 1994) as

$$E_{\text{int}}(V) = \frac{1}{n} \sum_{i=1}^{n} E_{\text{int}}(v_i)$$

$$= \frac{1}{n} \sum_{i=1}^{n} \frac{1}{l(V)} \|v_i - \alpha v_{i-1} - \beta v_{i+1}\|^2, \quad (2)$$

where α and β come from a shape matrix, which is an affine invariant representation of a shape of a snake, and

$$l(V) = \frac{1}{n} \sum_{i=1}^{n} \|v_i - v_{i-1}\|^2. \quad (3)$$

For the end points, we set $v_0 = v_n$, when $i = 1$, and $v_{n+1} = v_1$, when $i = n$. The shape matrix is initialized by a priori

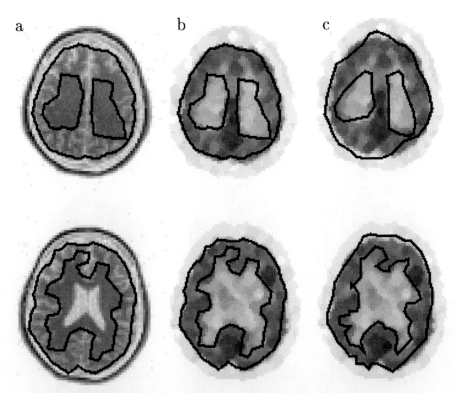

FIGURE 3. Delineation of subject 3, planes 15 and 18/35: (a) MRI and segmentation, (b) PET image and template snakes, and (c) delineation results with snakes from the corresponding FDG-PET image.

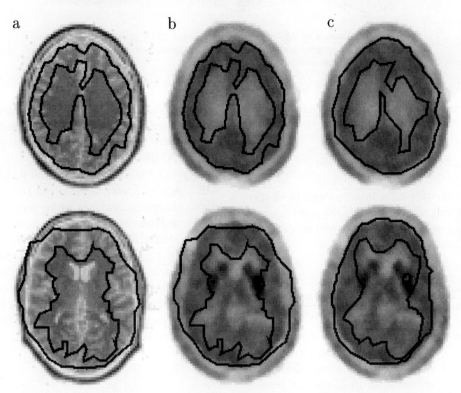

FIGURE 4. Delineation of subject 4, plane 6 and 9/15: (a) MRI and segmentation, (b) pet image and template snakes, and (c) delineation results with snakes from the corresponding FDOPA-PET image.

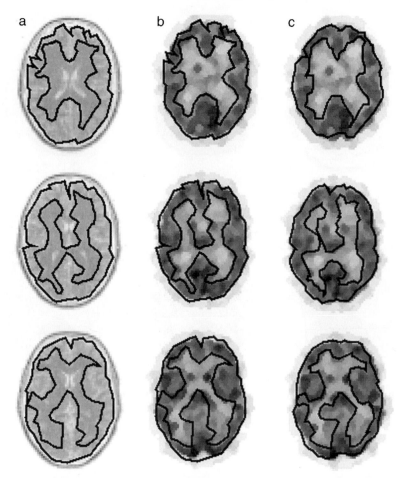

FIGURE 5.   Delineation of subject 5, planes 17, 18, and 20/35: (a) MRI segmentation, (b) PET images and template snakes, and (c) delineation results with snakes from the corresponding FDG-PET image.

information. The external energy E_{ext} is denoted (Lai, 1994) as

$$E_{ext}(V) = \frac{1}{n} \sum_{i=1}^{n} E_{ext}(v_i) = \frac{1}{n} \sum_{i=1}^{n} -\nabla f(x_i, y_i)^T \mathbf{n_i}, \quad (4)$$

where f returns an intensity value from an image to be segmented and $\mathbf{n_i}$ is a unit normal vector in the point (x_i, y_i). A set of two-dimensional snakes is used to find the expected three-dimensional target surface.

III. EXPERIMENTS AND RESULTS

Fluorodeoxyglucose (FDG)-PET scans were made with GE Advance (GE, Milwaukee, WI) and fluorodopa (FDOPA) PET scans with ECAT 931/08-12 (CTI/Siemens, Knoxville, TN). Sinograms were reconstructed to 128×128 resolution by the iterative median root prior (MRP) method (Alenius and Ruotsalainen, 1997). GE and ECAT scanners have

transaxial slice widths of 4.5 and 6.75 mm, respectively. MR images were obtained with 1.5 T Magneton (Siemens, Erlangen, Germany). PET images were coregistered to MR images by AMIR-fit (Ardekani, 1995).

A priori anatomic segmentation was obtained from an MR image by clustering (Jain and Dubes, 1988) tissues to the six clusters: gray matter, white matter, cerebrospinal fluid (CSF), and three background tissues. To get smooth boundaries, a studied cluster was postprocessed by using morphological operations (Russ, 1995). From results, the segments were extracted, and the number of points was deducted to get a template snake. The results were deformed by minimizing (1) with suitable parameter values. Implementation supports the use of multiresolution images and it was used with a part of the cases. Two regularization methods were used, a fixed λ for all points or a local minimum/maximum λ for each point in a snake. For the latter, (2) is

$$E_{int}(V) = \frac{1}{n} \sum_{i=1}^{n} \lambda_i E_{int}(v_i), \quad (5)$$

and (4) is

$$E_{\text{ext}}(V) = \frac{1}{n} \sum_{i=1}^{n} (1 - \lambda_i) E_{\text{ext}}(v_i). \qquad (6)$$

The minimization process takes a few seconds per snake to proceed. The cortex area was delineated for five subjects to evaluate the performance of the snakes. Part of the results is shown in Figs. 1–5.

IV. DISCUSSION

Snakes are able to transfer delineation from one modality to another having different resolution and dynamics, as we can see in Figs. 1–5, where brain structures are transformed from MR to PET. Structures differ because of differences in modality and registration errors.

As Fig. 1 shows, the snake is able to handle relatively large deformations in shape. In the bottom row in Fig. 2, the snake is able to smooth an initial segmentation error. The same effect can also be shown in Fig. 4, which is taken from another PET scanner with a different tracer.

The initialization of snakes is a critical issue and the use of the corresponding MR segmentation with anatomical structures is a solution. Therefore, a good MR image segmentation is required, such as presented in Garza-Jinich *et al.* (1999) and Davatzikos and Bryan (1996). The MR image segmentation was not our main interest at this stage, but it will certainly be an important part of further studies. The limitations of initialization give an opportunity to test the capabilities of snakes. In some cases, functional structures may not have an exact initialization at all. The functionality can be spread to several anatomic structures or it can be smaller than a single anatomic structure. The functional variance between individuals is also large. A suitable structurally driven method should be developed to handle these cases. Actually, this is what we have done by seeking a single functional target with a snake.

This study clearly shows the possibilities of snakes for the automated delineation, but still there is a long way to fully automated delineation. The complexity and variety of targets with functional structures are far too large. As other studies show, the best results are obtained under the guidance of a user.

To summarize this study, snakes can be used with automated delineation of the functional and anatomical targets from noisy and low-resolution PET images. The delineation with snakes can improve the registration alignment and correct errors in a priori information, especially with complex anatomical structures. The snake method with the template-driven PET delineation is certainly an interesting and promising subject for further study and it will be our next aim.

References

Alenius, S., and Ruotsalainen, U. (1997). Bayesian image reconstruction for emission tomography based on median root prior. *Eur. J. Nucl. Med.* **24**(3): 258–265.

Ardekani, B. A. (1995). *"Fusion of Anatomical and Functional Images of the Brain,"* Ph.D. thesis, University of Technology, Sydney.

Davatzikos, C., and Bryan, N. R. (1996). Using a deformable surface model to obtain a shape representation of the cortex. *IEEE Trans. Med. Imaging* **15**: 785–795.

Garza-Jinich, M., Meer, P., and Medina, V. (1999). Robust retrieval of three-dimensional structures from image stacks. *Med. Image Anal.* **3**(1): 21–35.

Huesman, R. H., Klein, G. J., Reutter, B. W., and Teng, X. (1998). Multiscale PET quantitation using three-dimensional volumes of interest. *In "Quantitative Functional Brain Imaging with Positron Emission Tomography"* (R. E. Carson, P. Herscovitch, and M. Daube-Witherspoon, Eds.), Chap. 8, pp. 51–58. Academic Press, San Diego.

Jain, A. K., and Dubes, R. C. (1988). *Algorithms for Clustering Data."* Prentice Hall, New York.

Kass, M., Witkin, A., and Terzopoulos, D. (1987). Snakes: Active contour models. *Int. J. Comput. Vision* **1**: 321–331.

Kelemen, A., Székely, G., and Gerig, G. (1998). Three-dimensional model-based segmentation of brain MRI. *In "Workshop on Biomedical Image Analysis"* (B. Vemuri, Ed.), pp. 4–13. IEEE Computer Society, Santa Barbara, CA.

Lai, K. F. (1994). *"Deformable Contours: Modeling, Extraction, Detection and Classification,"* Ph.D. thesis, University of Wisconsin–Madison.

McInerney, T., and Terzopoulos, D. (1995). Topologically adaptable snakes. *In "Proceedings of the Fifth International Conference on Computer Vision (ICCV'95),"* pp. 840–845. IEEE Computer Society Press, Santa Barbara, CA.

McInerney, T., and Terzopoulos, D. (1996). Deformable models in medical image analysis: A survey. *Med. Image Anal.* **1**(2).

Russ, J. C. (1995). *"The Image Processing Handbook,"* 2nd ed. CRC Press, Boca Raton, FL.

Analysis of Functional Imaging Data Sets via Functional Segmentation

ROGER N. GUNN, JOHN ASHBURNER,* JOHN ASTON, and VINCENT J. CUNNINGHAM

MRC Cyclotron Unit, Hammersmith Hospital, London, UK
Functional Imaging Laboratory, Queen Square, London, UK

A method for the analysis of functional imaging data sets including PET ligand displacement, PET-CBF, and fMRI studies is presented. The principle involves the functional segmentation of the data sets in the same stereotaxic space across subjects and conditions. No a priori information is required for the functional segmentation and the initial part of the analysis is purely data driven. Subsequently, simple statistical tests may be applied to the functionally segmented regions of interest.

I. INTRODUCTION

Functional imaging data sets in the brain are often analyzed across subjects and conditions. The analysis may be divided into two stages: first, registration (intrasubject) and spatial normalization (intersubject) of image volumes into a stereotaxic space, second, application of either data-driven (such as principal components analysis (PCA)) or hypothesis-driven (general linear model) methods at the voxel level for analysis.

Cluster analysis has recently been applied to the analysis of fMRI data sets (Baumgartner *et al.*, 1997; Moser *et al.*, 1997; Goutte et al., 1999). This has certain advantages over principal components analysis (Friston *et al.*, 1993) in terms of interpretation of identified components. The characteristic functions identified by cluster analysis are maintained in the same space as the original data, whereas eigenvectors derived from PCA are rotated. The application of independent component analysis (ICA; McKeown *et al.*, 1998) is also proving to be a useful tool in the analysis of fMRI data

and this also maintains the extracted components in the original space easing interpretation.

Here a technique is presented which uses cluster analysis to functionally segment the spatially normalized image volumes. Subsequently, the identified response functions are analyzed with simple statistical tests. This attempts to form a hybrid between the data-driven and hypothesis-driven methods alluded to by McKeown in his response to Friston (1998).

Simulated data is analyzed for a simple phantom and a ligand displacement study. Measured data sets from PET blood flow (single subject) and fMRI (single subject) are also analyzed.

II. MATERIALS AND METHODS

A. Functional Segmentation and Analysis

Cluster analysis was used to segment the stereotaxically normalized data sets across the different measured volumes. $N (=10)$ clusters were chosen which dictated that the data sets were segmented into N regions of interest. Characteristic data was extracted from each region of interest and statistically assessed using a t test. As only N statistical tests are performed a simple Bonferroni correction for N multiple comparisons was applied.

Cluster analysis has been implemented based on an adaptation of a mixture model (Hartigan, 1975; Ashburner *et al.*, 1996). The analysis is performed on the data sets by normalizing each individual response function such that the inte-

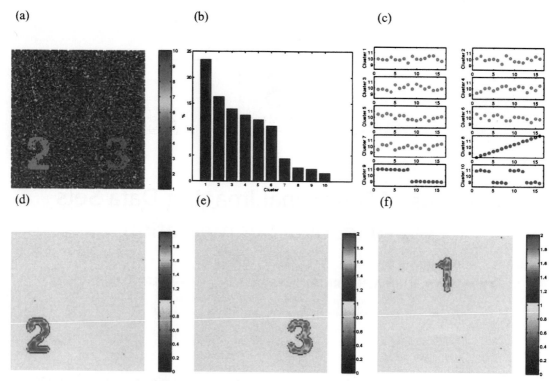

FIGURE 1. Functional segmentation of simulated data set. (a) Cluster segmentation into 10 response functions, (b) proportion of voxels contained in each cluster, (c) response functions for each cluster, (d) probability of voxel belonging to cluster 8 superimposed on normalized integral image of all scan volumes, (e) probability of voxel belonging to cluster 9 superimposed on normalized integral image of all scan volumes, (f) probability of voxel belonging to cluster 10 superimposed on normalized integral image of all scan volumes. (For d, e, f colorbar 1–2 represents *p* value 0–1.)

gral is equal to one. This allows for an analysis that simply interrogates the shape of the response functions and ignores their magnitude. The clustering is described in more detail by Ashburner *et al.* (1996), but briefly, iterations are performed and the cluster centers optimized (maximization of the log-likelihood function), with the gradual introduction of the full complement of clusters (N). The whole process yields N response functions each with an associated probability volume; hence all volumes contain values between 0 and 1. The *p* values at each voxel determine how likely it is that the voxel characteristics are described by the associated response function. A segmentation volume is generated by assigning each voxel to the cluster for whom its *p* value is greatest (examples of these segmentation volumes for various examples may be seen in Figs. 1a, 2a, 3a, and 4a). For presentation the clusters have been reordered such that they are ranked in terms of the proportion of voxels which they best characterize.

B. Data Sets

Four different data sets were investigated:

Data set 1: A single plane of data (128×128 voxels) was generated containing 16 frames of data. Three separate re-

gions were given different characteristic response functions and the background was maintained as a constant. Noise was then added to the whole data set (SNR = 1).

Data set 2: Simulated [^{11}C]SCH 23390 ligand displacement data set, consisting of six baseline binding potential (BP) image volumes and six activation (20% reduction in k_3 for 0–20 min for the left caudate) BP volumes. Dynamic data were generated using a plasma metabolite-corrected input function, a two-tissue compartment model and a segmented MRI consisting of 28 regions. Typical parameter values were assigned to the regions and dynamic PET data was generated using a PET simulator. This models the whole acquisition and reconstruction process (Ma and Evans, 1997). These data sets were analyzed using a basis function implementation of the simplified reference tissue model (Gunn *et al.*, 1997) to generate the BP volumes.

Data set 3: A single subject PET blood flow activation data set was downloaded from http://www.fil.ion.ucl.ac.uk/ spm/data. The data set consisted of 12 scans, odd scans were activation and even scans were baseline. In the activation scans the subject used their left hand to perform a finger opposition task which involved touching their thumb to their index finger, to their middle finger, to their ring finger, to

(a)

(b)

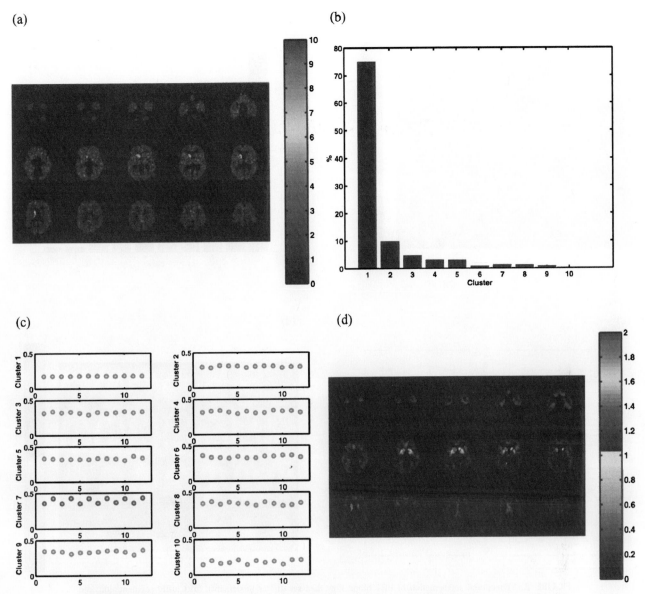

(c)

(d)

FIGURE 2. Functional segmentation of simulated PET ligand displacement data set. (a) Cluster segmentation into 10 response functions, (b) proportion of voxels contained in each cluster, (c) response functions for each cluster, (d) probability of voxels belonging to cluster 7 superimposed on normalized integral image of all scan volumes. (Colorbar 1–2 represents p value 0–1.)

their pinky, and then repeating this process. The subject performed this at a rate of 2 Hz, as guided by a visual clue. For baseline, there was no finger movement, but the visual cue was still present.

Data set 4: A single subject fMRI blood flow activation data set was downloaded from http://www.fil.ion.ucl.ac.uk/ spm/data. A subset of this data set was used using six blocks of six scans (36 scans), where odd blocks were activation and the even blocks were baseline. Each scan was acquired over 7 s and hence each block represented a duration of 42 s. The condition for successive blocks alternated between rest and

auditory stimulation, during which the subject was presented binaurally with bisyllabic words at a rate of 60 per minute.

III. RESULTS AND DISCUSSION

A successful segmentation of all the data sets was achieved. These analyses took at most a few minutes on an Ultra 1 Sun workstation.

Data set 1 (Fig. 1): The three distinct components were clearly identified for a signal-to-noise ratio of 1.

(a)   (b)

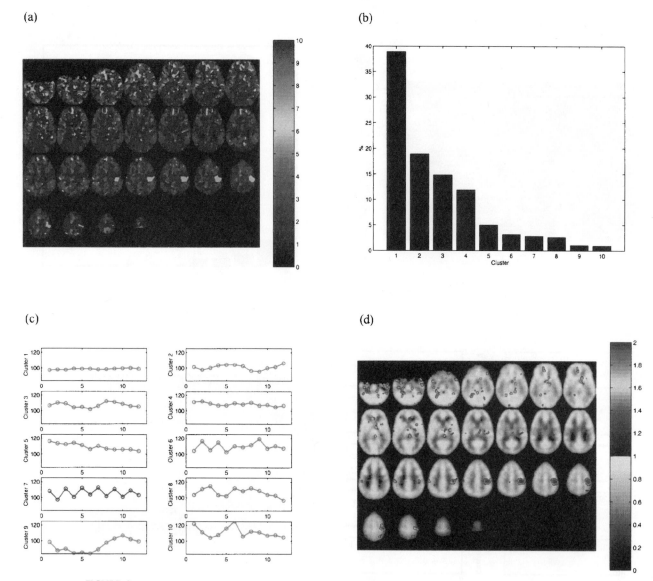

(c)   (d)

FIGURE 3. Functional segmentation of PET blood flow data set (finger movement). (a) Cluster segmentation into 10 response functions, (b) proportion of voxels contained in each cluster, (c) response functions for each cluster, (d) probability of voxels belonging to cluster 7 superimposed on normalized integral image of all scan volumes. (Colorbar 1–2 represents *p* value 0–1.)

Data set 2 (Fig. 2): The left caudate was identified as the only significant component ($p < 0.00005$ [paired *t*-test], magnitude, −16%). The response function can be seen as cluster 7 in Fig. 2c.

Data set 3 (Fig. 3): A single significant cluster was identified in the motor cortex ($p < 0.00013$ [*t*-test], magnitude, 9.3%), consistent with previous analyses using SPM. The response function can be seen as cluster 7 in Fig. 3c.

Data set 4 (Fig. 4): A single significant cluster was identified within the auditory cortex ($p < 0.000002$ [*t* test], magnitude, 1.3%). Furthermore, the time course of the hemodynamic response was directly characterized

(Fig. 4c, cluster 5) and thus would allow for more sophisticated modelling and statistical inference.

The method performed robustly on the data sets considered, but further testing is required to investigate if the clustering algorithm can fall into local minima. Further simulations and analyses are also required to characterize the ideal number of clusters and to consider the option of hierarchical clustering.

A generic method has been developed for the analysis of functional imaging data sets. The method involves two stages, the first a purely data-led functional segmentation of the data, followed by statistical inference across subjects and conditions.

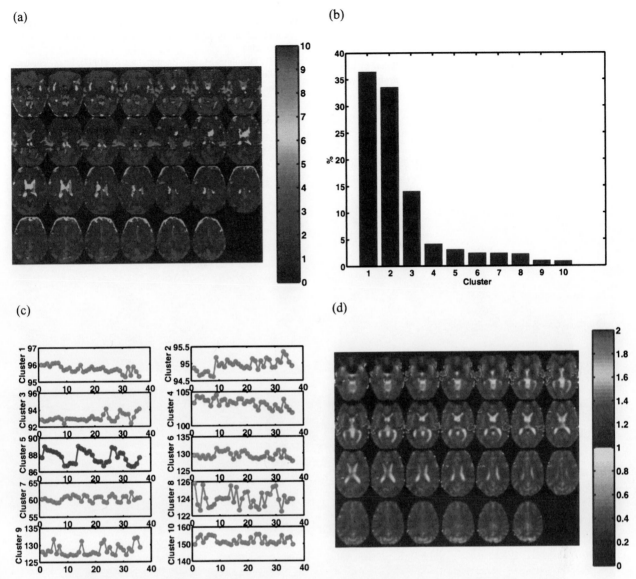

FIGURE 4. Functional segmentation of fMRI data set (auditory stimulation). (a) Cluster segmentation into 10 response functions, (b) proportion of voxels contained in each cluster, (c) response functions for each cluster, (d) probability of voxels belonging to cluster 5 superimposed on normalized integral image of all scan volumes. (Colorbar 1–2 represents p value 0–1.)

Acknowledgments

The majority of this work was performed while at the Montreal Neurological Institute. I thank the staff for their help and constructive discussions, in particular Alan Evans, Alain Dagher, Keith Worsley, and Peter Neelin.

References

Ashburner, J., Haslam, J., Taylor, C., Cunningham, V. J., and Jones, T. (1996). A cluster analysis approach for the characterization of dynamic PET data. *In "Quantification of Brain Function using PET"* (R. Myers, V. J. Cunningham, D. L. Bailey, and T. Jones, Eds.), pp. 301–306. Academic Press, San Diego.

Baumgartner, R., Scarth, G., Teichtmeister, C., Somorjai, R., and Moser, E. (1997). Fuzzy clustering of gradient-echo functional MRI in the human visual cortex. I. reproducibility. *J. Magn. Reson. Image.* **7**(6): 1094–1101.

Gunn, R. N., Lammertsma, A. A., Hume, S. P., and Cunningham, V. J. (1997). Parametric imaging of ligand–receptor binding in PET using a simplified reference region model. *NeuroImage* **6**: 279–287.

Hartigan, J. A. (1975). *"Clustering Algorithms,"* pp. 113–129. Wiley, New York.

Friston, K. J., Frith, C. D., Liddle, P. F., and Frackowiak, R. S. (1993). Functional connectivity: The principal-component analysis of large (PET) data sets. *J. Cereb. Blood Flow Metab.* **13**: 5–14.

Friston, K. J. (1998). Modes or models: A critique on independent component analysis for fMRI. *Trends Cognit. Sci.* **2**: 373–375.

Goutte, C., Toft, P., Rostrup, E., Nielsen, F. A., and Hansen, L. K. (1999). On clustering fMRI time series. *NeuroImage* **9**: 298–310.

McKeown, M. J., Makeig, S., Brown, G. G., Jung, T.-P., Kindermann, S. S., Bell, A. J., and Sejnowski, T. J. (1998). Analysis of fMRI data by blind separation into independent spatial components. *Hum. Brain Mapping* **6**: 160–188.

Ma, Y., and Evans, A. C. (1997). Analytical modeling of PET imaging with correlated functional and structural images. *IEEE Trans. Nucl. Sci.* **44**: 2439–2444.

Moser, E., Diemling, M., and Baumgartner, R. (1997). Fuzzy clustering of gradient-echo functional MRI in the human visual cortex. II. Quantification. *J. Magn. Reson. Image* **7**(6): 1102–1108.

Preliminary Studies of Computer Aided Ligand Design for PET

A. J. ABRUNHOSA,[*,†] **F. BRADY,**[*] **S. LUTHRA,**[*] **J. J. DE LIMA,**[†] **and T. JONES**[*]

[*]*MRC Cyclotron Unit, Hammersmith Hospital, London W12 ONN, UK*
[†]*IBILI, Faculty of Medicine, Az. de Sta Comba, Celas P-3000-354 Coimbra, Portugal*

The selection of a suitable radioligand for in vivo molecular imaging requires that contributions from nonselective, non-specific binding and the presence of radiolabeled metabolites are minimized. Several approaches have been made to relate in vivo behavior of positron emission tomography (PET) ligands with the in vitro and in vivo data and other experimental or calculated molecular properties. These approaches have been based on simple graphical analysis and the use of one or two measured parameters. Here we have attempted to generate statistically significant models that correlate some characteristics of the in vivo behavior of PET ligands with a combination of molecular properties derived solely from their chemical structures. Some of the most useful calculated properties include measures of lipophilicity, charge, bulkiness, and other parameters related to intermolecular interactions. Statistical methods are used to build models that relate the in vivo behavior of the compounds with their calculated molecular properties. Examples are given of such models derived for in vivo specificity (dopamine uptake site and dopamine D_2 receptor radioligands). This approach offers the potential to improve our understanding of the intermolecular interactions involved in the in vivo behavior of PET ligands. The models generated could also be used to assist in the selection of molecules for labeling even when minimal information about the compounds is available. The automation of this procedure could open the possibility of screening real or virtual libraries of compounds in search of new specific probes for in vivo molecular imaging.

I. INTRODUCTION

The criteria of a suitable radioligand for *in vivo* molecular imaging relies on the identification of compounds that exhibit suitable *in vivo* behavior. One of the most important aspects in this quest is to maximize the strength of the specific over the nonspecific signal (Fig. 1). To maximize the specific component of the signal, compounds are selected that exhibit good affinity for a particular molecular target, expressed in good concentration, in certain areas of the brain. To minimize the nonselective component, compounds are selected with minimal affinity for other related molecular targets. Additional interactions, more difficult to predict, play a vital role in the *in vivo* behavior of the compounds and many times impair their development as successful PET radioligands. In order to reach the molecular targets the compound has to cross the relevant biological barriers (e.g., the blood–brain barrier) and not bind strongly to non-specific components in plasma and in tissue. The formation of metabolites, especially when they can access the same specific compartments can have a very detrimental effect and the choice of labeling position can play a very important role here.

The nature of all these processes is nevertheless the same. It involves a series of molecular interactions between the radioligand and a variety of other molecules. Those interactions are ultimately a function of the molecule's physico-chemical properties. A prior knowledge of these properties and their role in the various processes can potentially provide very valuable tools in the selection and development of new PET radioligands.

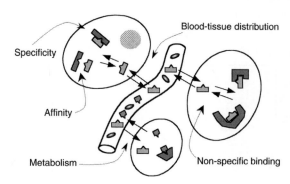

FIGURE 1. Simplified diagram of factors influencing the *in vivo* behavior of radioligands.

Several attempts have been made to relate some of the *in vitro* and *in vivo* behavior of candidate PET radioligands and some of their experimentally obtained (Zoghbi *et al.*, 1997), or calculated (Reichert *et al.*, 1997a) molecular properties. Although there are examples of the use of models for affinity (Reichert *et al.*, 1997b), most of the models derived have been directed toward maximising the *in vivo* signal (Kessler *et al.*, 1991; Stöcklin, 1992; Pike, 1993). These approaches have been based on simple graphical analysis and the use of one or two molecular properties. The log of the partition coefficient between *n*-octanol and water (log P) is one of the most cited. It has been related, among others, with brain penetration (Pardridge, 1998), specific to nonspecific binding (Coenen *et al.*, 1988), free fraction in tissue (Delforge *et al.*, 1996), and plasma protein binding (Osman, pers. commun.). Charge is another classical physicochemical property usually associated with polar intermolecular interactions and pK_a has been used either as a component of log P (to obtain log D) or on its own (Zoghbi *et al.*, 1997). Steric interactions have been limited to considerations about size (molecular mass) and its role in the ability of the compound to cross the blood–brain barrier (Pardridge, 1998). Nevertheless, no study has been made so far to assess which combination of individual molecular properties influence the various processes that make up the *in vivo* behavior of radioligands.

One of the problems is the availability of the data. We previously reported the work we have done in the creation of an in-house database of reported compounds labeled with positron emitters (Abrunhosa *et al.*, 1998). We have recently expanded this resource to include a collection of more than 40 calculated properties for all compounds targeted at the brain.

Our aim was to relate some of these calculated molecular properties to the characteristics that are desirable in a good *in vivo* radioligand.

As our starting example we chose to focus on the dopaminergic system. A substantial effort has been made in recent years on the design of radioligands for this system (for reviews, see Mazière *et al.*, 1992; Pike, 1993). We selected

two particular molecular targets in this system with sufficient *in vivo* data for the generation of models for specificity. The first one was a series of tropane analogs, labeled with positron emitters as radioligands for the dopamine uptake site. The second was a group of ligands for the dopamine D_2 receptor. All the generated models were validated using cross-validation techniques. The results obtained were very encouraging for the potential of this methodology as a valuable tool in the process of selection and development of radioligands for *in vivo* molecular imaging.

II. MATERIALS AND METHODS

A. Compound Selection

Molecular structures of a series of dopamine D_2 receptor and dopamine transporter radioligands, having *in vivo* specificity data from primates (striatum:cerebellum ratios) were taken from our in-house PET Chemistry database.

1. Radioligands for the Dopamine Uptake Site

For the studies with ligands for the dopamine uptake site, a series of 10 2β-3β-substituted-phenyl-tropane analogs with published data in nonhuman primates was selected. The selected compounds along with their parent structure are listed in Table 1 along with the reported striatum:cerebellum ratios at 60 min postinjection.

2. Radioligands for the Dopamine D_2 Receptor

For the studies with ligands for the dopamine D_2 receptor, a series of eight compounds with published striatum to cerebellum ratios in nonhuman primates was selected (Table 2). The compounds belong to three different chemical classes: 10 butyrophenones (parent structures Ia–Ic), 3 substituted benzamides (II), and the nonselective ligand clozapine (III).

B. Molecular Modeling

Energy minimized 3D structures of these compounds were computed using a mechanistic algorithm (method MM2, PersonalCache 2.1 Oxford Molecular) and a series of calculated molecular properties were derived from structure (Molecular Modelling Pro 3.1, ChemSW Inc). Calculated descriptors included general properties, e.g., hydrophilic/lipophilic parameters (e.g., clog P, clog D, hydrophilic surface area, hydrophilic/lipophilic balance), volumetric (e.g., polarizability, molar refractivity) solvent interaction (Hansen's 3D parameters), and electronic (e.g., charge, dipole moment) and hydrogen bonding (total, acceptor, and donor). Some local parameters were also calculated for specific series of analogs, e.g., interatomic distances,

TABLE 1. List of 2β-3β-Substituted-Phenyl-Tropane Analogs with *in Vivo* Striatum:Cerebellum (S/C) Values
Included in the Model for the Dopamine Uptake Site

Compound	R1(R4)	R2	R3	S/C
RTI-55/ß-CIT	I	CH3	CH3	5
WIN-35428/ß-CFT	F	CH3	CH3	4.5
ß-CIT-FE	I	CH3	(CH2)2F	9
ß-CIT-FP	I	CH3	(CH2)3F	8
RTI-131/CCT	Cl	CH3	CH3	3
CDCT	Cl+R4=Cl	CH3	CH3	2.3
FPCT	Cl	CH3	(CH2)3F	2.65
ß-CBT	Br	CH3	CH3	2.5
ß-CBT-FE	Br	CH3	(CH2)2F	12
ß-CBT-FP	Br	CH3	(CH2)3F	6

Note. Data taken from Farde *et al.*, 1994; Loc'h *et al.*, 1995; and Lunkvist *et al.*, 1995.

TABLE 2. List of Compounds Labeled for the Dopamine D_2 Receptor Included in the Model for Specificity, along with
Their Parent Structures

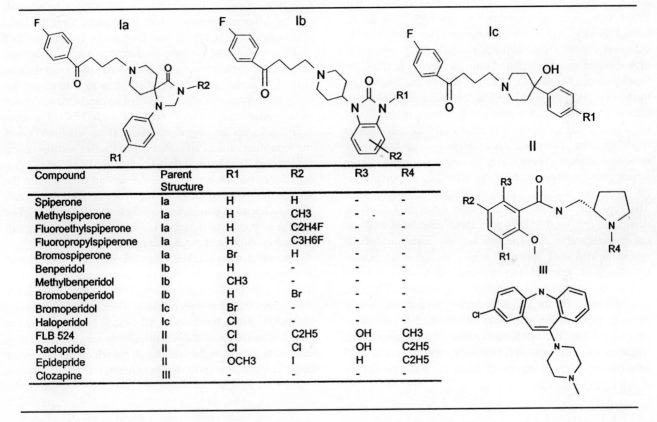

Compound	Parent Structure	R1	R2	R3	R4
Spiperone	Ia	H	H	-	-
Methylspiperone	Ia	H	CH3	-	-
Fluoroethylspiperone	Ia	H	C2H4F	-	-
Fluoropropylspiperone	Ia	H	C3H6F	-	-
Bromospiperone	Ia	Br	H	-	-
Benperidol	Ib	H	-	-	-
Methylbenperidol	Ib	CH3	-	-	-
Bromobenperidol	Ib	H	Br	-	-
Bromoperidol	Ic	Br	-	-	-
Haloperidol	Ic	Cl	-	-	-
FLB 524	II	Cl	C2H5	OH	CH3
Raclopride	II	Cl	Cl	OH	C2H5
Epidepride	II	OCH3	I	H	C2H5
Clozapine	III	-	-	-	-

bond angles, partial charges and calculated bond orders (using CNDO, a semiempirical quantum mechanics method).

Statistical analysis was performed using multiple linear regression (MLR) to try to relate their *in vivo* specificity to some of the molecular properties (Molecular Analysis Pro 3.52, ChemSW Inc). Due to the small number of observations ($n = 10$–14), linear models of up to only three descriptors were used. The equations were of the form:

$$R = c_1 P_1 + c_2 P_2 + c_3 P_3 + I, \qquad (1)$$

in which, R represents the response parameter; P_1, P_2, and P_3 the molecular properties, c_1, c_2 and c_3 their coefficients, and I the intercept.

Cross-validation of the models generated was carried out using predictive residual sum of squares (PRESS) analysis and models were considered acceptable if they showed a PRESS/sum of squares of the response variable (SSY) below 0.4. As an additional cross-validation procedure (model for the D_2 only), the calculated value for each one of the compounds was also predicted by models that exclude the response values for that specific compound in the calculations ("leave one out" analysis).

III. RESULTS AND DISCUSSION

A. Dopamine Transporter Radioligands

An attempt was made to relate the *in vivo* specificity of the selected tropane analogs with some of the calculated molecular properties. In this case, because all the compounds share the same parent structure (Table 1), some local properties were also included. This approach usually improves the accuracy of the model, but restricts the application of the equation to analogs of the compounds selected. Some of these local properties were shown in a previous study to be correlated ($r^2 = 0.98833$) with the reported affinity (K_i) of these compounds for the dopamine transporter (author's unpublished results). The *in vivo* property selected was the reported ratio of activity in the striatum, a receptor-rich area, over the one in the cerebellum (s/c) at 60 min postinjection. The data were collected from studies performed with nonhuman primates, as there were not enough studies reported in man to derive a consistent model and the primate data is usually close to the one observed in humans. There was no *a priori* assumption as to which physicochemical properties would be important and all the global and local properties calculated were used. The reported striatum to cerebellum ratio showed to be correlated ($r^2 = 0.93118$) with three calculated molecular properties (Fig. 2). The first was the capability of the molecule to donate hydrogen bonds ($H_{bond\rightarrow}$), a parameter important in membrane crossing and binding to other molecules (see discussion in the next section). The second property was the calculated magnitude of the dipole moment (μ), another parameter involved in intermolecular interactions. The last property was the deviation of the length of the substituent in R3 from the more favorable ethyl substructure (d_{R3}), a parameter taken from the model for

affinity. The variation of the length of substituents is often related to ligand selectivity for the specific molecular target.

B. Dopamine D2 Receptor Radioligands

A similar study was performed with compounds targeted at the dopamine D_2 receptor (Table 2). The response variable was again the reported striatum to cerebellum ratio as a measure of the specific over the nonspecific binding. A total of 14 compounds with reported values in nonhuman primates were used. The list of compounds reflects many years of radiopharmaceutical development for the dopaminergic system. It includes 3 compounds from the most recent class of reversible radioligands (substituted benzamides), 10 of the less selective and irreversible butyrophenones and 1 nonselective radioligand, clozapine. In an attempt to normalize for different kinetics, values were taken from the literature references at the 60-min time point. Due to the fact that the compounds belong to different parent structures, only global parameters were computed. The model derived (Fig. 3) correlates the reported striatum to cerebellum activity with three calculated physicochemical characteristics of the compounds ($r^2 = 0.91539$, $P < 0.0001$). The most correlated with the response variable ($r = -0.49483$), was the logarithm of the calculated octanol:water partition coefficient (log P). Also correlated well with the reported striatum to cerebellum activity was the second parameter in the model: the ability of the molecule to donate hydrogen bonds ($H_{bond\rightarrow}$) with a correlation coefficient of -0.47114. The third contribution to the model ($r = -0.16273$) was the calculated magnitude of the dipole moment, μ.

The striatum to cerebellum ratios obtained by the model constructed with the described parameters are compared, in Table 3, with the literature-reported values. The values in the third column are the ones obtained when all the compounds are used to build the model (full model) and are plotted in Fig. 3. The values in the fourth column are calculated with models that exclude the compound for which the values are being calculated ("leave one out" analysis). The results are

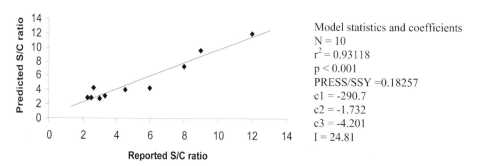

Model statistics and coefficients
N = 10
r^2 = 0.93118
p < 0.001
PRESS/SSY = 0.18257
c1 = -290.7
c2 = -1.732
c3 = -4.201
I = 24.81

FIGURE 2. Predicted vs reported striatum to cerebellum activity measured at 60 min postinjection in nonhuman primates for a series of dopamine transporter ligands. Parameters included in the model: $P_1 = H_{bond\rightarrow}$, $P_2 = \mu$, $P_3 = d_{R3}$.

very similar to the ones obtained with the full model, as we can observe from the corresponding residual (last column in Table 3). The low PRESS/SSY of 0.17210 is also a good indication that the model is not due to a chance correlation in the data.

Although the selection of parameters to include in the model was based only on the statistical significance and the cross-validation results, the included variables are quite relevant in view of the intermolecular interactions involved (Fig. 1). The negative contribution of log P in the model suggests that an increase in lipophilicity is unfavorable. This is in agreement with an increase in nonspecific binding of the molecules to lipids and proteins in the tissue due to an increase in lipophilic interactions. The negative effect of the ability to donate hydrogen bonds, also found in the dopamine transporter model, can be explained in a similar way. Hydrogen bonds play an important role in the process of binding small molecules to proteins. The third factor, also with a negative effect and in common with the tropane series, is again an important factor in intermolecular interactions. An increase in the magnitude of the dipole moment would increase dipole–dipole and dipole–ion interactions with the surrounding molecules. In conclusion, the greater the ability of the molecule to establish intermolecular interactions, the greater the probability that it will, *in vivo*, exhibit high nonspecific binding. This principle could, in theory, be applied to any radioligand, but, due to the small number of compounds included in the models, should be taken as an indication valid for compounds related with the modeled structures.

We think that the type of approach outlined in this work may constitute an important factor in the process of under-

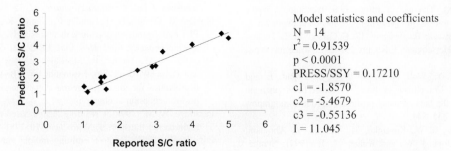

Model statistics and coefficients
$N = 14$
$r^2 = 0.91539$
$p < 0.0001$
PRESS/SSY $= 0.17210$
$c1 = -1.8570$
$c2 = -5.4679$
$c3 = -0.55136$
$I = 11.045$

FIGURE 3. Predicted vs reported striatum to cerebellum activity measured at 60 min postinjection in nonhuman primates for a series of dopamine D2 receptor ligands. Parameters included in the model: $P_1 = \log P$, $P_2 = H_{bond\rightarrow}$, $P_3 = \mu$.

TABLE 3. Striatum to Cerebellum Ratios Reported in the Literature for a Series of D_2 Receptor Ligands in Nonhuman Primates, Their Calculated Values According to the Model Indicated in Fig. 3, the Cross-Validation Results and the Predictive Residuals

Compound	Reported*	Predicted (full model)	Predicted (leave one out)	Predictive residual
Haloperidol	1.5	1.78	1.85	0.07
Bromobenperidol	1.15	1.16	1.16	0.00
Benperidol	3	2.75	2.72	−0.03
Bromoperidol	1.05	1.49	1.62	0.13
Bromospiperone	1.5	2.01	2.12	0.11
Clozapine	1.25	<1	<1	−0.29
Epidepride	4	4.07	4.07	0.00
Fluoroethylspiperone	4.8	4.74	4.71	−0.03
FLB 524	1.65	1.30	1.02	−0.28
Fluoropropylspiperone	5	4.44	4.14	−0.30
N-methyl-benperidol	1.6	2.06	2.27	0.21
N-methyl-spiperone	$2.9^{\dagger}$–4	2.68	2.43	−0.25
Raclopride	$2.5^{\dagger}$–5	2.47	2.45	−0.02
Spiperone	3.2	3.61	3.75	0.14

*Data taken from Arnett, 1985; Coenen, 1988; Farde, 1985; Hartvig, 1988; Kessler, 1993; Moerlein, 1986, and Suehiro, 1990.
† Value used in the model.

standing the molecular basis of the *in vivo* behavior of our radioligands, therefore an important tool in radioligand development.

The models derived, based only on calculated physicochemical properties, can be applied to virtually any molecule for which the molecular structure is known. The extension of this concept could ultimately allow the screening of real or virtual libraries of molecules in the search for new, more specific, radiopharmaceuticals.

Acknowledgments

This work was supported by a grant from the Portuguese Science and Technology Foundation (FCT PRAXIS XXI/4563/94).

References

Abrunhosa, A. J., Brady, F., Luthra, S. K., Morris, H., de Lima, J. J., and Jones, T. (1998). The use of Information Technology in the search for new PET tracers. *In "Quantitative Functional Brain Imaging with Positron Emission Tomography"* (R. E. Carson, M. E. Daube-Witherspoon, and P. Herscovitch, Eds.), pp. 267–271. Academic Press, San Diego.

Arnett, C. D., Shiue, C.-Y., Wolf, A. P., Fowler, J. S., Logan, J., and Watanabe, M. (1985). Comparison of three ^{18}F-labeled butyrophenone neuroleptic drugs in the baboon using positron emission tomography. *J. Neurochem.* **44**(3): 835–844.

Carroll, F. I., Mascarella, S. W., Kuzemko, M. A., Gao, Y., Abraham, P., Lewin, A. H., Boja, J. W., and Kuhar, M. J. (1994). Synthesis, ligand binding and QSAR (CoMFA and classical) study of 3β-(3′-substituted phenyl)-, 3β-(4′-substituted phenyl)-, and 3β-(3′,4′-disubstituted phenyl)tropane-2β-carboxylic acid methyl esters. *J. Med. Chem.* **37**: 2865–2873.

Coenen, H. H., Wienhard, K., Stöcklin, G., Laufer, P., Hebold, I., Pawlik, G., and Heiss, W.-D. (1988). PET measurement of D_2 and S_2 receptor binding of 3-N-([2′-^{18}F]fluoroethyl)spiperone in baboon brain. *Eur. J. Nucl. Med.* **14**: 80–87.

Delforge, J., Syrota, A., and Bendriem, B. (1996). Concept of reaction volume in the *in vivo* ligand-receptor model. *J. Nucl. Med.* **37**: 118–125.

Farde, L., Ehrin, E., Eriksson, L., Greitz, T., Hall, H., Hedström, C.-G., Litton, J.-E., and Sedvall, G. (1985). Substituted benzamides as ligands for visualisation of dopamine receptor binding in the human brain by positron emission tomography. *Proc. Natl. Acad. Sci. USA.* **82**: 3863–3867.

Farde, L., Halldin, C., Müller, L., Suhara, T., Karlsson, P., and Hall, H. (1994). PET study of [^{11}C]β-CIT binding to monoamine transporters in the monkey and human brain. *Synapse* **16**: 93–103.

Hartvig, P., Eckernäs, Å., Ekblom, B., Lindström, L., Lundqvist, H., Axelsson, S., Fasth, K. J., Gullberg, P., and Långström, B. (1988). Receptor binding and selectivity of three ^{11}C-labelled dopamine receptor antagonists in the brain of Rhesus monkeys studied with positron emission tomography. *Acta Neurol. Scand.* **77**: 314–321.

Kessler, R. M., Ansari, M. S., De Paulis, T., Schmidt, D. E., Clanton, J. A., Smith, H. E., Manning, R. G., Gillespie, D., and Ebert, M. H. (1991). High affinity dopamine D2 receptor radioligands. 1. Regional rat brain distribution of iodinated benzamides. *J. Nucl. Med.* **32**(8): 1593–1600.

Kessler, R. M., Votaw, J. R., Schmidt, D. E., Ansari, M. S., Holdeman, K. P., de Paulis, T., Clanton, J. A., Pfeffer, R., Manning, R. G., and Ebert, M. H. (1993). High affinity dopamine D2 receptor radioligands. 3. [^{123}I] and [^{125}I]Epidepride: *In vivo* studies in Rhesus monkey brain and comparison with *in vitro* pharmacokinetics in rat brain. *Life Sci.* **53**: 241–250.

Loc'h, C., Halldin, C., Hantraye, P., Swahn, C.-G., Lundqvist, C., Patt, J., Mazière, M., Farde, L., and Mazière, B. (1995). Preparation and PET evaluation of radiobrominated cocaine analogues for quantitation of monoamine uptake sites. *J. Lab. Comp. Radiopharm.* **37**: 64–65.

Lunkvist, C., Halldin, C., Swahn, C.-G., Müller, L., Ginovart, N., Nakashima, Y., Karlsson, P., Neumeyer, J. L., Wang, S., Milius, R. A., and Farde, L. (1995). Synthesis of [11]C- or ^{18}F-labelled analogues of β-CIT. Labelling in different positions and PET evaluation in cynomolgus monkeys. *J. Lab. Comp. Radiopharm.* **37**: 52–54.

Mazière, B., Coenen, H. H., Halldin, C., Någren, K., and Pike, V. W. (1992). PET radioligands for dopamine receptors and re-uptake sites: Chemistry and biochemistry. *Nucl. Med. Biol.* **19**(4): 497–512.

Moerlein, S. M., Laufer, P., Stöcklin, G., Pawlik, G., Wienhard, K., and Heiss, W.-D. (1986). Evaluation of [75]Br-labelled butyrophenone neuroleptics for imaging cerebral dopaminergic receptor areas using positron emission tomography. *Eur. J. Nucl. Med.* **12**: 211–216.

Pardridge, W. M. (1998). CNS drug design based on principles of blood-brain-barrier transport. *J. Neurochem.* **70**: 1781–1792.

Pike, V. W. (1993). Positron emitting radioligands for studies *in vivo*—Probes for human psychopharmacology. *J. Psychopharmacol.* **7**(2): 139–158.

Reichert, D. E., and Welch, M. J. (1997b). Comparative molecular force field analysis (CoMFA) studies on radiolabeled estradiol derivatives. *In "XII International Symposium on Radiopharmaceutical Chemistry,"* pp. 827–829. Uppsala, Sweden.

Reichert, D. E., Cutler, C. S., Anderson, C. J., Jones-Wilson, T. M., and Welch, M. J. (1997a). Molecular mechanics and semi-empirical studies of radiometal complexes. *In "XII International Symposium on Radiopharmaceutical Chemistry,"* pp. 825–826. Uppsala, Sweden.

Stöcklin, G. (1992). Tracers for metabolic imaging of brain and heart. Radiochemistry and radiopharmacology. *Eur. J. Nucl. Med.* **19**: 527–551.

Suehiro, M., Dannals, R. F., Scheffel, U., Stathis, M., Wilson, A. A., Ravert, H. T., Villemagne, V. L., Sanchez-Roa, P. M., and Wagner, H. N., Jr. (1990). *In vivo* labeling of the dopamine D2 receptor with N-^{11}C-methyl-benperidol. *J. Nucl. Med.* **31**: 2015–2021.

Zoghbi, S. S., Baldwin, R. M., Seibyl, J., Charney, D. S., and Innis, R. B. (1997). A radiotracer technique for determining apparent pKa of receptor-binding ligands. *In "XII International Symposium on Radiopharmaceutical Chemistry,"* pp. 136–138. Uppsala, Sweden.

S E C T I O N

II

GENERAL COMPARTMENT KINETICS

10

Correction of Partial Volume Effect on Rate Constant Estimation in Compartment Model Analysis of Dynamic PET Study

K. UEMURA,* H. TOYAMA,† Y. IKOMA,* K. ODA,‡ M. SENDA,‡ and A. UCHIYAMA*

*School of Science and Engineering, Waseda University, 3-4-1 Okubo Shinjuku-ku, Tokyo 169-0072, Japan
†National Institute of Radiological Sciences, 4-9-1 Anagawa Inage-ku Chiba 263-8555, Japan
‡Positron Medical Center, Tokyo Metropolitan Institute of Gerontology, 1-1 Naka-cho Itabashi-ku Tokyo 173-0022, Japan

The estimates for the rate constants in compartment model analysis of [^{18}F] Fluorodeoxyglucose (FDG)-positron emission tomography are affected by the fraction of tissue mixture in the region of interest (ROI), which depends both on the spatial resolution and on the size of the ROI. We have investigated the relationship among the estimates of [^{18}F]FDG rate constants, spatial resolution, and ROI size for the striatum and cerebral cortex, using dynamic digital brain phantom. The shape of the dynamic digital brain phantom with cortex, white matter, striatum, and cerebrospinal fluid space were manually extracted from T1-weighted MRI. The activity time course in each tissue was determined with a three-parameter model for [^{18}F]FDG. Noise was generated by Poisson process according to the collected count for each frame. Sinograms with various spatial resolutions were produced by forward projection, and the images were reconstructed by filtered backprojection with spatial resolution of 2–10 mm FWHM. The parameters (K_1, k_2, k_3, cerebral metabolic rate of glucose), were estimated by modified Marquardt method. The fractional tissue component (FTC) was determined as the percentage of the target tissue volume in the total volume of the ROI in the smoothed images. The values of estimates linearly correlated with the FTC for the ROIs with various sizes. The true value was estimated as an extrapolation to 100% FTC on the linear regression line. When this method was applied to the human data for three normal subjects, the estimates also linearly correlated with the FTC and the value at 100% FTC was obtained as a partial volume corrected parameter estimate.

I. INTRODUCTION

Rate constants estimated by a compartment model analysis for dynamic positron emission tomography (PET) studies are useful in evaluating the regional function of the target tissue. However, it is difficult to estimate the rate constants precisely, because the target tissue mixes with surrounding tissues by partial volume effect (PVE) due to the limited spatial resolution of the PET camera (Herholz and Patlak, 1989; Schmidt et al., 1991). O'Sullivan et al. used a mixture analysis model, which approximates pixelwise time–activity curves (TACs) in terms of linear combination of a number of underlying model subTACs, to correct for the effect of tissue mixture on rate constant estimation (O'Sullivan, 1993). We have investigated the relationship among the estimates of [^{18}F]FDG rate constants, spatial resolution, and ROI size for the striatum and cerebral cortex, using a "dynamic digital brain phantom" (Ikoma et al., 1998) reconstructed from a simulated sinogram, both noise-free and in noise added conditions. Then we have devised a technique to correct for PVE in the rate constant estimation.

II. MATERIALS AND METHODS

A. Simulation

1. Dynamic Digital Brain Phantom

A dynamic digital brain phantom simulating an [^{18}F] fluorodeoxyglucose (FDG) dynamic scan was generated

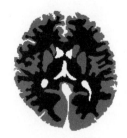

FIGURE 1.   Digital brain phantom.

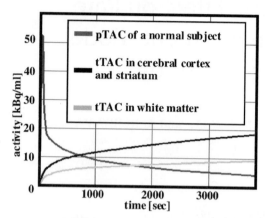

FIGURE 2.   Time-activity curves.

TABLE 1.   "True" Values of Rate Constants for Two-Tissue, Three-Parameter Model for [^{18}F]FDG

	K_1 (mL/g/min)	k_2 (L/min)	k_3 (L/min)
Cortical gray matter	0.182	0.130	0.062
White matter	0.954	0.109	0.045
Striatum	0.102	0.130	0.062
CSF space	0.000	0.000	0.000

by using a two-tissue-compartment, three-parameter (3K) model for [^{18}F]FDG. The boundaries between cortical gray matter, white matter, striatum, and CSF space were manually extracted from a T1-weighted MRI of a normal subject, as shown in Fig. 1. The whole brain phantom and the three segmented tissue phantoms (cortical gray matter, white matter, and striatum) were generated, respectively. Time-dependent radioactive tracer concentration for each tissue type (tTAC) was determined by using a measured plasma time–activity curve (PTAC) in a normal subject (Fig. 2) based on the kinetic model with "true" values of rate constants shown in Table 1. The scanning schedule of the dynamic digital brain phantom was determined according to the human scan (30 s × 2, 60 s × 4, 120 s × 4, 240 s × 8; total 45 min) in our institute.

The sinograms were produced for the simulated phantoms by forward projection with bin size of 1 mm and ma-

trix size of 256 × 256, assuming the spatial resolution of 2–10 mm FWHM, both noise-free and with noise level of 20% at the last frame. The noise was generated by Poisson process according to the collected count for each frame. Then the images were reconstructed by the filtered backprojection method with a Gaussian filter of 8 mm FWHM as shown in Fig. 3 (matrix size; 128 × 128; pixel size; 2.0 mm).

2. Data Analysis

Oval ROIs with various sizes were drawn over the striatum and over the frontal cortex. The fractional tissue components (FTCs) were calculated from the segmented tissue phantoms for each ROI drawn on each image with various resolution. For example, The FTC of stratium for a ROI is

$$
\begin{aligned}
&FTC\ of\ striatum \\
&= \frac{ROI(striatum)}{ROI(cortical) + ROI(white) + ROI(striatum)} \\
&\quad \times 100\%,
\end{aligned} \tag{1}
$$

where ROI (striatum), ROI (cortical), and ROI (white) are the volumes of segmented striatum and cortical gray and white matter phantoms, respectively, in the ROI.

The rate constants K_1, k_2, k_3, and cerebral metabolic rate of glucose (CMRGlc) ($= K_1 \cdot k_3/(k_2 + k_3) \cdot$ Glc/LC: Glc = 88.5 mg/100 mL LC = 0.52) were estimated by a modified Marquardt method for the TAC of each ROI with various values of FTC. The values of estimates vs FTC were plotted and the relationship was evaluated.

B. Human Data

1. Subjects

Three normal subjects had an MRI study on a 1.5 T superconducting magnet system (SIGNA, GE), and T1-weighted MRIs were acquired in transaxial sections at 3 mm pitch. The [^{18}F]FDG dynamic data were also acquired for 45 min after injection of [^{18}F]FDG. The PET study was performed with a HEADTOME IV (Shimadzu, Kyoto, Japan), providing 7 slices of tomographic images at 13-mm intervals with spatial resolution of 8.0 mm FWHM.

2. Data Analysis

The [^{18}F]FDG dynamic data of each subject were registered to the subject's MR images (Ardekani, 1995). The segmented images of cortical gray matter, white matter, and striatum were generated by the same method as the phantom from the MRI (Fig. 4c) and were smoothed with a 2D Gaussian filter to make the same resolution as the PET image. Then the concentric circular or oval ROIs with various sizes were manually drawn over the mesial frontal gray matter and over the striatum in the [^{18}F]FDG dynamic data as shown in Fig. 4a. The FTCs in the ROIs were estimated from

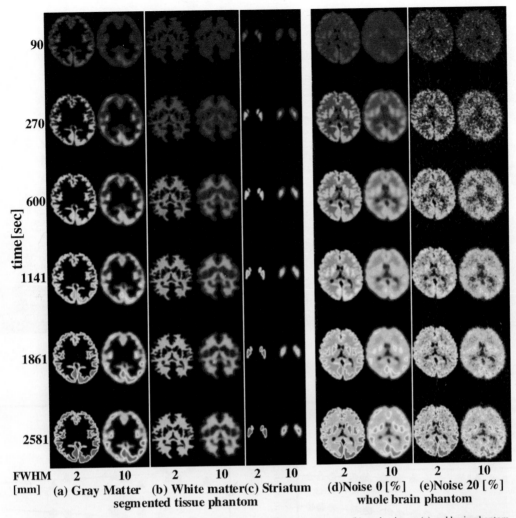

FIGURE 3. Segmented tissue phantom of cortical gray matter (a), white matter (b), and striatum (c) and brain phantom simulating an [^{18}F]FDG study reconstructed with spatial resolution of 2.0 (left) and 10.0 (right) mm FWHM, respectively, without noise (d) and with noise level of 20% at the last frame (e). As the spatial resolution increased, each segment spilled over into other segments and the peak count in a small region such as striatum decreased because of PVE.

the smoothed segmented images (Fig. 4d) and then the relationship between the parameter estimates (K_1, CMRGlc) and FTC for each ROI was examined.

III. RESULTS

In the simulation study, all the estimates correlated with FTC ($r > 0.99$) in the noise-free data and became close to the true values as FTC became 100% as shown in Fig. 5a. The PVE-corrected estimates were predicted from the regression line. The true values for K_1 and CMRGlc for the data with 20% noise were estimable by extrapolation of the regression line determined from the estimates for large ROIs with less noise, as shown in Fig. 5b. The regression lines were calculated for the estimates in the oval ROIs larger than

92 mm^2 to avoid noise effect. The noise contributed to the error in K values as much as PVE. On the other hand, PVE on CMRGlc were predicted from the striatal component even with noise. The effect of noise was minimal on the estimates of CMRGlc, moderate for K_1, and was prominent for k_2 and k_3 in determining the regression line to correct for PVE.

In the human data, the estimates of K_1 and CMRGlc correlated with the FTC ($r > 0.90$) and the values at 100% of FTC were estimated as shown in Fig. 6 and Table 2.

IV. DISCUSSION AND CONCLUSION

The mixture with surrounding tissues due to PVE affects the estimated values of rate constants. These estimates were linearly correlated with the fractional component of the

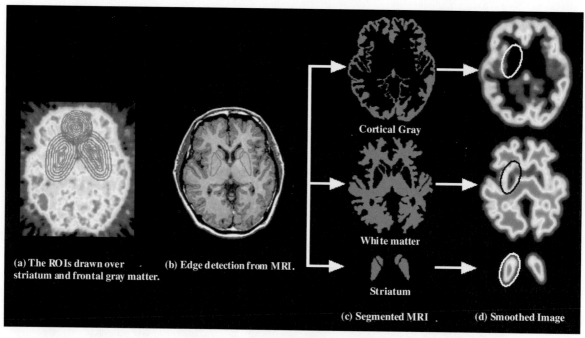

FIGURE 4. The method of analyzing the human data.

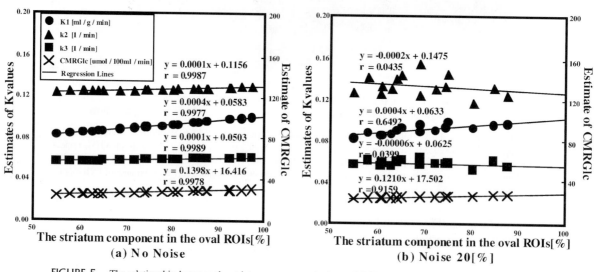

FIGURE 5. The relationship between the striatum component in the oval ROIs and the estimates of the rate constants for brain phantom without noise (a) and with 20% noise level (b).

TABLE 2.   The Values at 100% of FTC

	Subject 1		Subject 2		Subject 3	
	K_1(mL/g/min)	CMRGlc (μmol/100 mL/min)	K_1(mL/g/min)	CMRGlc (μmol/100 mL/min)	K_1(mL/g/min)	CMRGlc (μmol/100 mL/min)
Frontal gray	0.1693	43.3	0.1736	29.6	0.2385	50.2
Striatum	0.1491	44.0	0.1061	26.9	0.1778	32.6

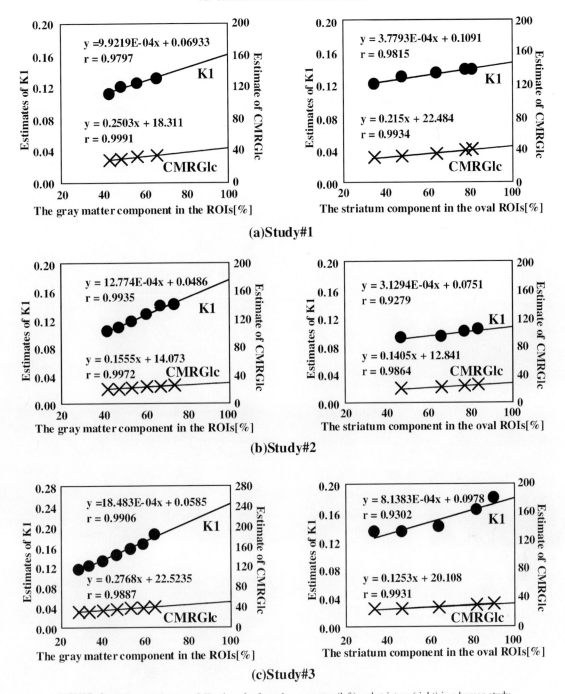

(a)Study#1

(b)Study#2

(c)Study#3

FIGURE 6. FTC vs estimates of *K* values for frontal gray matter (left) and striatum (right) in a human study.

target tissue (FTC) in the ROIs of various sizes. The PVE corrected value was estimable as the extrapolation to 100% FTC on the regression line. This study suggests that the effect of tissue mixture due to PVE on the rate constant estimation may be corrected for by using various sizes of ROIs and their FTC obtained from MRI.

The regression line of K_1 and CMRGlc on FTC was precisely determined, both for the noise-added simulation data and for the human data, and a nearly true value of K_1 and

CMRGlc was obtained by extrapolating the value at 100% FTC.

Acknowledgments

This work was supported by Grant-In-Aid for Scientific Research (c) 11670653 from the Ministry of Education, Science, Sports, and Culture, Japan. This work was supported by Individual Research of Waseda University Grant for Special Research Projects 98A-848.

References

Herholz, K., and Patlak, C. S. (1989). The influence of tissue heterogeneity on results of fitting nonlinear model equations to regional tracer uptake curves with an application to compartmental models used in positron emission tomography. *IEEE Trans. Med. Imaging* **MI-8**: 337–343.

Schmidt, K., Mies, G., and Sokoloff, L. (1991). Model of kinetic behavior of deoxyglucose in heterogeneous tissue in brain: A reinterpretation of the significance of parameters fitted to homogeneous tissue models. *J. Cereb. Blood Flow Metab.* **11**: 10–24.

O'Sullivan, F. (1993). Imaging Radiotracer Model Parameters in PET: A mixture analysis approach. *IEEE. Trans. Med. Imaging* **12**(3): 399–412.

Ikoma, Y., Toyama, H., Yamada, T., Uemura, K., *et al.* (1998). Creation of a dynamic digital phantom and its application to a kinetic analysis. *Kaku Igaku* **35**(5): 293–303. [in Japanese]

Ardekani, A, B., Braun, M., Hutton, F. B., Kanno, I., and Iida, H. (1995). A fully automatic multimodality image registration algorithm. *J. Comput. Assist. Tomogr.* **19**(4): 615–623.

11

Partial Volume Effect Correction: Methodological Considerations

MARK SLIFSTEIN, OSAMA MAWLAWI, and MARC LARUELLE

Department of Psychiatry and Radiology, Columbia University College of Physicians and Surgeons, NewYork, New York 10032

Several authors have proposed methods for correction of partial volume effect (PVE) in PET data that assume a linear model of the point spread function of the scanner. We have tested our own refinements of the methods of Labbé (1998) and Rousset (1998) in various settings and report on their performance. We tested their accuracy in the context of sphere phantom experiments. We tested their robustness to poor coregistration by simulating misregistration between the PET and MRI images. We tested their noise response characteristics by simulation. Finally, we tested their effect on estimated total volume of distribution (V_T) in cortical regions using data from tracer studies with the D_1 radioligand [^{11}C]NNC-112. Our results show that both methods are accurate, at least in the case of simple geometry, such as that of sphere phantoms. Labbé's method has slightly superior noise characteristics when the noise is additive and gaussian, but the methods are comparable for other noise models. The relative error incurred from poorly coregistered data appears to be comparable for both methods to that seen with no PVE correction. Both methods lead to sizable increases in V_T in cortical regions; the increase with Labbé's method is slightly larger than that with Rousset's.

I. INTRODUCTION

PET measurement of tracer concentration in a region is generally confounded by partial volume averaging effect (PVE), the phenomenon of blurring in which the activity measured in a region includes a measure of activity from the surrounding volume (Kessler *et al.*, 1984; Hoffman *et al.*, 1979). If the surrounding volume has a lower concentration of activity (for example, white matter or cerebrospiral fluid (CSF) surrounding gray matter), the measured activity in the region is smaller than the actual value. PVE is most significant for small structures comparable in size to the FWHM of the point spread function (PSF) of the PET scanner and for regions with high surface to volume ratio. In particular, tracer studies in which high contrast between cortical gray matter and white matter is expected are good candidates for PVE correction procedures.

Several methods have been proposed for PVE correction (Labbé *et al.*, 1998; Rousset *et al.*, 1998; Meltzer *et al.*, 1996). In this study, we compared the methods of Labbé and Rousset in several settings using our own implementation of the model PSF and report on their properties. In particular, we have examined them in terms of their accuracy and their robustness to various confounds such as statistical noise and poor registration.

II. MATERIALS AND METHODS

A. Theory

Both methods make similar assumptions. First, they assume that the distribution of tracer in the brain is well approximated by several regions of interest (ROIs) within which tracer concentration is roughly homogeneous and so can be characterized by a regional mean T. Second, they also assume that the PSF is a linear operator acting on the space of images, i.e., it has the property of superposition. Let $ROI_j(x)$ be the indicator image of the jth region evaluated at voxel x. That is, $ROI_j(x) = 1$ if x is in the jth region and $ROI_j(x) = 0$ if x is outside the jth region. Then, by the homogeneity assumption above, the true activity distribution can be represented as a sum of images,

$$\text{PET}_{\text{TRUE}}(x) = \sum_{j=1}^{P} T_j ROI_j(x),$$

where T_j is the true activity in the jth region and P is the total number of regions. If we denote the action of the PSF on an image I by $\text{PSF}(I)$, then the linearity of the PSF implies

$$\text{PSF}(\text{PET}_{\text{TRUE}})(x) = \text{PSF}\left(\sum_{j=1}^{P} T_j ROI_j\right)(x)$$

$$= \sum_{j=1}^{P} T_j \cdot \text{PSF}(ROI_j)(x).$$

Labbé's model equation is $\text{PET} = XT + \varepsilon$, where all images are represented as column vectors, PET is the observed PET image, column j of the design matrix X is $\text{PSF}(ROI_j)$, T is the vector of true regional values, and ε is a noise term. This is a classical linear least squares problem, and the estimate $\hat{T}$

of T is given by $\hat{T} = (X^T X)^{-1} X^T \cdot \text{PET}$. Rousset's model equation is $t = \text{GTM} \cdot T + \varepsilon$, where t is a vector of measured regional averages, the ij element of the $p \times p$ matrix GTM (geometrical transfer matrix) is spillover from region j into region i, and T and ε are as before. The solution is $\hat{T} = \text{GTM}^{-1} \cdot t$. The matrices X and GTM can be seen to be related in the following way. Let A (for "averaging") be the $p \times N$ matrix (N is the total number of voxels in the image) defined by $\text{row}_j(A) = ROI_j^T / (\text{vol}(ROI_j))$, where vol(Image) is the total number of nonzero voxels in the image. Then, as $t = A \cdot \text{PET}$, GTM can be derived as $A \cdot X$. Figure 1 shows a schematic representation of the procedure for the Labbé model.

B. Procedures

The following set of preliminary procedures are applicable to both correction methods.

1. ROIs are drawn on a structural MRI of the subject (or phantom). Each MRI with an ROI drawn on it is saved as a separate image.

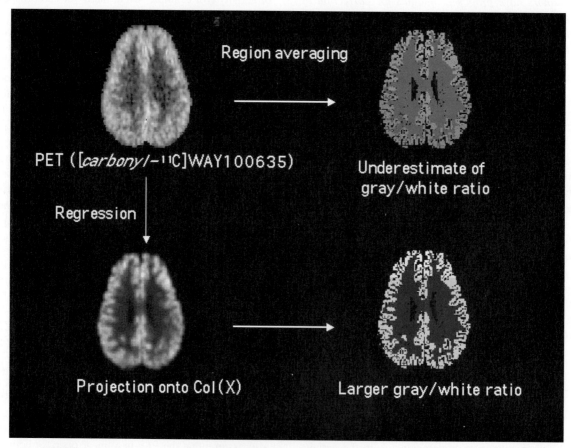

FIGURE 1. Schematic representation of the Labbé method. Upper left: PET image (human brain with [*carbonyl* - [11]C] WAY-100635). Upper right: Mean regional activity map of the same slice with no PVE correction. Lower left: the slice after regression of the PET image onto the design matrix X. Lower right: Mean regional activity map after PVE correction (Labbé method, same gray scale as upper right).

2. The ROI images are segmented according to tissue type (gray, white, or CSF).
3. The segmented images are converted into indicator images by setting them to a value of 1 in the ROI gray matter and zero outside. White matter and CSF are saved separately and eventually combined into one large image for each tissue type.
4. The indicator images are acted on by the model PSF (see discussion below regarding formation of PSF).
5. The matrices X or GTM are formed from the transformed ROI images.
6. PET images are coregistered to the original MRI.

At this point the correction procedures can be implemented as described under Theory, above.

III. DETERMINATION OF THE PSF

One aspect of the procedure that is the same in theory for both methods but different in practice is the implementation of the PSF. If the FWHM of the PSF were constant across the field of view (FOV) then the action of the PSF on the PET image could be represented by convolution with a gaussian kernel, a process which is relatively inexpensive in terms of computation. Unfortunately, the FWHM varies across the FOV, increasing in both axial and transverse components as x moves away from the center of the transverse plane. Explicit reproduction of the PSF would entail a different formula at each voxel, a procedure that would be prohibitively expensive with current technology. On the other hand, loss of accuracy is incurred by convolving with a kernel having a single average FWHM. We have devised a compromise

by taking advantage of the symmetry and seperability of the PSF function. The FWHM in the transverse plane and the axial direction are set to constants on concentric rings about the center of the FOV in the transverse plane, the constants being determined by the manufacturer's specifications for the scanner (HR+ in our case). We note that Labbé convolves with three different constant kernels, with FWHMs obtained at various distances from the center of FOV, and then interpolates the three according to position (Labbé *et al.*, 1996). Rousset convolves the entire image with a constant kernel. In our tests, we used our own implementation with both methods. The relative merits of the various PSF implementations is a topic for further investigation.

IV. TEST PROCEDURES

A. Sphere Phantoms

A cylindrical phantom (Data Spectrum Corp., Hillsborough NC) with six hollow spheres was used. The internal diameters of the spheres were 34, 28, 21, 12 and 9 mm. A hot sphere warm background configuration was used with approximately 3.78×10^3 Bq/cc activity concentration in the spheres and 1.54×10^3 Bq/cc in the background. The two test procedures were applied to the resulting PET image. The mean activities in the spheres were also measured without PVE correction for comparison (Fig. 2).

B. Simulated Misregistration

To simulate the effect of poor ROI drawing and/or poor coregistration of PET to MRI, the largest of the six spheres

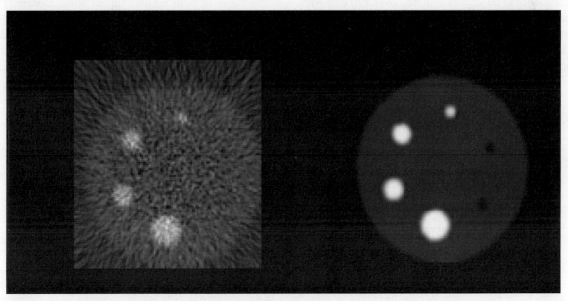

FIGURE 2. The sphere phantom. Left side is the PET image of the phantom. Right side is the MRI.

was intentionally misaligned by 1, 2, 3, and 4 mm radially. The correction methods, with the resulting inaccurate design matrices, were then applied to the PET image. Activity in the misaligned sphere was also estimated without PVE. The relative error due to misregistration was assessed by computing the ratio of the measured activity for a given offset to the activity measured by the same method with no offset. The ratio is expressed in percentages (Table 1).

C. Noise Analysis

A test image was created by drawing an ROI around the dorsolateral prefrontal cortex of a human brain MRI. The resulting ROI image was then segmented into gray and white matter. Three noise simulations were performed. In the first, gray and white matter were assigned the arbitrary respective values of 3 and 2.5, respectively. The image was then blurred by the simulated PSF. Gaussian random noise with mean = 0 and standard deviation = 1 was added independently to each voxel in the ROI image, and the PVE correction methods were applied to the resulting "noisy" image. The procedure was repeated 1000 times. The mean and standard deviation of the gray and white matter estimates were computed. For comparison, two other noise structures were tested. In one, each voxel in the image was assigned a Poisson random variable with mean = 3 in gray matter and mean = 2.5 in white matter. The final simulation used a combination of a Poisson variable and additive gaussian noise.

D. Time–Activity Curves

Data from a human brain PET study were reanalyzed using the PVE correction methods. Five studies, each consisting of a sequence of 21 frames acquired over 120 min with the D_1 radioligand [^{11}C]NNC-112 were analyzed with PVE correction. Twenty-seven regions (bilateral regions were treated as one) were drawn on the MRI to create the design matrices. This included areas of the brain not typically

thought of as ROIs, such as vasculature, because activity in these regions can spill over into true ROIs. The resulting time–activity curves (TACs) were analyzed by standard tracer kinetic modeling and the estimated total volume of distribution was computed. This was compared to previous analysis done without PVE correction.

V. RESULTS AND DISCUSSION

A. Sphere Phantom

Both methods returned within 10% of the true activity for all six spheres and close to 100% for the largest sphere (Fig. 3). Note that the percentage of true activity recovered decreases with decreasing sphere size for the four largest spheres, but then increases for the 12- and 9-mm spheres. This may be a "cancellation of errors" effect due to the fact that with voxel dimensions of $1.7162 \times 1.7162 \times 2.425 \, mm^3$, spheres of these sizes cannot be accurately reproduced in indicator images.

B. Simulated Misregistration

All three methods (the two PVE corrections and no correction) show decreasing measured activity with increasing extent of misregistration. The relative error (activity recovered for a given offset as a percentage of the activity recovered by the same method with no offset) was similar for all three methods (Table 1).

C. Noise Simulation

Both methods returned mean activity within 0.5% of the true value in both regions for all noise models (Table 2). For additive gaussian noise, the Labbé method had slightly better variance characteristics, but this is totally in accord with standard statistical theory (Scheffé, 1959). If the noise is additive, independent across voxels, and homoscedastic,

TABLE 1.   Results of Misregistration Experiment

Offset (mm)	Labbé (%)	Δ Labbé (%)	Rousset (%)	Δ Rousset (%)	No corr. (%)	Δ No corr (%)
0	99	100	98	100	84	100
1	99	100	98	100	84	100
2	99	100	97	99	83	99
3	96	97	95	97	82	97
4	93	94	90	92	79	93

Note. Results of misregistration experiment. Offset refers to amount by which the sphere was misregistered in the radial direction. The columns with method names contain the percent of the true activity recovered. The columns with Δ contain the ratio of the recovered activity for the given offset to the recovered activity for zero offset with the same method. Note that the Δ columns for the correction methods are similar to the Δ column with no correction.

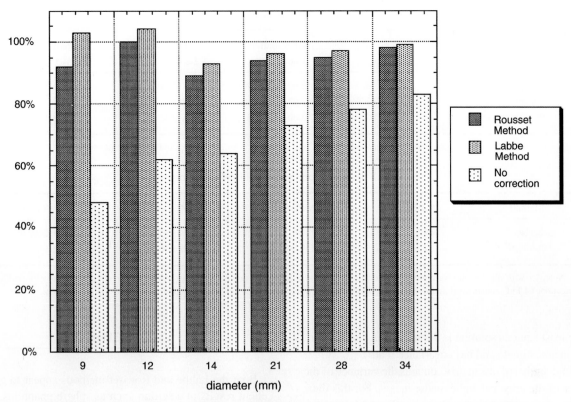

FIGURE 3.   Results from the sphere phantom experiment. Percent of true activity recovered is shown for each of three methods (Labbé, Rousset, and no correction) and each of six spheres.

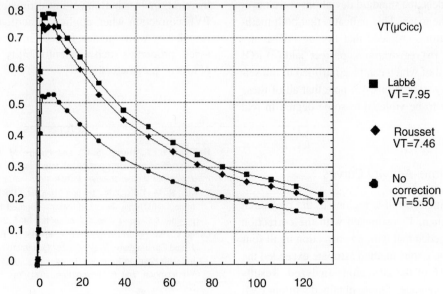

FIGURE 4.   Time activity curves in the DLPFC ([^{11}C]NNC-112, in human) units are μCi vs. minutes. The PVE correction methods attribute more activity to the region than the uncorrected estimate in each frame, resulting in higher estimates of V_T. Note that the Labbé method is consistently slightly higher than the Rousset method.

TABLE 2.   Noise Simulation Results

Noise structure	Gray matter, Labbé	Gray matter, Rousset	White matter, Labbé	White matter, Rousset
Gaussian	3.00 ± 0.0354	3.00 ± 0.0554	2.50 ± 0.0308	2.50 ± 0.0508
Poisson	3.00 ± 0.0329	3.00 ± 0.0308	2.50 ± 0.0271	2.50 ± 0.0253
Poisson + Gaussion	3.00 ± 0.0530	3.00 ± 0.0693	2.50 ± 0.0359	2.50 ± 0.0533

Note. All results are estimated mean ± standard deviation. True means are 3 for gray matter and 2.5 for white matter.

TABLE 3. V_T as percentage of No Correction

Method	DLPFC (%)	MPFC (%)	OPFC (%)	PUT (%)
Labbé	$147 \pm 12\%$	$143 \pm 26\%$	$198 \pm 72\%$	$144 \pm 22\%$
Rousset	$137 \pm 12\%$	$139 \pm 17\%$	$180 \pm 43\%$	$142 \pm 9\%$

Note. V_T with K_1/k_2 ratio constrained to cerebellum value ($n = 5$). Means and deviations expressed as percent of outcome with no PVE correction. DLPFC, dorsolateral prefrontal cortex; MPFC, medial prefrontal cortex; OPFC, orbito prefrontal cortex; PUT, putamen.

then the least squares solution is the minimum variance unbiased linear estimate and has covariance matrix $(X^T X)^{-1}$. The main diagonal of this matrix contains the variance of the estimates of the gray and white matter means. For this data, the square roots of the first and second diagonal elements of the matrix are equal to 0.0350 and 0.0303, respectively (compare to Table 2, line 1). For the same noise, the theoretical covariance matrix of the Rousset model is $(AX)^{-1} A A^T (AX)^{-T}$. The square roots of the main diagonal of this matrix are 0.0595 and 0.0513 (compare to Table 2, line 1). The other noise models had slightly different variance characteristics, but for all models, the standard deviations were never more than 2.6% of the mean. This indicates that both methods are good estimators, provided that (a) the PSF is accurately reproduced, (b) registration is perfect, and (c) ROI drawing is perfect, all of which can be controlled in the setting of pure computer simulations. We note that all of these conditions are likely to be violated to some degree in real studies.

D. Time–Activity Curves

Results are presented for three regions in the prefrontal cortex and the putamen. V_T estimated with the correction methods always exceeded that with no correction in all four regions (Table 3). The Labbé method estimate exceeded the Rousset method in 16 of the 20 regions analyzed. Results are presented as a percentage of those obtained without PVE correction. Note that V_T in the putamen was increased by approximately the same proportion as in the cortical regions (Fig. 4).

VI. CONCLUSION

Both the Labbé and Rousset methods appear to give excellent results in a setting such as sphere phantoms, where the geometry is simple and very accurate coregistration is possible. The error incurred due to poor registration or inaccurate region drawing, taken as a fraction of the estimated activity with good registration, seems to be no worse than that seen without correction, but it remains for future work to quantify how these errors will propagate in terms of estimation of kinetic parameters. Both methods lead to sizable increases in distribution volume estimates as compared to no PVE correction when applied to cortical gray regions surrounded by white matter. It also remains for future work to assess properties such as identifiability and reproducibility of kinetic parameters.

References

Hoffman, E., J., Huang, S.-C., and Phelps, M. E. (1979). Quantitation in positron emission computed tomography. 1. Effect of object size. *J. Comput. Assist. Tomogr.* **3**: 299–308.

Kessler, R. M., Ellis, J. R., Jr., and Eden, M. (1984). Analysis of emission tomographic scan data: Limitations imposed by resolution and background. *J. Comput. Assist. Tomogr.* **8**: 514–522.

Labbé, C., Koepp, M., Ashburner, J., Spinks, T., Richardson, M., Duncan, J., and Cunningham, V. (1998). *In* "Quantitative Functional Brain Imaging with Positron Emission Tomography" (R. E. Carson, M. E. Daube-Witherspoon, and P. Herscovitch, Eds.), pp. 59–66. Academic Press, San Diego.

Labbé, C., Froment, J. C., Kennedy, A., Ashburner, J., and Cinotti, L. (1996). Positron emission tomography metabolic data corrected for cortical atrophy using magnetic resonance imaging. *Alzheimer Dis. Assoc. Diorders* **10**: 141–170.

Meltzer, C. C., Zubieta, J. K., Links, J. M., Brakeman, P., Stumpf, M. J. and Frost, J. (1996). MR-based correction of brain PET measurements for heterogeneous gray matter radioactivity distribution. *J. Cereb. Blood Flow Metab.* **16**: 650–658.

Rousset, O. G., Yilong, M., and Evans, A. C. (1998). Correction for partial volume effects in PET: Principle and validation. *J. Nucl. Med.* **39**: 904–911.

Scheffé, H. (1959). "The Analysis of Variance." Wiley, New York.

12

Venous Sinuses vs On-Line Arterial Sampling as Input Functions in PET

M.-C. ASSELIN,* V. J. CUNNINGHAM,[†] N. TURJANSKI,[†] L. M. WAHL,[‡] P. M. BLOOMFIELD,[†] R. N. GUNN,[†] and C. NAHMIAS[§]

*Department of Physics and Astronomy, McMaster University, Hamilton, Ontario, Canada
[†]MRC Cyclotron Unit, Hammersmith Hospital, London, UK
[‡]Theoretical Biology, Institute for Advanced Study, Princeton, New Jersey
[§]Department of Nuclear Medicine, McMaster University Medical Centre, Hamilton, Ontario, Canada

The purpose of the present work was to investigate a noninvasive method for acquiring an input function from dynamic positron emission tomography (PET) scans without arterial cannulation. The method is based on regions of interest drawn around the venous sinuses visible in the PET images. Two subjects, one normal and one Parkinsonian, were studied in a high-resolution, high-sensitivity PET scanner after the intravenous injection of [^{18}F]6-fluoro-L-meta-tyrosine (FmT). The time course of radioactivity in the cerebral blood, corrected for partial volume and spillover, was compared to that in the arterial blood sampled on line. The results of compartmental and graphical analyses of FmT using the directly sampled and the image-derived input functions were also compared. Shape differences were observed in the time to peak, peak height and the area under the two blood curves. The use of the venous sinus input function enabled the discrimination of normal and Parkinsonian subjects and the identification of the affected and unaffected sides of the parkinsonian subject as indicated by the model parameter reflecting presynaptic dopaminergic metabolism and the graphically determined influx constant. This method of noninvasively obtaining an input function for quantitative analysis of PET brain data is general enough to be applicable to other tracers and simple enough to be used with the clinical population.

I. INTRODUCTION

The radiotracer [^{18}F]6-fluoro-L-meta-tyrosine (FmT) is a substrate for the enzyme aromatic amino-acid decarboxylase (AADC) and is used in conjunction with positron emission tomography (PET) to study the activity of dopaminergic cells in the human brain (Nahmias et al., 1995). In order to perform a quantitative analysis of the PET images, an accurate measurement of the time course of the radiotracer in the blood is needed.

Arterial sampling is the recognized method used to obtain the blood input function. It is not used universally, however, because of the risks associated with an arterial cannulation. Furthermore, the detection system used to count the blood radioactivity needs to be carefully calibrated against the PET scanner. This is done to ensure compatibility between the blood and tissue data as is required by the subsequent mathematical modeling.

We have investigated a noninvasive method for obtaining an input function from regions of interest (ROIs) drawn around the venous sinuses visible in the PET images. The time course of radioactivity in the cerebral blood was corrected for partial volume and spillover. The method was validated by comparing the results of compartmental and graphical analyses of [^{18}F]FmT using on-line arterial sampling with those using the image-derived input function.

II. MATERIALS AND METHODS

A. Data Acquisition

Two subjects, one normal (47-year-old man) and one Parkinsonian (60-year-old woman) affected on the right side (Hoehn and Yahr stage 1), underwent a 2-h PET scan (EXACT3D, CTI/Siemens, Knoxville, TN) (Spinks *et al.*, 1996) after the intravenous injection of 3 mCi (110 MBq) of [^{18}F]FmT. Arterial blood was counted continuously on-line (Ranicar *et al.*, 1991) for the entire duration of the study. In addition, eight blood samples were drawn to measure the exchange between plasma and red blood cells and the peripheral metabolism of [^{18}F]FmT. The images were acquired using list mode so that the frame durations could be defined postacquisition using the whole head time–activity curve (typically 1 frame at 50 s per frame, 8 frames at 5 s per frame, 3 frames at 10 s per frame, 2 frames at 30 s per frame, 4 frames at 60 s per frame, 4 frames at 120 s per frame, 5 frames at 300 s per frame, and 8 frames at 600 s per frame). This enabled the accurate definition of the early time course in the venous sinus. The PET images were reconstructed using the reprojection algorithm (Kinahan and Rogers, 1989) with a ramp filter at Nyquist cutoff frequency and using the model-based scatter correction developed by Watson *et al.* (1996). Correction for attenuation was performed using the reconstructed single-photon transmission images segmented with the local threshold technique (Xu *et al.*, 1994).

B. Data Analysis

Cluster analysis (Ashburner *et al.*, 1996) was carried out on the dynamic PET images to segment the venous sinuses based on the characteristic shape of the time–activity curve (TAC). The PET image of the first 3 min summed is compared with the blood cluster image in Fig. 1. Circular ROIs (diameter = 5 mm) were drawn manually around the voxels having the highest probability of belonging to the blood cluster. The time course of radioactivity (in kBq/mL) in the venous sinuses was produced after applying these ROIs to the dynamic PET images. The use of the cluster image reduced the observer's variations in the placement of the ROIs around the small cerebral blood vessels. Further ROIs were drawn manually on the summed PET images from the last hour of the study. In order to correct for the partial volume and spillover effects relating to the venous sinus, a background ROI was delimited as an annulus outside the venous sinus in the occipital lobe (see below). Elliptical ROIs were drawn around the cerebellum and the left and right putamina that served as target tissues for the compartmental and graphical analyses. All these ROIs were overlaid on the dynamic PET images to generate the time course of the radioactivity concentration in these tissues.

C. Correction for Partial Volume and Spillover

A simple correction (Wahl *et al.*, 1999) was applied to the venous sinuses TAC to overcome the nonnegligible effects of partial volume and spillover. The method used recovery coefficients (RCs) (Hoffman *et al.*, 1979) to correct for the activity that spilled in and out of the region of interest from and to the surrounding area. As indicated by Eq. (1) (Kessler *et al.*, 1984), the true activity concentration in the venous sinus (C_{true}) at each mid-frame time t is obtained after correcting both the activity concentration measured in the region of interest around the venous sinus, (C_{ROI}) and the activity concentration in the background region (C_{bkg}):

$$C_{\text{true}}(t) = \frac{1}{RC}\big(C_{\text{ROI}}(t) - C_{\text{bkg}}(t)\big) + C_{\text{bkg}}(t). \quad (1)$$

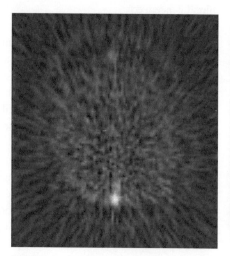 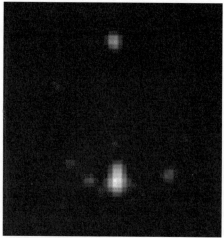

FIGURE 1. Comparison of the summed PET image from the first 3 min (left) with blood cluster image (right).

Note that the diameter of the inner circle of the annulus defining the background region was set to the sum of the vessel diameter and the FWHM of the PET scanner to avoid spillover from the blood vessel into the background region. The recovery coefficients were estimated for a given ROI radius by smoothing a circular blood vessel of known size (diameter = 5 mm) with a two-dimensional Gaussian line spread function (LSF) having a full width at half maximum equal to the measured transaxial spatial resolution of the PET scanner (FWHM = 5 mm) (Spinks *et al.*, 1996). For vessel and ROI radius in the range 2–4 mm, the RC values varied between 0.2 and 0.6.

D. Comparison of Input Functions

The peak height, time to peak, and the total integral of the venous sinus TAC were compared with those of the arterial sampling TAC. The arterial sampling and the venous sinus TACs were then corrected for plasma to blood partition and radiolabeled metabolites in order to be used as input functions to the cerebellum and left and right putamina. The effects of using either the sampled or the ROI-based input functions on the results of both compartmental and graphical analyses were then evaluated. Successive compartmental models of increasing complexity were compared using an *F* test (DiStefano *et al.*, 1984). For the cerebellum TAC, a two-tissue, three-rate-constant compartment model adequately described the data. The cerebral blood volume was included as an additional parameter and the delay between the blood and tissue TACs was fixed to the values presented in Table 1 (see Results). For the left and right putamina TACs, fits were improved by fixing the K_1/k_2 ratio to that of the cerebellum

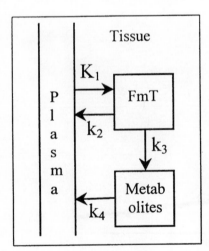

FIGURE 2. Compartmental model: K_1 and k_2 represent the forward and reverse transport rate constants of [^{18}F]FmT from the plasma to the tissue, k_3 is the rate constant of decarboxylation of [^{18}F]FmT in tissue by the enzyme aromatic amino-acid decarboxylase (AADC), and k_4 is the rate constant of clearance of the [^{18}F]FmT metabolites out of the tissue to the plasma.

and by inclusion of a fourth rate constant (see Fig. 2). It was assumed that k_4 represented loss of labeled metabolites from the tissues to the blood pool, but this being an apparent effect of tissue heterogeneity cannot be excluded (Schmidt *et al.*, 1991, 1992). For graphical analysis, the method proposed by Wong *et al.* (1986) was used.

III. RESULTS

A. Direct Comparison of Input Functions

The corrected venous sinus TAC is compared with the arterial sampling TAC in Fig. 3. As shown in Table 1, the venous sinus curve peaked earlier than the arterial sampling curve for both the normal and hemi-Parkinsonian (HPD) subjects. Note that the delay values for the arterial sampling TACs are consistent with the values typically used in compartmental analysis. The peak height of the venous sinus curve was larger in the normal subject but smaller in the HPD subject than that of the arterial sampling one. The area under the venous sinus curve was smaller than under the arterial sampling curve for both the normal and the HPD subjects.

B. Compartmental Analysis

The results of the compartmental modeling are presented in Table 2 and illustrated in Fig. 4. The compartmental analysis yielded values for the forward transport rate constant K_1 that are similar with both input functions. The difference in reverse transport rate constant k_2, however, was inversely related to the difference in peak height between the two input functions. Using either input function, the metabolic rate constant k_3 is lower for the affected side of the HPD subject than the unaffected side, and the k_3 of both sides of the HDP subject are lower than the normal subject.

C. Graphical Analysis

The Patlak plots of the cerebellum and the left and right putamina are displayed in Fig. 5. Note that these plots are in agreement with the results of the compartmental analysis. The apparent accumulation in the left and right putamina is consistent with the small value of k_4 (see Table 2). The near zero slope of the cerebellum curve is likewise consistent with the small value of k_3 (again see Table 2). In Table 3, the influx constant K_i of either putamina is lower for the HPD subject than the normal one. The affected side of the HPD subject has a lower influx constant than the unaffected one. These differences between the two subjects and between the affected and unaffected side of the HPD subject are observed using either input function.

Normal

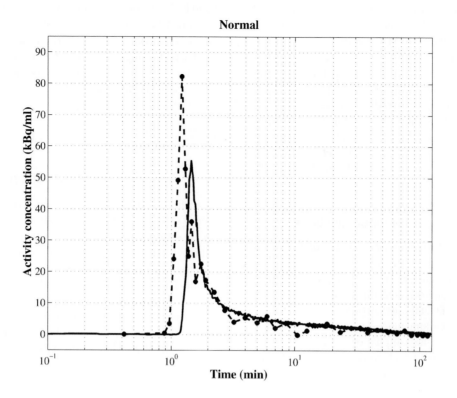

HPD

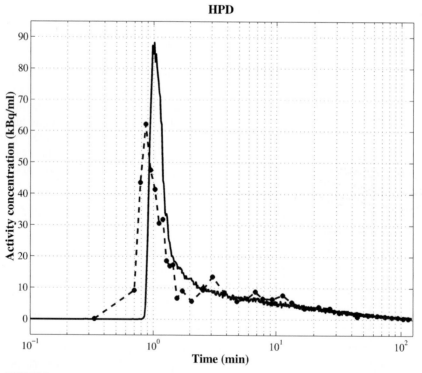

FIGURE 3. Venous sinus curve after correction for partial volume and spillover (-•-) compared to arterial sampling curve (—) for the normal (top) and the Parkinsonian (bottom) subjects. Note that the arterial sampling curves were plotted without correction for delay.

Normal

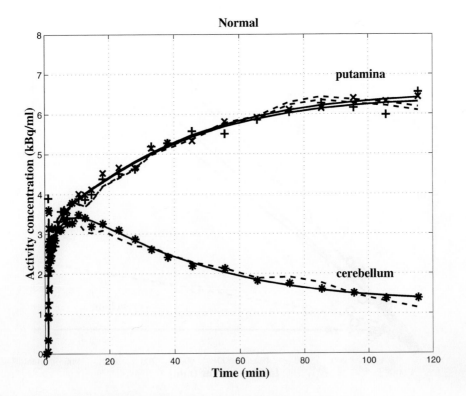

HPD

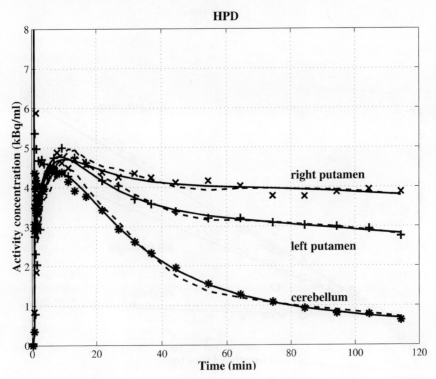

FIGURE 4. Fitted curves to the experimental data points for cerebellum (*), and left (+) and right (x) putamina using the sampled (—) and ROI-based (– –) input functions for the normal (top) and Parkinsonian (bottom) subjects. Note that the normal subject was given 150 mg of Carbidopa 1 h prior to the start of the scan. The Parkinsonian patient was withdrawn from medication 4 h before the scan started.

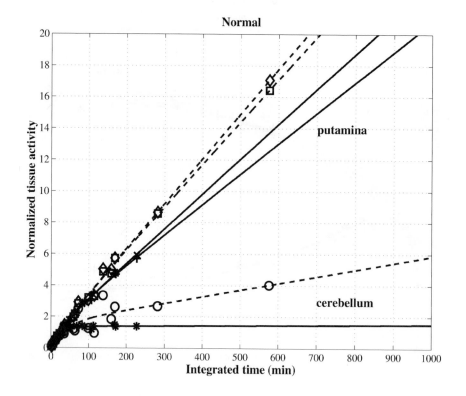

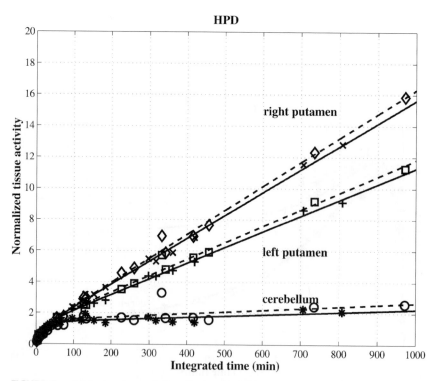

FIGURE 5.   Patlak plots for cerebellum (*, ○) and left (+, □) and right (×, ◇) putamina using the sampled (— and crosses) and ROI-based (– – and open symbols) input functions for the normal (top) and Parkinsonian (bottom) subjects. Note again that the normal subject was given 150 mg of Carbidopa 1 h prior to the start of the scan. The Parkinsonian patient was withdrawn from medication 4 h before the scan started.

TABLE 1. Results of Comparison of the Shape of the Arterial Sampling and the Corrected Venous Sinus Time–Activity Curves

Subject	Input function	Peak position (s)	Delay[†] (s)	Peak height (kBq/mL)	Difference (%)	Total integral Mdis/mL	Difference (%)
Normal	Arterial sampling	87	14	55.4		13.9	
	Venous sinus	73	0	82.2	+48	9.94	−28
HPD	Arterial sampling	61	11	88.2		16.1	
	Venous sinus	52	0	62.2	−30	14.9	−7.5

[†] Delay = input function peak position − venous sinus peak position.

TABLE 2. Results of Compartmental Analysis Using the Sampled and the ROI-Based Input Functions

		Arterial sampling as input function				Venous sinus as input function			
Subject	Tissue	K_1^a (mL/mL/min)	k_2^a (min^{-1})	k_3^a (min^{-1})	k_4^a (min^{-1})	K_1^a (mL/mL/min)	k_2^a (min^{-1})	k_3^a (min^{-1})	k_4^a (min^{-1})
Normal	Cerebellum	0.0526	0.0638	0.0023		0.0490	0.0531	0.0049	
		±0.0011	±0.0017	±0.0002	n/a	±0.0019	±0.0033	±0.0006	n/a
		(2.1)	(2.7)	(8.7)		(3.9)	(6.2)	(12)	
	R. putamen	0.0563	0.0704	0.0888	0.0043	0.0505	0.0561	0.109	0.0018
		±0.0020	±0.0025	±0.0066	±0.0003	±0.0025	±0.0028	±0.017	±0.0004
		(3.6)	(3.6)	(7.4)	(7.0)	(5.0)	(5.0)	(16)	(22)
	L. putamen	0.0556	0.0695	0.0890	0.0045	0.0504	0.0560	0.106	0.0019
		±0.0029	±0.0036	±0.010	±0.0005	±0.0029	±0.0032	±0.019	±0.0005
		(5.2)	(5.2)	(11)	(11)	(5.8)	(5.7)	(18)	(26)
HPD	Cerebellum	0.0542	0.0681	0.0014	n/a	0.0548	0.0748	0.0025	n/a
		±0.0010	±0.0015	±0.0001		±0.0031	±0.0047	±0.0004	
		(1.8)	(2.2)	(7.1)		(5.7)	(6.3)	(16)	
	R. putamen	0.0578	0.0723	0.0337	0.0026	0.0623	0.0890	0.0372	0.0023
		±0.0025	±0.0031	±0.0022	±0.0006	±0.0042	±0.0060	±0.0026	±0.0007
		(4.3)	(4.3)	(6.5)	(23)	(6.7)	(6.7)	(7.0)	(30)
	L. putamen	0.0612	0.0765	0.0224	0.0034	0.0667	0.0953	0.0262	0.0033
		±0.0025	±0.0031	±0.0014	±0.0008	±0.0036	±0.0051	±0.0015	±0.0007
		(4.1)	(4.1)	(6.3)	(24)	(5.4)	(5.4)	(5.7)	(21)

[a] Model parameters expressed as least-square estimates ± standard error (coefficient of variation in %).

D. Discussion

The results of the compartmental and graphical analyses of FmT using the image-derived input function were similar to those obtained using the directly sampled input function, despite the shape differences of the two input functions. The use of the venous sinus input function enabled the discrimination of normal and HPD subjects and the identification of the affected and unaffected sides of the HPD subject as indicated by the values of rate constants k_3 and K_i.

The precision on the parameter estimates was slightly poorer when using the ROI-based input function compared to the sampled input function. When designing the scanning protocol, consideration should be given to the noise in addition to the accurate definition of the early time course of radioactivity in the venous sinus.

The difference in the shape of the venous sinus input function compared to the arterial sampling input function probably reflects in large part errors in the correction for partial volume and spillover. To be more accurate, the correction should account for the heterogeneity of the tissues surrounding the venous sinus (skull and occiput) and use the actual size of the venous sinus (Strul and Bendriem, 1999).

The arterial sampling input function is confounded by the dispersion and delay during the pumping and counting of the radioactive blood. Its accuracy is also dependent upon cross-calibration between the PET scanner and the detection system used to count the radioactive blood. These problems are circumvented by extracting the time course of radioactivity in the blood from the PET images, hence ensuring compatibility between the blood and tissue data.

TABLE 3. Results of Graphical Analysis Using the Sampled and the ROI-Based Input Functions

Subject	Tissue	Arterial sampling as input function		Venous sinus as input function	
		K_i^a (min^{-1})	Intercept^a (mL/mL)	K_i^a (min^{-1})	Intercept^a (mL/mL)
Normal	Cerebellum	0.00017 ±0.00049 (5.4)	1.361 ±0.074	0.0043 ±0.0013 (30)	1.54 ±0.41 (27)
	R. putamen	0.02276 ±0.00055 (2.4)	0.821 ±0.084 (10)	0.02894 ±0.00051 (1.8)	0.48 ±0.13 (27)
	L. putamen	0.0204 ±0.0015 (7.4)	1.15 ±0.27 (23)	0.02764 ±0.00053 (1.9)	0.62 ±0.14 (23)
HPD	Cerebellum	0.00086 ±0.00020 (23)	1.391 ±0.080 (5.8)	0.00115 ±0.00041 (36)	1.51 ±0.19 (13)
	R. putamen	0.01473 ±0.00019 (1.3)	0.896 ±0.073 (8.1)	0.01546 ±0.00028 (1.8)	0.86 ±0.12 (14)
	L. putamen	0.01014 ±0.00017 (1.7)	1.146 ±0.066 (5.8)	0.01061 ±0.00029 (2.7)	1.21 ±0.13 (11)

^{a} Model parameters expressed as least-square estimates ± standard error (coefficient of variation in %).

The venous sinus was chosen over the internal carotid artery (Chen *et al.*, 1998) for a number of reasons. The venous sinus has a larger diameter, there is less intersubject size variations, and the tissues surrounding it are less heterogeneous. Another important factor is that the venous sinus is isolated from other blood vessels. This removes the need for an additional correction to contend for the spill of the radioactivity between the two adjacent internal carotid arteries.

This alternative means of obtaining an input function is not restricted to [^{18}F]FmT and thus its use could be extended to other radiotracers. Ultimately, our goal is to develop a noninvasive method for quantitative analysis of PET brain data that is applicable to the clinical population.

Acknowledgments

This work was conducted at the MRC Cyclotron Unit. The authors thank the chemists D. Brown and S. K. Luthra for the synthesis of FmT, Dr. S. Osman for the metabolites analysis, D. Griffiths and A. Blyth for their assistance in the data collection, and L. Schnorr for the reconstruction of the PET images. M.C.A. is supported by the Natural Sciences and Engineering Research Council of Canada and holds an E.S. Garnett bursary in Medical Imaging.

References

Asburner, J., Haslam, J., Taylor, C., *et al.* (1996). A Cluster Analysis Approach for the Characterization of Dynamic PET Data. *In* "Quantification of Brain Function Using PET" (R. Myers, V. J. Cunningham, D. L. Bailey, and T. Jones, Eds.), pp. 301–306. Academic Press, San Diego.

Chen, K., Bandy, D., Reiman, E., *et al.* (1998). Noninvasive quantification of the cerebral metabolic rate for glucose using positron emission tomography, ^{18}F-fluorodeoxyglucose, the Patlak method, and an image-derived input function. *J. Cereb. Blood Flow Metab.* **18**: 716–723.

DiStefano, J. J., III, and Landaw, E. M. (1984). Multiexponential, multicompartmental, and noncompartmental modeling. I. Methodological limitations and physiological interpretations. *Am. J. Physiol.* **246**: R651–R664.

Hoffman, E. J., Huang, S. C., and Phelps, M. E. (1979). Quantitation in positron emission computed tomography: Effect of object size. *J. Comput. Assist. Tomogr.* **3**: 299–308.

Kessler, R. M., Ellis, J. R., and Eden, M. (1984). Analysis of emission tomographic scan data: Limitations imposed by resolution and background. *J. Comput. Assist. Tomogr.* **8**: 514–522.

Kinahan, P. E., and Rogers, J. G. (1989). Analytic 3D image reconstruction using all detected events. *IEEE Trans. Nucl. Sci.* **36**: 964–968.

Nahmias, C., Wahl, L. M., Chirakal, R., *et al.* (1995). A probe for intracerebral aromatic amino-acid decarboxylase activity: Distribution and kinetics of [^{18}F]6-fluoro-L-m-tyrosine in the human brain. *Mov. Dis.* **10**: 298–304.

Ranicar, A. S. O., Williams, C. W., Schnorr, L., *et al.* (1991). The on-line monitoring of continuously withdrawn arterial blood during PET studies using a single BGO/photomultiplier assembly and non-stick tubing. *Med. Progr. Tech.* **17**: 259–264.

Schmidt, K., Lucignani, G., Moresco, R. M., *et al.* (1992). Errors introduced by tissue heterogeneity in estimation of local cerebral glucose utilization with current kinetic models of the [^{18}F]fluorodeoxyglucose method. *J. Cereb. Blood Flow Metab.* **12**: 823–834.

Schmidt, K., Mies, G., and Sokoloff, L. (1991). Model of kinetic behavior of deoxyglucose in heterogeneous tissues in brain: A reinterpretation of the significance of parameters fitted to homogeneous tissue models. *J. Cereb. Blood Flow Metab.* **11**: 10–24.

Spinks, T. J., Bailey, D. L., Bloomfield, P. M., *et al.* (1996). Performance of a new 3D-only PET scanner—the EXACT3D. *In "IEEE Nuclear Science Symposium Conference Record"* (A. Del Guerra, Ed.), Vol. 2, pp. 1275–1279.

Strul, D., and Bendriem, B. (1999). Robustness of anatomically guided pixel-by-pixel algorithms for partial volume effect correction in positron emission tomography. *J. Cereb. Blood Flow Metab.* **19**: 547–559.

Wahl, L. M., Asselin, M.-C., and Nahmias, C. (1999). Regions of interest in the venous sinuses as input functions for quantitative positron tomography. *J. Nucl. Med.* **40**: 1666–1675.

Watson, C. C., Newport, D., and Casey, M. E. (1996). A single scatter simulation technique for scatter correction in 3D PET. *In "Three-Dimensional Image Reconstruction in Radiation and Nuclear Medicine"* (P. Grangeat, and J.-L. Amans, Eds.), pp. 255–268, Kluwer Academic Publishers, Dordrecht, The Netherlands.

Wong, D. F., Gjedde, A., and Wagner, H. N., Jr. (1986). Quantification of neurotransmitters in the living human brain. I. Irreversible binding of ligands. *J. Cereb. Blood Flow Metab.* **6**: 137–146.

Xu, M., Luk, W. K., Cutler, P. D., *et al.* (1994). Local threshold for segmented attenuation correction of PET imaging of the thorax. *IEEE Trans. Nucl. Sci.* **41**: 1532–1537.

13

Identification of Linear Compartmental Systems That Can Be Analyzed by Spectral Analysis of the Sum of all Compartments

K. C. SCHMIDT

Laboratory of Cerebral Metabolism, National Institute of Mental Health, Bethesda, Maryland 20892

General linear time-invariant compartmental systems were examined to determine which systems are structured so that they meet the conditions necessary for application of the spectral analysis technique to the sum of the concentrations in all compartments. Spectral analysis can be used to characterize the reversible and irreversible components of the system and to estimate the minimum number of compartments, but it applies only to systems in which the measured data can be expressed as a positively weighted sum of convolution integrals of the input function with an exponential function that has real-valued nonpositive decay constants. The conditions are shown to be met by noncyclic strongly connected compartmental systems that have exchange of material with the environment confined to a single compartment, as well as by certain noncyclic systems with traps and all non-interconnected collections of such systems. These compartmental structures may prove to be useful for kinetic modeling of positron emission tomographic radiotracers in which unmetabolized tracer delivered by the blood exchanges with a pool of unmetabolized tracer in the tissue before undergoing parallel series of metabolic changes and/or binding to various receptors in the tissue.

I. INTRODUCTION

Linear compartmental systems have been used extensively to describe the kinetics of radiotracers used with positron emission tomography (PET). As it is not possible with the PET camera to measure separately compartments that are smaller than the spatial resolution of the instrument nor to distinguish among the various chemical species, only the sum of radioactivities in the compartments in the field of view is measured. Usually a compartmental model is specified *a priori* for kinetic analyses, but this may lead to significant errors if the specified model fails to account for heterogeneity of the tissues included in the voxel or region of interest (Schmidt *et al.*, 1992). The spectral analysis technique provides an alternative to the use of a fixed kinetic model (Cunningham and Jones, 1993; Turkheimer *et al.*, 1994). It does not require that the number of compartments be fixed *a priori*; instead the technique itself provides an estimate of the minimum number of compartments in the system. Since the spectral analysis technique applies to heterogeneous as well as homogeneous tissues, it is particularly useful for the analysis of tracer kinetics in brain with PET as the limited spatial resolution of the scanner assures that most measurements include activities from a heterogeneous mixture of gray and white matter tissues. Spectral analysis also provides an estimate of the rate constant of trapping of tracer in the tissue as well as the amplitudes and decay constants of reversible components (Cunningham and Jones, 1993). This information can be used for subsequent specification of a kinetic model, or it can be used to estimate selected parameters of the system that do not depend on the specific model configuration, such as the total volume of distribution of the tracer (Cunningham and Jones, 1993; Turkheimer *et al.*, 1998).

II. CONDITIONS NECESSARY FOR APPLICATION OF SPECTRAL ANALYSIS TO TOTAL TISSUE RADIOACTIVITY

For the purposes of the current study we consider only extravascular radioactivity, as the modification to add an intravascular component to the total activity is straightforward. For application of the spectral analysis technique it is assumed that the total concentration of tracer in the voxel or region of interest, C_T, can be described by the equation,

$$C_T(T) = \sum_{j=1}^{n} \gamma_j \int_0^T C_p(t) e^{\lambda j(T-t)} \, dt, \qquad (1)$$

where T is the time of the measurement, n is the number of detectable compartments in the tissue, $C_p(t)$ is the arterial input function, γ_j and λ_j are parameters that depend on the intercompartmental transfer rate constants. The exponents λ_j are assumed to be nonpositive real numbers and the coefficients γ_j to be nonnegative real numbers. Identification of systems meeting these conditions will be made within the framework of the general linear compartmental model described below.

III. THE TIME-INVARIANT LINEAR COMPARTMENTAL MODEL

The general n-compartmental system is described by the state equation,

$$\dot{x}(t) = Ax(t) + Bu(t), \qquad x(0) = x_0, \qquad (2)$$

where $x(t)$ is the n-vector of state variables, $u(t)$ is a p-vector of input functions, and x_0 is the initial state of the system. For the time-invariant system, B is the $n \times p$ matrix of influx rate constants from each input to each compartment, and the state transition matrix $A = [a_{ij}]$ consists of rate constants k_{ij} (transfer from compartment j to compartment i) and k_{0j} (transfer from compartment j to the environment) as follows:

$$a_{ij} = k_{ij}, \qquad i, j = 1, 2, \ldots, n, \qquad i \neq j,$$
$$a_{jj} = -k_{0j} - \sum_{\substack{i=1 \\ i \neq j}}^{n} k_{ij}, \qquad j = 1, 2, \ldots, n.$$

The elements of A and B are assumed to be constant with respect to time during the experimental period, but their values may change under differing experimental conditions.

Solution of Eq. (2) yields

$$x(t) = e^{At} x_0 + \int_0^t e^{A(t-\tau)} Bu(\tau) \, d\tau,$$

where e^{At} is the matrix exponential whose entries are linear combinations of $t^k e^{\lambda_j t}$, k an integer ≥ 0, and λ_j the distinct eigenvalues of A. If A has n linearly independent eigenvectors, then all entries in e^{At} are linear combinations of $e^{\lambda_j t}$.

The m observed outputs, $y(t)$, are linear combinations of the state and input variables,

$$y(t) = Cx(t) + Du(t),$$

where C is the $m \times n$ matrix of measurement gains, and D is an $m \times p$ matrix describing the direct contribution of each input to the observed output.

IV. CHARACTERISTICS OF PET DATA

In the system there is initially no radioactivity ($x_0 = 0$), and at all times only a single input, i.e., $p = 1$ and $u(t) = C_p(t)$. The matrix $B = [k_{10}, k_{20}, \ldots, k_{n0}]^T$ contains the influx rate constants from the arterial blood or plasma into compartments $1, 2, \ldots, n$. Each measurement consists of one value ($m = 1$), namely the total activity in the region, which is the sum of the concentrations in each of the tissue compartments. Therefore, $C = [1 \ 1 \ldots 1]$, $D = 0$, and $y(t) = C_T(t) = \sum_{j=1}^{n} x_j(t)$.

V. STRONGLY CONNECTED SYSTEMS THAT MEET THE SPECTRAL ANALYTIC CONDITIONS

Systems in which it is possible for material to reach every compartment from every other compartment, but not possible for material to pass from a given compartment through two or more other compartments back to the starting compartment, are called strongly connected and noncyclic. They include catenary and mammillary systems as well as arbitrary strongly connected noncyclic combinations of these systems (Fig. 1). The state transition matrix of a noncyclic strongly connected system is sign-symmetrical; i.e., $a_{ij}a_{ji} \geq 0$, $i \neq j$; $a_{ij} = 0 \Leftrightarrow a_{ji} = 0$, and it possesses the important property of diagonal symmetrizability; i.e., it is similar to a symmetric matrix via a diagonal similarity transform (Hearon, 1963).

A. Linearly Independent Eigenvectors

A matrix with distinct eigenvalues has linearly independent eigenvectors. Since any coefficient matrix taken at random is likely to have distinct eigenvalues, linear independence of eigenvectors is often simply assumed. Diagonally symmetrizable matrices, however, always have linearly independent eigenvectors whether or not there are repeated eigenvalues (Hearon, 1963), and entries in the matrix exponential are of the form $e^{\lambda_j t}$. Therefore, total tissue activity for strongly connected noncyclic compartmental system can be expressed by Eq. (1).

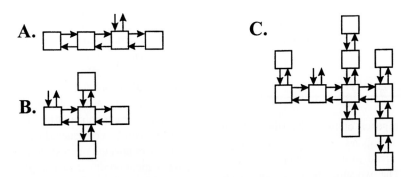

FIGURE 1. Strongly connected compartmental systems that do not contain cycles: (A) catenary, (B) mammillary, and (C) arbitrary noncyclic combination of catenary and mammillary systems. Each of the systems illustrated has exchange with the environment confined to a single compartment and therefore meets the conditions necessary for applying spectral analysis to total tissue activity irrespective of the values of the intercompartmental transfer rate constants.

B. Real-Valued Nonpositive Eigenvalues

The eigenvalues of a compartmental system may, in general, be either real or complex numbers, but the real part of every complex eigenvalue is always nonpositive (Hearon, 1963). Similarity of the state transition matrix A to a symmetric matrix implies that the eigenvalues of A are real. Hence the eigenvalues are nonpositive and real. Furthermore, a strongly connected system that is open (has efflux to the environment) does not contain any traps, so zero is not an eigenvalue of the system (Fife, 1972). Therefore, for strongly connected open noncyclic compartmental systems, the exponents λ_j of Eq. (1) are negative real numbers.

C. Nonnegative Coefficients

Nonnegativity of coefficients has been shown to occur in the content of a *single* compartment of a noncyclic, strongly connected system if that single compartment is the only compartment initially loaded (Hearon, 1979). In the notation of the current study (with the symbol e_i denoting the vector $[0, \ldots, 0, 1, 0, \ldots, 0]^T$, where the nonzero entry is in the ith position) the theorem states the following.

1. Monotonicity Theorem (Hearon, 1979)

Let A be diagonally symmetrizable, and let $x(t)$ be the solution of $\dot{x}(t) = Ax(t)$, $x(0) = ce_i$, for arbitrary i and $c > 0$. Then $x_i(t) = \sum_{j=1}^n \alpha_j \exp(\lambda_j t)$, where $\alpha_j \geq 0$.

Equivalently, if the system is initially in the zero state and a single compartment receives input from the environment, then that compartment will exhibit nonnegative coefficients of its component terms as stated below.

2. Monotonicity Theorem 2

Let the state transition coefficient matrix A be diagonally symmetrizable and $x(t)$ be the solution of $\dot{x}(t) = Ax(t) + Bu(t)$, $x(0) = \mathbf{0}$, and $B = ce_i$, for arbitrary i and $c > 0$.

Then

$$x_i(t) = \sum_{j=1}^n \alpha_j \int_0^t u(\tau) \exp[\lambda_j(t - \tau)]\, d\tau, \quad \text{where } \alpha_j \geq 0.$$

Note that nonnegativity of coefficients applies *only* to compartment i.

Now consider a strongly connected compartmental system that not only restricts environmental input to compartment i, but also has efflux to the environment only from compartment i. The change in concentration in the sum of the compartments of this system is the difference between the influx into compartment i and the efflux from compartment i:

$$\dot{y}(t) = k_{i0}u(t) - k_{0i}x_i(t). \tag{3}$$

If the system is closed, $k_{0i} = 0$ and integation of Eq. (3) produces

$$y(t) = k_{i0} \int_0^t u(\tau)\, d\tau,$$

which clearly has only positive coefficients. On the other hand, if the system is open, i.e., $k_{0i} > 0$, then all λ_j are real and negative, and Monotonicity Theorem 2 can be applied to obtain

$$x_i(t) = \sum_{j=1}^n \alpha_j \int_0^t u(\tau) \exp[\lambda_j(t - \tau)]\, d\tau, \tag{4}$$

where $\alpha_j \geq 0$. Substituting Eq. (4) into Eq. (3) and integrating yields

$$y(t) = K \int_0^t u(\tau)\, d\tau + \sum_{j=1}^n \gamma_j \int_0^t u(\tau) \exp[\lambda_j(t - \tau)]\, d\tau, \tag{5}$$

where $\gamma_j = -k_{0i}\alpha_j/\lambda_j$ and $K = k_{i0} - \Sigma_j\gamma_j$. It follows immediately from $\lambda_j < 0$, $-k_{0i} < 0$, and $\alpha_j \geq 0$ that $\gamma_j \geq 0$. The coefficient K is the steady-state total activity in the system following a unit impulse input; because the strongly connected open system contains no traps, $K = 0$.

VI. SYSTEMS WITH TRAPS

Consider a system composed of two subsystems: S_1, a noncyclic strongly connected open system that has exchange with the environment confined to a single compartment i, and S_2, a trap (Fig. 2A). The two subsystems are connected unidirectionally, i.e., material can be transfered from any compartment in S_1 to any compartment in S_2, but material cannot be transfered from any compartment in S_2 either to the environment or to S_1. Since the subsystem S_1 is strongly connected, its state transition matrix is diagonally symmetrizable; therefore, it has linearly independent eigenvectors and the matrix exponential consists of linear combinations of $e^{\lambda_j t}$ where the eigenvalues λ_j are real and nonpositive. Furthermore, S_1 contains no traps so $\lambda_j < 0$. Since S_1 has influx only into compartment i, by Monotonicity Theorem 2, $x_i(t)$ can be described by Eq. (4) (with n replaced by n_1, the number of compartments in S_1). Because compartment i is the only compartment of the combined system $S = S_1 \oplus S_2$ that exchanges material with the environment, the rate of change in the sum of the concentrations in all compartments of S is given by Eq. (3). Substituting Eq. (4) into Eq. (3) and integrating again produces Eq. (5) (with n replaced by n_1). It follows from $\lambda_j < 0$, $-k_{0i} < 0$, $\alpha_j \geq 0$ that $\gamma_j \geq 0$. The steady-state total concentration in the combined system following a unit impulse input is strictly positive ($K > 0$) since the system contains a trap, namely S_2.

Note that since the eigenvalues of S_2 do not appear in the measured sum of concentrations in the system as a whole, the internal structure of S_2 is not restricted; e.g., it may contain cycles (Fig. 2B). Adding additional traps to the system does not alter the characteristics of the total measured output; namely, eigenvalues remain real and nonpositive and coefficients remain nonnegative. Thus any compartmental system that can be constructed by adding traps to a noncyclic strongly connected open system that has exchange with the environment confined to a single compartment satisfies the spectral analytic conditions.

VII. DISCUSSION

The compartmental structures shown in the present study to satisfy the spectral analytic conditions include the catenary, mammillary, and various noncyclic combinations of these systems when input from the environment and efflux to the environment are confined to the same single compartment of the system. Any system in which tracer delivered by the blood exchanges with a single compartment in the tissue and then undergoes one or more series of metabolic and/or binding transformations (without feedback) satisfies these conditions. Examples include the one-compartment kinetic model (Fig. 3A) used for measurement of cerebral blood flow in a homogeneous tissue with $H_2^{15}O$ (Herscovitch $et\ al.$, 1983) and the two-compartment model (Fig. 3B) for measurement of glucose metabolism in homogeneous tissues with $[^{18}F]$fluorodeoxyglucose ($[^{18}F]$FDG) (Reivich et

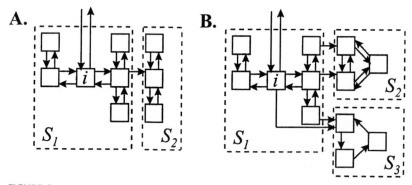

FIGURE 2. Compartmental systems formed by adding traps to a noncyclic strongly connected open system. The subsystem S_1 (without the links to subsystem S_2 or S_3) has exchange of material with the environment confined to compartment i. It satisfies the conditions for application of spectral analysis, and, as it contains no traps, all the eigenvalues are negative. Adding traps S_2 (A) or S_2 and S_3 (B) adds a term with a zero eigenvalue to the equation describing the total tissue activity; i.e., it adds a term containing the integrated input function. The combined systems, therefore, also meet the conditions necessary for application of spectral analysis. The number of visible negative eigenvalues is less than or equal to the number of compartments in S_1. Note that since the eigenvalues of the traps (subsystems S_2 and S_3) do not appear in the equation for total activity, there are no restrictions on their structure, e.g., they may contain cycles as illustrated in (B).

al., 1979). The three-compartment model (Fig. 3C) used in many receptor ligand binding studies does not contain any cycles and includes the assumption that there is a single pool of free ligand in the tissue that exchanges with the plasma. Hence it satisfies the spectral analytic conditions.

The models discussed thus far were designed for application in homogeneous tissues. A collection of replicates of a compartmental system designed for homogeneous tissues can be used to model the activity in a heterogeneous mixture of tissues (Fig. 3D). Due to the additivity of the components in Eq. (1), any collection of systems that are not interconnected and in which each independent system in the collection satisfies the spectral analytic conditions also satisfies the spectral analytic conditions. This property renders the spectral analysis technique particularly useful for the analysis of tracer kinetics in PET studies in brain as several kinetically distinct tissues may be included in each voxel or region of interest.

Spectral analysis may not be appropriate for analyzing models that contain feedback loops, that have efflux from more than one compartment, or that consist of connected collections of subsystems, even if the subsystems in the collection individually meet the conditions for application

of spectral analysis (Fig. 4). It is, however, only the general application of spectral analysis that is precluded; spectral analysis may still be applicable to these systems if the rate constants are known to satisfy some constraints. For example, the three-compartment model that describes deoxyglucose kinetics and accounts for intracellular compartmentation of glucose phosphatase contains a feedback loop (Schmidt *et al.*, 1989) (Fig. 4A). For some combinations of values of the rate constants, the eigenvalues of the system are complex-valued; this indicates that oscillations in the system are possible. Unless one of the rate constants in the cycle is zero, or the system is closed ($k_2 = 0$), this model is not a candidate for application of spectral analysis. The model of Huang *et al.* (1991) that describes the kinetics of L-3,4-dihydroxy-6-[^{18}F]fluorophenylalanine ([^{18}F]FDOPA) and L-3,4-dihydroxy-6-[^{18}F]fluoro-3-*O*-methylphenylalanine ([^{18}F]3-OMFD) in the striatum is given in Fig. 4B. It consists of two unconnected catenary subsystems, so the eigenvalues of the system are real-valued. Spectral analysis cannot be applied to the total tissue activity in this system as configured, however, since positivity of coefficients is not assured unless the rate constants satisfy certain constraints; e.g., k_4, loss from the fluorodopa metabolite compartment,

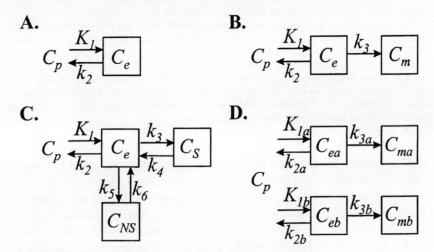

FIGURE 3. Compartmental models that satisfiy the conditions for application of spectral analysis to total tissue activity irrespective of the values of the rate constants. (A) Model for measurement of cerebral blood flow in a homogeneous tissue with $H_2^{15}O$. C_p and C_e represent the concentration of tracer in the arterial blood and in the exchangeable pool in the tissue, respectively, K_1 the blood flow per unit weight of tissue (mL/g/min), and $k_2 = K_1/\lambda$, where λ is the tissue: blood partition coefficient. (B) Model for measurement of cerebral glucose utilization with [^{18}F]FDG in a homogeneous tissue. C_p and C_e represent the concentration of tracer in the arterial plasma and the unmetabolized tracer in the exchangeable pool in the tissue, respectively; C_m represents the concentration of metabolites in the tissue. The rate constants are labeled K_1 through k_3. (C) Model of receptor ligand distribution in a homogeneous tissue. C_p and C_e represent the concentration of free tracer in the arterial plasma and in the tissue, respectively; C_S is the concentration of ligand bound to specific receptors, and C_{NS} is the concentration of ligand that is nonspecifically bound. Transfer rate constants are labeled K_1 through k_6. (D) Model to describe the kinetics of [^{18}F]FDG in heterogeneous mixture of two tissues. Symbols are the same as in (B), with the additional subscripts *a* and *b* indicating concentrations and rate constants in tissue *a* and tissue *b*.

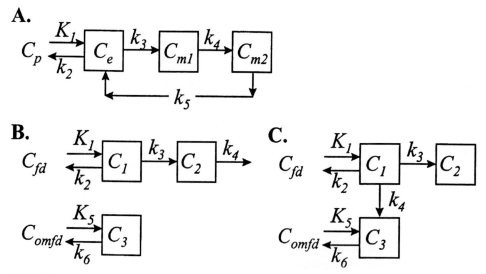

FIGURE 4. Compartmental models that do not *a priori* meet the conditions for application of spectral analysis. (A) Three-compartmental model to describe the kinetic behavior of deoxyglucose (DG) in a homogeneous tissue of the brain that accounts for the intracellular compartmentation of glucose 6-phosphatase (Schmidt *et al.*, 1989). C_p and C_e represent the concentration of unmetabolized tracer in the arterial plasma and in the exchangeable pool in the tissue, respectively. The concentration of DG-6-phosphate in the cytosol, where it is formed, and in the cisterns of the endoplasmic reticulum, where the phosphatase resides, are represented by C_{m1} and C_{m2}, respectively. In this model k_4 represents transport and/or diffusion of DG-6-phosphate from the cytosol into the cisterns of the endoplasmic reticulum, and k_5 represents the combined hydrolysis of DG-6-phosphate and return of DG to the tissue precursor pool. Unless k_2, k_3, k_4, or $k_5 = 0$, the existence of the cycle from C_e to C_{m1} to C_{m2} and back to C_e implies that the system may not meet the conditions necessary for application of spectral analysis. (B) Three-compartmental model to describe the kinetics of [18F]FDOPA and [18F]3-OMFD in the striatum (Huang *et al.*, 1991). C_{fd} and C_{omfd} represent the concentrations of [18F]FDOPA and [18F]3-OMFD, respectively, in the arterial plasma, and C_1 and C_3 represent the corresponding concentrations in the tissue. The concentration of 6-[18F]fluorodopamine ([18F]FDA) and its metabolites is represented by C_2. Only if the rate constants satisfy certain constraints, e.g., $k_4 = 0$ or $k_2 > k_4$, are the conditions necessary for application of spectral analysis to this system met. (C) Three-compartmental model of Gjedde *et al.* (1991) for transport and metabolism of [18F]FDOPA in brain. All symbols except k_4 are as in (B). This model does not include loss of [18F]FDA metabolites; Instead, k_4 represents the methylation of [18F]FDOPA in the brain tissue. Conditions for application of spectral analysis to this system are met if the rate constants satisfy $k_4 = 0$ or $k_2 > k_6$.

is zero, or $k_2 > k_4$. Two subsystems are also used in the model of Gjedde *et al.* (1991) to describe [18F]FDOPA and [18F]3-OMFD kinetics (Fig. 4C). In this model it is assumed that there is no loss of fluorodopa metabolites during the experimental period, so both subsystems in this model meet the conditions necessary for application of spectral analysis. Because this model includes the possibility of conversion of [18F]FDOPA to [18F]3-OMFD in the tissue, however, the subsystems are connected and the complete system may not satisfy the conditions necessary for application of spectral analysis to total tissue activity. (The system does meet the conditions if $k_4 = 0$ or $k_2 > k_6$).

VIII. CONCLUSION

Spectral analysis is potentially a useful tool for the analysis of dynamic PET data as it applies to heterogeneous as well as homogeneous tissues. It does not require that the ki-

netic model of the system be completely specified *a priori*, but instead provides an estimate of the minimum number of compartments needed to describe the system and characterizes the system components. It is based on a general linear compartmental system but does not apply to all such systems. The present study has identified a class of compartmental systems to which the analysis can be generally applied.

References

Cunningham, V. J., and Jones, T. (1993). Spectral analysis of dynamic PET studies. *J. Cereb. Blood Flow Metab.* **13**: 15–23.

Fife, D. (1972). Which linear compartmental systems contain traps? *Math. Biosci.* **14**: 311–315.

Gjedde, A., Reith, J., Dyve, S., Léger, G., Guttman, M., Diksic, M., Evans, A., and Kuwabara, H. (1991). Dopa decarboxylase activity of the living human brain. *Proc. Natl. Acad. Sci. USA* **88**: 2721–2725.

Hearon, J. Z. (1963). Theorems on linear systems. *Ann. NY Acad. Sci.* **108**: 36–68.

Hearon, J. Z. (1979). A monotonicity theorem for compartmental systems. *Math. Biosci.* **46**: 293–300.

Herscovitch, P., Markham, J., and Raichle, M. E. (1983). Brain blood flow measured with intravenous $H_2^{15}O$. I. Theory and error analysis. *J. Nucl. Med.* **24**: 782–789.

Huang, S.-C., Yu, D.-C., Barrio, J. R., Grafton, S., Melega, W. P., Hoffman, J. M., Satyamurthy, N., Mazziotta, J. C., and Phelps, M. E. (1991). Kinetics and modeling of L-6-[18F]fluoro-DOPA in human positron emission tomographic studies. *J. Cereb. Blood Flow Metab.* **11**: 898–913.

Reivich, M., Kuhl, D., Wolf, A., Greenberg, J., Phelps, M., Ido, T., Casella, V., Fowler, J., Hoffman, E., Alavi, A., Som, P., and Sokoloff, L. (1979). The [18F]fluorodeoxyglucose method for the measurement of local cerebral glucose utilization in man. *Circ. Res.* **44**: 127–137.

Schmidt, K., Lucignani, G., Mori, K., Jay, T., Palombo, E., Nelson, T., Pettigrew, K., Holden, J. E., and Sokoloff, L. (1989). Refinement of the kinetic model of the 2-[14C]deoxyglucose method to incorporate effects of intracellular compartmentation in brain. *J. Cereb. Blood Flow Metab.* **9**: 290–303.

Schmidt, K., Lucignani, G., Moresco, R. M., Rizzo, G., Gilardi, M. C., Messa, C., Colombo, F., Fazio, F., and Sokoloff, L. (1992). Errors introduced by tissue heterogeneity in estimation of local cerebral glucose utilization with current kinetic models of the [18F]fluorodeoxyglucose method. *J. Cereb. Blood Flow Metab.* **12**: 823–834.

Turkheimer, F., Moresco, R. M., Lucignani, G., Sokoloff, L., Fazio, F., and Schmidt, K. (1994). The use of spectral analysis to determine regional cerebral glucose utilization with positron emission tomography and [18F]Fluorodeoxyglucose: theory, implementation and optimization procedures. *J. Cereb. Blood Flow Metab.* **14**: 406–422.

Turkheimer, F., Sokoloff, L., Bertoldo, A., Lucignani, G., Reivich, M., Jaggi, J. L., and Schmidt, K. (1998). Estimation of component and parameter distributions in spectral analysis. *J. Cereb. Blood Flow Metab.* **18**: 1211–1222.

14

Evaluation of the Reliability of Parameter Estimates in the Compartment Model Analysis by Using the Fitting Error

Y. IKOMA,* H. TOYAMA,[†,‡] K. UEMURA,* Y. KIMURA,[‡] M. SENDA,[‡] and A. UCHIYAMA*

*Waseda University, Tokyo, Japan
[†]National Institute of Radiological Sciences, Chiba, Japan
[‡]Positron Medical Center, Tokyo Metropolitan Institute of Gerontology, Tokyo, Japan

In order to evaluate the reliability of parameter estimates for the human [^{11}C]flumazenil data, the relationship among the fitting error, noise, and reliability of parameter estimates was investigated by means of computer simulation with various sets of k values according to the one-tissue-compartment, two-parameter model. Rate constants were estimated by means of a modified Marquardt method pixel by pixel in simulated flumazenil scans. The relationship among the fitting error, noise, and reliability of parameter estimates depended only on k_2, not on K_1. An empirical equation for reliability estimation was generated using this relationship between the fitting error and the value of k_2. Applying this empirical equation, the noise level and reliability of parameter estimates for human flumazenil data were evaluated. As the size of the region of interest decreased or the administration dose decreased, the noise level increased and the reliability decreased reasonably.

I. INTRODUCTION

In a quantitative analysis of the tracer kinetics in a PET dynamic study, the precision in rate constant estimation is affected by the noise. Although estimation of the reliability of parameter estimates is valuable in a clinical study, it is not easy for human data, in which the true noise level is not precisely known (Feng *et al.*, 1996; Millet *et al.*, 1996).

We have investigated the relationship among the fitting error, noise, and reliability of parameter estimates by means of computer simulation and have developed a method for estimating the noise level and the reliability for human data from the fitting error obtained in the model analysis. The method was applied to the 2-parameter model for [^{11}C]-flumazenil.

II. MATERIALS AND METHOD

A. Computer Simulation

Simulation data with various noise levels were generated for 12 sets of "true" k values: various K_1 ($K_1 = 0.20, 0.25, 0.28, 0.30, 0.40, 0.45$ mL/mL/min, and $k_2 = 0.07$ min^{-1}) and various k_2 ($k_2 = 0.03, 0.05, 0.07, 0.10, 0.15, 0.20$ min^{-1}, and $K_1 = 0.28$ mL/mL/min), and 1000 simulated data sets were generated for each k and noise level. A dynamic tracer concentration for [^{11}C]flumazenil was derived from the one-tissue-compartment two-parameter model (Koeppe *et al.*, 1991) and a measured input function according to the human PET imaging protocol (30 s × 12, 60 s × 4, 120 s × 24, total about 60 min, 40 frames). The unmetabolized plasma time–activity curve of a normal subject was used as an input function, for which the data sampling was performed at 22 points for 62 min.

The collected count $N(t)$ and noise (%) for each frame were determined by the following equations:

$$N(t) = \int_{t-t_d/2}^{t+t_d/2} C_t(t') \cdot \exp\left(-\frac{\ln 2}{T} t'\right) dt' \cdot factor, \quad (1)$$

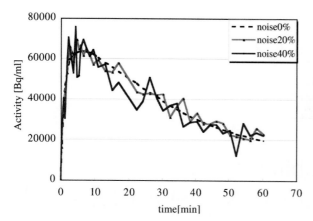

FIGURE 1. Typical tissue time activity curves for $K_1 = 0.28$, $k_2 = 0.07$ without noise and with 20% noise and 40% noise.

$$\text{Noise}(\%) = \frac{\sqrt{N(t)}}{N(t)} \times 100 = \frac{1}{\sqrt{N(t)}} \times 10e, \quad (2)$$

where $C_t(t)$ is the nondecaying tissue radioactivity concentration derived from the k values and the input function, t_d is the data collection time, and T is the physical half-life of the radionuclide. The "factor" is a scaling factor representing the sensitivity of the measurement system and is introduced here to adjust the noise level. The noise was generated with random numbers based on Poisson distribution and added to $C_t(t)$, i.e., decay corrected tissue activity for each frame (Fig. 1). The level of the noise for the dynamic data was expressed as the percentage of noise to the count for the last frame ($t_{\text{last}} = 58$ min). Seven noise levels (1, 3, 5, 7, 10, 20, and 40% at the last frame) were studied, and the "factor" was determined for each (Ikoma et al., 1998).

It should be noted that the effect of sampling period (t_d) and physical decay are incorporated in the noise determination. It should also be noted that this model assumes that the noise, which is added to the time–activity curve, is determined by the curve itself, i.e., K_1 and k_2 of the region. In fact, the noise is determined by the total count of the slice which depends on the K_1 and k_2 of other regions.

B. Reliability of Parameter Estimates

In a simulation study, the rate constants (K_1, k_2) and the distribution volume (DV = K_1/k_2) were estimated by a modified Marquardt method (Taguchi et al., 1997), and the fitting error was calculated by

$$\sqrt{\frac{\sum(C_{\text{simulated},t} - C_{\text{estimated},t})^2}{N-2}} \cdot 100 \Big/ C_{\text{max}}, \quad (3)$$

where $C_{\text{simulated},t}$ is the simulated value at time t, $C_{\text{estimated},t}$ is the estimated value at time t with the modified Marquardt method, C_{max} is the maximum value of the simulated time curve, N is the number of frames. Parameter estimates were

considered invalid if either K_1 or k_2 was outside the range, $0.0 < K_1 < 1.0, 0.0 < k_2 < 1.0$. To account for asymmetric distribution of the estimation error, deviation of the estimates from the true value was calculated separately for lower and higher range by:

$$R_\pm = \sqrt{\frac{\sum(k_{\text{estimated},i} - k_{\text{true}})^2}{n}} \cdot 100 \Big/ k_{\text{true}}, \quad (4)$$

where $k_{\text{estimated},i}$ is the estimated k value for ith simulation, k_{true} is the true value of k, n is the number of simulations in which estimated k value is lower or higher than the true value, R_+ is the reliability of estimates for higher range, and R_- is the reliability of estimates for lower range.

The relationship among the fitting error, noise level, and reliability were investigated by the following processes.

(a) The relationship between the fitting error and noise level was investigated for various k values.
(b) The relationship between the noise level and reliability of parameter estimates was investigated for various k values.
(c) The empirical equations were derived from the relationship among the fitting error, noise, and reliability.

C. Analysis of Human Data with [^{11}C]flumazenil

Two human PET data sets with [^{11}C]flumazenil on normal volunteers were analyzed, with the administration dose of 155 MBq (Subject 1) and 615 MBq (Subject 2). Data were acquired with a PET camera, HEADTOME IV(Shimadzu, Kyoto, Japan) according to the time schedule of 30 s × 12, 60 s × 4, 120 s × 24, i.e., a total of 40 frames and about 1 h. Arterial blood sampling was performed over 60 min after intravenous injection.

Using an empirical equation derived from the computer simulation, the reliability of parameter estimates for cortical gray matter, white matter, and cerebellum were calculated with the tissue time–activity curve in regions of interest (ROIs) of various sizes.

III. RESULTS

A. Computer Simulation

The value of K_1 did not affect the relationship between the noise level and the fitting error, but the value of k_2 did, as shown in Fig. 2. Their relationship was expressed empirically by a quadratic equation:

$$\text{Noise}(\%) = a \times (\text{fitting error})^2 + b \times (\text{fitting error}). \quad (5)$$

The relationships between the value of k_2 and the coefficients a and b were also investigated and was expressed by

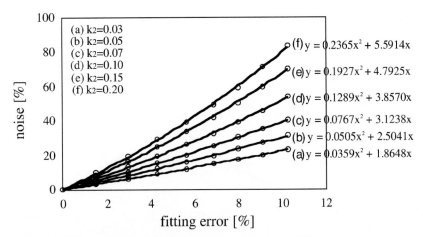

FIGURE 2. The relationship between the fitting error and the noise level for the various values of k_2.

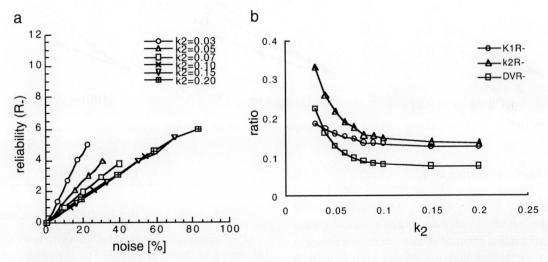

FIGURE 3. (a) Relationship between the noise level and R_- for DV. (b) Relationship between the value of k_2 and the ratio of reliability R_- to noise.

polynomial equations:

$$a = -45.042 \times k_2^3 + 13.942 \times k_2^2$$
$$+0.0945 \times k_2 + 0.0196,$$
$$b = -61.728 \times k_2^2 + 35.817 \times k_2 + 0.8719. \quad (6)$$

Using Eqs. (5) and (6), the noise level was described empirically with the fitting error and k_2.

Parameter estimates distributed widely and asymmetrically as the noise increased in the simulation as shown in Table 1. The measure of reliability, $R_\pm$, grew roughly in a proportional manner with the noise level as shown in Fig. 3a. Their ratio also depended only on k_2, not on K_1. When the relationship between the value of k_2 and the ratio of reliability to noise was examined, the ratio became constant when the value of k_2 was higher than 0.09, as shown in Fig. 3b. Using these results, the reliability was also described with the fitting error and k_2.

TABLE 1. Reliability of Parameter Estimates in the Simulation

		Noise 10%	Noise 20%	Noise 40%
K_1	R_-	1.41	2.83	5.98
	R_+	1.37	2.73	5.13
k_2	R_-	1.67	3.40	7.17
	R_+	1.85	3.67	7.25
DV	R_-	0.97	1.98	3.79
	R_+	0.83	1.61	3.64

Note. R_- or R_+ of K_1, k_2, and DV with the noise level of 10, 20 and 40% when the k values were $K_1 = 0.28$, $k_2 = 0.07$.

To verify the empirical equations, the values of noise and reliability were estimated in the simulation data for three

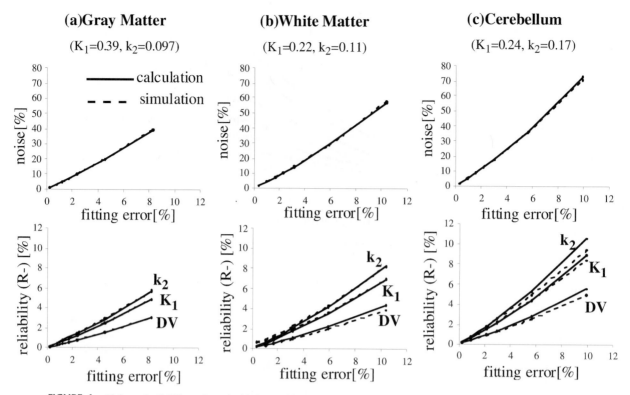

FIGURE 4.   Noise and reliability estimated with the empirical equation and simulation versus fitting error in the simulation data.

sets of k values representing the gray matter ($K_1 = 0.39$, $k_2 = 0.097$), white matter ($K_1 = 0.22$, $k_2 = 0.11$), and cerebellum ($K_1 = 0.24$, $k_2 = 0.17$). As a result, the noise and reliability estimated with the empirical equations almost agreed with the values obtained in the simulation as shown in Fig. 4. In the cerebellum, however, estimated reliability with the empirical equation somewhat differed from the value obtained in the simulation.

B. Human Data

When the empirical equations were applied to human data, estimated reliability depended on the administration dose. For subject 1, with administration dose of 155 MBq, estimated noise was 35% at the last frame, and R_- of DV was about 4.3% in the gray matter. For subject 2, with administration dose of 615 MBq, estimated noise was 20%, and R_- of DV was about 2.0%.

The noise and reliability for tissue activity curves obtained from ROIs of various sizes were estimated by the empirical equations and shown in Fig. 5. The estimated noise level became lower and parameter estimates became more reliable as the ROI size became larger. However, limitation was observed beyond which the reliability did not get better by increasing the ROI size.

IV. DISCUSSION

A. Validity of the Empirical Equations

Parameter estimates deviated from symmetry as the noise increased, as shown in Table 1. This is probably because the model equation is nonlinear. If the model equation is linear and the noise is Gaussian, the distribution of parameter estimates will be symmetric, and the standard deviation may be calculated from the residual error without simulation. In fact, the distribution was asymmetric to the true value, and we calculated the deviation separately for lower and higher range to true value as Eq. (4), and the reliability of parameter estimates was derived empirically from the computer simulation. The validity of the equation was verified as shown in Fig. 4, and the noise level estimated with the empirical equation agreed well with the value obtained in the simulation in gray matter, white matter, and cerebellum. The reliability of parameter estimates also agreed well, especially in the gray matter. The empirical approach presented here may be useful for estimating the reliability in human study where the true noise is not known.

B. Effect of ROI Size in the Model Analysis

As shown in Fig. 5, the estimated noise became smaller as the ROI size became larger up to a certain point due to

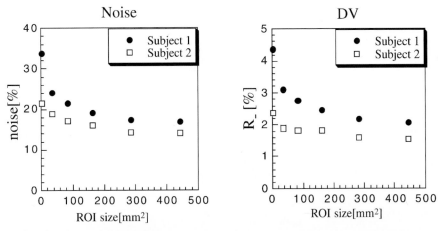

FIGURE 5. Reliability estimates of the gray matter in various ROI sizes for human data.

noise reduction by pixel averaging. Beyond that point, however, the estimated noise computed from the fitting error did not become smaller, probably because the k_2 value within the ROI became heterogeneous and the model did not describe the time curve of ROIs. This result also suggests that there is an optimum ROI size as a tradeoff between noise and heterogeneity.

V. CONCLUSION

The noise level and the reliability was described with the empirical equations using the fitting error and k_2. The reliability of parameter estimates for human data was evaluated from the fitting error. The estimated noise level and the reliability calculated with the empirical equation responded reasonably to the ROI size and administration dose.

Acknowledgments

This work was supported in part by Grants in Aid for Scientific Research No. 11670653 from the Ministry of Education, Science and Culture, Japan.

References

Feng, D., Huang, S.-C., Wang, Z., and Ho, D. (1996). An unbiased parametric imaging algorithm for nonuniformly sampled biomedical system parameter estimation. *IEEE Trans. Med. Image.* **15**: 512–518.

Ikoma, Y., Toyama, H., Yamada, T., Uemura, K., Kimura, Y., Senda, M., and Uchiyama, A. (1998). Creation of a dynamic digital phantom and its application to a kinetic analysis. *Kaku Igaku* **35**(5): 293–303. [in Japanese]

Koeppe, R. A., Holthoff, V. A., Frey, K. A., Kilbourn, M. R., and Kuhl, D. E. (1991). Compartmental analysis of [11C]flumazenil kinetics for the estimation of ligand transport rate and receptor distribution using positron emission tomography. *J. Cereb. Blood Flow Metab.* **11**: 735–744.

Millet, P., Delforge, J., Pappata, S., Syrota, A., and Cinotti, L. (1996). Error analysis on parameter estimates in the Ligand-receptor model: Application to parameter imaging using PET data. *Phys. Med. Biol.* **41**: 2739–2756.

Taguchi, A., Toyama, H., Kimura, Y., Senda, M., and Uchiyama, A. (1997). Comparison of the number of parameters using nonlinear iteration methods for compartment model analysis with 18F-FDG brain PET. *Kaku Igaku* **34**: 25–33. [in Japanese]

15

A Linear Solution for Calculation of K_1 and k_2 Maps Using the Two-Compartment CBF Model

P.-J. TOUSSAINT and E. MEYER

Positron Imaging Laboratories, McConnell Brain Imaging Centre, Montreal Neurological Institute, McGill University Montréal, Québec, Canada

We have previously demonstrated that a solution of the two-compartment model, which accounts for residual intravascular radioactivity (V_0) and explicitly includes correction for tracer arrival delay (Δt) and dispersion (τ) of the input function, provides accurate estimates of cerebral water clearance (K_1, an index of cerebral blood flow) and of the washout rate of metabolized water from brain to blood ($K_1 = k_2(V - V_0)$, where V is the total distribution volume). However, calculation of parametric images from high-resolution dynamic positron emission tomography data using nonlinear least-squares fitting has proven to be time consuming. We have therefore derived a linear solution to accelerate the pixel-by-pixel computation of parametric maps. Simulations showed that estimates of K_1 and k_2 can be obtained with less than 2% bias using our linear solution of the two-compartment model at all flow values in regions with low to average vascularity ($V_0 \sim 2$ mL/100 g), and with less than 14% error in highly vascularized areas ($V_0 = 10$ mL/100 g).

I. INTRODUCTION

Quantification of cerebral blood flow (CBF) *in vivo* using [^{15}O]water positron emission tomography (PET) studies requires measurement of the uptake and washout of the tracer in tissue as a function of time (tissue time–activity curve $C(t)$), as well as its delivery to the tissue (arterial input function $C_a(t)$) (Herscovitch *et al.*, 1983; Raichle *et al.*, 1983). Direct measurement of the radioactivity concentration in arterial blood requires arterial cannulation and either manual or automatic sampling.

A two-compartment model has been described recently (Ohta *et al.*, 1996) which allows for the quantification of CBF and also provides a measure of the initial distribution volume of the tracer (V_0, the second compartment). This second compartment had been introduced to correct for residual vascular radioactivity, which caused erroneously high blood flow values to be observed with the one-compartment model (Kety, 1951). The method requires that appropriate corrections for delay (Δt) and dispersion (τ) of the sampled blood data be applied (Meyer, 1989; Iida *et al.*, 1986). In this method presented by Ohta and colleagues (1996), delay corrections are performed on a slice-by-slice basis and a fixed value is assumed for dispersion (4 s). These corrections may not be appropriate for all subjects, particularly in disease situations where Δt and τ may vary from one brain area to the next. Therefore, we have initially introduced a nonlinear solution of the two-compartment model which implicitly includes the correction terms for delay and dispersion (Toussaint and Meyer, 1996, 1998). This approach fits five parameters (K_1, k_2, V_0, τ, and Δt). Estimation of the main parameters of interest, cerebral water clearance and washout rate of metabolized water (K_1 and k_2), has proven to be robust using the nonlinear equation, but this solution provides imprecise estimates of the other three parameters, including V_0. In addition, convergence rates are slow and care must be taken to avoid solutions which correspond to local minima rather than to the true best fit parameter values, thereby increasing the overall analysis time.

In the present study, we propose a linear solution of the two-compartment model which increases convergence rates while retaining the accuracy of the main parameters of interest.

Copyright © 2001 by Academic Press
All rights of reproduction in any form reserved.

II. THEORY

Derivation of the linear expression starts from the general solution of the two-compartment model (see Ohta *et al.* (1996), for derivation of the two-compartment model three-weighted integration approach),

$$C(t) = (K_1 + k_2 V_0) \int_0^t C_a(u)\, du \\ - k_2 \int_0^t C(u)\, du + V_0 \cdot C_a(t), \tag{1}$$

where $C(t)$ is the tissue time–activity concentration and $C_a(t)$ the radioactivity concentration in blood. Here, the true input function, $C_a(t)$, is related to the measured delayed and dispersed blood activity, $g(t)$, by

$$C_a(t) = g(t) + \tau \frac{dg(t)}{dt}. \tag{2}$$

Substituting Eq. (2) into Eq. (1), we obtain

$$C(t) = a \cdot \int_0^t g(u + \Delta u)\, du + b \cdot g(t + \Delta t) \\ + c \cdot \frac{dg(t + \Delta t)}{dt} - d \cdot \int_0^t C(u)\, du, \tag{3}$$

where $g(t + \Delta t)$ is the measured input function that has been corrected for delay, and the linear parameters (a, b, c, d) are related to the original nonlinear parameters (K_1, k_2, V_0, τ) through the following equations: $a = K_1 + k_2 V_0$, $b = \tau K_1 + (\tau k_2 + 1) V_0$, $c = \tau V_0$, and $d = k_2$.

III. METHODS

For simulations, typical measurements of arterial blood radioactivity concentration obtained from a dynamic PET study with ^{15}O-labeled water were used to generate a bolus type input function by approximation to a sum of four gamma variates (Toussaint and Meyer, 1996). Various levels of Gaussian noise (5, 10, and 20%) were added to tissue data sets created using the two-compartment model Eq. (1) with the baseline parameters $K_1 = 25, 50, 75,$ and $100\ mL/100\ g/min$, $V_0 = 2$ and $10\ mL/100\ g$, $K_1/k_2 = 0.9\ mL/g$, $\Delta t = 10\ s$ and $\tau = 5\ s$. Using these tissue data for several combinations of baseline parameter sets, together with the simulated bolus input function, $g(t + \Delta t)$, we have carried out simulations to assess the performance of the linear solution of the two-compartment model with multiparameter least-squares fitting.

In order to test our solution on real data, an $H_2^{15}O$ PET study was performed on a normal subject using the ECAT EXACT HR+ whole body tomograph (CTI/Siemens) operating in a 3D acquisition mode. Written informed consent was obtained from the subject prior to participation in the study whose protocol was approved by the Research Ethics Committee of the Montreal Neurological Institute and Hospital. Immobilization of the subject's head was achieved by means of a customized head contention device (Vac-Lock, MED-TECH). A short indwelling catheter was placed into the left radial artery for automatic blood sampling. Following bolus injection of about 370 MBq of $H_2^{15}O$, activity images were acquired over 3 min as the subject was lying supine in the scanner while fixating a cross-hair at the center of a television monitor in a baseline condition (Vafaee *et al.*, 1999).

The arterial radioactivity concentration was corrected for external delay and dispersion due to automatic blood sampling and calibrated with respect to the tomograph. Data were reconstructed and cerebral blood flow was calculated using the two-compartment model linear least-squares solution, as well as the three-weighted integration method for comparison. The parametric images were normalized to the mean global values and average measurements were obtained for K_1 and k_2 in selected regions of interest.

IV. RESULTS

Using the linear solution reduced the analysis time by two orders of magnitude compared to the nonlinear solution. Fitting all 5 parameters, as expected, gave fits that were indistinguishable from the nonlinear fit upon visual inspection (Fig. 1). The linear parameter a was recovered within 1% with a coefficient of variation (COV) between ± 1.6 and $\pm 8.1\%$ for small V_0, and these increased to at most $6.7 \pm 4.7\%$ at large V_0. However, estimates of c were either much smaller (at small V_0 and low to average K_1) or larger (at large V_0 and K_1 values) than the expected value. Estimates of b were similarly affected at small V_0, but were recovered with less than 16% bias at large V_0 values (Fig. 2, top left). These inaccuracies in the linear parameters introduced a large bias in the estimates of V_0, τ, and Δt (Fig. 2, top right), whereas absolute values of K_1 were within 1.9% of expectation on average for all baseline flow values at $V_0 = 2\ mL/100\ g$, with a COV ranging from ± 3.6 to $\pm 3.9\%$ (Fig. 2, bottom left). Estimates of k_2 were within 0.4% on average, with a COV of ± 4.3 to $\pm 10.3\%$.

At larger V_0, estimates of K_1 were within 4.5% for average to high flow values (COV between ± 3.4 and $\pm 4.1\%$) and corresponding k_2 values were within 4.6% (COV between ± 4.9 and $\pm 6.5\%$). K_1 and k_2 were overestimated for low baseline flow at large V_0 (at most $+10.5\%$ for K_1 and $+14.3\%$ for k_2) due to considerable underestimation of V_0 from the fit.

As the time of integration (total study duration used in Eq. (3)) was increased above $120\ s$, K_1 and k_2 estimates were recovered accurately at average to high flow for all V_0 values (COV $< \pm 12.9\%$ for k_2 and COV $< 5.2\%$ for K_1) (Fig. 2, bottom right). With $V_0 = 10\ mL/100\ g$, both K_1 and k_2 were slightly overestimated at average flow ($50\ mL/100\ g/min$),

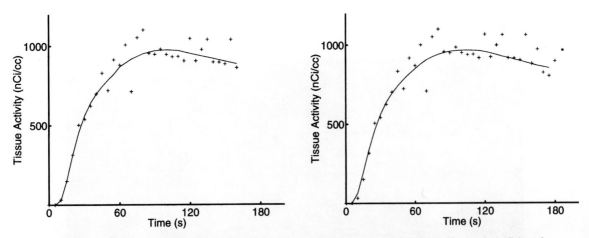

FIGURE 1. Noisy (10% Gaussian) simulated tissue data (+) and fitted curve from nonlinear least-squares fitting of $(K_1, k_2, V_0, \tau, \Delta t)$ (left) and linear least-squares fitting of $(a, b, c, d, \Delta t)$ (right). Baseline $K_1 = 50$ mL/100 g/min, $V_0 = 2$ mL/100 g.

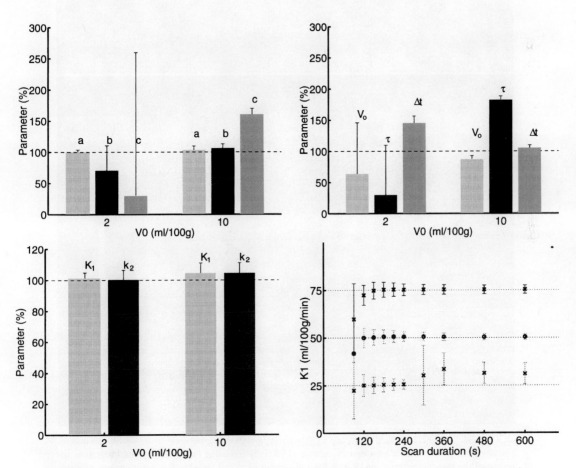

FIGURE 2. Mean estimates from five-parameter least-squares fitting of the linear two-compartment model equation, expressed as a percentage of their expected values (baseline $K_1 = 50$ mL/100 g/min, 10% Gaussian noise, $n = 100$). Estimates of a (top left) are within 3% error. Inaccuracies in (c), and thus in (b) (see equations), caused errors in V_0, τ, and Δt (top right), which in most instances canceled each other's effect on estimates of K_1 and k_2 (bottom left). These were within 2% of expectation at average to high initial K_1 for integration over 120 s and beyond (bottom right).

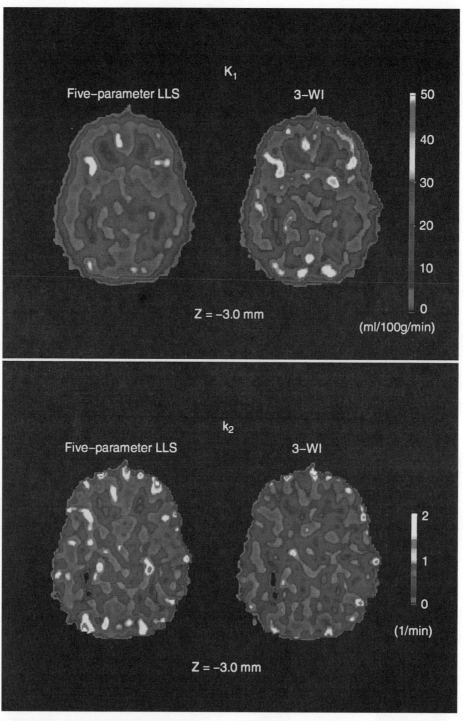

FIGURE 3. K_1 (mL/100 g/min) (top) and k_2 (1/min) images (bottom), obtained by fitting five parameters pixel-by-pixel with the linear least-squares solution (LLS, left) and using the three-weighted integration approach of the two-compartment model with fixed dispersion (4 s) and slice-by-slice delay correction (right) (Ohta *et al.*, 1996).

but were more accurate at higher flow values for a total study duration of more than 150 s. Both parameters were overestimated at low flow values for all baseline values of V_0, but more so for large V_0, particularly for integration times longer than 240 s, due to inaccuracies in estimates of V_0 and τ.

Maps of K_1 and k_2 from real dynamic [^{15}O]water PET data were also derived from five-parameter fitting of the linear solution of the two-compartment model (Fig. 3, left). These were very similar to images obtained by fitting the same data using the three-weighted integration approach of Ohta and colleagues (Fig. 3, right).

V. DISCUSSION

The two-compartment model of Ohta *et al.* (1996) allows for accurate quantification of K_1, provided that delay and dispersion corrections are appropriate (Toussaint and Meyer, 1996). The technique has been successfully used in vibrotactile and visual stimulation studies (Fujita *et al.*, 1997; Vafaee *et al.*, 1999). The nonlinear solution reported previously (Toussaint and Meyer, 1996, 1998) implicitly performs corrections for Δt and τ, and provides accurate estimates of the main parameters of interest, K_1 and k_2, and of the differences in K_1 and V_0 between a baseline and a stimulation condition.

The main assumption of the two-compartment model is that the intravascular signal constitutes a nonnegligible second compartment, called V_0. This model has been discussed in detail by Ohta and colleagues (1996). In the present study, no substantial differences between the linear and the nonlinear fits were found, indicating that the linear method retains the sensitivity (and specificity) of the nonlinear solution, at least for estimation of K_1 and k_2. In addition, like the nonlinear solution, our linear solution does not suffer from the so-called falling flow phenomenon (FFP) which is characteristic of the one-compartment model (Iida *et al.*, 1989). This was confirmed by the absence of a decrease in estimated flow values as the time of integration was increased above 120 s. Discrepancies between estimates and expectation values of K_1 and k_2 at integration times below 120 s may be linked to the greater correlation between some of the model parameters (V_0, Δt, and τ) in the nonlinear solution at these lengths of study duration (Toussaint and Meyer, 1996), as well as to the larger statistical error due to the reduced number of counts.

An important observation was that the results of the linear solution were entirely independent of the fit starting values, unlike the nonlinear approach. For the linear solution, the data analysis time was decreased by two orders of magnitude as compared to the nonlinear approach. The insensibility to starting values and the greater stability (consistent parameter estimates) of the linear solution allows for fitting tissue data

with relatively higher noise levels, which can be important in those situations where the amount of radioactivity which can be injected during a study is limited. In addition to the faster, linear approach, it is possible to perform voxel-by-voxel fits of high-resolution dynamic PET data, thereby creating functional image volumes of all model parameters (K_1, k_2, V_0, τ, for nonlinear solution and a, b, c, d for linear solution, as well as Δt for both), within a few hours. The feasibility of this approach was suggested by the stability of the linear fits during simulations (for a single pixel) and was verified using $H_2^{15}O$ and dynamic PET-CBF in normal human volunteers.

In conclusion, although more accurate and precise estimates of K_1 and k_2 are obtained with the nonlinear equation, the shorter analysis time required by the linear solution makes it the preferred approach for pixel-by-pixel calculation of parametric images from high-resolution dynamic PET data.

Acknowledgments

This work was supported by MRC (Canada) Grant SP-30, the Isaac Walton Killam Fellowship Fund of the Montreal Neurological Institute and by the McDonnell-Pew Program in Cognitive Neuroscience.

References

Fujita, H., Meyer, E., Reutens, D. C., Kuwabara, H., Evans, A. C., and Gjedde, A. (1997). Cerebral [^{15}O] water clearance in humans determined by positron emission tomography. II. Vascular responses to vibrotactile stimulation. *J. Cereb. Blood Flow Metab.* **17**: 73–79.

Herscovitch, P., Markham, J., and Raichle, M. E. (1983). Brain blood flow measured with intravenous $H_2^{15}O$. I. Theory and error analysis. *J. Nucl. Med.* **24**: 782–789.

Iida, H., Kanno, I., Miura, S., Murakami, M., Takahashi, K., and Uemura, K. (1989). A determination of the regional brain/blood partition coefficient of water using dynamic positron emission tomography. *J. Cereb. Blood Flow Metab.* **9**: 874–885.

Iida, H., Kanno, I., Miura, S., Murakami, M., Takahashi, K., and Uemura, K. (1986). Error analysis of a quantitative cerebral blood flow measurement using O-15H$_2$O autoradiography and positron emission tomography, with respect to the dispersion of the input function. *J. Cereb. Blood Flow Metab.* **6**: 536–545.

Kety, S. S. (1951). The theory and application of the exchange of inert gas at the lungs and tissues. *Pharmacol. Rev.* **3**: 1–41.

Meyer, E. (1989). Simultaneous correction for tracer arrival delay and dispersion in CBF measurements by the $H_2^{15}O$ autoradiographic method and dynamic PET. *J. Nucl. Med.* **30**: 1069–1078.

Ohta, S., Meyer, E., Fujita, H., Reutens, D. C., Evans, A., and Gjedde, A. (1996). Cerebral [^{15}O]water clearance in humans determined by PET. I. Theory and normal values. *J. Cereb. Blood Flow Metab.* **16**: 765–780.

Raichle, M. E., Martin, W. R. W., Herscovitch, P., Mintun, M. A., and Markham, J. (1983). Brain blood flow measured with intravenous $H_2^{15}O$. II. Implementation and validation. *J. Nucl. Med.* **24**: 790–798.

Toussaint, P.-J., and Meyer, E. (1996). A sensitivity analysis of model parameters in dynamic blood flow studies using $H_2^{15}O$ and PET. *In* "*Quantification of Brain Function using PET*" (R. Myers, V. J. Cunningham, D. L. Bailey, and T. Jones, Eds.), pp. 196–200. Academic Press, San Diego.

Toussaint, P.-J., and Meyer, E. (1998). Simultaneous estimation of perfusion (K_1) and vascular (V_0) responses in [^{15}O]water PET activation studies. *In "Quantitative Functional Brain Imaging with Positron Emission Tomography"* (R. E. Carson, M. E. Daube-Witherspoon, and P. Herscovitch, Eds.), pp. 367–370. Academic Press, San Diego, 1998.

Vafaee, M. S., Meyer, E., Marrett, S., Paus, T., Evans, A. C., and Gjedde, A. (1999). Frequency-dependent changes in cerebral metabolic rate of oxygen during activation of human visual cortex. *J. Cereb. Blood Flow Metab.* **19** 272–277.

S E C T I O N

III

SPECIFIC COMPARTMENT KINETICS

SECTION

III

SPECIFIC COMPARTMENT
KINETICS

Do *In Vitro* Enzyme Kinetics Predict *In Vivo* Radiotracer Kinetics?

SCOTT E. SNYDER, THEODORE BRYAN, NEERAJA GUNUPUDI, ELIZABETH R. BUTCH,
and MICHAEL R. KILBOURN

Division of Nuclear Medicine, Department of Internal Medicine, University of Michigan School of Medicine, Ann Arbor, Michigan 48109

Optimizing in vivo radiotracer pharmacokinetics requires either the synthesis and evaluation of multiple radiolabeled candidate compounds or predicting in vivo pharmacokinetic properties from the easier to obtain in vitro values (binding affinities, kinetic rates). We have examined here the ability to predict, using in vitro values for rates of hydrolysis by purified acetylcholinesterase and butyrylcholinesterase, the in vivo rates of substrate hydrolysis by these two cholinesterases in the primate brain. A series of N-methylpiperidinyl and N-methylpyrrolidinyl esters with various length acid groups were synthesized in carbon-12 and carbon-11 forms. In vitro hydrolysis rates were determined using a standard colorimetric assay and in vivo hydrolysis rates estimated using imaging with positron emission tomography (PET). The in vitro hydrolysis rates correctly predict both the relative in vivo rates of hydrolysis of substrates and also the selectivity of different ester chain lengths for the two cholinesterases. It is thus possible to use the in vitro rates of enzyme hydrolysis to "fine-tune" the in vivo radiotracer pharmacokinetics, and thus optimize radiopharmaceuticals for PET or single photon emission computed tomography measures of cholinesterases in the brain.

I. INTRODUCTION

A large number of radiotracers labeled with gamma-emitting or positron-emitting radionuclides have now been prepared for *in vivo* studies of neurochemical processes in the mammalian brain. A significant fraction of these have been used to successfully "image," using autoradiography, single photon emission computed tomography (SPECT), or positron emission tomography (PET), the regional distribution of high affinity binding sites throughout the brain; these sites may be on receptors, transporters, enzymes, or other biological macromolecules. However, many of these radiotracers exhibit poor *in vivo* pharmacokinetic properties which have made it difficult (or at times impossible) to accurately quantify the numbers of binding sites or enzymes present or to reliably measure small to medium changes in the numbers of sites as a result of physiological or pharmacological challenges or disease. Thus, of the many radiotracers which have been reported in the literature, only a handful have proven of widespread utility for *in vivo* quantification of neurochemistry.

Improving the properties of such radiotracers is, of course, of great interest, but the mechanisms to do so are less apparent. Attempts to correlate *in vitro* measures of binding affinities with *in vivo* radiotracer properties have met with limited success: in most cases, the comparison of an equilibrium value (such as an *in vitro* binding affinity) with a dynamic value (regional brain pharmacokinetics) has failed to accurately predict a better radiotracer. Synthesizing a family of derivatives, in the hope of finding one with an improved pharmacokinetic profile, can be a time- and resource-consuming task for the radiopharmaceutical chemist.

The use of radiolabeled substrates to study the *in vivo* rates of enzyme reactions provides a different, and perhaps more promising, area in which radiotracer pharmacokinetics might be better "optimized." In many cases, either relatively homogeneous or purified enzyme preparations are available which can be used to examine the *in vitro* rates of reaction with a series of potential substrates, without the need to synthesize each compound in radiolabeled form. As an example, we have examined here the two related enzymes, acetylcholinesterase (AChE) and butyrylcholinesterase (BuChE), both found in the mammalian brain and both of interest in Alzheimer's disease (Kuhl *et al.*, 1999; Perry *et al.*, 1978). Both of these enzymes are available in highly purified form, and there are well established *in vitro* assays for enzymatic hydrolysis of compounds by these two cholinesterases (Rappaport *et al.*, 1959). This provided an excellent system for testing the ability to predict *in vivo* pharmacokinetic rates for cholinesterase action from the corresponding *in vitro* enzyme rates of hydrolysis.

II. MATERIALS AND METHODS

A. Chemicals and Radiochemicals

The series of *N*-methylpiperidinyl and *N*-methylpyrrolidinyl esters shown in Fig. 1 were synthesized by literature methods (Nguyen *et al.*, 1998; Snyder *et al.*, 1998; Bryan *et al.*, 1999). Radiochemical purities of carbon-11 compounds were routinely >95%, with specific activities >18.5 GBq/µmol (500 Ci/mmol). Acetylcholinesterase (electric eel), butyrylcholinesterase (purified horse serum), acetylcholine chloride, and butyrylcholine chloride were obtained from Sigma Chemical Co.

B. *In Vitro* Enzyme Assays

Rates of hydrolysis of the esters were determined at 37°C using a colorimetric method (Rappaport *et al.*, 1959). Piperidinyl or pyrrolidinyl ester substrates were used in excess, and rates of hydrolysis were expressed relative to the preferred substrates for each enzyme (acetylcholine for AChE, butyrylcholine for BuChE).

C. *In Vivo* Enzyme Hydrolysis Rates

Tissue time–activity curves were determined following bolus intravenous injections of carbon-11-labeled esters into a nemistrina macaque monkey (female, 4.6 kg). The animal was anesthetized throughout each study (ketamine, xylazine). PET imaging was done using a TCC PCT4600A tomograph, which had been modified with additional septa to increase the resolution to 7.5 mm FWHM at the center of the field of view. Scan intervals increased from 1 min immediately after radiotracer injection to 10-min scans from 40 to 80 min. Regions of interest were placed on the left and right basal ganglia and several regions over the cortex, and the entire tissue time–activity curves were determined. These were analyzed using a curve-shape program which calculates the apparent combined forward rate constant $k_3 = k_{\text{hydrolysis}}[\text{ChE}]$. This combined forward rate constant has been demonstrated to be a sensitive method for detecting changes in cholinesterase enzymatic activity in the primate (Callahan *et al.*, 1998) and human brain (Koeppe *et al.*, 1999).

III. RESULTS AND DISCUSSION

The *in vitro* assays of substrate hydrolysis by the two cholinesterases produced results consistent with the expected selectivity of these two enzymes. For acetylcholinesterase, the rates of ester hydrolysis decreased with increasing size of the carboxylic acid; any esters with a chain length of four or more were not appreciably cleaved by AChE (Table 1). In contrast, BuChE was capable of hydrolyzing nearly every single ester (Table 2), consistent with the larger substrate binding pocket proposed for this rather non-specific enzyme (Vellom *et al.*, 1993).

It is not possible to directly compare hydrolysis rates of the same compound with the two enzymes: the rates shown above are expressed per unit of enzyme used in the assay,

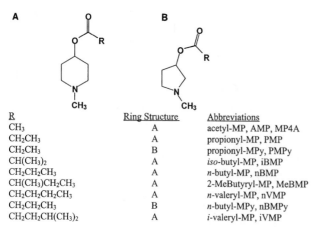

R	Ring Structure	Abbreviations
CH₃	A	acetyl-MP, AMP, MP4A
CH₂CH₃	A	propionyl-MP, PMP
CH₂CH₃	B	propionyl-MPy, PMPy
CH(CH₃)₂	A	*iso*-butyl-MP, iBMP
CH₂CH₂CH₃	A	*n*-butyl-MP, nBMP
CH(CH₃)CH₂CH₃	A	2-MeButyryl-MP, MeBMP
CH₂CH₂CH₂CH₃	A	*n*-valeryl-MP, nVMP
CH₂CH₂CH₃	B	*n*-butyl-MPy, nBMPy
CH₂CH₂CH(CH₃)₂	A	*i*-valeryl-MP, iVMP

FIGURE 1. Structures and abbreviations for *N*-methylpiperidin-4-yl (A) and *N*-methylpyrrolidin-3-yl (B) esters.

TABLE 1. Actual (Change in Absorbance/Unit Enzyme/min/L) and Relative Rates of Ester Hydrolysis by Purified AChE *In Vitro*

Substrate		Rate	Relative rate
Acetylcholine	ACh	0.47	100
Acetyl-MP	AMP (MP4A)	0.22	47
Propionyl-MP	PMP	0.065	13
Propionyl-MPy	PMPy	0.065	13
i-Butyryl-MP	iBMP	0.014	3

Note. Butyrylcholine and the *n*-Butyryl, 2-Methylbutyryl, *n*-Valeryl, and *i*-Valeryl esters of *N*-methylpiperidin-4-ol are not hydrolyzed. Abbreviations as per Fig. 1.

and the values for one unit of AChE and one unit of BuChE hydrolytic activity are defined using different substrates.

These *in vitro* data would support that the substrate with fastest *in vivo* hydrolysis by AChE should be the acetyl ester, and the best BuChE substrate should be the butyryl or valeryl ester. Furthermore, *in vivo* regional hydrolysis rates of substrates for AChE should reflect the known higher concentration of that enzyme in the striatum relative to the cortex, whereas the substrates for BuChE should produce regional hydrolysis rates that reflect the essentially uniform distribution of that enzyme in mammalian brain.

The *in vivo* imaging studies done in the monkey (Table 3) agreed completely with these predictions. The acetyl ester shows high rates of hydrolysis with proportionately more in the striatum; the propionyl ester also shows selectivity for AChE as evidenced by the relative hydrolysis rates in striatum vs cortex. For the two potential BuChE substrates, the *n*-butyl and *n*-valeryl esters, the longer chain es- ter (*n*-valeryl) shows a higher *in vivo* k_3, consistent with the slightly higher *in vitro* hydrolysis rate, but both esters produce a similar rate in striatum and cortex, consistent with BuChE but not AChE. The *iso*-butyl ester, however, is an interesting compound. The rate of hydrolysis *in vivo*, and the excellent ratio between striatum and cortex rates, would suggest that it is predominantly an *in vivo* AChE substrate. The *in vitro* data, however, clearly supports hydrolysis of this *iso*-butyl ester by BuChE. These data could result from a high AChE/BuChE concentration ratio, particularly in the striatum. It is not, at this point, possible to clearly identify which enzyme (or if both) is responsible for *in vivo* hydrolysis of the *iso*-butyl compound, as it is not known what the proportions of each enzyme are in the brain tissues of the nemistrina monkey.

IV. CONCLUSIONS

Using *in vitro* assays of enzyme hydrolysis for a series of esters, it was possible to predict both the relative *in vivo* rates of hydrolysis and the enzyme specificity. Increasing the size of the carboxylic acid group increases the rate of hydrolysis of such esters by BuChE, and complete specificity for AChE was shown for esters longer than propionyl. With AChE, moving from acetyl to propionyl to *iso*-butyl esters progressively reduces the *in vitro* rate of hydrolysis, which is mirrored in the *in vivo* rates of enzyme hydrolysis. For BuChE, the rates of hydrolysis both *in vitro* and *in vivo* increase as the acid chain length grows; we did not determine in this study if rates would continue to increase with even larger acid chains, although the data from the *iso*-valeryl ester suggests some limits on the size and/or shape of longer alkyl chains.

TABLE 2. Actual (Change in Absorbance/Unit Enzyme/min/L) and Relative Rates of Ester Hydrolysis by Purified BuChE *in Vitro*

Substrate		Rate	Relative rate
Acetylcholine	ACh	0.56	57
Acetyl-MP	AMP (MP4A)	0.021	2
Propionyl-MP	PMP	0.048	5
Propionyl-MPy	PMPy	0.039	4
i-Butyryl-MP	iBMP	0.026	3
n-Butyryl-MP	nBMP	0.086	9
2-MeButyryl-MP	MeBMP	0.019	2
n-Valeryl-MP	nVMP	0.10	11
n-Butyryl-MPy	nBMPy	0.077	8
i-Valeryl-MP	iVMP	0.006	<1
Butyrylcholine	BuCh	0.97	100

Note. Abbreviations as per Fig. 1.

TABLE 3. Regional *in Vivo* Rates of Hydrolysis of N-[^{11}C]Methylpiperidinyl Esters in the Monkey Brain

Substrate		k_3 (min^{-1})	
		STR	CTX
Acetyl-MP	AMP (MP4A)	0.117	0.85
Propionyl-MP	PMP	0.075	0.045
i-Butyl-MP	iBMP	0.044	0.017
n-Butyl-MP	nBMP	0.054	0.047
n-Valeryl-MP	nVMP	0.067	0.073

Note. The value k_3 represents the forward rate constant, $k_{hydrolysis}$[enzyme]; STR, striatum; CTX, cortex. Abbreviations as per Fig. 1.

Acknowledgments

This work was supported by grants form the Department of Energy (DE-FG02–87ER60561) and the National Institutes of Health (NS 24896 and CA 09015). The authors thank Phil Sherman and Kyle Kuszpit for technical assistance and Dr. Robert Koeppe for helpful discussions.

References

Bryan, T. A., Snyder, S. E., Sherman, P. S., Kuhl, D. E., and Kilbourn, M. R. (1999). Synthesis of a carbon-11 labeled pyrrolidine ester as a potential *in vivo* substrate for acetylcholinesterase and butyrylcholinesterase. *J. Labeled Compd. Radiopharm.* **42** (Suppl. 1): S207–S209.

Callahan, M. J., Schwarz, R. D., Sherman, P., Koeppe, R. A., Kilbourn, M. R., and Frey, K. A. (1998). Correlation of behavioral effects with PET quantification of AChE inhibition in rhesus monkey brain. *Soc. Neurosci. Abstr.* **24**: 1219.

Frey, K. A., Koeppe, R. A., Kilbourn, M. R., Snyder, S. E., Schwarz, R. D., Callaghan, M. J., and Kuhl, D. E. (1997). PET quantification of acetylcholinesterase activity in monkey brain without blood sampling: Methodology and effect of THA. *J. Cereb. Blood Flow Metab.* **17** (Suppl. 1): S328.

Kilbourn, M. R., Nguyen, T. B., Snyder, S. E., and Koeppe, R. A. (1998). One for all or one for each? Matching radiotracers and brain pharmacokinetics. *In "Quantitative Functional Brain Imaging with Positron Emission Tomography"* (R. E. Carson, M. E. Daube-Witherspoon, and P. Herscovitch, Eds.), pp. 261–265. Academic Press, San Diego.

Kilbourn, M. R., Nguyen, T. B., Snyder, S. E., and Sherman, P. (1998). N-[^{11}C]methylpiperidine esters as acetylcholine substrates: An *in vivo* structure–activity study. *Nucl. Med. Biol.* **25**: 755–760.

Koeppe, R. A., Frey, K. A., Snyder, S. E., Kilbourn, M. R., and Kuhl, D. E. (1997). Evaluation of two distinct kinetic analyses for use with [C-11]PMP: an irreversible tracer for mapping AChE activity. *J. Cereb. Blood Flow Metab.* **17** (Suppl. 1): S329.

Koeppe, R. A., Frey, K. A., Snyder, S. E., Meyer, P., Kilbourn, M. R., and Kuhl, D. E. (1999). Kinetic modeling of N-[^{11}C]methylpiperidin-4-yl propionate: Alternatives for analysis of an irreversible PET tracer for measurement of acetylcholinesterase activity of human brain. *J. Cereb. Blood Flow Metab.* **19**: 1150–1163.

Kuhl, D. E., Koeppe, R. A., Minoshima, S., Snyder, S. E., Ficaro, E. P., Foster, N. L., Frey, K. A., and Kilbourn, M. R. (1999). *In vivo* mapping of cerebral acetylcholinesterase activity in aging and Alzheimer's disease. *Neurology* **52**: 691–699.

Nguyen, T. B., Snyder, S. E., and Kilbourn, M. R. (1998). Synthesis of carbon-11 labeled piperidine esters as potential *in vivo* substrates for acetylcholinesterase. *Nucl. Med. Biol.* **25**: 761–768.

Perry, E. K., Perry, R. H., Blessed, G., and Tomlinson, B. E. (1978). Changes in brain cholinesterases in senile dementia of the Alzheimer type. *Neuropathol. Appl. Neurobiol.* **4**: 273–277.

Rappaport, F., Fischil, J., and Pinto, N. (1959). An improved method for the estimation of cholinesterase activity in serum. *Clin. Chim. Acta* **4**: 227–230.

Snyder, S. E., Tluczek, L., Jewett, D. M., Nguyen, T. B., Kuhl, D. E., and Kilbourn, M. R. (1998). Synthesis of 1-[^{11}C]methylpiperidin-4-yl propionate ([^{11}C]PMP) for *in vivo* measurements of acetylcholinesterase activity. *Nucl. Med. Biol.* **25**: 751–754.

Vellom, D. C., Radic, Z., Li, Y., Pickering, N. A., Camp, S., and Taylor, P. (1993). Amino acid residues controlling acetylcholinesterase and butyrylcholinesterase specificity. *Biochemistry* **32**: 12–17.

17

Accuracy of FDG Images as a Measure of Cerebral Glucose Metabolic Rate: Influence of Rate Constant Heterogeneity

MICHIO SENDA,* HINAKO TOYAMA,*,† YOKO IKOMA,*,‡ YUICHI KIMURA,* KOJI UEMURA,*,‡ and KENJI ISHII*

*Tokyo Metropolitan Institute of Gerontology, Tokyo, Japan
†National Institute of Radiological Sciences, Chiba, Japan
‡Waseda University, Tokyo, Japan

To see how the fluorodeoxyglucose (FDG) images accurately reflect glucose metabolic rate, the relationship between influx rate constant $K (= K_1 k_3/(k_2 + k_3))$ and fractional uptake $U(=$ FDG images divided by the integrated input) was examined. Allowing for the noise, K and U showed a strong linear relationship both within and between subjects, mainly due to a positive correlation among K_1, k_2, and k_3. The K–U line had a slope of nearly unity and a positive U intercept, indicating that the use of proportionally scaled FDG image as a relative measure of cerebral metabolic rate of glucose (CMRGlc) is subject to overestimation in hypometabolic subjects and regions. The results also suggest that CMRGlc may be readily estimable from FDG images if the integrated input is estimated less invasively.

I. INTRODUCTION

An orthodox way of measuring the regional cerebral metabolic rate of glucose (CMRGlc) with [18F]fluorodeoxyglucose ([18F]FDG) is to estimate the regional rate constants (K_1, k_2, k_3) from the tissue–activity curve $C_t(t)$ and the plasma curve $C_p(t)$ based on the tracer kinetic model and to compute the influx rate constant K, the product of K and plasma glucose divided by the lumped constant yielding CMRGlc (Huang *et al.*, 1980). Recently, a simplified method has been frequently used, in which the radioactivity images acquired around 45–60 min postinjection $(= T)$ and normalized by the global value or by the value at a ref-

erence region (e.g., pons) are used as a measure of relative CMRGlc. This approximation is accurate if $C_t(T)$ is proportional to K across the brain, which is assumed by many investigators and appears acceptable as long as clinical diagnosis is concerned. Meanwhile, the so-called *in vivo* autoradiographic method with standard rate constant values (Phelps *et al.*, 1979) has also been used, in which the operational equation assumes $C_t(T)$ to be linearly related, though not proportional, to K. It has been pointed out that the autoradiographic method provides a good approximation of K in normal subjects (Heiss *et al.*, 1984). Theoretically, however, the proportionality between $C_t(T)$ and K may not hold depending on the regional k_2 and k_3. The validity of uniform $K/C_t(T)$ assumption has not been investigated in detail, especially in the patients where the rate constants may take values far from the standard. The present study deals with the heterogeneity of rate constant values in the brain of patients with neurodegenerative disorders and how it affects the $K/C_t(T)$ relationship.

II. METHODS

The subjects were 3 normal volunteers and 13 patients with neurodegenerative disorders (ages 36–70) including 8 patients with Alzheimer-type dementia (AD) and 3 with corticobasal degeneration (CBD). The dedicated PET camera was a 2D scanner Headtome-IV (Shimadzu, Japan). Under fasting conditions, the subjects underwent an 18-frame time-sequential scan for 45 min, starting at the injection of 120–220 MBq of FDG followed by a single-frame scan for 12 min centering at 56–58 min ($= T$), together with sequential arterial blood sampling. Physical decay was corrected to obtain the tissue–activity curve $C_t(t)$ and the plasma–activity curve $C_p(t)$.

Parametric images of K_1, k_2, k_3, and K were generated from $C_t(t)$ and $C_p(t)$ for the initial 45 min with Marquardt algorithm based on the kinetic model containing the three rate constants plus blood volume, while the fractional uptake U image was computed as $C_t(T)$ divided by the integral of $C_p(t)$ up to T. The within-subject relationship between K and U and its intersubject variation was examined on the K/U ratio image as well as with a principal component analysis on (U, K) scatter plots. The within-subject correlation between the rate constants (K_1, k_2, k_3) was also examined.

Apart from it, simulation was performed to estimate the effect of the noise on the K–U relationship for selected subjects. The root mean square error (RMSE) for the rate constants estimation was computed and averaged across the brain. The noise level of the human data was estimated using the relationship between noise and RMSE obtained from another simulation. Then the effect of the noise on K and U was simulated using the mean rate constant values of the subject.

III. RESULTS AND DISCUSSION

The K images were less noisy than the images of K_1, k_2, or k_3. The K/U ratio appeared homogeneous across the brain including the white matter and degenerate areas in every subject. The mean K/U ranged from 0.81 to 0.93 and

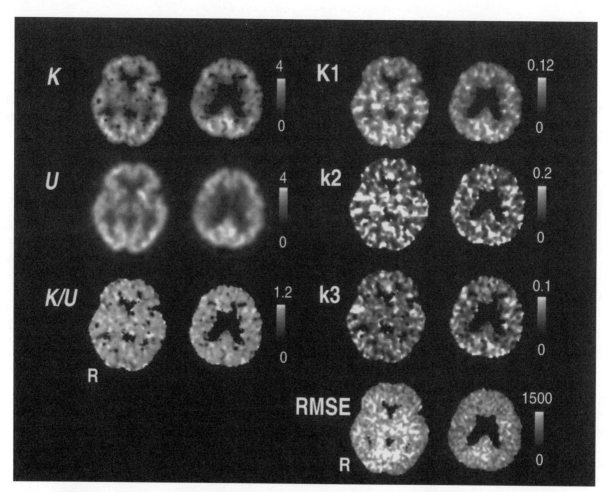

FIGURE 1. FDG parametric images of a 64-year-old patient with corticobasal degeneration. Influx rate constant ($K = K_1k_3/(k_2 + k_3)$) and fractional uptake (U = radioactivity at 56 min divided by integrated input) are displayed in mL/min/100 mL, their ratio (K/U) is an absolute number, three rate constants (K_1, k_2, and k_3) are in 1/min, and root mean square error for the rate constants estimation (RMSE) is in Bq/mL. Note apparently homogeneous K/U ratio across the brain including hypometabolic regions in the right striatum and in the right Rolandic area.

its within-subject variation was 10.3–17.9% (SD/mean) depending on the subject. The (U, K) plots were scattered along and just below the line of identity with a correlation coefficient of 0.83–0.95 in each subject. The first principal component explained 91.8–97.2% of the total variation, and the slope (K/U) of the eigen vector was 0.94–1.13, which was slightly larger than the K/U ratio. The three rate constants (K_1, k_2, k_3) were positively correlated with each other in each subject, the correlation coefficient ranging 0.63–0.77, 0.28–0.51, and 0.46–0.72 for K_1 vs k_2, K_1 vs k_3, and k_2 vs k_3, respectively. The RMSE image was homogeneous except for the image center being higher than the periphery.

Figure 1 illustrates the parametric images of a patient with CBD. The K image appeared similar to the U image, resulting in the K/U image being apparently homogeneous and visualizing no contrast between gray matter and white matter. The homogeneity covered pathologically hypometabolic areas, i.e., the right Rolandic area and the right striatum. In another patient with AD (not shown here), severely hypometabolic temporoparietal cortex showed a slightly lower K/U than the preserved primary sensorimotor cortex, and the scalp and other extracerebral regions had very low K/U values.

Figure 2A shows the (U, K) plots of the patient presented in Fig. 1. The plots were distributed along and just below the line of identity. The K/U ratio for the subject was 0.91 ± 0.12 (mean $\pm$ SD, SD/mean = 14%), and the correlation coefficient was 0.93. The first principal component explained 97% of the total variation, the slope of the eigen vector was 1.04, and the U intercept was 0.24 when the eigen vector axis was extrapolated. Because the scatter

plots include the variation caused by the noise, its contribution was estimated by simulation for the data of the patient. The mean RMSE was 1100 Bq/ml, corresponding to a noise level of 7.0% (percentage of the activity in the 18th frame). Using the mean rate constant values ($K_1 = 0.07$, $k_2 = 0.11$, $k_3 = 0.05$) for the patient and the estimated noise, the contribution of the noise to the (U, K) plots was computed and presented in Fig. 2B. The variation of K, U, and K/U was 8, 5, and 11% (SD/mean) in Fig. 2B, while they were 29, 30, and 14% in Fig. 2A, respectively. It is notable that K and U are not correlated in Fig. 2B because they are derived from different data. Although the noise and its effect depend on the rate constants and other factors, simply subtracting the variation due to noise from the observed variation resulted in the "true" K/U variation of 8.7% and correlation coefficient of nearly unity, as a rough estimation. This suggests that the true U and K have an extremely strong within-subject correlation and are almost linearly related.

Figure 3A presents the correlation between K_1, k_2, and k_3 within the patient shown in Fig. 1. Positive correlation was observed between any two of the three rate constants. The correlation coefficient was 0.68, 0.51, and 0.69, for K_1 vs k_2, K_1 vs k_3, and k_2 vs k_3, respectively. A simulation similar to Fig. 2B indicated a positive correlation between the estimation error of the three rate constants due to noise as shown in Fig. 3B, in which the correlation coefficient was 0.91, 0.66, and 0.87, for K_1 vs k_2, K_1 vs k_3, and k_2 vs k_3, respectively. It is difficult to estimate the true variation of the rate constants and their true correlation from this simulation, because the effect of noise depends on the rate constant values and that the observed distribution is deviated from Gaussian. How-

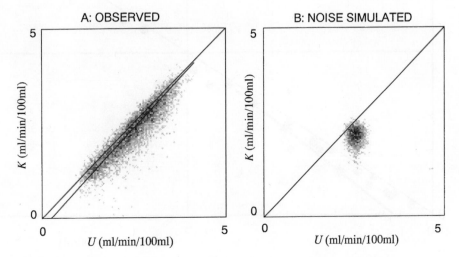

FIGURE 2. Within-subject correlation between U and K. (A) Scatter plots of pixel values for the patient presented in Fig. 1. The eigenvector derived from a principal component analysis is drawn together with the line of identity. (B) Variation due to noise in the (U, K) plots that has been simulated by using the estimated noise level and assuming that K_1, k_2, and k_3 are uniform and take the observed mean values in A.

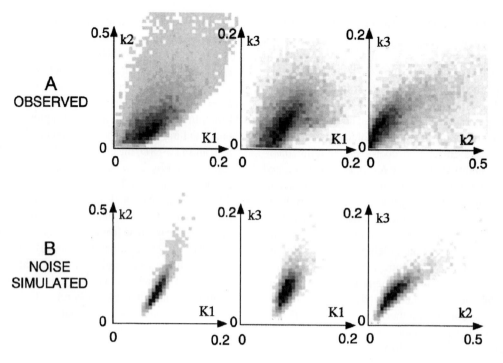

FIGURE 3. Within-subject correlation between K_1, k_2, and k_3. (A) Scatter plots of pixel values for the patient presented in Fig. 1. (B) Correlation due to noise simulated in the same way as in Fig. 2B.

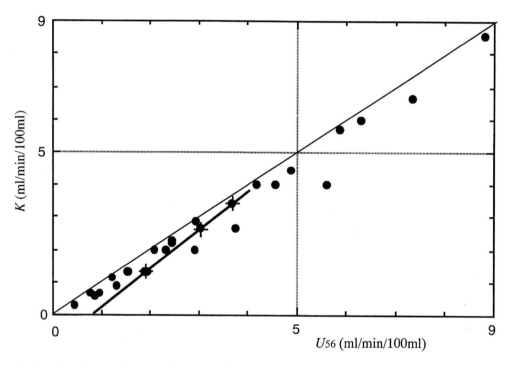

FIGURE 4. Noise-free simulation for the relationship between influx rate constant (K) and fractional uptake at 56 min (U_{56}), using 27 sets of three rate constants: $K_1 = 0.04$, 0.08, or 0.12; $k_2 = 0.04$, 0.10, or 0.25; $k_3 = 0.02$, 0.05, or 0.10. Cross marks represent results when K_1, k_2, and k_3 each assumes the lowest, middle, or highest of the designated three values.

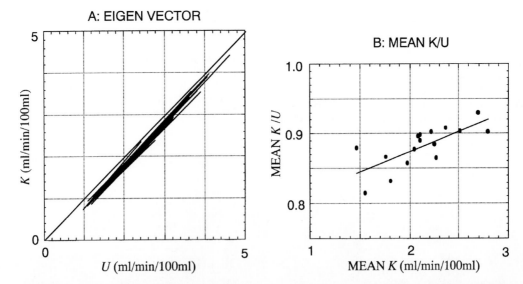

FIGURE 5. Between-subjects variation in K–U relationship. (A) Long axis of the 95% distribution ellipse for (U, K) estimated from the principal component analysis for each of the 16 subjects drawn on the K–U chart. (B) Individual variation in the mean K/U ratio and its dependency on the mean K.

ever, a positive correlation among K_1, k_2, and k_3 remained after simply subtracting the noise from the observed variation. According to Michaelis–Menten kinetics, regional K_1, k_2, and k_3 cannot assume independent values, but a positive correlation arises between K_1 and k_2 as well as between k_2 and k_3 (Kuwabara *et al.*, 1990). Where the concentration of glucose transporter is reduced, both K_1 and k_2 are decreased. Where hexokinase activity (k_3) is reduced, intracerebral glucose is increased, resulting in decreased intracerebral vacant transporter and decreased k_2. In degenerate areas, K_1 and k_3 are also more likely to decrease together than to change toward opposite directions. Therefore, it is reasonable to speculate that a positive correlation exists among K_1, k_2, and k_3 in most subjects.

To assess the effect of rate constant heterogeneity on the K–U relationship, noise-free simulation was performed using designated values: $K_1 = 0.04$, 0.08, or 0.12; $k_2 = 0.04$, 0.10, or 0.25; $k_3 = 0.02$, 0.05, or 0.10. They nominally represent the lowermost, average, and uppermost of the frequently observed values obtained in the present study. The U for $T = 56$ min and K computed from the 27 sets of (K_1, k_2, k_3) and a typical input function showed a positive correlation as shown in Fig. 4. This suggests that the positive correlation in Fig. 2A is explained even if K_1, k_2, and k_3 are randomly distributed within the observed range. Furthermore, when K_1, k_2, and k_3 each assumed the lowest, middle or the highest of the designated three values, the resulting three plots were located along a straight line. This suggests that the possible strong linear relationship between the true U and K deduced from Fig. 2 is explained by the presence of a positive correlation among K_1, k_2, and k_3 as speculated above.

Figure 5 indicates the between-subjects variation in the K–U relationship. When the long axis of the 95% distribution ellipse derived from the principal component analysis for each of the 16 subjects was drawn on the K–U chart, minimal individual variation was observed. Because the line had a slope of nearly unity and a positive U intercept, the K/U ratio depended on the mean K and had smaller values in hypometabolic subjects and regions.

IV. CONCLUSION

(i) $C_t(T)$ is almost linearly related, though not proportional, to $K = K_1 k_3/(k_2 + k_3)$ within subjects, mainly due to positive correlation between K_1, k_2, and k_3. Because the noise-induced error for U is smaller than that for K, a CMRGlc image may be generated more precisely from the $C_t(T)$ image and the (U, K) regression than from pixelwise estimation of the rate constants alone. (ii) The relationship between K and U ($= C_t(T)$ divided by integrated input) is fairly consistent between subjects under constant T, and the K–U line may be determined empirically from a large number of subjects. Therefore, if the integrated input is estimated less invasively, e.g., from injected dose and body surface area or by one-point blood sampling, then the absolute value of CMRGlc may be estimated from $C_t(T)$. (iii) K/U is nearly, but not exactly, uniform within or across subjects. Use of proportionally normalized $C_t(T)$ as relative CMRGlc overestimates K in hypometabolic subjects and regions due to free tissue [^{18}F]FDG.

References

Heiss, W.-D., Pawlik, G., Herholz, K., Wagner, R., Göldner, H., and Wienhard, K. (1984). Regional kinetic constants and cerebral metabolic rate for glucose in normal human volunteers determined by dynamic positron emission tomography of $[^{18}F]$-2-fluoro-2-deoxy-D-glucose. *J. Cereb. Blood Flow Metab.* **4**: 212–223.

Huang, S. C., Phelps, M. E., Hoffman, E. J., Sideris K., Selin, C. J., and Kuhl, D. E. (1980). Noninvasive determination of local cerebral metabolic rate of glucose in man. *Am. J. Physiol.* **238**: E69–E82.

Kuwabara, H., Evans, A. C., and Gjedde, A. (1990). Michaelis–Menten constraints improved cerebral glucose metabolism and regional lumped constant measurements with $[^{18}F]$fluorodeoxyglucose. *J. Cereb. Blood Flow Metab.* **10**: 180–189.

Phelps, M. E., Huang, S. C., Hoffman, E. J., Selin, C., Sokoloff, L., and Kuhl, D. E. (1979). Tomographic measurement of local cerebral glucose metabolic rate in humans with (F-18)2-fluoro-2-deoxy-D-glucose: Validation of method. *Ann. Neurol.* **6**: 371–388.

18

Regional FDG Lumped Constant in the Normal Human Brain

MICHAEL M. GRAHAM, MARK MUZI, ALEXANDER M. SPENCE, FINBARR O'SULLIVAN, THOMAS K. LEWELLEN, JEANNE M. LINK, and KENNETH A. KROHN

Division of Nuclear Medicine, Department of Radiology, and Department of Neurology, University of Washington, Seattle, Washington

The lumped constant (LC) is a correction factor used to infer glucose metabolic rate (MRGlc) from fluorodeoxyglucose (FDG) metabolic rate (MRFDG). LC was determined in normal brain by measuring regional MRGlc and MRFDG independently using 1-[^{11}C]glucose and FDG with dynamic positron tomographic imaging, arterial blood sampling, and region-of-interest time–activity curve analysis with appropriate compartmental models. Ten subjects (4 male, 6 female) were studied. The mean LC ($\pm$ SD) for normal brain was 0.89 ± 0.08. The value for cerebellum was slightly lower, 0.78 ± 0.11 ($P = 0.006$, two tailed paired t test). LC values for individual subjects tended to be similar across the brain but differed significantly between subjects, suggesting that the LC may not be truly constant. The LC values determined in this study are considerably higher than older values in the literature, probably because of methodological differences, but agree with a recent study by Hasselbalch.

I. INTRODUCTION

The lumped constant (LC) is a term proposed by Sokoloff, *et al.* (1977) to describe the ratio of deoxyglucose metabolic rate (MR) to glucose MR in the rat brain. The concept has been extended to fluorodeoxyglucose (FDG) in human studies. Its value has been estimated by measuring whole brain extraction fractions, by comparison to whole brain glucose MR, either literature values or in the same subjects, and by comparison with oxygen MR. In this study we used a more direct approach, i.e., we independently determined FDG and glucose MR regionally using positron emis-

sion tomography (PET) imaging and 1-[^{11}C]D-glucose injected followed by FDG.

II. MATERIALS AND METHODS

A. Human studies

Normal volunteers (4 males and 6 females) were recruited with ages ranging from 28 to 47. Potential subjects were rejected if there was any history of head injury with loss of consciousness, migraine headaches, seizures, or other neurologic problems. No subjects were taking any medications at the time of the study and none was diabetic. The subjects fasted overnight prior to the PET study. The study was approved by the University of Washington Human Subjects Committee and all subjects documented their informed consent.

A foam headrest was molded to the back of the head and a thermoplastic face mask was fitted. An arterial catheter was inserted into the radial artery and was connected to an automated blood sampler (Graham and Lewellen, 1993). A venous catheter was placed in the contralateral arm for injection. Subjects were awake, but with minimal visual, auditory, or tactile stimulation.

The FDG and 1-[^{11}C]glucose syntheses are described in Spence *et al.* (1998). The fraction of fluorodeoxymannose in the FDG was less than 5%. The range of FDG activity injected was 8.4 to 9.8 mCi. One patient received only 3.5 mCi 1-[^{11}C]glucose, but the others received from 14.4 to 18.9 mCi. The data set associated with the lower dose showed more noise than the other studies, but yielded similar results for parameter estimates and therefore was included in all analyses.

B. Blood Sampling, Counting, and Calibration

During the PET studies arterial blood was sampled at 20-s intervals initially, followed by progressively longer intervals. The samples were centrifuged and 0.5 mL of plasma was pipeted into plastic tubes for counting. Activity was decay corrected to the time the blood sample was obtained. Both the plasma and the tissue data were in units of μCi/mL for subsequent data analysis. Blood glucose was determined using a calibrated Beckman Glucose Analyzer II at several times during the study.

C. PET Imaging

The tomograph was a General Electric Advance whole body tomograph providing 35 image planes of data over a 15-cm axial field of view. After positioning the subject in the tomograph, an attenuation image was obtained with rotating Ge-68 rod sources. Both attenuation imaging and emission imaging were done in 2D mode. At $t = 0$ the isotope infusion began and was continued for 1 min using a syringe pump (Harvard). The image acquisition sequence, starting at $t = 0$ was four 15-s, four 30-s, four 1-min, four 3-min, and eight 5-min images. The same sequence was used for both 1-[^{11}C]glucose, and then for FDG, which began at $t = 90$ to 100 min. Following acquisition, PET data were reconstructed with attenuation correction and a Hanning filter yielding images with a resolution of approximately 6 mm. T1-weighted magnetic resonance (MR) images were acquired on the same day with a GE Signa 1.5 T system. T1-weighted images were acquired at an interval of 6.5 mm and were reoriented to align with the PET images.

D. Data Analysis

Image manipulation, including summation and region-of-interest (ROI) placement, was done on a Macintosh computer with the program Alice (Parexel Software). FDG images, summed from 45 to 75 min, were used for placement of ROIs. Using the MRI images for reference, ROIs were placed over the following areas: frontal cortex, temporal cortex, parietal cortex, occipital cortex, caudate, putamen, thalamus, cerebellum, and white matter. All structures were well resolved on the FDG PET studies. Partial volume effects were minimized by placing ROIs at least 5 mm from the edge of each structure. Following placement of the ROIs, time–activity curves (TACs) were generated for all ROIs. Subsequent manipulation of the TAC data was performed

with EXCEL (Microsoft). The ROI TACs were not decay corrected. The plasma TACs were smoothed and interpolated with a program designed for the task (Graham, 1997).

The parameter optimization program included models for both FDG and 1-[^{11}C]glucose. The models were set up as differential equations and were solved numerically. Each model included rate constants K_1, k_2, k_3, and k_4 as well as delay (time shift of the tissue data relative to the plasma data). A common blood volume term was applied to both tracers. Data points were weighted proportional to the square root of the product of activity times image duration.

The FDG model was identical to that described by Phelps et al. (1979) as an extension of Sokoloff et al. (1977) original model (Fig. 2A). The differential equations were

$$\frac{dC_e}{dt} = K_1^* C_p - (k_2^* + k_3^*)C_e + k_4^* C_m, \tag{1}$$

$$\frac{dC_m}{dt} = k_3^* C_e - k_4^* C_m, \tag{2}$$

where C_p is the plasma activity, C_e is the extravascular FDG, and C_m is the tissue FDG-6-PO4.

The 1-[^{11}C]glucose model (Fig. 1B) was similar to the model proposed by Blomqvist et al. (1990). It is like the FDG model, except in the way k_4 is handled. Once glucose is phosphorylated to glucose-6-PO$_4$ it continues to be metabolized and is not dephosphorylated. The k_4 term represents all the possibilities for loss of label from tissue associated with the metabolism of glucose. This includes loss primarily as lactate and CO_2, but also includes loss of other labeled metabolic products. (Blomqvist et al., 1990) discussed the potential limitations of this approach.

The differential equations for the 1-[^{11}C]glucose model were

$$\frac{dC_e}{dt} = K_1 C_p - (k_2 + k_3)C_e, \tag{3}$$

$$\frac{dC_m}{dt} = k_3 C_e - k_4 C_m. \tag{4}$$

In both models radioactive decay was included. Thus tissue data were not decay corrected prior to analysis. The plasma data were decay corrected to the time each blood sample was obtained. To account for 1-[^{11}C]glucose activity that was still present at the time the FDG was injected, the calculated glucose activity was extrapolated and added to the calculated FDG activity.

Since glucose is metabolized in the rest of the body, as well as in the brain, there is a steady accumulation of metabolites in the plasma. We showed previously that the appearance rate of ionic metabolites was linear (in terms of fraction of total activity in plasma) and reached 18% at 1 h (Spence et al., 1998). In that study we found no detectable [^{11}C]CO$_2$ in the arterial plasma. The concept of linear appearance of metabolites, passing through 18% at 1 h, was incorporated into the analysis model. It was assumed that these metabolites remained in the blood, thus contributing to total tissue activity, but there was no uptake of metabolites into brain.

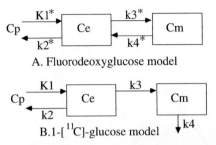

FIGURE 1. Models for describing the kinetic behavior of FDG (A) and 1-[^{11}C]glucose (B). The kinetic models also include a blood volume term which was constrained to be identical for both tracers and delay terms to allow temporal shifting of the tissue activity relative to blood activity.

III. RESULTS

Ten subjects were studied, 4 male, 6 female. Average plasma glucose concentration during the PET scans was 4.98 mM (range from 4.46 to 5.55) and did not change significantly over the course of the combined imaging study. Typical tissue time–activity curves for several tissues, along with model output, are shown in Fig. 3. Note that the quality of the data was excellent, i.e., low noise, and that the model output accurately fit the observed data.

The LC estimates are shown in Table 1. The mean LC calculated for normal brain using PET imaging of 1-[^{11}C]glucose and FDG is 0.89 ± 0.08 SD and agrees with our recent studies on contralateral brain in glioma patients (Spence *et al.*, 1998). The value for cerebellum (0.78±0.11) is slightly lower than that for the rest of the brain and the value for white matter (0.92 ± 0.12) is slightly higher. The

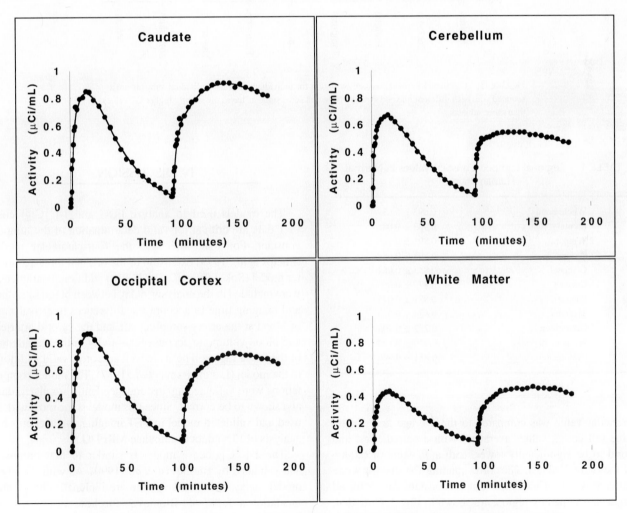

FIGURE 2. Typical data sets from both radiopharmaceutical studies in four different regions in four different subjects. In each case 1-[^{11}C]glucose was injected at $t = 0$ and FDG was injected at $t \approx 90$ min. The filled circles are tissue activity and the line is model output after the parameters of the model have been optimized to achieve a best fit.

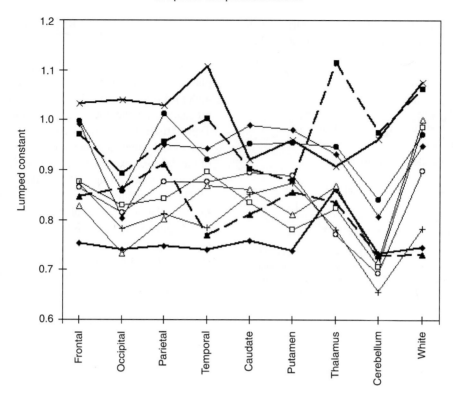

FIGURE 3. Calculated lumped constant by region for individual subjects. Individual subjects are associated with each different line and symbol. Note that some subjects have consistently higher LC values than other subjects.

TABLE 1. Regional Lumped Constant Values in Normal Human Brain

Whole brain	0.893 ± 0.084
Frontal	0.903 ± 0.085
Temporal	0.891 ± 0.106
Parietal	0.894 ± 0.089
Occipital	0.835 ± 0.084
Caudate	0.877 ± 0.065
Putamen	0.871 ± 0.075
Thalmus	0.884 ± 0.095
Cerebellum	0.782 ± 0.106
White matter	0.920 ± 0.120
Gray matter	0.881 ± 0.096

cerebellar value was compared to the average gray matter value (all cortex values averaged) with a paired *t* test and found to be significantly lower, with a *P* value of 0.0006 for the two-tailed test. The corresponding *P* value for white matter was 0.15. Calculated lumped constant values for all subjects, grouped by region, are shown in Fig. 3. Note that, although there is some variation, the LC values for individual subjects tended to group together. Thus the average LC seems to be somewhat different between subjects.

IV. DISCUSSION

The models used to analyze FDG and 1-[^{11}C]glucose PET data are critical for valid determination of the lumped constant. For FDG this was the four-parameter model (Phelps *et al.*, 1979) extension of the original three-parameter model (Sokoloff *et al.*, 1977). Two additional parameters were included in the analysis: delay between blood sampling and imaging time to account for differences in arrival time of blood at the artery sampling site and the carotid arteries; and blood volume to account for activity in the vasculature of the tissue region. These are well accepted, valid additions to the model (Lammertsma *et al.*, 1987). The model computations were validated against independently simulated data and shown to be correct. Since this model is the most widely used and validated model in PET imaging, we accept it for analysis of FDG data to calculate MRFDG.

The 1-[^{11}C]glucose model is similar to that proposed and validated by Blomqvist *et al.* (1990). As with the FDG model, delay and blood volume are included. Our implementation differs from Blomqvist's in how we approached loss of metabolites from tissue. We propose that the rate of loss is proportional to the amount of metabolized tracer in the tissue. This is a broad assumption, in that glucose

TABLE 2.   Literature Values for the Lumped Constant

First author	Year	Method	N	LC
Phelps	1979	MRFDG vs literature MRGlc	12	0.43
Gjedde	1985	Distribution vol + MM constants	4	0.45
Reivich	1985	AV difference	9	0.52
Brooks	1987	MRFDG vs literature MRGlc	27	0.50
Lammertsma	1987	MRFDG vs MRO_2	3	0.75
Hasselbalch	1998	MRFDG vs Fick MRGlc	15	0.80
Spence	1998	MRFDG vs [^{11}C]glucose MRGlc	40	0.80
Present study	1999	MRFDG vs [^{11}C]glucose MRGlc	10	0.89

metabolism and label loss occurs via several metabolic pathways. Blomqvist measured this loss term explicitly by determining cerebral blood flow and the arteriovenous (A-V) difference for metabolites. This explicit approach is more invasive and is not feasible for regional analysis. Thus we use a linear loss term, directly proportional to the amount of phosphorylated tracer in the tissue. The resultant cerebral MRGlc estimates are essentially identical to Blomqvist's, suggesting that our simpler approach to dealing with tracer loss is reasonable.

Previous studies estimating values for the LC are summarized in Table 2. There is a general trend toward higher gray matter FDG metabolic rates in more recent studies compared to that seen in older studies. This is likely due to improved resolution in modern tomographs and therefore less partial volume effect. Another reason for the different values in this study compared to older values is that most early studies did not include k_4^*. Finally many of the early studies used FDG with significant fluorodeoxymannose (FDM) contamination that would result in lower estimates for K_1^* and k_3^* (Wienhard et al., 1991).

The only previous work in which both 1-[^{11}C]glucose imaging and FDG imaging were done in the same normal subjects was by Blomqvist et al. (1994). The underlying hypothesis was that the difference in behavior between FDG and 1-[^{11}C]glucose (corrected only for $^{11}CO_2$ loss) resulted from nonoxidative glucose consumption. A critical underlying assumption was that the true lumped constant for FDG in normal brain was 0.52, taken from Reivich et al. (1985). Given this assumption, Blomqvist et al. found, in the resting brain, that MRGlc was linearly related to MRFDG with a slope of 0.65. This could be interpreted either as evidence for substantial nonoxidative glycolysis in the normal resting brain or that the real lumped constant is higher; 0.52/0.65 = 0.80. Lactate production in the normal resting human brain averages 0.112 mol per mole of glucose consumed (Lying-Tunell et al., 1980). This means that approximately 5.6% of the glucose is metabolized via glycolysis and the balance is via oxidative metabolism. This low fractional lactate production by the normal resting brain suggests that the expla-

nation of Blomqvist's results is more likely a higher lumped constant.

Previous approaches used to estimate the FDG LC can be categorized as follows:

1. PET measure of FDG uptake compared with literature values for glucose metabolism (Brooks et al., 1987; Phelps et al., 1979);
2. determination of arteriovenous extraction fractions for both FDG and glucose (Reivich et al., 1985);
3. derivation of the LC from kinetic parameters of FDG, given certain assumptions regarding FDG and glucose kinetics (Gjedde et al., 1985);
4. PET measure of FDG uptake compared with glucose metabolism inferred from oxygen metabolism determined using [^{15}O]oxygen in the same subjects (Lammertsma et al., 1987);
5. PET measure of FDG uptake compared with Fick measures of glucose uptake in the same subjects (Hasselbalch et al., 1998).

The attempts to estimate the FDG LC by comparison of FDG uptake with literature values of cerebral glucose metabolic rate are probably in error since the older literature estimates of brain GMR are on the high side as shown recently by (Hasselbalch et al., 1998).

The A-V differences approach (Reivich et al., 1985) has a significant problem in that FDG extraction fraction steadily decreases due to dephosphorylation, which causes the extraction fraction of FDG to decrease with time. In addition, their FDG was contaminated with up to 20% FDM for which the extraction fraction is lower than for FDG (Wienhard et al., 1991). This would lower the measured extraction fraction and underestimate LC.

Gjedde et al. addressed the LC question using a number of assumptions regarding values for the brain water volume, the phosphorylation ratio (k_3^*/k_3), and the Michaelis constant for glucose transport (Gjedde et al., 1985). The values chosen were selected with the assumption that the whole brain FDG LC was 0.42 (Phelps et al., 1979). Because of this assumption, the value of the LC is really not addressed, although they showed that the LC became relatively elevated in areas of cerebral strokes in some patients, presumably due to increased nonoxidative glycolysis.

Lammertsma et al. (1987) studied 3 subjects with FDG and with [^{15}O]oxygen. The ^{15}O study was analyzed in a well validated manner to estimate regional cerebral oxygen metabolism. Glucose metabolism was estimated from the MRO_2 by assuming a stoichiometric uptake ratio for MRO_2/MRGlc of 5.6. The FDG study was analyzed in a number of ways. The most sophisticated methods estimated the four parameters of the FDG model ($K_1^*, k_2^*, k_3^*, k_4^*$) to calculate metabolic rate. They estimated an FDG LC of 0.75 ± 0.08 (SD). Overall, the results of Lammertsma's study support the concept that the FDG LC is higher than the early values of 0.43 to 0.52.

Hasselbalch *et al.* recently estimated the FDG LC by measuring FDG uptake with PET imaging, compared against Fick determination of whole brain glucose uptake (Hasselbalch *et al.*, 1998). Their study emphasized careful determination of cerebral blood flow using ^{133}Xe and found that global cerebral blood flow was somewhat lower than previously reported. This had the effect of yielding lower glucose MR so the resulting LC was 0.81 ± 0.15. The only significant limitation with their approach was that they used the three-parameter model for FDG, ignoring the impact of dephosphorylation. Although there is controversy regarding k_4^*, we have found consistently better fits of the model output to the FDG data when k_4^* was included. We also found that the estimated FDG MR was consistently lower when k_4^* was zero. If k_4^* had not been included in our analysis, then the average LC would have been 0.75 ± 0.11, very close to that reported by Hasselbalch *et al.*

In addition to finding that the value for the LC is higher than previous estimates, our results show that the LC may vary between subjects. Figure 3 shows that values for the LC are tighter for individual subjects than for the group as a whole. This concept of individual variation in the LC was supported by Blomqvist *et al.* in comparing MRGlc determined with 1-[^{11}C]glucose to MRFDG (Blomqvist *et al.*, 1994). They state, "even if MRglc and MRFDG are on the average proportional within a subject, the factor of proportionality varies considerably between the subjects." Also the LC values for cerebellum are consistently lower than for the remainder of the brain. All but one of our LC values for cerebellum were lower than the values for average gray matter in each subject, $P = 0.0006$.

The value of 0.89 for the LC in normal human brain is consistent with the work from several investigators and with our LC of 0.86 in contralateral brain in 40 glioma-bearing subjects (Spence *et al.*, 1998). The only lingering doubt relates to the performance of the glucose model. As Blomqvist *et al.* (1994) commented, the model probably does not adequately account for nonoxidative glycolysis, with rapid loss of label as lactate.

In the meantime, we recommend that when MRGlc is calculated from MRFDG, a value of 0.89 be used for the LC for normal brain. As we have previously reported (Spence *et al.*, 1998), the LC for brain tumor varies so much that no LC can be assigned, and so studies that use only FDG tracer should refer to the result only as MRFDG.

Acknowledgments

The authors thank Barbara Lewellen, Scott Freeman, and Aaron Charlop for technical assistance with PET imaging and Kenneth Maravilla and Vernon Terry for assistance with MR imaging. Supported by NIH Grant CA42045.

References

Blomqvist, G., Seitz, R. J., Sjogren, I., Halldin, C., Stone-Elander, S., Wid'en, L., Solin, O., and Haaparanta, M. (1994). Regional cerebral oxidative and total glucose consumption during rest and activation studied with positron emission tomography. *Acta Physiol. Scand.* **151**: 29–43.

Blomqvist, G., Stone-Elander, S., Halldin, C., Roland, P. E., Wid'en, L., Lindqvist, M., Swahn, C. G., Langstrom, B., and Wiesel, F. A. (1990). Positron emission tomographic measurements of cerebral glucose utilization using [1–11C]D-glucose. *J. Cereb. Blood Flow Metab.* **10**: 467–483.

Brooks, R. A., Hatazawa, J., Di-Chiro, G., Larson, S. M., and Fishbein, D. S. (1987). Human cerebral glucose metabolism determined by positron emission tomography: A revisit. *J. Cereb. Blood Flow Metab.* **7**: 427–432.

Gjedde, A., Wienhard, K., Heiss, W. D., Kloster, G., Diemer, N. H., Herholz, K., and Pawlik, G. (1985). Comparative regional analysis of 2-fluorodeoxyglucose and methylglucose uptake in brain of four stroke patients. With special reference to the regional estimation of the lumped constant. *J. Cereb. Blood Flow Metab.* **5**: 163–178.

Graham, M. M. (1997). Physiologic smoothing of blood time–activity curves for PET data analysis. *J. Nucl. Med.* **38**: 1161–1168.

Graham, M. M., and Lewellen, B. L. (1993). High-speed automated discrete blood sampling for positron emission tomography. *J. Nucl. Med.* **34**: 1357–1360.

Hasselbalch, S. G., Madsen, P. L., Knudsen, G. M., Holm, S., and Paulson, O. B. (1998). Calculation of the FDG lumped constant by simultaneous measurements of global glucose and FDG metabolism in humans. *J. Cereb. Blood Flow Metab.* **18**: 154–160.

Lammertsma, A. A., Brooks, D. J., Frackowiak, R. S., Beaney, R. P., Herold, S., Heather, J. D., Palmer, A. J., and Jones, T. (1987). Measurement of glucose utilisation with [18F]2-fluoro-2-deoxy-D-glucose: A comparison of different analytical methods. *J. Cereb. Blood Flow Metab.* **7**: 161–172.

Lying-Tunell, U., Lindblad, B. S., Malmlund, H. O., and Persson, B. (1980). Cerebral blood flow and metabolic rate of oxygen, glucose, lactate, pyruvate, ketonebodies and amino acids. *Acta Neurol. Scand.* **62**: 265–275.

Phelps, M. E., Huang, S. C., Hoffman, E. J., Selin, C., Sokoloff, L., and Kuhl, D. E. (1979). Tomographic measurement of local cerebral glucose metabolic rate in humans with (F-18)2-fluoro-2-deoxy-D-glucose: Validation of method. *Ann. Neurol.* **6**: 371–378.

Reivich, M., Alavi, A., Wolf, A., Fowler, J., Russell, J., Arnett, C., MacGregor, R. R., Shiue, C. Y., Atkins, H., Anand, A., *et al.* (1985). Glucose metabolic rate kinetic model parameter determination in humans: the lumped constants and rate constants for [18F]fluorodeoxyglucose and [11C]deoxyglucose. *J. Cereb. Blood Flow Metab.* **5**: 179–192.

Sokoloff, L., Reivich, M., Kennedy, C., Des-Rosiers, M. H., Patlak, C. S., Pettigrew, K. D., Sakurada, O., and Shinohara, M. (1977). The [14C]deoxyglucose method for the measurement of local cerebral glucose utilization: Theory, procedure, and normal values in the conscious and anesthetized albino rat. *J. Neurochem.* **28**: 897–916.

Spence, A. M., Muzi, M., Graham, M. M., O'Sullivan, F., Krohn, K. A., Link, J. M., Lewellen, T. K., Lewellen, B., Freeman, S. D., Berger, M. S., and Ojemann, G. A. (1998). Glucose metabolism in human malignant gliomas measured quantitatively with PET, 1-[11C]glucose and FDG: Analysis of the FDG lumped constant. *J. Nucl. Med.* **39**: 440–448.

Wienhard, K., Pawlik, G., Nebeling, B., Rudolf, J., Fink, G., Hamacher, K., Stocklin, G., and Heiss, W. D. (1991). Estimation of local cerebral glucose utilization by positron emission tomography: Comparison of [18F]2-fluoro-2-deoxy-D-glucose and [18F]2-fluoro-2-deoxy-D-mannose in patients with focal brain lesions. *J. Cereb. Blood Flow Metab.* **11**: 485–491.

Validation of Noninvasive Quantification Technique for Neurologic FDG-PET Studies

KOON-PONG WONG,*,†,‡ **DAGAN FENG,***,† **STEVEN R. MEIKLE,**†,‡ and **MICHAEL J. FULHAM**†,‡

**Center for Multimedia Signal Processing, Department of Electronic and Information Engineering, The Hong Kong Polytechnic University, Hong Kong*
†*Biomedical and Multimedia Information Technology (BMIT) Group, Basser Department of Computer Science, The University of Sydney, NSW 2006, Sydney, Australia*
‡*Departments of PET and Nuclear Medicine, Royal Prince Alfred Hospital, Camperdown, NSW 2050, Sydney, Australia*

Dynamic imaging with positron emission tomography (PET) is widely regarded as the "gold standard" for the in vivo measurement of regional cerebral metabolic rate for glucose (rCMRGlc) with $[^{18}F]$ fluorodeoxy-D-glucose ($[^{18}F]$ FDG) and is used for the clinical evaluation of neurological disease. However, in addition to the acquisition of dynamic images, continuous arterial blood sampling is required to obtain the tracer time–activity curve in blood for the estimation of rCMRGlc. The insertion of arterial lines and the subsequent collection and processing of multiple blood samples are impractical for clinical PET studies for various reasons. We recently proposed a cascaded modeling method that estimates the input function and determines the kinetic model parameters simultaneously from multiple regions of interest with one or more late venous blood samples for calibration. The aims of this study were (i) to validate the original method in dynamic neurologic FDG-PET studies, and (ii) to propose and validate an improved method so that the estimation results could be even more reliable in the cases that the measurement noise is very high.

I. INTRODUCTION

Positron emission tomography (PET) with $[^{18}F]$fluorodeoxy-D-glucose ($[^{18}F]$FDG) has been widely used to study regional brain glucose metabolism in humans. However, the major drawback of the method is the need to measure the input function (IF) for the compartmental model of $[^{18}F]$FDG (Phelps *et al.*, 1979; Huang *et al.*, 1980). The input function is usually obtained by frequent blood sampling at the radial artery or arterialized vein (a-v) (Phelps *et al.*, 1979; Huang *et al.*, 1980). Arterial blood sampling is impractical in clinical studies because it is cumbersome for the patients and staff involved. Effort has been directed toward reducing or obviating the need for arterial blood sampling. Based on the information embedded in the dynamic PET images, we recently developed a theory for a practical noninvasive measurement technique which can simultaneously estimate the kinetic parameters from the PET images and recover the input function without the need of invasive arterial blood sampling (Feng *et al.*, 1997). This study validated the proposed method with dynamic neurologic FDG-PET data obtained from humans. In addition, we proposed and validated an improved method, so that the parameter estimates could be more reliable when the measurement noise is very high.

II. MATERIALS AND METHODS

A. Kinetic Model

The model is based on the three-compartment model originally developed by Sokoloff *et al.* (1977) with the correction for the dephosphylation reaction as proposed by Phelps *et al.* (1979) and Huang *et al.* (1980). Details of this compartmental model can be found elsewhere (Phelps *et al.*,

1979; Huang *et al.*, 1980; Hawkins *et al.*, 1986). The actual tissue activity, $c_T^*(t)$, measured by PET can be expressed as

$$c_T^*(t) = (1 - CBV) \cdot c_i^*(t) + CBV \cdot c_a^*(t), \qquad (1)$$

where $c_i^*(t)$ is the total tissue activity, $c_a^*(t)$ is the measured arterial IF, and CBV is the fractional cerebral blood volume which accounts for the contribution of the activity in the vascular space of tissue (Hawkins *et al.*, 1986). The regional cerebral metabolic rate for glucose (rCMRGlc) is calculated according to Phelps *et al.* (1979) and Huang *et al.* (1980)

$$rCMRGlc = \frac{k_1^* k_3^*}{k_2^* + k_3^*} \frac{C_{glc}}{LC} \equiv K \frac{C_{glc}}{LC}, \qquad (2)$$

where k_1^*–k_4^* are the rate constants for the $[^{18}F]$FDG model, C_{glc} is the glucose concentration in plasma, and LC is the lumped constant which embodies the difference between $[^{18}F]$FDG and glucose in transportation and phosphorylation (Phelps *et al.*, 1979; Huang *et al.*, 1980). The macroparameter, K, is proportional to the uptake of the $[^{18}F]$FDG and thereby proportional to rCMRGlc. In this study, rCMRGlc was used as one of the main criteria to evaluate the performance of the proposed method.

B. Invasive Method

In the invasive (conventional) approach, the rate constants in the kinetic model are determined by nonlinear least squares (NLLS) fitting to the tissue time–activity curve (TAC) which is conventionally obtained by manually delineation of regions of interest (ROIs) over the PET images (Huang *et al.*, 1980). However, the model equation that describes the tissue response requires knowledge of the input function, which is usually obtained from the arterial blood samples.

C. Proposed Modeling Method and Its Modification

The theoretical basis and Monte-Carlo simulation results of the proposed method have been reported previously (Feng *et al.*, 1997). Only a brief summary of the method is presented here. The basic assumptions of the proposed method (called simultaneous estimation, SIME, in the sequel) are that all the tissue TACs are driven by the same input function and that the information of the input function is embedded in the dynamic PET images. It is, therefore, possible to recover the input function by minimizing the errors between the model predicted tissue responses and the measurements of TACs obtained from two or more ROIs, and thus the kinetic parameters can be estimated simultaneously using NLLS (Feng *et al.*, 1997) if the input function model is properly accounted into the overall fitting procedure. In order to reduce the uncertainties in the parameter estimates and

to reduce the computational complexity of the fitting procedure for fitting too many data points, an input function model proposed in (Feng *et al.*, 1993) was used to model the arterial IF, $c_a^*(t)$, instead of fitting the measurements directly by the proposed method. To achieve absolute quantification of the kinetic parameters and the input function, at least one venous blood sample is required for calibration.

The accuracy of the parameter estimate is reflected by the asymptotic standard deviation (SD), which can be computed using the information matrix approximated by NLLS. Theoretically, accurate parameter estimates can be obtained with SIME. Due to high nonlinearity of the parameter space and high noise level in the PET data, the information matrix in the NLLS will be unstable because the stability of the Jacobian matrix is disrupted by the numerical error (noise) and the linear approximation to the curvature might not valid. Subsequent estimation of the SDs for the parameter estimates will be very poor, though the values of the estimates are close to those obtained with the invasive method, and the shape and magnitude of the recovered IF agreed with the measured arterial IF. We have developed a technique that is applied after SIME for the situations as mentioned above. We refer to this method as simultaneous estimation with post-estimation (SIMEP), which is based on the assumption that the input function recovered by SIME can globally minimize the errors between the measured data and the predicted tissue TACs in the least squares sense. The parameters in the individual tissue TACs can be estimated using conventional compartmental model fitting with the recovered IF as the input function. The SDs of the parameters can be greatly reduced due to the reduction in dimensionality and nonlinearity of the parameter space.

D. Human Studies

Dynamic FDG-PET studies were performed on three human subjects with a PC4096-15WB PET tomograph (GE/Scanditronix). All subjects were fasted overnight before the study. After performing a 15-min transmission scan, 370 MBq of $[^{18}F]$FDG was injected intravenously. Nineteen arterial blood samples were collected every 15 s from 0 to 2 min; every 30 s from 2 to 3.5 min; and at 7, 10, 15, 20, 30, 60, 90, and 120 min, respectively, to form the measured arterial IF for comparison. Raw PET data were acquired following the tracer injection over a period of 2 h (10×12-s scans; 2×30-s scans; 2×1-min scans; 1×1.5-min scan; 1×3.5-min scan; 2×5-min scans; 1×10-min scan; 3×30-min scans). Tomographic images were reconstructed using filtered backprojection with a Hann filter (cutoff frequency = 0.5 of the Nyquist frequency). Five ROIs were put over the PET images to obtain the tissue TACs in gray matter, white matter, occipital lobe, parietal lobe, and the average of the whole brain. Two late venous blood samples were taken at

60 and 120 min for calibration and to accelerate the convergence as well as to improve the numerical identifiability. Kinetic rate constants were then estimated using NLLS for the compartmental model fitting, SIME, and SIMEP, respectively.

III. RESULTS AND DISCUSSION

It was found that there were very good agreements between the measured and the estimated (recovered) IFs, and between the measured and the predicted TACs from both SIME and SIMEP, for all three studies. Figure 1 shows the

fitting results obtained from one of the subjects. As seen from Fig. 1, the TACs predicted by the parameters obtained from SIME and SIMEP are almost identical to the measured TACs. Table 1 shows the comparison of the estimated rate constants in the gray matter and white matter, obtained by the conventional (invasive) method, SIME, and SIMEP (using three ROIs), respectively. The results indicated that the main difference between SIME and SIMEP is in the parameter variances rather than the values themselves. Linear regression on the areas under the curves (AUCs) covered by both IFs at different times ($n = 19$) yielded a straight line with slope = 1.033 and intercept = 0.639 ($r = 0.99$, $p < 0.001$). Good estimates of rCMRGlc were obtained

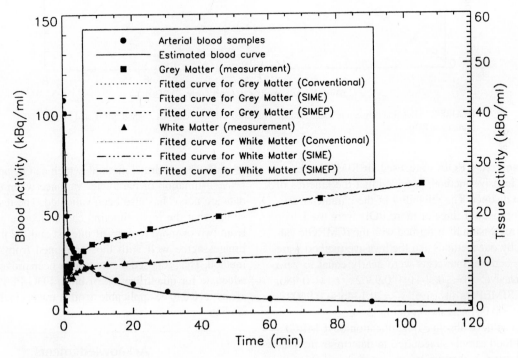

FIGURE 1. Arterial blood samples, estimated blood curve, and predicted tissue TACs corresponding to gray matter and white matter, respectively, estimated by SIME with two regions (gray matter and white matter), SIMEP, and the conventional (invasive) method.

TABLE 1. Estimated Rate Constants in Gray Matter and White Matter Using Different Methods (Estimate ± Standard Deviation)

Parameters	Gray matter			White matter		
	Invasive	SIME	SIMEP	Invasive	SIME	SIMEP
k_1^*	0.0836 ± 0.0149	0.0776 ± 0.0569	0.0810 ± 0.0079	0.0727 ± 0.0082	0.0726 ± 0.0335	0.0725 ± 0.0085
k_2^*	0.1901 ± 0.0444	0.1776 ± 0.1045	0.1977 ± 0.0369	0.2556 ± 0.0554	0.2708 ± 0.1095	0.2763 ± 0.0585
k_3^*	0.0614 ± 0.0042	0.0609 ± 0.0303	0.0654 ± 0.0081	0.0504 ± 0.0130	0.0511 ± 0.0322	0.0517 ± 0.0124
k_4^*	0.00071 ± 0.00031	0.00053 ± 0.00044	0.00084 ± 0.00024	0.00751 ± 0.00331	0.00698 ± 0.00440	0.00707 ± 0.00304
CBV	0.0378 ± 0.0076	0.0474 ± 0.0349	0.0452 ± 0.0068	0.0152 ± 0.0082	0.0176 ± 0.0143	0.0170 ± 0.0082
rCMRGlc	7.098 ± 0.379	6.892 ± 4.589	7.003 ± 0.353	4.165 ± 0.613	4.008 ± 2.611	3.974 ± 0.560

Note. The units for the rate constants k_1^*–k_4^* and rCMRGlc are 1/min, and mg/min/100 mL, respectively.

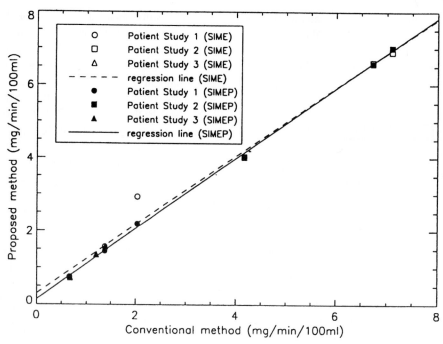

FIGURE 2. Relation between rCMGlc estimates obtained with conventional method, SIME and SIMEP using three ROIs.

even when only two ROIs were used in SIME, as compared to the invasive method, provided that the kinetics of the TACs are distinct. The reliability of the parameter estimates was improved if three or more ROIs were used. The slopes of the regression lines formed with the rCMRGlc values obtained by our methods and the invasive method were close to one, and the intercepts were nearly equal to zero (SIME vs invasive: $y = 0.3041 + 0.9362x$, $r = 0.986$, $p < 0.001$; SIMEP vs invasive: $y = 0.1446 + 0.9600x$, $r = 0.994$, $p < 0.001$), as shown in Fig. 2.

In order to achieve absolute quantification of rCMRGlc, at least one blood sample is required to determine the glucose concentration in plasma. It was found that if this one or two late venous blood samples are used, the performance of our method is generally improved (Feng *et al.*, 1997). Furthermore, venous sampling is much preferred to arterial sampling as it is less invasive and is much more tolerated by the subject while the potential difficulties associated with arterial catheterization can be avoided. In addition, the radiation exposure associated with rapid continuous blood sampling is significantly reduced.

IV. SUMMARY

A noninvasive modeling approach for the calculation of rCMRGlc has been validated by clinical studies. The method requires one or more late venous blood samples, which is minimally invasive and markedly reduces radiation exposure

to staff. The modified method, which can be applied to improve estimation of parameter variance when the raw PET data are noisy, has also been validated. The results demonstrated that the input function can be recovered accurately from two or more regions of interest and the parameter estimates agree well with those obtained from the invasive method. The proposed methods have been proven to be more adequate for quantification of brain FDG-PET studies and are expected to be applicable to other tracers in dynamic PET studies.

Acknowledgments

This research was supported by NH&MRC, ARC, and RGC research grants. The authors are grateful to the National PET/Cyclotron Center, Taipei Veterans General Hospital, Taiwan, for providing the FDG data sets used in this study.

References

Feng, D., Huang, S. C., and Wang, X. (1993). Models for computer simulation studies of input functions for tracer kinetic modeling with positron emission tomography. *Int. J. Biomed. Comput.* **32**: 95–110.

Feng, D., Wong, K. P., Wu, C. M., and Siu, W. C. (1997). A technique for extracting physiological parameters and the required input function simultaneously from PET image measurements: Theory and simulation study. *IEEE Trans. Info. Tech. Biomed.* **1**: 243–254.

Hawkins, R. A., Phelps, M. E., and Huang, S. C. (1986). Effects of temporal sampling, glucose metabolic rates, and disruptions of the blood-brain barrier on the FDG model with and without a vascular compartment: Studies in human brain tumors with PET. *J. Cereb. Blood Flow Metab.* **6**: 170–183.

Huang, S. C., Phelps, M. E., Hoffman, E. J., Sideris, K., Selin, C., and Kuhl, D. E. (1980). Noninvasive determination of local cerebral metabolic rate of glucose in man. *Am. J. Physiol.* **238**: E69-E82.

Phelps, M. E., Huang, S. C., Hoffman, E. J., Selin, C., Sokoloff, L., and Kuhl, D. E. (1979). Tomographic measurement of local cerebral glucose metabolic rate in humans with (F-18)2-fluoro-2-deoxy-D-glucose: Validation of method. *Ann. Neurol.* **6**: 371–388.

Sokoloff, L., Reivich, M., Kennedy, C., DesRosiers, M. H., Patlak, C. S., Pettigrew, K. D., Kakurada, D., and Shinohara, M. (1977). The [^{14}C]deoxyglucose method for the measurement of local cerebral glucose utilization: Theory, procedure and normal values in the conscious and anesthetized albino rat. *J. Neurochem.* **28**: 897–916.

20

Effect of Population k_2 Values in Graphical Estimation of DV Ratios of Reversible Ligands

J. E. HOLDEN,* V. SOSSI,[†] G. CHAN,[†] D. J. DOUDET,[‡] A. J. STOESSL,[‡] and T. J. RUTH[†]

*University of Wisconsin, Madison, Wisconsin
[†]UBC/TRIUMF, Vancouver, Canada
[‡]UBC, Vancouver, Canada

The graphical method of Logan with brain tissue time courses as input functions provides a robust and accurate estimate of the ratio of the total distribution volumes (DV) of reversibly binding ligands in two brain tissues. A reference tissue time course is used to estimate the running integral of the concentration of ligand in plasma. This estimator includes a term dependent on k_2, the compartmental backflux rate constant for the ligand in the reference tissue. As the goal of the method is to allow the ratio estimation without measuring the plasma time course, a population average k_2 value is used. We studied the effect of population average k_2 values for three reversible ligands (SCH23390, methylphenidate, dihydrotetrabenazene, all labeled with ^{11}C) by comparing DV ratios estimated using the plasma-input function graphical method (pl) with those estimated by the tissue-input method performed three different ways: no k_2 dependent term (no), term calculated using individual k_2 values (ind), and term calculated using population average k_2 values (av). Regressions were performed for no vs pl, ind vs pl, av vs pl, and av vs ind. DV ratio estimates from average k_2 values were highly correlated with both those derived from plasma and those from individual k_2 values for all ligands. Small systematic discrepancies among the methods for methylphenidate appear to arise from the slow equilibration between tissue and plasma for that ligand. Our results provide strong support for the use of average k_2 values for these three ligands.

I. INTRODUCTION

The graphical method introduced by Logan et al. (1996) using brain tissue time courses as input functions provides a robust and accurate estimate of the ratio of the total distribution volumes of reversibly binding ligands in two brain tissues. If one of the tissues (reference) is devoid of specific binding sites, but is otherwise similar to the other (target), the ratio can be further reduced to yield the binding potential B/K_d in the target tissue. The method is based on the use of the reference tissue time course to estimate the running integral of the concentration of ligand in plasma. This estimator divided by the concentration in the target tissue gives the abscissa values for the graphical points to be fitted. The second term in the abscissa calculation is $(1/k_2)(R(t)/T(t))$, where $R(t)$ and $T(t)$ are the reference and target tissue time courses, respectively, and k_2 is the compartmental backflux rate constant for the ligand in the reference tissue. As the point of the method is to allow the ratio estimation to be performed without requiring the plasma time course necessary for the evaluation of k_2, a population average value is used. We have investigated the effect of the use of a population average value for k_2 for three reversible ligands by comparing distribution volume ratios estimated using the plasma-input function graphical method with those estimated by the tissue-input function method, with the abscissa values estimated three different ways: no second term, second term calculated using k_2 values specific to the particular study, and second term calculated using population average k_2 values.

II. METHODS

A. PET Studies

Studies of the D1 dopamine receptor binding ligand [^{11}C]SCH23900 ([^{11}C]SCH), the type 2 monoamine vesicular transporter ligand [^{11}C]dihydrotetrabenazene ([^{11}C]DTBZ), and the dopamine uptake transporter ligand [^{11}C]methylphenidate ([^{11}C]MP) were performed in 10 normal volunteer subjects. The scanning protocol was the same for all of the tracers and consisted of 4×60-s, 3×120-s, 8×300-s, 1×600-s frames for a total of 1 h. Twenty-three arterial blood samples were also taken and 5 of these analyzed for metabolites. The metabolite-corrected plasma time courses were estimated in all cases over the entire scanning period.

B. Analysis

1. Region of Interest (ROI) Placement

For each scan seven consecutive axial image planes containing the striatum were summed to produce a composite image. Four circular (61 mm^2) ROIs were placed on each striatum, one on the caudate and three on the putamen, and six circular ROIs (297 mm^2) on the occipital cortex. In addition, two consecutive slices containing the cerebellum were summed and one oval ROI (556 mm^2) was placed on the cerebellum. These ROIs were replicated on images obtained from each time frame. Time–activity curves were thus obtained for the left and right caudate, left and right putamen, occipital cortex, and cerebellum.

2. Calculations

Cerebellar (CER) and occipital cortical (COR) time courses were fitted with a one-tissue compartment model to provide the reference tissue k_2 values for each study. These were then averaged to provide the required average values. Distribution volume ratios (DVR) for striatal brain regions versus both reference tissues were estimated using four distinct approaches:

- The individual total distribution volumes were estimated in both target and reference tissues by the Logan graphical method with plasma-input function and the DVR (pl) calculated as their ratio.
- The DVR was determined directly by the Logan graphical method with tissue time course as input function. The k_2-dependent term in the abscissa calculation was dealt with in three different ways: it was neglected entirely (no), it was estimated using the individual k_2 value determined from compartmental fitting (ind), and it was estimated using the average value for all subjects (av), determined independently for each reference tissue.

The fitted time period was 15 to 60 min postinjection in all cases.

C. Statistical Comparisons

Regressions were performed for no k_2 term versus plasma (no vs pl), individual k_2 value versus plasma (ind vs pl), average k_2 value versus plasma (av vs pl), and average k_2 value versus individual k_2 value (av vs ind). Relative qualities of the correlations were assessed using the correlation coefficient r.

III. RESULTS

The results for anterior putamen as target tissue are presented in Tables 1 and 2. These are representative of those from other striatal regions. Slopes, intercepts, and correlation coefficients are presented for the four regressions performed. Results are presented separately for cerebellum (Table 1) and for occipital cortex (Table 2) as reference tissue. The first columns (no vs pl) indicate the relative importance

TABLE 1.   Correlation Results: Cerebellum as Reference Tissue

Ligand		No vs pl	Ind vs pl	Av vs pl	Av vs ind
SCH	Slope	1.01	0.97	0.99	1.02
	Intercept	−0.13	0.007	−0.039	−0.047
	r	0.97	0.99	0.99	1.0
DTBZ	Slope	0.78	0.96	0.93	0.97
	Intercept	0.24	0.043	0.074	0.038
	r	0.97	0.99	0.99	0.99
MP	Slope	0.65	0.94	1.03	1.06
	Intercept	0.28	0.074	−0.13	−0.14
	r	0.82	0.98	0.96	0.96

TABLE 2.   Correlation Results: Occipital Cortex as Reference Tissue

Ligand		No vs pl	Ind vs pl	Av vs pl	Av vs ind
SCH	Slope	0.95	0.96	0.98	1.02
	Intercept	0.018	0.037	0.034	−0.031
	r	0.96	0.99	0.99	1.0
DTBZ	Slope	0.84	0.93	0.91	0.95
	Intercept	0.21	0.099	0.14	0.10
	r	0.96	0.98	0.99	0.98
MP	Slope	0.52	0.91	1.07	1.16
	Intercept	0.67	0.14	−0.21	−0.36
	r	0.82	0.99	0.98	0.98

of the k_2-dependent term. The second columns (*ind* vs *pl*) show the relationship between the tissue-input and plasma-input methods when complete information (individual k_2 values) is available. The last two columns indicate the effect of using population average values rather than individual values for k_2. Interpretation of these results is provided in the following section.

IV. DISCUSSION

The inclusion of the k_2-dependent term in the estimation of abscissa values in the tissue-input graphical estimation of distribution volume ratios provides a considerable improvement in the statistical precision of the results by allowing the slope estimation to be performed over a significantly increased time range. As one of the primary advantages of the tissue-input method is avoiding the technical burden of arterial blood sampling, the estimation of the actual k_2 value in each individual study is not a practical option. Use of a population average value for k_2 may result in additional variability of the DVR values due to variation of k_2 between subjects. We have investigated this question for three reversible binding ligands.

For all three tracers DVR values estimated with the tissue-input method implemented with the actual individual k_2 values were highly correlated with those estimated with the plasma-input method, thus confirming the practical equivalence of the two approaches (Tables 1 and 2, second columns). Correlation coefficients were in excess of 0.98 regardless of the reference tissue chosen. The relative importance of the k_2-dependent term was established in each case by comparing DVR values derived without the term with values derived with the plasma-input method (Tables 1 and 2, first columns). Neglecting the k_2-dependent term resulted in the systematic underestimation of DVR, with the effect ranging from nearly insignificant for $[^{11}C]$SCH to extremely significant for MP. Finally, the potential variability and bias arising from the use of an average population k_2 value was assessed by comparing DVR values determined with average k_2 values with values determined with individual k_2 values (Tables 1 and 2, last columns). The two approaches were statistically indistinguishable for $[^{11}C]$SCH and $[^{11}C]$DTBZ. For $[^{11}C]$MP the regression shows a small amount of residual scatter, together with a positive bias in excess of 10%, arising from the use of average k_2 values.

Our observations can be largely explained in terms of the constancy of the ratio $R(t)/T(t)$ over the fitted time period. If this ratio is constant, then the inclusion of the second term, or the value of its coefficient when included, has no effect on the fitted slope value. This is particularly important in the case of $[^{11}C]$SCH, which is known to have significant specific binding site density in occipital cortex but not in cerebellum. The DVR for COR was indeed significantly lower than that for CER, presumably reflecting reference tissue-specific binding. Despite this, the correction term using a one-compartment assumption for $R(t)$ performed equally well for both tissues. Specific binding in the reference tissue does not violate the assumptions required by the tissue-input Logan method. In that case, the coefficient of the second term in the abscissa estimation becomes a more complicated combination of the compartmental parameters k_2, k_3, and k_4. Because the two-tissue-compartment fits of these ligands are so unstable, we chose to study the performance of the tissue-input approach implemented with the assumption of a single compartment in the reference tissue. The equivalence of the average k_2 method to the individual k_2 method for $[^{11}C]$SCH or for $[^{11}C]$DTBZ does not demonstrate that the one-compartment assumption was met or that variation of k_2 between subjects did not occur. Rather, the constancy of the ratio $R(t)/T(t)$ observed for these two compounds meant that these considerations became unimportant.

The high sensitivity of $[^{11}C]$MP to the second term and to the value of its coefficient arises from the relatively slow equilibration of this ligand. The ratio $R(t)/T(t)$ declined rapidly at the start of the fitting range and continued to decline slowly over the entire study duration. Despite this sensitivity, the values derived from the *av* and *ind* approaches were highly correlated ($r = 0.98$) for occipital cortex as reference tissue. Occipital cortex is the reference tissue used for $[^{11}C]$MP studies in our laboratory.

Our results strongly support the use of average k_2 values and the one-compartment assumption for these ligands. Caution must be used in the analysis of $[^{11}C]$MP in cases where significant k_2 changes are possible.

References

Logan, J., Fowler, J. S., Volkow, N. D., Wang, G.-J., Ding, Y.-S., and Alexoff, D. L. (1996). Distribution volume ratios without blood sampling from graphical analysis of PET data. *J. Cereb. Blood Flow Metab.* **16**: 834–840.

21

Measuring the BP of Four Dopaminergic Tracers Utilizing a Tissue Input Function

V. SOSSI,* J. E. HOLDEN,† G. CHAN,* M. KRZYWINSKI,* A. J. STOESSL,‡ and T. J. RUTH*

UBC/TRIUMF, Vancouver, Canada
†*University of Wisconsin, Madison, Wisconsin*
‡*UBC, Vancouver, Canada*

The presynaptic tracers [^{11}C] methylphenidate and [^{11}C] dihydrotetrabenazine ([^{11}C] DTBZ) and the postsynaptic tracers [^{11}C] raclopride and [^{11}C] Schering 23990 ([^{11}C] SCH) are used to study the integrity of the dopaminergic system with positron emission tomography. These are reversible tracers where the binding potential $BP = B_{max}/K_d$ is often used to quantify the amount of their specific binding to the sites of interest. The kinetics of these tracers make them possible candidates for simplified analysis methods that do not require blood input function, thus eliminating the need for arterial blood sampling. We have compared the BP values obtained for the four tracers using the Logan graphical tissue method and the Lammertsma reference tissue method (RTM). The BP estimates obtained with the two methods were nearly identical in most cases with similar reliability and reproducibility. R_1 estimated by the RTM method proved to have very low inter- and intrasubject variability. k_2, also estimated by the RTM method, had good reliability only for [^{11}C] SCH with cerebellar input function and [^{11}C] DTBZ with occipital input function.

I. INTRODUCTION

[^{11}C]Methylphenidate ([^{11}C]MP), [^{11}C]dihydrotetrabenazine ([^{11}C]DTBZ), [^{11}C]raclopride ([^{11}C](RAC), and [^{11}C]Schering 23990 ([^{11}C]SCH) provide complementary information on the integrity of the dopaminergic system using PET.

Methylphenidate inhibits dopamine reuptake and enhances synaptic dopamine levels. One of its isomers, D-*threo*-methylphenidate, has been labeled with ^{11}C ([^{11}C]MP) (Ding *et al.*, 1994) for positron emission tomography (PET). Its binding in the human brain is reversible, highly reproducible and saturable and thus MP is deemed an appropriate PET ligand to measure dopamine transporter availability (Volkow *et al.*, 1995).

Dihydrotetrabenazine is one of the metabolites of tetrabenazine, a high affinity inhibitor of the type 2 vesicular monoamine transporter (VMAT2). Dihydrotetrabenazine binds to the VMAT2 and blocks the storage of monoamine neurotransmitters in presynaptic vesicles (Pletscher *et al.*, 1962; Henry and Scherman, 1989). [^{11}C]DTBZ is used in PET to examine *in vivo* the integrity of striatal presynaptic monoaminergic terminals (Koeppe *et al.*, 1995, 1996) and to estimate neuronal losses in aging (Frey *et al.*, 1996) and neurodegenerative disease (Frey *et al.*, 1996; Gilman *et al.*, 1996). (±)-α-DTBZ was used in these studies.

The postsynaptic tracer [^{11}C]SCH has been widely used as a ligand to study dopamine D1 receptor function using PET (Farde *et al.*, 1987; Suhara *et al.*, 1991; Farde *et al.*, 1992; Shinitoh *et al.*, 1993). Likewise [^{11}C]RAC, a D2 receptor antagonist is commonly used to study dopamine D2 receptor function (Farde *et al.*, 1985).

All of these tracers are reversible, although with different time courses, and several analysis methods have been proposed. They range from standard compartmental analysis, which requires dynamic scanning and a plasma input function (IF) to simplified ratio methods where only a static image at a certain time interval after injection and the definition of a reference region devoid of specific binding are required. Several comparisons of the various methods have already been reported (Lammertsma *et al.*, 1996; Chan *et al.*, 1998, 1999).

A very common measure that quantifies the tracer binding is the binding potential, BP. BP = k_3/k_4, where k_3 is the rate constant for transfer from free to bound compartment and k_4 is the rate constant for transfer from bound to free compartment. In this study we have compared BP values obtained by two tissue input methods, the Logan graphical approach (Logan *et al.*, 1996) and the Lammertsma simplified reference tissue method (Lammertsma and Hume, 1996) with the Gunn implementation (Gunn *et al.*, 1997) (RTM), for all four tracers. Both methods are very attractive in that they do not require a plasma input function and they consequently reduce the burden of the scanning procedure. By its nature, the Logan method is an equilibrium method, thus requiring tracer equilibration during the scanning procedure, while the RTM method is a kinetic method with constraints on the number of compartments. Therefore there could be, a priori, differences in BP estimates by the two methods as a consequence of the kinetic behaviors of the tracers.

In addition to BP, the RTM method provides an estimate of the rate constant for transfer from free to plasma compartment, k_2, and the ratio of the rate constants for transfer from plasma to free compartment (k_1) between target and reference region, $R_1 = k_1/k'_1$. The reproducibility and accuracy of these two parameters was also estimated. This study is thus divided into four parts: examination of the kinetics of the four tracers, comparison of the BP values, estimation of reliability and reproducibility of the BP values provided by each method, and estimation of the reliability and reproducibility of R_1 and k_2.

II. METHODS

Ten normal volunteers were scanned using [^{11}C]DTBZ, [^{11}C]SCH, and [^{11}C]RAC, mean age 54.9 ± 17.9, 50.5 ± 18.4, and 51.3 ± 16.4, respectively. Seven normal volunteers were scanned using [^{11}C]MP, mean age 47.6 ± 13.7. The subjects recruited for the [^{11}C]SCH, [^{11}C]DTBZ, or [^{11}C]MP study underwent a repeat scan approximately 14 days after the first scan to evaluate the reproducibility of the measurement and the analysis method. The scanning protocol was the same for all of the tracers and consisted of 4×60, 3×120, 8×300, 1×600 s. For [^{11}C]DTBZ, [^{11}C]MP, and [^{11}C]SCH 23 blood samples were also taken and five analyzed for metabolites. The plasma data were used to perform a standard compartmental analysis to complement the comparison of these methods. All scans were taken on the ECAT 953B (Spinks *et al.*, 1992) in 2D mode, except for the

[^{11}C]RAC scans, which were performed in 3D mode. Details of the processing of the 3D data can be found elsewhere (Sossi *et al.*, 1998). All scans were corrected for attenuation using a ^{68}Ge transmission scan.

III. ANALYSIS

A. Region of Interest (ROI) Placement

For each scan seven consecutive axial image planes containing the striatum were summed to produce a composite image. Four circular (61 mm^2) ROIs were placed on each striatum, one on the caudate and three without overlap along the longitudinal axis of the putamen, and six circular ROIs (297 mm^2) were placed on the occipital cortex (OC). In addition, two consecutive slices containing the cerebellum were summed and one oval ROI (556 mm^2) was placed on the cerebellum. These ROIs were replicated on images obtained from each time frame. Time–activity curves were thus obtained for the left and right caudate, left and right putamen (average of the three ROI values), occipital cortex, and cerebellum.

B. Analysis Methods

The Logan method determines the distribution volume ratio according to

$$\int_0^T C_t(t)\, dt / C_t(T)$$

$$= \text{DVR}\left[\left(\int_0^T C_r(t)\, dt + C_r(T)/k'_2\right)/C_t(T)\right] + \text{int}, \quad (1)$$

where $C_t(t)$ is the tracer concentration in the target region, $C_r(t)$ is the tracer concentration in the reference region, and k'_2 is the population derived k'_2 value. This method requires the tracer kinetics to satisfy the following assumptions: steady state must be reached during the course of the study, the ratio of the influx/efflux constants is the same in the target and reference regions ($k_1/k_2 = k'_1/k'_2$), and a population k'_2 value can be used. This is a two-parameter method, with the variable of interest being BP and no specific assumption on the number of compartments. The k'_2 values used in this study were derived from compartmental analysis: [^{11}C]SCH 0.061 (OC IF) 0.101 (cerebellar IF), [^{11}C]MP 0.039 (OC IF) 0.055 (cerebellar IF), [^{11}C]DTBZ 0.073 (OC IF) 0.086 (cerebellar IF), [^{11}C]RAC 0.3545 (OC IF) 0.163 (cerebellar IF).

The RTM method requires these assumptions to be satisfied: the free and the bound tracer in the target region can be described by a single compartment and $k_1/k_2 = k'_1/k'_2$, while no assumption on equilibrium is necessary. The model provides three parameters: BP, the local rate of delivery $R_1 = k_1/k'_1$, and k_2, using Eq. (2), where $C_t(t)$ is the concentration of the free and bound tracer in the target region:

$$C_t(t) = R_1 C_r(t) + \{k_2 - R_1 k_2/(1 + BP)\}C_r(t)^*$$
$$\times \exp\{-k_2 t/(1 + BP)\}. \quad (2)$$

The standard one-compartment approach was also applied to [11C]DTBZ, [11C]MP, and [11C]SCH data and the BP values were compared to those obtained from the Logan and RTM methods. BP was estimated from the distribution volumes using the indirect method $BP = (DV_t - DV_r)/DV_r$. The BP values were compared numerically for all four tracers and linear correlation coefficients were estimated. For [11C]DTBZ, [11C]MP, and [11C]SCH standard deviations between and within subjects (STDB, STDW) and reliability $R(R = STDB^2/(STDB^2 + STDW^2))$ were calculated using a one-way ANOVA (Scheffe, 1959). STDW was used as a measure of reproducibility.

R_1 values were compared to the equivalent parameter determined from the compartmental model, k_1/k_1', using linear regression. In addition, STDB and STDW and reliability were calculated.

There is no equivalent of the RTM k_2 using the compartmental model approach, therefore only STDB, STBW, and

reliability R were used in the assessment of the k_2 measurement precision.

IV. RESULTS

A. Tracer Kinetics

The decay-corrected time–activity curve of the specifically bound tracer $C_b(t)$ is shown in Fig. 1 for each tracer for a representative target region. $C_b(t)$ has been obtained by subtracting the reference region activity from the target region activity, which contains both free and bound tracer. The specific binding estimate is dependent on the choice of the reference region and a detailed investigation of the appropriate choice is beyond the scope of the present study. The specifically bound tracer as estimated using either the cerebellum or the occipital cortex input function clearly reaches equilibrium for [11C]RAC, [11C]SCH, and [11C]DTBZ. MP, however, appears to barely reach equilibrium during the course of the study.

The compartmental model analysis performed on the [11C]DTBZ, [11C]MP, and [11C]SCH showed that a one-compartment model describes the kinetics of [11C]MP and

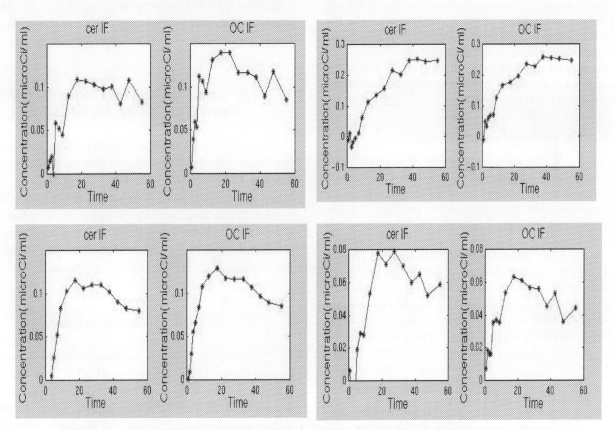

FIGURE 1. Time–activity course for the specifically bound tracer with cerebellum as reference region and occipital cortex as reference region. Top left, [11C]DTBZ, top right, [11C]MP, bottom left, [11C]RAC; and bottom right, [11C]SCH.

TABLE 1. Comparison of BP Estimates between Various Methods and Correlation Coefficients r

Tracer	Region	BP_{Log}/BP_{RTM}	r(Log, RTM)	BP_{Log}/BP_{comp}	r(Log, comp)	BP_{RTM}/BP_{comp}	r(RTM, comp)
				CER IF			
[11]C]MP	cau	1.01 ± 0.11	0.859	0.84 ± 0.08	0.894	0.84 ± 0.14	0.685
	put	1.02 ± 0.10	0.853	0.85 ± 0.07	0.854	0.84 ± 0.13	0.632
[11]C]DTBZ	cau	0.96 ± 0.05	0.968	0.88 ± 0.06	0.959	0.93 ± 0.09	0.888
	put	0.92 ± 0.06	0.968	0.89 ± 0.08	0.968	0.97 ± 0.14	0.889
[11]C]SCH	cau	1.03 ± 0.05	0.970	0.73 ± 0.08	0.881	0.71 ± 0.07	0.907
	put	1.02 ± 0.05	0.984	0.73 ± 0.08	0.892	0.72 ± 0.06	0.940
[11]C]RAC	cau	1.02 ± 0.01	0.994	NA	NA	NA	NA
	put	1.02 ± 0.01	0.998				
				OC IF			
[11]C]MP	cau	0.11 ± 0.11	0.845	0.79 ± 0.06	0.923	0.72 ± 0.09	0.782
	put	1.12 ± 0.03	0.980	0.80 ± 0.04	0.932	0.71 ± 0.03	0.942
[11]C]DTBZ	cau	1.00 ± 0.03	0.986	0.90 ± 0.04	0.960	0.86 ± 0.03	0.972
	put	0.99 ± 0.03	0.994	0.86 ± 0.04	0.983	0.88 ± 0.07	0.964
[11]C]SCH	cau	1.00 ± 0.02	0.990	0.78 ± 0.07	0.834	0.79 ± 0.08	0.796
	put	0.97 ± 0.03	0.988	0.78 ± 0.06	0.913	0.80 ± 0.06	0.921
[11]C]RAC	cau	0.99 ± 0.01	0.998	NA	NA	NA	NA
	put	0.99 ± 0.01	0.999				

[11]C]SCH very well. Visual inspection of the model fit to the data showed that ^{11}C]DTBZ kinetics are better described by a two-compartment model as already previously determined (Koeppe *et al.*, 1996). [11]C]DTBZ might, therefore, not satisfy the requirements of the RTM model. The ^{11}C]RAC kinetics are known to follow the one-compartment model well (Lammertsma and Hume, 1996).

B. Binding Potential Comparison

Table 1 shows the relative ratios of the BP values obtained with the different methods and the correlation coefficients. Excellent agreement between the numerical values of the BP obtained with RTM and Logan is observed for ^{11}C]SCH, ^{11}C]RAC, and ^{11}C]DTBZ with the OC IF and ^{11}C]MP with the cerebellar IF. The Logan method gives an approximately 5% lower values for ^{11}C]DTBZ with the cerebellar IF and 10% higher values for ^{11}C]MP with the OC IF. The correlation coefficient is >0.96 for all tracers except for ^{11}C]MP, where it is approximately 0.85. The compartmental model BP estimates were consistently higher that any of the tissue input methods, indicating that there might be some degree of specific binding in the reference regions.

C. BP Reliability and Standard Deviations

Table 2 shows the BP values together with the standard deviation between and within subject (STDB, STDW). For all four tracers the STDB and STDW are comparable for the two methods with the exception of the DTBZ putamen BP

value obtained with the cerebellar input function. The lower reliability of the RTM method was in this case due to a single outlier where the program failed to properly fit the data.

D. R_1 Correlation with the Compartmental Model, STDB, STDW, and Reliability

Results are shown in Table 3. There is very good correlation between the R_1 values obtained from the compartmental model (a one-compartment model fit was used for all three tracers). The STDB and STDW are always <10%, indicating that R_1 can be measured very accurately; the reliability, however, tends to be low with the exception of the values obtained for ^{11}C]SCH with cerebellar IF and caudate R_1 values obtained for MP with either input function.

E. k_2 STDB, STDW, and Reliability

k_2 cannot be determined directly from the compartmental model approach; therefore only STDB, STDW, and reliability were used as figures of merit. Table 4 shows the results. The STDB and STDW are generally higher compared to their values obtained for R_1. The reliability tends to be low with the exception of ^{11}C]SCH with the cerebellar IF and ^{11}C]DTBZ with the occipital cortex IF.

TABLE 2.   BP Estimate Reliability R, STDB, and STDW for the Two Methods (See Text)

Tracer	Method	Region	First scan	Second scan	STDB (%)	STDW (%)	R
				<CER IF>			
[11C]MP	Logan	cau	1.36	1.33	0.117	0.094	0.61
		put	1.06	1.03	0.092	0.100	0.46
	RTM	cau	1.37	1.35	0.200	0.129	0.70
		put	1.06	1.04	0.139	0.106	0.63
[11C]DTBZ	Logan	cau	0.545	0.527	0.140	0.137	0.51
		put	0.471	0.448	0.241	0.154	0.71
	RTM	cau	0.570	0.590	0.143	0.108	0.64
		put	0.510	0.604	0.239	0.468	0.21
[11C]SCH	Logan	cau	1.14	1.12	0.151	0.052	0.89
		put	0.860	0.885	0.234	0.054	0.95
	RTM	cau	1.11	1.09	0.182	0.062	0.90
		put	0.848	0.870	0.247	0.055	0.95
				<OCC IF>			
[11C]MP	Logan	cau	1.32	1.35	0.152	0.073	0.81
		put	1.03	1.05	0.143	0.073	0.79
	RTM	cau	1.19	1.19	0.132	0.102	0.63
		put	0.92	0.97	0.143	0.055	0.87
[11C]DTBZ	Logan	cau	0.792	0.772	0.110	0.078	0.67
		put	0.707	0.681	0.215	0.098	0.93
	RTM	cau	0.787	0.779	0.111	0.073	0.70
		put	0.714	0.690	0.205	0.096	0.82
[11C]SCH	Logan	cau	0.734	0.722	0.147	0.102	0.67
		put	0.532	0.564	0.272	0.124	0.85
	RTM	cau	0.738	0.724	0.136	0.117	0.57
		put	0.546	0.574	0.258	0.121	0.81

TABLE 3. R_1 STDB, STDW, and Reliability R

Tracer	IF	Region	R	STDB (%)	STDW (%)	R
[11C]MP	Cer	cau	0.92	0.074	0.038	0.79
		put	0.89	0.046	0.058	0.39
	OC	cau	0.93	0.078	0.028	0.88
		put	0.98	0.032	0.047	0.32
[11C]DTBZ	Cer	cau	0.97	0.044	0.084	0.22
		put	0.95	0.047	0.071	0.30
	OC	cau	0.95	0.037	0.074	0.20
		put	0.95	0.060	0.071	0.41
[11C]SCH	Cer	cau	0.82	0.081	0.032	0.96
		put	0.87	0.091	0.029	0.91
	OC	cau	0.73	0.033	0.069	0.19
		put	0.80	0.069	0.052	0.63

TABLE 4. k_2 STDB, STDW, and Reliability R

Tracer	IF	Region	STDB (%)	STDW (%)	R
MP	Cer	cau	0.086	0.135	0.29
		put	0.043	0.131	0.10
	OC	cau	N.A.	0.180	0.0
		put	0.026	0.138	0.03
[11C]DTBZ	Cer	cau	0.169	0.159	0.53
		put	0.111	0.154	0.34
	OC	cau	0.255	0.179	0.67
		put	0.260	0.148	0.76
[11C]SCH	Cer	cau	0.141	0.080	0.76
		put	0.161	0.102	0.71
	OC	cau	0.092	0.185	0.20
		put	N.A.	0.310	0.0

V. DISCUSSION

The BP values determined from both methods were in very good agreement for all four tracers, indicating that all tracers adequately satisfy the kinetic assumptions required by each model. This finding included [11C]DTBZ in spite of the fact that a two compartment model provided a better fit of the [11C]DTBZ data then the one-compartment model. MP with the occipital cortex input function showed the largest discrepancy in the BP values, with the Logan method giving about 10% higher estimates. This could be potentially due to the late equilibration of this tracer. The BP estimates were also comparable in terms of reproducibility and reliability. The RTM method has one additional degree of freedom compared to the Logan method and is thus expected to be more sensitive to the noise in the data. Nevertheless, it failed to give an accurate BP value in an isolated case only.

R_1, the second parameter determined by RTM was found to vary little between and within patients. This could mean that there is indeed limited variability between subjects or that the method is not sensitive enough to changes in R_1. R_1 is potentially an important parameter, since it provides information on the relative rate of tracer delivery from plasma to the free compartment between target and reference region. The rate of tracer delivery depends on blood flow and ligand extraction from plasma across the blood–brain barrier into the brain. Therefore, any local alterations of either blood flow or blood–brain barrier permeability that would alter R_1 could have an effect on the BP estimate. This study has been performed on normal volunteers, where no alterations in either blood flow or blood–brain barrier were present; more studies in situations where such alteration is known to occur are needed to establish the sensitivity of the method to changes in R_1 and their impact on the determination of the BP.

The last parameter determined by RTM, the tracer efflux rate k_2, showed higher variability between and within subjects, often with low reliability. The reliability was particularly low for [^{11}C]MP, which, again, could reflect the late equilibration of this tracer and consequent difficulty in accurately determining k_2. In the case of [^{11}C]SCH, where the reliability is high when the cerebellum is used as input function and low when the occipital cortex is used, the result could be explained by the fact that the cerebellum has a lower density of dopamine receptors compared to the occipital cortex and is therefore a more suitable area to be used as reference region. A similar explanation might be valid for [^{11}C]DTBZ where the k_2 reliability is higher when the OC time activity course is used as input function: the fact that the BP values are higher when the OC IF is used than when to the cerebellar IF is used might indicate that the degree of specific binding is higher in the cerebellum. The tissue input model requirement that the reference region should be devoid of specific binding is satisfied to a lesser degree with the cerebellar IF, thus potentially rendering the k_2 estimate less reliable.

VI. CONCLUSION

Tissue input data analysis methods are clearly preferred to those requiring plasma input, since they reduce the inva-

siveness and the complexity of the scanning procedure. The requirements imposed by the tissue input methods are however, generally more stringent then the plasma input ones. In this study we have examined the suitability of four reversible tracers, [^{11}C]MP, [^{11}C]DTBZ, [^{11}C]RAC, and [^{11}C]SCH, to satisfy the assumptions required by the Logan graphical method and by the reference tissue method by examining the tracer time–activity course and comparing the BP estimates obtained from the two methods in terms of values, standard deviations, and reliability. All four tracers were found to adequately satisfy the requirements of both methods with very good agreement between BP estimates from both methods. Standard deviations and reliabilities were comparable for both methods.

Investigation of the other two parameters provided by RTM R_1 and k_2 showed that R_1 had limited intra- and intersubject variability potentially reflecting the fact that the ratio of the target to reference influx constant does not vary greatly between subjects. k_2 however, was found to have a greater variability, with higher reliability only for [^{11}C]SCH with cerebellar input function and [^{11}C]DTBZ with occipital cortex IF.

Acknowledgments

The authors are very thankful to Dr. M. Schulzer and E. Mak for their help with the statistical analysis. We also are grateful for the radiotracer preparations by the PET chemistry group at TRIUMF and the scanning team at UBC. This research was supported the MRC Canada, the National Parkinson Foundation, and the Pacific Parkinson's Research Institute.

References

Chan, G., Holden, J. E., Stoessl, A. J., Doudet, D. J., Wang, Y., Dobko, T., Morrison, K. S., Huser, J., English, C., Legg, B., Schulzer, M., Calne, D. B., and Ruth, T. J. (1998). Reproducibility of the distribution of carbon-11-SCH 23390, a dopamine D1 receptor tracer, in normal subjects. *J. Nucl. Med.* **39**: 792–797.

Chan, G., Holden, E. J., Stoessl, A. J., Samii, A., Doudet, D. J., Dobko, T., Morrison, K. S., Adam, M., Schulzer, M., Calne, D. B., and Ruth, T. J. (1999). Reproducibility studies with 11C-DTBZ, a monoamine vesicular transporter inhibitor in healthy human subjects. *J. Nucl. Med.* **40**: 283–289.

Ding, Y. S., Sugano, Y., Fowler, J. S., and Salata, C. (1994). Synthesis of the racemate and individual enantiomers of [11C]methylphenidate for studying presynaptic dopaminergic neuron with positron emission tomography. *J. Label. Comp. Radiopharmaceut.* **34**: 989–997.

Farde, L., Ehrin, E., Erriksson, L., Greitz, T., Hall, H., Hedstrom, C. G., Litton, J. E., and Sedvall, G. (1985). Substituted benzamides as ligands for visualization of dopamine receptor binding in the human brain by positron emission tomography. *Proc. Natl. Acad. Sci. USA* **82**: 3863–3867.

Farde, L., Halldin, C., Stone-Elander, S., and Sedvall, G. (1987). PET analysis of human dopamine receptor subtypes using ^{11}C-SCH and ^{11}C-raclopride. *Psychopharmacology* **92**: 278–284.

Farde, L., Nordstrom, A. L., Wiesel, F.-A., Pauli, S., Halldin, C., and Sedvall, G. (1992). PET analysis of central D1 and D2 dopamine receptor occupancy in patients treated with classical neuroleptics and clozapine. *Arch. Gen. Psychiatry* **49**: 538–544.

Frey, K. A., Koeppe, R. A., Kilbourn, M. R., *et al.* (1996). Presynptic monoainergic vesicles in Parkinson's disease and normal aging. *Ann. Neurol.* **40**: 873–884.

Gilman, S., Frey, K. A., Koeppe, R. A., *et al.* (1996). Decreased striatal monoaminergic terminals in olivopontocerebellar atrophy and multiple system atrophy demonstrated with positron emission tomography. *Ann. Neurol.* **40**: 885.

Gunn, R. N., Lammertsma, A. A., Hume, S. P., and Cunningham, V. J. (1997). Parametric imaging of ligand-receptor binding in PET using a simplified reference region model. *Neuroimage* **6**: 279–286.

Henry, J. P., and Scherman, D. (1989). Radioligands of the vesicular monoamine trasporter and their use as markers of monoamine storage vesicles. *Biochem. Pharmacol.* **38**: 2395–2404.

Koeppe, R. A., Frey, K. A., Vander Borgth, T. M., *et al.* (1995). Kinetic evaluation of alpha-[C-11]dihydrotetrabenazine (DTBZ): A PET ligand for assessing the vesicular monoamine transporter. *J. Nucl. Med.* **36**: 118P. [abstract].

Koeppe, R. A., Frey, K. A., Vander Borght, T. M., *et al.* (1996). Kinetic evaluation of [^{11}C]dihydrotetrabenazine by dynamic PET: A measurement of vesicular monoamine transporter. *J. Cereb. Blood Flow Metab.* **16**: 1288–1299.

Lammertsma, A. A., Bench, C. J., Hume, S. P., Osman, S., Gunn, K., Brooks, D. J., and Frackoviak, R. S. J. (1996). Comparison of methods for analysis of clinical [11C] Raclopride Studies. *J. Cereb. Blood Flow Metab.* **16**: 42–52.

Lammertsma, A. A., and Hume, S. P. (1996). Simplified reference tissue model for PET receptor studies. *Neuroimage* **4**: 153–158.

Logan, J., Fowler, J. S., Volkow, N. D., Wang, G.-J., Ding, Y.-S., and Alexoff, D. L. (1996). Distribution volume ratios without blood sampling from graphical analysis of PET data. *J. Cereb. Blood Flow Metab.* **16**: 834–840.

Pletscher, A., Brossi, A., and Grey, K. F. (1962). Benzoquinoline derivatives: a new class of monoamine decreasing drugs with psychotropic action. *Int. Rev. Neurobiol.* **4**: 275–306.

Scheffe, H. (1959). "*The Analysis of Variance,*" pp. 221–260. Wiley, New York, NY.

Shinotoh, H., Inoue, O., Hirayama, K., *et al.* (1993). Dopamine D1 receptors in Parkinson's disease and striatonigral degeneration: A positron emission tomography study. *J. Neurol. Neurosurg. Psychiatry* **56**: 467–472.

Sossi, V., Oakes, T. R., and Ruth, T. J. (1998). A phantom study evaluating the quantitative aspect of 3D PET imaging in the brain. *Phys. Med. Biol.* **43**: 2615–2630.

Spinks, T. J., Jones, T., Bailey, D. J., Townsend, D. W., Grootoonk, S., Bloomfield, P. M., Gilardi, M.-C., Casey, M. E., Sipe, B., and Reed, J. (1992). Physical performance of a positron tomograph for brain imaging with retractable septa. *Phys. Med. Biol.* **37**: 1637–1655.

Suhara, T., Fukuda, H., Inoue, O., Suzuki, K., Yamasaki, T., and Tateno, Y. (1991). Age-related changes in human D1 receptors measured by positron emission tomography. *Psychopharmacology* **103**: 43–45.

Volkow, N. D., Ding, Y. S., and Fowler, J. S. (1995). Carbon 11-d-threomethylphenidate: A new PET ligand for the dopamine transporter. II: Studies in the human brain, *J. Nucl. Med.* **36**: 2162–2168.

22

Tracer Conversion Rate and Accuracy of Compartmental Model Parameter in Irreversibly Trapped Radiotracer Method

SHIN-ICHIRO NAGATSUKA,[*,†] **KIYOSHI FUKUSHI,**[*] **HIROKI NAMBA,**[*,‡] **HITOSHI SHINOTOH,**[*,§]
NORIKO TANAKA,[*,#] **SHUJI TANADA,**[*] and **TOSHIAKI IRIE**[*]

[*]*Advanced Technology for Medical Imaging, National Institute of Radiological Sciences, Chiba, Japan*
[†]*ADME/TOX Research Institute, Daiichi Pure Chemicals, Co., Ltd., Ibaraki, Japan*
[‡]*Department of Neurological Surgery, Hamamatsu University School of Medicine, Shizuoka, Japan*
[§]*Department of Neurology, Chiba University School of Medicine, Chiba, Japan*
[#]*Department of Neurological Surgery, Tokyo Women's Medical College, Tokyo, Japan*

Blood flow limitation of tracer delivery to tissues often affects accuracy and precision of kinetic parameter estimates, particularly in the k_3 parameter (tracer conversion rate) of irreversible tracers in tissues with high k_3. As an example of such irreversible radiotracers, $[^{11}C]$ N-methylpiperidinyl-4-acetate, which is metabolically trapped in the brain, was used in the present study. The extent of PET data error in the scanning conditions employed was estimated by analyzing relationships in radioactivity count, standard error of pixel data, and observed PET data error in regions of interest taken in emission images of a homogenous phantom and healthy human subjects. In several cerebral regions with different k_3, standard time–radioactivity curves were obtained from one particular subject using three-compartment kinetic model analysis with nonlinear least-squares method. The standard time–radioactivity curve and the estimated PET data error were used to generate 100 different time–radioactivity data sets in each region. Two analytical methods for kinetic parameter estimation, i.e., conventional nonlinear least-squares method and shape analysis that estimates k_3 without using input function, were then applied to the simulated data sets to evaluate the accuracy and precision of k_3 estimates. Both methods gave higher accuracy and precision of k_3 in regions with lower k_3. The dynamic range of k_3 estimation was slightly wider in the nonlinear least-squares method than in the shape analysis.

I. INTRODUCTION

Irreversibly binding radioligands or metabolically trapped radiosubstrates are widely used to measure receptor or enzyme density in living human tissue using positron emission tomography (PET) or single photon emission computed tomography. $[^{11}C]$N-Methylpiperidinyl-4-acetate ($[^{11}C]$ MP4A) used to map cerebral regional acetylcholinesterase (AChE) activity is an example of such irreversible radiotracers (Irie *et al.*, 1994, 1996; Iyo *et al.*, 1997; Namba *et al.*, 1998; Nagatsuka *et al.*, 1998a; Namba *et al.*, 1999; Shinotoh *et al.*, 1999) (Fig. 1). By parametric error estimation using covariance matrices derived from theoretical time–radioactivity curves (TAC) (Nagatsuka *et al.*, 1998b), we have shown that the accuracy and precision of conversion rate (k_3) of the irreversible tracer are severely affected by flow-limitation when k_3 is much higher than the efflux rate constant (k_2).

In the present study, we have estimated the extent of PET data error in our scanning conditions by analyzing emission images from a homogenous phantom and five normal volunteers. Kinetic parameters of irreversible trapping were obtained from one particular subject using nonlinear least-squares (NLLS) analysis based on the three-compartment model shown in Fig. 1. The regression equation obtained was used as the standard TAC, which was then given Gaussian random error based on the estimated PET data error to gen-

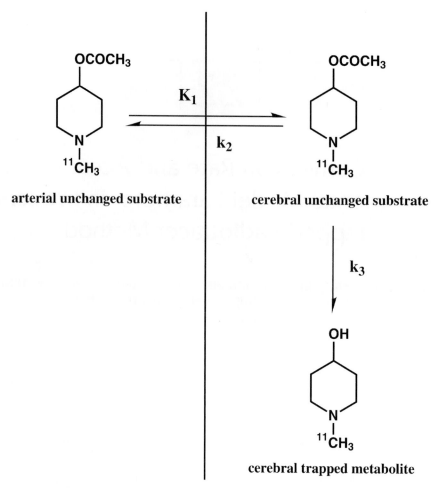

FIGURE 1.   Irreversible trapping of radioactive enzyme substrate in the brain.

erate 100 different time–radioactivity data sets for simulation study. NLLS analysis and shape analysis without using input function (Koeppe *et al.*, 1999) were then performed using each data set to evaluate the accuracy and precision of the k_3 parameter in several regions with different k_3/k_2 ratios.

II. MATERIALS AND METHODS

A Siemens ECAT EXACT47 scanner (Knoxville, TN) was used for two-dimensional PET scanning after 10 min transmission correction using ^{68}Ge–^{68}Ga external sources. Reconstruction of images using Hanning filter with cutoff intensity of 0.5 gave spatial resolution (full-width at half-maximum) of 6.0 mm transaxially and 5.4 mm axially, respectively, at the center of the field of view.

[^{11}C]MP4A was prepared as reported previously (Iyo *et al.*, 1997). Specific radioactivity of [^{11}C]MP4A was more than 18 TBq/mmol with radiochemical purity of more than 98%.

A cylindrical phantom supplied from Siemens for scanner maintenance (Ø 20 × 19 cm) was filled with ca. 0.123 MBq/mL of ^{11}C radioactivity in water followed by PET scanning according to the following 40 time frame sequence; 34×5 min, 1×4 min, 1×3 min, 1×2 min, 1×1 min, 1 × 30 s, and 1 × 10 s. Circular ROIs with ca. 3 cm diameter were taken over two planes (13 × 2 planes = 26). The center of each ROI was 6.5 cm away from the center of the phantom. The volume of ROI was 5.84 mL.

Five healthy male volunteers (ages 49–76 years) participated in the human PET study with written informed consent. Each subject was intravenously infused [^{11}C]MP4A (ca. 740 MBq/5 mL saline) over 1 min using an infusion pump. A series of PET scans composed of 14 time frames (3 × 20 s, 3 × 40 s, 1 × 1 min, 2 × 3 min and 5 × 6 min) was acquired over 40 min. ROIs were taken in the following 24 regions; superior frontal cortex, inferior frontal cortex, sensorimotor cortex, parietal cortex, temporal cortex, occipital cortex, hippocampus, amygdala, thalamus, striatum and cerebellar cortex in right and left hemisphere, respectively, and cingulate cortex and pons over both hemisphere. The ROI volume ranged from 1.08 to 7.40 mL.

In each ROI obtained from these studies, total disintegration number, Nd (Bq·s) was calculated from radioactivity concentration, ROI volume, tracer decay factor, and frame duration. The Nd was then converted to the reduced value, X (referred as count index), which is proportional to the coefficient of variation (COV) of radioactivity count in the ROI:

$$X = \frac{4000}{\sqrt{Nd}}.$$

In the homogenous phantom study, actual PET data error (E_{obs}) was observed as COV of 26 ROI data at each of 40 time points. The relationship between E_{obs} and count index was analyzed by linear regression analysis, giving the regression line of $E_{obs} = C1 \cdot X$.

For both phantom and human ROIs, another definition of PET data error, i.e., the standard error of pixel data (E_{pixel}), was calculated from the COV of pixel values in a ROI,

$$E_{pixel} = \frac{pixel\ COV}{\sqrt{pixel\ number}}.$$

By linear regression analysis as desribed above, two regression lines were obtained as $E_{pixel} = C2 \cdot X$ for phantom ROIs and $E_{pixel} = C3 \cdot X$ for human ROIs. The $C3/C2$, which shows the ratio of human ROI error to phantom ROI error, was used as the correction factor of the slope $C1$ obtained from the homogenous phantom study. As a practical error estimate for human ROI including both physical and biological errors, the error value E was calculated as follows and used in the simulation study;

$$E = \frac{C1 \times C3}{C2} X.$$

Kinetic parameters of tracer uptake (K_1), washout (k_2), and irreversible trapping (k_3) for left hemispheric temporal cortex, frontal cortex, hippocampus, thalamus, striatum, and cerebellum in one particular subject were obtained by nonweighted NLLS analysis of arterial input function and cerebral TAC as reported previously (Namba *et al.*, 1999). The NLLS regression equation for each region was used as the standard TAC to generate 100 different time–radioactivity data sets by giving Gaussian random error derived from the estimated PET data error; i.e.,

$$C(t) = C_s(t) \times (100 + E \times R)/100,$$

where $C(t)$ is generated time–radioactivity data for the time point t, $C_s(t)$ is time–radioactivity data of the standard TAC for corresponding time point, E is the estimated PET data error, R is normalized Gaussian random value (mean = 0, SD = 1). Each of 100 data sets was then subjected to nonweighted NLLS and shape analyses to estimate the accuracy and precision of kinetic parameters. Shape analysis was performed after the linear interpolation of eight data points be-

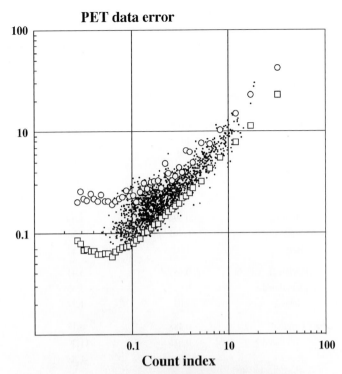

PET data error

FIGURE 2. Relationship between PET data error and count index. Larger open circles and open squares show E_{obs} and E_{pixel} obtained from the homogenous phantom study. Small dots show E_{pixel} obtained from the human PET study ($n = 1680$).

tween adjacent two time frames according to the following model (Koeppe *et al.*, 1999),

$$C(t_i) = C_s(t_i) + C_m(t_i)$$

$$C_m(t_i) = C_m(t_i - 1) + C_s(t_i - 1) \times \Delta t \times k_3,$$

where $C(t_i)$, $C_s(t_i)$, and $C_m(t_i)$ show cerebral radioactivity, unchanged substrate, and metabolite at time frame t_i, and Δt is duration of time frame. In this model, cerebral radioactivity at the first time frame was assumed to be solely unchanged substrate; i.e., $C(t_1) = C_s(t_1)$, $C_m(t_1) = 0$. The iterative calculations were done with changing k_3 until meeting the termination condition that cerebral radioactivity at the last time frame is solely metabolite without unchanged substrate, i.e., $C(t_n) = C_m(t_n)$, $C_s(t_n) = 0$.

III. RESULTS AND DISCUSSION

A. Estimation of PET Data Error

In the homogenous phantom study, actual PET data error, E_{obs} can be obtained by calculating COV of several independent ROI data taken in the wide homogenous area. Meanwhile, it is impossible to take many independent ROI in one particular region in the human PET study. One possible information relating to the extent of PET data error

TABLE 1. Kinetic Parameters of Tracer Uptake (K_1), Washout (k_2), and Irreversible Trapping (k_3) in the Original Subject Used to Lead the Standard TAC and 100 Times Simulation Study Using NLLS Analysis

Region	K_1 uptake		k_2 washout		k_3 trapping	
	mL/min/g	COV (%)	min^{-1}	COV (%)	min^{-1}	COV (%)
TC Original	0.532	1.84	0.135	7.91	0.092	7.74
Simulated	0.536	2.88	0.139	9.94	0.094	7.16
Ratio	1.01	1.57	1.03	1.26	1.02	0.93
FC Original	0.416	1.89	0.086	9.98	0.072	10.9
Simulated	0.420	3.34	0.090	13.9	0.074	11.5
Ratio	1.01	1.77	1.05	1.39	1.03	1.06
HP Original	0.394	4.36	0.093	24.4	0.095	23.8
Simulated	0.401	5.71	0.103	26.7	0.101	19.5
Ratio	1.02	1.31	1.11	1.09	1.06	0.82
TH Original	0.587	5.41	0.134	35.5	0.227	27.1
Simulated	0.603	7.43	0.160	41.1	0.243	24.1
Ratio	1.03	1.37	1.19	1.16	1.07	0.89
ST Original	0.488	6.18	0.112	70.8	0.429	48.5
Simulated	0.516	11.3	0.208	110	0.520	55.6
Ratio	1.06	1.83	1.86	1.55	1.21	1.15
CE Original	0.481	3.89	0.111	54.0	0.575	35.5
Simulated	0.515	11.8	0.307	136	0.662	57.4
Ratio	1.07	3.03	2.77	2.52	1.16	1.62

Note. The ratio shows the average value obtained from simulated data sets per the value of the standard TAC. Regions and their ROI volumes are FCl; frontal cortex (3.33 mL); TC, temporal cortex (5.15 mL); HP, hippocampus (1.64 mL); TH, thalamus (1.82 mL); CE, cerebellum (5.93 mL); ST, striatum (3.11 mL).

is the variation of pixel values, E_{pixel}. Despite the gap between E_{obs} and E_{pixel}, both error values are supposed to be correlated with each other because sources of variation for both values are mostly in common. Based on the assumption that PET data error is mainly governed by radioactivity count in the ROI, the relationship between these error values (E_{obs} and E_{pixel}) and count index was determined in the homogenous phantom and human PET studies (Fig. 2). The E_{obs} and E_{pixel} obtained from the homogenous phantom study were nearly proportional to the count index with a small intercept. This intercept is supposed to be due to the noise of reconstruction filter. The E_{pixel} obtained from the human PET study was proportional to the count index with almost negligible intercept. In the phantom study, the E_{pixel} was about one half of the E_{obs}. This difference might be explained by E_{pixel} depending on quite local data, i.e., the variation of pixel values in a small ROI region. While the E_{obs} shows the variation of 26 ROI data taken over the wide area of the homogenous phantom, which includes radial noise of the reconstruction filter. The count index in the human PET study ranged from 0.5 to 20. In this range, all error values could be approximated to be proportional to the count index. The slope for the E_{obs} obtained from the phantom

study, $C1$ was estimated to be 1.302 ± 0.024 ($r^2 = 0.984$, $p < 0.001$). This slope was corrected by using the ratio of slopes for the E_{pixel} obtained from the homogenous phantom and human PET studies. The slopes for the E_{pixel} were $C2 = 0.682 \pm 0.005$ ($r^2 = 0.997$, $p < 0.001$) in the phantom study and $C3 = 0.934 \pm 0.007$ ($r^2 = 0.798$, $p < 0.001$) in the human PET study. The slope for the E_{pixel} in the human PET study is about 1.37 times higher than that in the phantom study. The variation of the regression in the human PET study ($r^2 = 0.798$) was also larger than that in the phantom study ($r^2 = 0.922$). These results suggest that E_{pixel} in the human PET study includes effects of partial volume, heterogeneous radioactivity distribution or other error sources. In the present study, the practical PET data error was estimated as $E = C1 \cdot C3/C2X = 1.783X$.

B. Simulation study

Table 1 shows the kinetic parameters of tracer uptake (K_1), washout (k_2), and irreversible trapping (k_3) for the temporal cortex, frontal cortex, hippocampus, thalamus, striatum, and cerebellum in the original subject used to lead the standard TAC (original) and 100 times simulation study

TABLE 2. The Rate of Irreversible Trapping (k_3) in the Original Subject Used to Lead the Standard TAC and 100 Times Simulation Studies Using NLLS and Shape Analyses

Region	k_3/k_2 ratio	Original k_3 value		NLLS simulation		Shape simulation	
		min^{-1}	COV (%)	min^{-1}	COV (%)	min^{-1}	COV (%)
TC	0.67	0.092	7.74	0.094 (1.02)	7.16	0.086 (0.93)	7.27
FC	0.84	0.072	10.9	0.074 (1.03)	11.5	0.072 (1.00)	9.71
HP	1.02	0.095	23.8	0.101 (1.16)	19.5	0.090 (0.95)	15.0
TH	1.69	0.227	27.1	0.243 (1.07)	24.1	0.128 (0.56)	32.4
ST	3.83	0.429	48.5	0.520 (1.21)	55.6	0.383 (0.89)	159
CE	5.18	0.575	35.5	0.662 (1.15)	57.4	0.249 (0.43)	144

Note. The values in parentheses show accuracy rate calculated as the average k_3 obtained from simulated data sets over the k_3 of the original subject. Regions and their VOI volumes are the same as those shown in the legend for Table 1.

(simulated) using NLLS analysis. The COV of each original parameter was determined by the parametric method using the covariance matrix normalized by statistically expected average data variance, SS/$(n - m)$ (SS, sum of square error; n, number of time frame = 14; m, number of kinetic parameter = 3) (Carson, 1986). From the radioactivity concentration derived from the standard TAC, count index at each time frame was calculated to obtain the estimate of PET data error (Fig. 3). The estimated PET data error for the designated time point of the standard TAC was used to derive Gaussian random error to generate 100 different data sets of time–radioactivity for each of cerebral regions. The simulated data sets were subjected to nonweighted NLLS analyses to obtain the average and COV of kinetic parameters. The simulation study gave almost identical kinetic parameters and their COV as compared with the results from nonweighted NLLS analysis followed by calculation of the covariance matrix in the original subject. The estimation of PET data error in the present study was, therefore, valid and useful for nonparametric estimation of the accuracy and precision of kinetic parameters.

The shape analysis is a nonparametric method of estimating the rate of irreversible trapping of radiotracers. By using the 100 simulated data sets for each of cerebral regions, the shape analysis was also performed to obtain estimates of k_3 parameter and its COV. The results were compared with those obtained from the NLLS analysis of the original subject and simulation study using the NLLS analysis (Table 2). The accuracy rate of k_3 parameter was obtained as k_3 estimate of simulation study over k_3 of the standard TAC. The precision was evaluated by COV of k_3 obtained from the simulation study. In both simulations using NLLS and shape analyses, the accuracy and precision of k_3 parameter were worse in regions with higher k_3/k_2 ratio, suggesting that the flow-limitation of tracer distribution affected k_3 parameter estimation of irreversible radiotracers. The NLLS analysis gave good accuracy up to the thalamus with k_3/k_2 ratio of 1.69, while the shape analysis was valid up to the hippocampus with k_3/k_2 ratio of 1.02. The COV of k_3 in cortical regions and hippocampus was less than 20% in both NLLS and shape analyses. The NLLS analysis gave wider dynamic range of k_3 estimation as compared with the shape analysis. The computation of k_3 parameter by the shape analysis is, however, far easier than the NLLS analysis and quite reliable in cerebral cortical regions and hippocampus. Furthermore, the shape analysis, which does not require arterial input function, is very easy to apply and thought to be useful for the determination of k_3 parameter in regions with low k_3/k_2 ratio.

In conclusion, PET data error, which is indispensable for nonparametric estimation of kinetic parameter error by simulation study, could be roughly estimated from radioactivity

Radioactivity (nCi/mL)

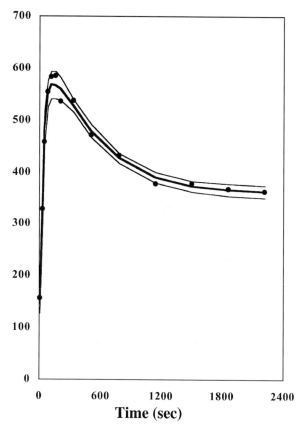

FIGURE 3. Standard time–radioactivity curve for the frontal cortex. Closed circles show observed PET data in the original subject used to lead the standard TAC. Thin lines above and below the TAC show range of standard deviation estimated from radioactivity count.

count in the ROI. The NLLS analysis of simulated data sets gave almost identical k_3 parameter and its COV as compared with the parametric calculation. The simulation studies using NLLS and shape analyses showed lower accuracy and precision of k_3 parameter in regions with higher k_3/k_2 ratio due to flow-limitation of tracer delivery. The dynamic range of k_3 estimation was slight wider in the NLLS analysis as compared with the shape analysis. The shape analysis was, however, thought to be of use for regions with lower k_3/k_2 ratio because of its simplicity. In irreversible radiotracers, it seems to be important to define their applicable range by examining the accuracy and precision of kinetic parameters.

Acknowledgments

The authors are grateful to Dr. T. Hasegawa, Kitasato University, and Drs. H. Murayama and H. Matsuura, National Institute of Radiological Sciences, for their help in homogenous phantom study. Dr. T. Nakajima, National Institute of Radiological Sciences is also acknowledged for his assistance in PET data acquisition.

References

Carson, R. E. (1986). Parameter estimation in positron emission tomography. *In* "*Positron Emission Tomography and Autoradiography: Principles and Applications for the Brain and Heart*" (M. Phelps, J. Mazziotta, and H. Schelbert, Eds.), pp. 347–390. Raven Press, New York.

Irie, T., Fukushi, K., Akimoto, Y., Tamagami, H., and Nozaki, T. (1994). Design and evaluation of radioactive acetylcholine analogs for mapping brain acetylcholinesterase (AchE) in vivo. *Nucl. Med. Biol.* **21**: 801–808.

Irie, T., Fukushi, K., Namba, H., Iyo, M., Tamagami, H., Nagatsuka, S., and Ikota, N. (1996). Brain acetylcholinesterase activity: Validation of a PET tracer in a rat model of Alzheimer's disease. *J. Nucl. Med.* **37**: 649–655.

Iyo, M., Namba, H., Fukushi, K., Shinotoh, H., Nagatsuka, S., Suhara, T., Sudo, Y., Suzuki, K., and Irie, T. (1997). Measurement of acetylcholinesterase by positron emission tomography in the brains of healthy controls and patients with Alzheimer's disease. *Lancet* **349**: 1805–1809.

Koeppe, R. A., Frey, K.A., Snyder, S. E., Meyer, P., Kilbourn, M. R., and Kuhl, D. E. (1999). Kinetic modeling of N-[^{11}C]methylpiperidin-4-yl propionate: Alternatives for analysis of an irreversible positron emission tomography tracer for measurement of acetylcholinesterase activity in human brain. *J. Cereb. Blood Flow Metab.* **19**: 1150–1163.

Nagatsuka, S., Namba, H., Iyo, M., Fukushi, K., Shinotoh, H., Suhara, T., Sudo, Y., Suzuki, K., and Irie, T. (1998a). Quantitative measurement of acetylcholinesterase activity in living human brain using a radioactive acetylcholine analog and dynamic PET. *In* "*Quantitative Functional Brain Imaging with Positron Emission Tomography*" (R. E. Carson, M. E. Daube-Witherspoon, and P. Herscovitch, Eds.), pp. 393–399. Academic Press, San Diego.

Nagatsuka, S., Fukushi, K., Namba, H., Iyo, M., Shinotoh, H., Tanada, S., and Irie, T. (1998b). Precision of kinetic parameters in irreversibly trapped radiotracers. *NeuroImage* **7**: A61.

Namba, H., Iyo, M., Fukushi, K., Shinotoh, H., Nagatsuka, S., Suhara, T., Sudo, Y., Suzuki, K., and Irie, T. (1999). Human acetylcholinesterase activity measured with PET: Procedure, normal value and effect of age. *Eur. J. Nucl. Med.* **25**: 135–143.

Namba, H., Iyo, M., Shinotoh, H., Nagatsuka, S., Fukushi, K., and Irie, T. (1998). Preserved acetylcholinesterase activity in aged cerebral cortex. *Lancet* **351**: 881–882.

Shinotoh, H., Namba, H., Yamaguchi, M., Fukushi, K., Nagatsuka, S., Iyo, M., Asahina, M., Hattori, T., Tanada, S., and Irie, T. (1999). Positron emission tomographic measurement of acetylcholinesterase activity reveals differential loss of ascending cholinergic systems in Parkinson's disease and progressive supranuclear palsy. *Ann. Neurol.* **46**: 62–69.

SECTION IV

PARAMETRIC IMAGING

A Reliable Unbiased Parametric Imaging Algorithm for Noisy Clinical Brain PET Data

DAGAN FENG,[*,†] **WEIDONG CAI,**[*,†] **and ROGER FULTON**[†,‡]

[*]Center for Multimedia Signal Processing, Department of Electronic and Information Engineering, Hong Kong Polytechnic University
[†]Biomedical and Multimedia Information Technology (BMIT) Group, Basser Department of Computer Science, The University of Sydney, Australia
[‡]Department of PET and Nuclear Medicine, Royal Prince Alfred Hospital, Australia

The generalized linear least squares (GLLS) algorithm for parameter estimation of nonuniformly sampled biomedical systems is a computationally efficient and statistically reliable way to generate parametric images for tracer dynamic studies with positron emission tomography (PET). However, for very noisy clinical PET data, parameter estimation using GLLS may not converge for every pixel. In this paper, we proposed an improved GLLS algorithm which can guarantee that the estimation converges for noisy pixel curves and the parameters estimated are within the physiological and pathological ranges. It has been investigated via clinical fluorodeoxyglucose-PET studies. The results showed that the parametric image generated by the improved GLLS is more smooth and reliable than that by the standard GLLS algorithm. Therefore, this new algorithm can provide more accurate parametric images in dynamic clinical brain PET studies.

I. INTRODUCTION

Biomedical parametric imaging, which requires the estimation of parameters for certain biosystems on a pixel-by-pixel basis, is an important technique providing image-wide quantification of physiological and biochemical function and visualization of the distribution of these functions corresponding to anatomic structures. A number of parametric imaging algorithms have been developed previously (Carson *et al.*, 1986; Huang *et al.*, 1980, 1987; Patlak *et al.*, 1983; Phelps *et al.*, 1986; Subramanyam *et al.*, 1978; Feng

et al., 1995). The steady-state method (Subramanyam *et al.*, 1978) uses a constant input of tracer, allowing the radioactivity concentrations in blood and tissue to reach an equilibrium. The autoradiographic approach (ARA) (Phelps *et al.*, 1986) has the advantage of using only one tissue concentration measurement although still requires a fully sampled arterial input function to estimate usually one parameter. In both the steady state and the ARA methods, parameter estimation is based on some assumptions which reduce the estimation accuracy, particularly when abnormal pathology is present. In the dynamic protocols, more than one unknown parameter can be estimated from a single input/single output (SISO) experiment. The nonlinear least squares (NLLS) algorithm (Huang *et al.*, 1980) can provide parameter estimates of optimum statistical accuracy. However, this NLLS method requires considerable computation time and good initial parameter values (without a good initial guess, NLLS will not converge). Therefore it is impractical for estimation of imagewide parameter estimates. Several alternative rapid parameter estimation schemes for certain specific dynamic PET data or model types have been proposed. For example, the integrated projection method can simultaneously estimate cerebral blood flow and distribution volume from the decay-uncorrected and -corrected PET data in a very efficient way (Huang *et al.*, 1982). The well-known Patlak graphical approach (PGA) can estimate the combination of the model rate constants, which allows for the determination of cerebral metabolic rate of glucose, when a unidirectional transfer process is dominant during the experimental period; i.e., the returning rate constant for the model used must

---

be assumed to be zero (Patlak *et al.*, 1983). Among these schemes, the weighted integration method (WIM) (Carson *et al.*, 1986) algorithm is more generally applicable. However, to increase the estimation reliability by predetermining the optimal sets of weighting functions for every pixel in the functional image is difficult. A generalized linear least squares (GLLS) algorithm (Feng *et al.*, 1996) has been proposed by our research group and is useful in image-wide parameter estimation to generate parametric images with PET. We have found that, compared with the existing algorithms, the GLLS algorithm: (1) can directly estimate continuous model parameters; (2) does not require initial parameter values; (3) is generally applicable to a variety of models with different structures; (4) can estimate individual model parameters as well as physiological parameters; (5) requires very little computing time; and (6) can produce unbiased estimates (Feng *et al.*, 1996). However, like most of the other parameter estimation algorithms (e.g., NLLS (Huang *et al.*, 1980)), due to the very noisy clinical PET data, parameter estimation using the GLLS may not converge to reasonable values within the expected physiological and pathological ranges for every pixel (noisy curve). Therefore, in this paper, we proposed an improved GLLS algorithm which can guarantee that the estimation converges for all pixels (noisy pixel curves) and the parameters estimated are within the physiological and pathological ranges.

II. MATERIALS AND METHODS

A. The Improved GLLS Algorithm

For simplicity and clarity, we developed the new algorithm using dynamic PET [^{18}F]-*2-fluoro-2-deoxy-D-glucose* ([^{18}F]FDG) images as a motivative example, although the algorithm can be used for other biomedical functional imaging. In this paper, the algorithm is based on an optimal image sampling schedule involving a smaller number of image frames with a five-parameter [^{18}F]FDG model (Cai *et al.*, 1998). The GLLS estimator can be derived as

$$\hat{\theta} = [\mathbf{Z}']^{-1}\mathbf{r}', \qquad (1)$$

where

$$\mathbf{r}' = [\psi_1 \otimes C_i^*(t_1'), \psi_1 \otimes C_i^*(t_2'), \psi_1 \otimes C_i^*(t_3'), \psi_1$$
$$\otimes C_i^*(t_4'), \psi_1 \otimes C_i^*(t_5')]^{\mathbf{T}}, \qquad (2)$$

$\theta = [P_1, P_2, P_3, P_4, P_5]^{\mathbf{T}}$, and $C_p^*(t)$ is the [^{18}F]FDG concentration in plasma represented by the plasma time–activity curve (PTAC), $C_i^*(t)$ is the total [^{18}F]FDG concentration in tissue, or tissue time–activity concentration curve (TTAC). Once estimates of the macroparameters are obtained, the microparameters, i.e., the rate constants for the [^{18}F]FDG model, can be calculated as

$$\hat{k}_1^* = \frac{P_1 P_4 + P_2}{1 - P_1}, \qquad \hat{k}_2^* = -\frac{P_1 P_5 + P_3}{P_1 P_4 + P_2} - P_4,$$

$$\hat{k}_3^* = -(\hat{k}_2^* + \hat{k}_4^* + P_4), \qquad \hat{k}_4^* = -\frac{P_5}{\hat{k}_2^*}, \qquad \hat{k}_5^* = P_1. \quad (4)$$

Then the local cerebral metabolic rate of glucose (*LCMRGlc*) in the human brain can be calculated from

$$LCMRGlc = (C_p/LC)(k_1^* k_3^*)/(k_2^* + k_3^*), \qquad (5)$$

where *LC* denotes the lumped constant, and C_p is the "cold" glucose concentration in plasma (Feng *et al.*, 1995). Details of the GLLS can be found in (Feng *et al.*, 1996; Cai *et al.*, 1998).

In this new algorithm, a clustering technique (Ho *et al.*, 1997) was employed to determine the lower/upper bounds of parameter estimation at various pixels. Many clustering algorithms have been developed (Bow, 1984; Ciaccio *et al.*, 1994), which can be divided into direct (constructive) or indirect (optimization) ones, depending on whether a criterion measure is used during cluster analysis. The direct algorithms perform clustering without the necessity of a criterion measure, whereas indirect algorithms use the criterion measure to optimize clustering. Clustering algorithms can be further classified as agglomerative or divisive, according to whether classification is in a top-down or bottom-up direction. In this paper, we use an indirect agglomerative clustering algorithm based on the traditional Euclidean distance criterion measure. In general, a time–activity curve (TAC) can be obtained from each pixel. However, many TACs have similar kinetics. The clustering technique can be used to

$$\mathbf{Z}' = \begin{bmatrix} \psi_1 \otimes C_p^*(t_1') & \psi_2 \otimes C_p^*(t_1') & \frac{1}{\lambda_1 \lambda_2}\int_0^{t_1'} C_p^*(\tau)\,d\tau + \psi_3 \otimes C_p^*(t_1') & \psi_2 \otimes C_i^*(t_1') & \frac{1}{\lambda_1 \lambda_2}\int_0^{t_1'} C_i^*(\tau)\,d\tau + \psi_3 \otimes C_i^*(t_1') \\ \psi_1 \otimes C_p^*(t_2') & \psi_2 \otimes C_p^*(t_2') & \frac{1}{\lambda_1 \lambda_2}\int_0^{t_2'} C_p^*(\tau)\,d\tau + \psi_3 \otimes C_p^*(t_2') & \psi_2 \otimes C_i^*(t_2') & \frac{1}{\lambda_1 \lambda_2}\int_0^{t_2'} C_i^*(\tau)\,d\tau + \psi_3 \otimes C_i^*(t_2') \\ \psi_1 \otimes C_p^*(t_3') & \psi_2 \otimes C_p^*(t_3') & \frac{1}{\lambda_1 \lambda_2}\int_0^{t_3'} C_p^*(\tau)\,d\tau + \psi_3 \otimes C_p^*(t_3') & \psi_2 \otimes C_i^*(t_3') & \frac{1}{\lambda_1 \lambda_2}\int_0^{t_3'} C_i^*(\tau)\,d\tau + \psi_3 \otimes C_i^*(t_3') \\ \psi_1 \otimes C_p^*(t_4') & \psi_2 \otimes C_p^*(t_4') & \frac{1}{\lambda_1 \lambda_2}\int_0^{t_4'} C_p^*(\tau)\,d\tau + \psi_3 \otimes C_p^*(t_4') & \psi_2 \otimes C_i^*(t_4') & \frac{1}{\lambda_1 \lambda_2}\int_0^{t_4'} C_i^*(\tau)\,d\tau + \psi_3 \otimes C_i^*(t_4') \\ \psi_1 \otimes C_p^*(t_5') & \psi_2 \otimes C_p^*(t_5') & \frac{1}{\lambda_1 \lambda_2}\int_0^{t_5'} C_p^*(\tau)\,d\tau + \psi_3 \otimes C_p^*(t_5') & \psi_2 \otimes C_i^*(t_5') & \frac{1}{\lambda_1 \lambda_2}\int_0^{t_5'} C_i^*(\tau)\,d\tau + \psi_3 \otimes C_i^*(t_5') \end{bmatrix}. \quad (3)$$

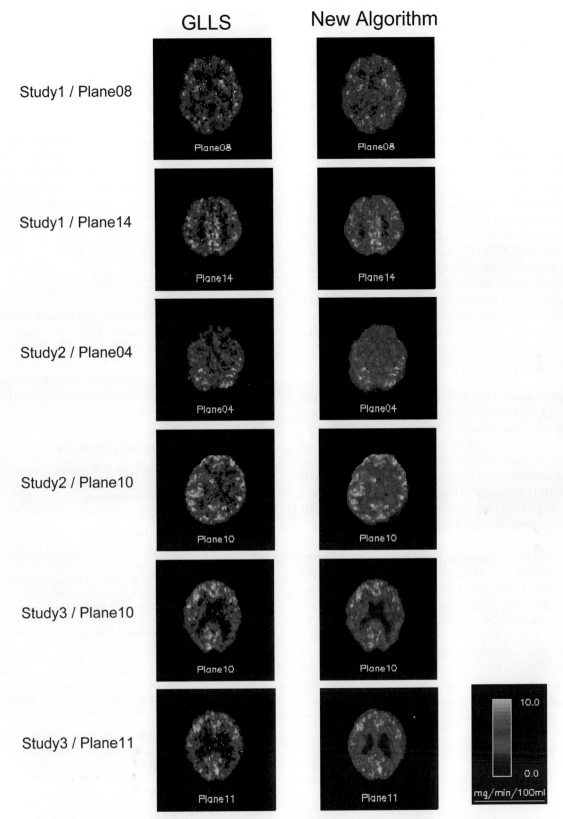

FIGURE 1. Illustration of the difference between the parametric image of local cerebral metabolic rates of glucose (LCMRGlc) generated from the GLLS algorithm and from the new algorithm. The unit of LCMRGlc is mg/min/100 mL.

classify image-wide TACs, $C_i(t)$ (where $i = 1, 2, \ldots, R$, and R is the total number of image pixels), into S cluster groups C_j (where $j = 1, 2, \ldots, S$, and $S \ll R$) by measurement of the magnitude of natural association (similarity characteristics). It is expected that TACs with high degrees of natural association will belong to the same cluster groups, and conversely, TACs with low degrees will belong to different groups (Feng *et al.*, 1996; Cai *et al.*, 1998). For clustering to be valid, each TAC must be assigned uniquely to a cluster group (i.e., no TAC is allowed to belong to two different groups). The indexed image will map each pixel into a particular cluster. The respective temporal information for each cluster group will be contained in a look-up table (LUT). The LUT will be sequentially indexed by cluster group and each index will contain the mean TAC cluster values for that group. For each TAC group, the maximum and minimum values of parameter estimates can be precalculated according to certain rules based on the mean TAC, including the consideration of the physiological and pathological ranges. For each pixel parameter within the group, the estimation is forced to be converged within the upper and lower bounds.

B. Clinical FDG-PET Studies

Clinical dynamic FDG-PET studies were carried out at the National PET/Cyclotron Center, Taipei Veterans General Hospital, Taiwan, using a PC4096-15WB PET scanner (GE/Scanditronix) which has 8 detector rings and 15 slices. This scanner contains 4096 detectors with axial and transaxial resolutions of 6.5 mm full width at half maximum (FWHM). Between 200 and 400 MBq (approximately 0.5 mg) of [^{18}F]FDG was injected intravenously and serial arterial blood samples (each 2–3 ml) were taken at 0.25, 0.5, 0.75, 1, 1.25, 1.5, 1.75, 2, 2.5, 3, 3.5, 7, 10, 15, 20, 30, 60, 90, and 120 min postinjection. These samples were immediately placed on ice and the plasma was separated for the determination of plasma [^{18}F]FDG and glucose concentrations. PET scanning was performed according to a schedule which consisted of 22 temporal frames: 10×0.2-, 2×0.5-, 2×1-, 1×1.5-, 1×3.5-, 2×5-, 1×10-, and 3×30-min scans. The scanning was completed within 120 min of tracer injection. The PET data were corrected for attenuation and decay-corrected to the time of injection and then reconstructed using filtered backprojection with a Hanning filter. The reconstructed images were 128×128 with pixel size of 2×2 mm.

III. RESULTS AND DISCUSSION

The parametric image of the local cerebral metabolic rates of glucose (LCMRGlc) generated by the improved GLLS algorithm was compared with that generated by the standard GLLS. The resultant physiological parametric images for six different planes from three clinical [^{18}F]FDG-PET studies are shown in Fig. 1. From the images, we can see the parametric images generated by the improved GLLS algorithm are more smooth and reliable that those by the standard GLLS algorithm, in which there are some pixels with diverged estimates. In contrast with other kinds of images, dynamic PET images have a consistent general structure consisting of an approximately oval region containing almost all of the information of interest. In our new algorithm, background pixels were neglected, which greatly reduced computing time of parameter estimation, and removed background noise. Therefore, this new algorithm can provide more accurate parametric images in dynamic clinical brain PET studies.

IV. CONCLUSIONS

An improved GLLS algorithm, which can generate more smooth and reliable parametric images, has been presented. It has been investigated via clinical [^{18}F]FDG-PET studies. Our results demonstrated that this new parametric imaging algorithm can provide more accurate parametric images and is potentially very useful in dynamic clinical brain PET studies.

Acknowledgments

This research is supported by the RGC and ARC Grants. The authors are grateful to the National PET/Cyclotron Center, Taipei Veterans General Hospital, Taiwan, for their kind assistance.

References

Bow, S. T. (1984). *"Pattern Recognition: Applications to Large Data-Set Problems."* Dekker, New York.

Cai, W., Feng, D., and Fulton, R. (1998). A fast algorithm for estimating FDG model parameters in dynamic PET with an optimised image sampling schedule and corrections for cerebral blood volume and partial volume. *"Proceedings of the 20th Annual International Conference of IEEE Engineering in Medicine and Biology Society,"* Oct. 29–Nov. 1, 1998, Hong Kong, pp. 767–770.

Carson, R. E., Huang, S. C., and Green, M. E. (1986). Weighted integration method for local cerebral blood flow measurements with positron emission tomography, *J. Cereb. Blood Flow Metab.* **6**: 245–258.

Ciaccio, E. J., Dunn, S. M., and Akay, M. (1994). Biosignal pattern recognition and interpretation systems: Methods of classification. *IEEE Eng. Med. Biol.* **13**: 129–135.

Feng, D., Huang, S. C., Wang, Z., and Ho, D. (1996). An unbiased parametric imaging algorithm for non-uniformly sampled biomedical system parameter estimation. *IEEE Trans. Med. Imaging.* **15**: 512–518.

Feng, D., Ho, D., Chen, K., Wu, L. C., Wang, J. K., Liu, R. S., and Yeh, S. H. (1995). An evaluation of the algorithms for determining local cerebral metabolic rates of glucose using positron emission tomography dynamic data. *IEEE Trans. Med. Imaging.* **14**: 697–710.

Ho, D., Feng, D., and Chen, K. (1997). Dynamic image data compression in spatial and temporal domains: Theory and algorithm. *IEEE Trans. Info. Tech. Biomed.* **1**(4): 219–228.

Huang, S. C., Phelps, M. E., Hoffman, E. J., Sideris, K., Selin, C. J., and Kuhl, D. E. (1980). Non-invasive determination of local cerebral metabolic rate of glucose in man. *Am. J. Physiol.* **238**: E69–E82.

Huang, S. C., Carson, R. E., and Phelps, M. E. (1982). Measurement of local blood flow and distribution volume with short-lived isotopes: A general input technique. *J. Cereb. Blood Flow Metab.* **2**: 99–108.

Patlak, C. S., Blasberg, R. G., and Fenstermacher, J. (1983). Graphical evaluation of blood to brain transfer constants from multiple-time uptake data. *J. Cereb. Blood Flow Metab.* **3**: 1–7.

Phelps, M. E., Mazziotta, J., and Schelbert, H. (1986). *"Positron Emission Tomography and Autoradiography: Principles and Applications for the Brain and Heart."* Raven Press, New York.

Subramanyam, R., Alpert, N. M., Hoop, B., Brownell, G. L., and Taveras, J. M. (1978). A model for regional cerebral oxygen distribution during continuous inhalation of ^{15}O, $C^{15}O$ and $C^{15}O_2$. *J. Nucl. Med.* **19**: 48–53.

Smooth Variance Maps Using Parameter Projections

R. P. MAGUIRE, K. L. LEENDERS, and N. M. SPYROU

Groningen NeuroImaging Program, University and University Hospital Groningen, The Netherlands
Department of Physics, University of Surrey, Guildford, UK

Previously, we have used the concept of parameter projections to show that parametric images of physiological parameters can be calculated directly from raw sinogram data, using a linear pharmacokinetic model, as opposed to the activity image based approach which requires reconstruction of activity concentration images. As with all parameter estimation techniques, it is important to have an estimate of the inaccuracy of the estimated variables. This paper shows an application of variance estimation techniques drawn from standard literature on tomographic variance estimation to the estimation and reconstruction of parametric images using parameter projections. The method is shown to be more precise, in the estimation of the individual pixel variance, than either the method based on reconstructed images or variance calculation based on comparing pixels between reconstructed sets of parametric images. A method based solely on a reconstructed image cannot recover information, contained in neighboring pixels, which is necessary in order to make a good estimate of the variance. This information is suppressed during reconstruction. The method has the same computational advantages for the calculation of variances as does the original method for the calculation of parametric images.

I. INTRODUCTION

The spatial and kinetic aspects of PET measurements are succinctly summarized in parametric images. These are normally calculated from reconstructed images of activity concentration. However, we have shown previously (Maguire et al., 1997) that it is possible with certain restrictions to apply linear pharmacokinetic models to the projection data. This has obvious advantages for computational efficiency, since only one set of images per parameter of interest needs to be reconstructed, instead of an entire time series of activity concentration images.

In this paper we address the characteristics of parameter estimation on projection data, with respect to the estimation of the individual pixel variance of parametric images. These variance estimates are crucial to the application of image statistical analyses to determine loci of differences between respective studies. At the outset it is noted that the literature (Huesman, 1984) has stressed the requirement to take into account the covariance between neighboring pixels when estimating variance in a region of interest. However, we are equally aware of methods for correcting for interpixel correlations, and the aim of this paper will be to address the issue of estimating smooth pixel variance maps, which are required for that technique.

II. THEORY

The concept of parameter projections can be briefly summarized as follows. Consider any linear pharmacokinetic model, and note that linearity may be achieved by transforming the data (Gjedde, 1982). It can be written as

$$\mathbf{C} = \mathbf{XP}, \qquad (1)$$

where $\mathbf{C}$ is a vector of activity concentrations in the subject, at times t_i, $\mathbf{X}$ is a matrix of basis functions $f_j(t_i)$, and $\mathbf{P}$ is a vector of parameters. If the radon transform of the activity concentration at a time t is considered, then for a projection in a rotated coordinate system $[x_r, y_r]$ we can write

$$\lambda(x_r, t) = \int_l C_t(\mathbf{r}, t) dy_r, \qquad (2)$$

where $\mathbf{r}$ is a vector locating an element of tissue in the subject in space.

Integrating Eq. (1) along direction y_r and substituting the left-hand side into the right hand side of (2), the following can be derived,

$$\Lambda = X\Pi, \qquad (3)$$

where Λ is a vector of the radon transforms of the object at times t_i, and Π is a vector of parameter transforms, that is radon transforms of the parameters of the model, which can be backprojected to form parameteric images.

The parameter transforms can be estimated from the measured data by performing linear regression analysis on the projection data and may be reconstructed using any technique. It is also possible to estimate the variance of the parameter projections, using standard techniques in linear regression, and these can be backprojected using methods described in the literature for the estimation of individual pixel variance in emission computed tomography (Alpert *et al.* 1982), using filtered backprojection,

$$\sigma^2_{c(\mathbf{r})} = \left(\frac{\pi}{m}\right)^2 \int d\phi \left(\{\sigma^2_{\lambda(x_r)} A^2(x_r)\} \otimes h^2(x_r)\right)_{x_r = |\mathbf{r}| \cos(\theta) - \phi}, \qquad (4)$$

where $\sigma^2_{c(\mathbf{r})}$ and $\sigma^2_{\lambda(x_r)}$ are the pixel variance estimation and the projection variance estimation, respectively, and m is the number of projection angles. Further symbolism is elucidated in the original paper. In the original application of this backprojection technique, it is assumed that a noiseless attenuation correction factor $\mathbf{A}$, is known for each projection. The technique can, however, be applied using any filter $\mathbf{h}$.

III. METHOD

An implementation of Eq. (4) was explored using simulations of an object 20 cm in diameter with two inserts, 8 and 2 cm in diameter (Fig. 1). The activity concentration in the central compartment was governed by the following equation,

$$C(\mathbf{r}, t_i) = P_1(\mathbf{r}, t_i) t_i C_p + P_2(\mathbf{r}, t_i) C_p, \qquad (5)$$

where P_1 and P_2 are model parameters, and C_p can be thought of as analogous to the activity concentration level of a continuous infusion.

The activity concentration in the two inserts "A" and "B" was 20% higher than that in compartment "C," so that the parameters $[P_1, P_2]$ were also increased by 20% in those areas.

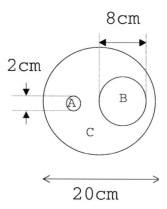

FIGURE 1. Simulated object, containing two inserts, 8 and 2 cm in diameter.

A set of 100 time series of acquisition data was simulated. Each time series, equivalent to a dynamic acquisition, consisted of a set of 10 images with $t = 1, 2, 3 \ldots 10$. C_p was chosen such that the full projection set contained 1.5×10^6 counts at t_5. P_1 and P_2 were fixed at 0.1 and 0.5, in the central compartment.

Simulations were repeated, without any difference in the activity concentrations in the three compartments, to allow a comparison between the two configurations—with and without a contrast between the compartments. The comparison was done by creating statistical parametric maps for the difference of the mean images from each configuration for different numbers of simulations. The reconstruction filter was varied to assess the impact of smoother reconstructed images.

Parametric images were estimated using both linear regression of individual pixels in reconstructed activity images, and by applying estimating parameter projections. Variance estimates were similarly performed on reconstructed images, or while estimating the parameter projections. Additionally the individual pixel variances were estimated by comparing across the 100 sets of parametric images as a direct verification of the other methods.

IV. RESULTS

To compare the estimates of the individual pixel estimates for one simulation—equivalent to one PET study—a 3-mm-wide, 1-pixel-width profile was calculated in the variance images estimated from *one* run (selected randomly), these are displayed in Fig. 2.

The estimates based on parameter projections for a single run are spatially smoother than the estimates based on images. This is also true for the estimated variance of the mean of 100 runs as shown in Fig. 3; additionally here the direct estimate across runs (group estimate) is given as a validation of the accuracy of the estimates.

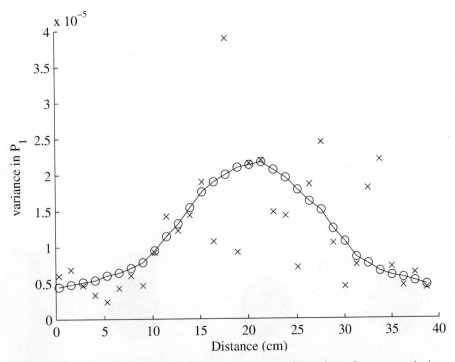

FIGURE 2. Profiles 3 mm through the individual pixel estimates of the variance of parameter projections of P_1. Estimates based on parameter projections ($\bigcirc$) and on reconstructed images ($\times$).

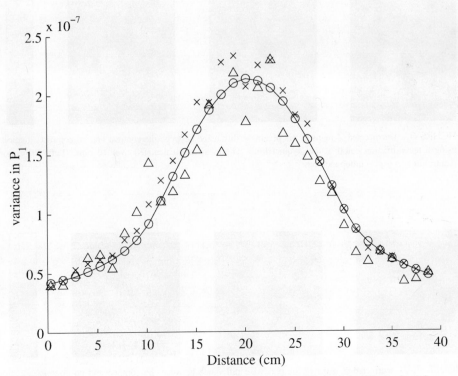

FIGURE 3. Estimates of the variance of the mean of P_1 taken over 100 simulations for the dame 3-mm profile as in Fig. 2. Group estimates ($\triangle$) parameter projection estimates ($\bigcirc$) and image estimates ($\times$).

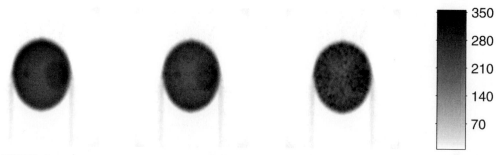

FIGURE 4. Estimated local signal-to-noise ratio for the mean image (P_1) of a set of 100 runs. (Left) Projection estimation; (center) image estimation; (right) group estimation.

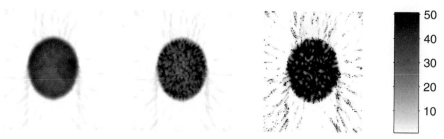

FIGURE 5. Estimated local signal-to-noise ratio for the mean image (P_1) of a set of two runs. (Left) Projection estimation; (center) image estimation; (right) group estimation.

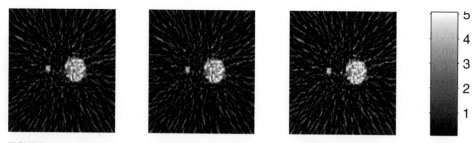

FIGURE 6. Uncorrected Z-maps of the pixelwise difference between the contrast and no contrast situations, using a reconstruction cutoff at the Nyquist limit. The mean was calculated over 10 runs. (Left) Projection; (center) image; (right) group methods.

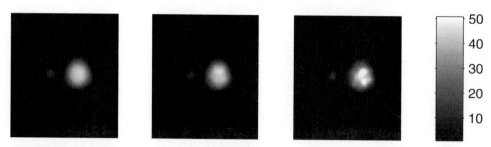

FIGURE 7. Uncorrected Z-maps of the pixelwise difference between the contrast and no contrast situations, using a reconstruction cutoff at 0.2 times the Nyquist limit. The mean was calculated over 10 runs. (Left) Projection; (center) image; (right) group methods.

Figures 4 and 5 show the estimated local signal-to-noise ratio for images from each of the three methods, reconstructed with a hanning filter, with the cutoff set at 0.5 Nyquist limit. In Fig. 4, the SNR of the mean of 100 simulations is considered, and in Fig. 5, the mean of only 2 is considered. Clearly the precision of the projection- based estimates is highest, although it is not biased with respect to the other estimation methods.

Surprisingly, the group method, which is a direct estimate of the SNR in the parametric images, yields the poorest results. Finally, in Figs. 6 and 7, Z-maps of the difference between the phantom with contrast and without are given for a grouping of 10 simulations. In the image reconstructed with the lower signal-to-noise ratio (reconstruction up to the Nyquist limit) the differences between the Z-maps from the three methods are not apparent. However, if the cutoff is reduced to 0.2 Nyquist, so that the images have smoothness on the order of 10 mm FWHM, then the higher precision estimates of the projection method result in a much smoother map.

V. DISCUSSION

Calculation of both the parametric images, and their associated variances using projection data has the advantage of avoiding the reconstruction of large time series of data, which is especially important in 3D imaging. Applying standard literature methods to estimate the variance of a parameter projection and backprojecting using methods described in the literature allows this.

The estimates of variance calculated in this way are in agreement with those calculated using reconstructed image data and those estimated by comparing between a large series of parametric images. However, the estimates are precise. This is because the method inherently uses all of the information available in the calculation.

In this work we have used the assumption of a perfectly calculated attenuation estimate. This is not always available, although contour fitting methods may give a noiseless approximation. The issue of transmission scan correction has not been addressed here.

When calculating Z-maps, the differences between the methods are most apparent when the signal-to-noise ratio of the parametric images is high and when the number of runs is low. It should be recalled that this method is still applicable when there is only one measurement. Naturally we have not included any intersubject variability in our simulations; however, this is a very important factor, and perhaps the dominant factor, in the estimation of variance in human studies. It remains to be seen how the method can contribute to improved parametric map estimation in the context of high intersubject variability.

References

Alpert, N. M., Chesler, D. A., Correia, J. A., Ackerman, R. H., Chang, J. Y., Finkelstein, S., and Davis, S. M. (1982). Estimation of the local statistical noise in emission computed tomography. *IEEE Trans. Med. Imaging* **MI-1**: 142–146.

Gjedde, A. (1982). Calculation of cerebral glucose phosphorylation from brain uptake of glucose analogs in vivo: A re-examination. *Brain Res.* **257**: 237–274.

Huesman, R. H. (1984). A new fast algorithm for the evaluation of regions of interest and statistical uncertainty in computed tomography. *Phys. Med. Biol.* **29**: 543–552.

Maguire, R. P., Calonder, C., and Leenders, K. L. (1997). An investigation of multiple time/ Graphical analysis applied to projection data: Theory and validation. *J. Catal.* **21**: 327–331.

Model Fitting with Spatial Constraint for Parametric Imaging in Dynamic PET Studies

YUN ZHOU, SUNG-CHENG HUANG, and MARVIN BERGSNEIDER

Division of Nuclear Medicine and Biophysics, Department of Molecular & Medical Pharmacology, UCLA School of Medicine, Los Angeles, California 90095

In PET dynamic studies, the parametric images estimated by conventional nonlinear least square (NLS) method usually have high variability. Ridge regression with spatial constraint (RRSC) was proposed in the present study. The penalty of spatial variation of parameters was added to the cost function of NLS for RRSC. Ridge regression theory was applied at each iteration step. For RRSC evaluation, a three-compartment, five-parameter fluorodeoxyglucose (FDG) model was used in computer simulation and human study. Results of computer simulations show that the mean square error of estimates can be decreased 50–90% compared to that seen with NLS by applying RRSC methods at middle or high noise level (6 or 10 million counts per plane over a 60-min dynamic study). Applied to real data, RRSC generated good quality parametric images in FDG dynamic studies, K_i images estimated by RRSC and graphical method (i.e., Patlak plot) have high correlation ($R^2 = 0.98$). RRSC, thus, can significantly increase the reliability of the estimated model parameters, and is ideally suited for generating parametric images.

I. INTRODUCTION

In PET dynamic studies, the measured PET kinetic data includes spatial and temporal information, which can be used to evaluate the biochemical and physiological status of tissue. The parameters which describe the delivery, transport, and biochemical transformation of a tracer can be estimated by fitting the model to the tissue kinetics. Due to the high noise level of pixel kinetics in PET dynamic studies, pixelwise microparameters of tracer kinetics estimated by

conventional nonlinear least square (NLS) method usually have high variability. Therefore, linear regression methods, such as Patlak plot (Patlak *et al.*, 1983), weighted integration (Carson *et al.*, 1986), and generalized linear least square (GLLS) (Chen *et al.*, 1996), have been proposed to generate parametric images based on some linearity approximation. We have proposed a spatially coordinated imagewise model fitting (SCMF) (Huang and Zhou, 1998) method to incorporate spatial constraint into the imagewise nonlinear regression process. In that algorithm, Marquardt algorithm was used to minimize the cost function and a simple linear spatial smoothing is applied to estimates at each iteration. In this study, ridge regression with spatial constraint (RRSC) is performed in each iteration, so spatial smoothing on parameters is less stringent. This method was evaluated with computer simulation and human fluorodeoxyglucose (FDG) dynamic studies.

II. MATERIALS AND METHODS

A. Theory

At each iteration of RRSC, the cost function to determine the step ΔP is based on minimizing the cost function,

$$Q_{\text{RRSC}}(\Delta P | \lambda)$$
$$= (Y - F(P_0 + \Delta P))'W(Y - F(P_0 + \Delta P))$$
$$+ \lambda(P_0 + \Delta P - SP_0)'(P_0 + \Delta P - SP_0), \quad (1)$$

where $Y(n \times 1)$ is output measurement, $P(m \times 1)$ is the parameter vector to be estimated. W is diagonal matrix and its diagonal element $w_{ii} = $ (duration of ith frame of PET dynamic scanning). SP_0 is spatially smoothed P_0, where P_0 is the estimated result in a previous iteration step. $F(P_0 + \Delta P)$ is the predicted value from the tracer kinetic model and is approximated by first order Taylor's expansion as

$$F(P_0 + \Delta P) \approx F(P_0) + \frac{dF}{dP}\Delta P, \qquad (2)$$

where $dF/dP(n \times m)$ is the gradient of F at P_0 on the parameter space. λ is a ridge parameter that is equal to C^*h, where h is equal to Sh_0 with h_0 being a value determined based on simple ridge regression with spatial constraint (Zhou et al., IEEE MIC'99 conference record). Sh_0 is a spatially smoothed h_0, and C is a constant over all iterations. So, the updating formula for ΔP at each iteration can be derived by substituting Eq. (2) to Eq. (1) and then minimizing it w.r.t. ΔP. The result is

$$\Delta P = \left(\left(\frac{dF}{dP}\right)' W \frac{dF}{dP} + \lambda I_m \right)^{-1}$$
$$\times \left(\left(\frac{dF}{dP}\right)' W(Y - F(P_0)) + \lambda(SP_0 - P_0) \right). \qquad (3)$$

The initial estimates were estimated from the whole brain average time–activity curve by a Marquardt algorithm. The estimates of RRSC are close to those of NLS as C tends to zero. It is also easy to see that estimates of RRSC are close to SCMF as C tends to infinity. The smoothing filter used in the present study was [1 2 4 2 1]/10 in both x- and y-directions.

B. Computer Simulation Study

A three-compartment, five-parameter (K_1 (the forward transport from plasma across the blood–brain barrier (BBB)), k_2 (the reversed transport across BBB), k_3 (the phosphorylation rate constant), and k_4 (the dephosphorylation rate constant), k_5 (the vascular volume)) FDG compartment model (Huang et al., 1980) was used to simulate the tissue FDG kinetics. The parameter vectors of K_g ([0.102, 0.13, 0.062, 0.004, 0.05]) for gray matter and K_w([0.04, 0.08, 0.029, 0.003, 0.03]) for white matter were used in the simulation. A plasma FDG time–activity curve from a human PET study was used as the input function. Spatial configuration of gray

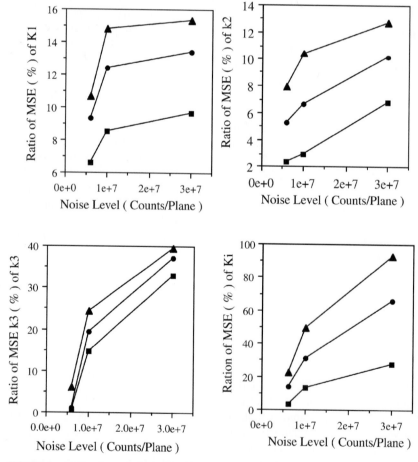

FIGURE 1. Ratio of average MSE (RRSC/NLS) increases as noise level decreases. The constant C was fixed at 5 for the ridge parameter λ. The solid triangles, squares, and circles are the average MSE for gray matter, white matter, and whole brain, respectively.

and white matter on the simulated images followed the Hoffman phantom. The scanning sequence of (4×0.5, 4×4, 10×5 min) was the same as for real dynamic FDG study at UCLA. Pseudorandom noise (normally distributed with its variance proportional to its mean) was added to each simulated data point in the sinogram. Three different noise levels which correspond to total counts of (6×10^6, 10×10^6, 30×10^6) were simulated. Hanning filter with cutoff at Nyquist frequency was used for image reconstruction (image matrix size 128×128, pixel size 0.125 cm). Twenty realizations for each noise level were obtained for evaluating the noise level of parametric images. True parametric images were then reconstructed. Mean, variance, and mean square error (MSE) were calculated for each parameter and for each pixel. The value of the parameter k_4 was fixed at its true value for model fitting, because the total scanning time of dynamic study was only 60 min, which is not long enough to provide reliable estimate of k_4. A set of values for the constant C (0.5, 1, 5, 10, 50, 100) was selected to eval-

uate the effects of the ridge parameter on model parameter estimation.

C. Human FDG PET Study

After a bolus of FDG ($\sim$150 MBq) was injected intravenously, a sequence of (4×0.5, 4×2, 10×5 min, total 60 min, 18 frames) dynamic human FDG-PET scan was started simultaneously on Siemens/CTI exact HR+ scanner in 3D (septa out) mode. Arterial blood samples were taken during the scan and counted in a well counter to give the plasma FDG time–activity curve as the input function. Dynamic images (128×128, pixel size = 0.1446 cm, plane separation 0.2425 cm) were reconstructed using a Ramp-0.5 filter. The FDG model (see section B above) was then applied to the dynamic PET images. RRSC and NLS were applied separately to the dynamic images (k_4 fixed at 0.0068/min). Results from RRSC were also compared to values obtained from fitting region of interest (ROI) kinetics. Pixelwise com-

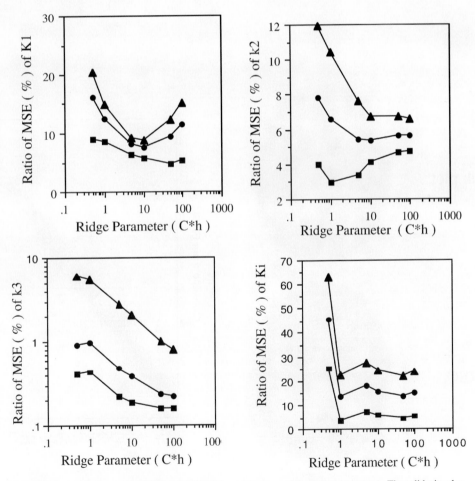

FIGURE 2. Ratio of average MSE (RRSC/NLS) as a function of the ridge parameter λ. The solid triangles, square, circle are the average MSE for gray matter, white matter, and whole brain, respectively. K_i gets its lowest MSE at the $\lambda = h$. K_1 gets its lowest MSE for λ around 10^*h. As λ becomes larger than 50^*h, MSE of k_2 and k_3 tend to become smaller.

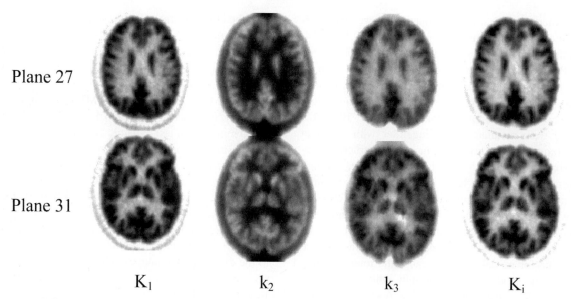

Plane 27

Plane 31

K_1 k_2 k_3 K_i

FIGURE 3. Parametric images generated by RRSC method in human FDG dynamic study. Ridge parameter λ was selected to be $50*h$ (i.e., $C = 50$) for all planes and all pixels. $K_i = K_1^*k_3/(k_2 + k_3)$.

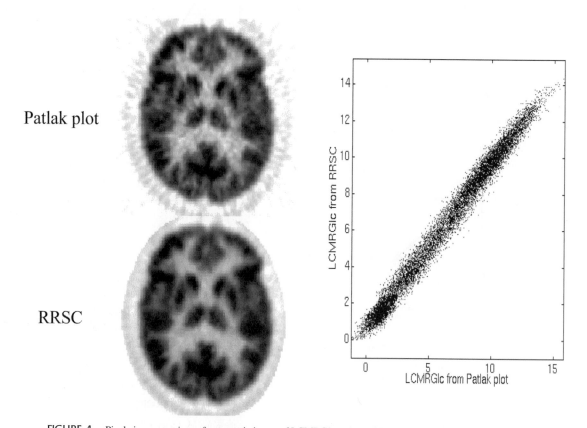

Patlak plot

RRSC

FIGURE 4. Pixelwise comparison of parametric image of LCMRGlc estimated by Patlak plot and RRSC. The correlation between the two estimates is 0.99. There is a good linear relationship between the two estimates (RRSC (LCMRGlc) = 0.93*Patlak (LCMRGlc) +0.54 with $R^2 = 0.98$). The K_i image by RRSC is shown to have a higher signal-to-noise ratio.

parison of local cerebral metabolic rate of glucose (LCMR-Glc) between RRSC and Patlak plot was also examined. The same set of C values used in computer simulation study was also applied to the human study for evaluating the ridge parameter.

III. RESULTS AND DISCUSSION

In Fig. 1, the value of C was fixed at 5. The computer simulation results show that RRSC can decrease MSE of the estimates by 50–90% compared to that seen with the NLS method at high and middle noise levels. Due to higher noise level of white matter regions than in gray matter, the MSE reduction due to RRSC in white matter region is more than in gray matter. Effect of ridge parameter λ on the estimated parameters was shown in Fig. 2. For the linear case, it was shown that the MSE criteria that resulted in optimal ridge parameter was given as h_0 (Hoerl and Kennard, 1976). For the nonlinear situation, our computer simulation results show that the MSE of K_i was minimized at h_0. The optimal ridge parameter for K_1 image is around 10^*h. Overall the MSE of k_2, k_3, and K_i are close to their minimum when λ is between 50^*h and 100^*h. This means that the range of ridge parameter h for K_1 and K_i should be in $[h\ 10^*h]$, and in the range of $[50^*h\ 100^*h]$ for k_2 and k_3. In general, the initial estimates can affect the results of a nonlinear regression, if the cost function has multiple local minima. The sensitivity of the RRSC algorithm to initial estimates of the FDG model parameters is currently under investigation.

Consistent with the simulation results, the value of 50^*h for λ is a good compromise among all parameters to generate parametric images in human FDG-PET dynamic studies. Figure 3 shows the parametric images estimated by RRSC method. The parametric images for different parameters are comparable in image quality. We found that the variability of k_2 and k_3 increases as λ decreases, and estimates of the nonlinear parameters k_2 and k_3 are more sensitive to the ridge parameter than the linear parameters K_1 and K_i (in the case of k_4 close to 0). The parametric image of local cerebral metabolic rate of glucose (LCMRGlc) as estimated by RRSC and Patlak plot methods are shown in Fig. 4. The correlation of all the pixel values between the two parametric images is 0.99. The linear regression line is:

$$RRSC(LCMRGlc) = 0.93^*Patlak(LCMRGlc) + 0.54,$$
and $R^2 = 0.98$.

This shows that RRSC gives LCMRGlc estimates as reliable as Patlak plot does but with better image quality.

In conclusion, at high or middle noise level, RRSC significantly increases the reliability of the estimated parameters and is ideally suited for generating parametric images.

Acknowledgments

This work was partially supported by DOE Contract DE-FC03-87ER60605 and NIH Grant NS30308-06A2.

References

Carson, R. E., Huang, S. C., and Green M. V. (1986). The weighted integration method for local cerebral blood flow measurements with positron emission tomography. *J Cereb. Blood Flow Metab.* **6**: 667–678.

Chen, K., Lawson, M., Reiman, E., Cooper, A., Feng, D., Huang, S.-C., Bandy, D., and Ho, D. (1998). Generalized linear least squares method for fast generation of myocardial blood flow parametric images with N-13 Ammonia PET. *IEEE Trans. Med. Imaging* **17**: 236–243.

Hoerl, A. E., and Kennard, R.W. (1976). Ridge regression: Iterative estimation of the biasing parameter. *Commun. Statis.-Theor. Methods* **A5**(1): 77–88.

Huang, S. C., Phelps, M. E., Hoffman, E. J., *et al.* (1980). Noninvasive determination of local cerebral metabolic rate of glucose in man. *Am. J. Physiol.* **238**: E69–E82.

Huang, S. C., and Zhou Y. (1998). Spatially-coordinated regression for image-wise model fitting to dynamic PET data for generating parametric images. *IEEE Trans. Nucl. Sci.* **45**: 1194–1199.

Patlak, C. S., Blasber, R. G., and Fenstermacher, J. D. (1983). Graphical evaluation of blood-to-brain transfer constants from multiple-time uptake data. *J Cereb. Blood Flow Metab.* **3**: 1–7.

Zhou, Y., Huang, S.-C., Hser, Y.-I., and Bergsneider, M. (1999). Generalized ridge regression versus simple ridge regression for generation of kinetic parametric images in PET, "*IEEE MIC'99 Conference Record*".

A Less-Sensitive to Noise Spectral Analysis Method

A. BERTOLDO and C. COBELLI

Department of Electronics and Informatics, University of Padova, Italy

The spectral analysis (SA) method proposed by Cunningham and Jones in 1993 aims to identify the kinetic components of the tissue tracer activity without specific model assumptions, e.g., presence or absence of FDG dephosphorylation or homogeneity in the tissue. However, this technique bears some nonnecessary assumptions and is sensitive to noise in the data. A generalization of the classical SA method is presented which is also statistically robust. This new approach relaxes the constraint on exponential amplitudes being positive, uses parsimony criteria to identify the model order, i.e., the number of exponentials in the data, and provides estimates of both exponential amplitudes and time constants with their precision. This novel SA method gives a more accurate and clearer model-independent picture of the system under study.

I. INTRODUCTION

Spectral analysis (SA) was introduced by Cunnigham and Jones (1993a) as a model-independent technique for quantitating PET tracer kinetics in heterogeneous as well as homogeneous tissues. SA has been used in several PET studies including glucose metabolism (Turkheimer *et al.*, 1994; Bertoldo *et al.*, 1998), blood flow (Cunningham *et al.*, 1993b), ligand–receptor binding (Richardson *et al.*, 1996; Tadokoro *et al.*, 1993), and pharmacokinetic parameters (Meikle *et al.*, 1998a). It is also sufficiently fast to be used for parametric image reconstruction of PET projection data

(Cunningham *et al.*, 1993b; Meikle *et al.*, 1998b). SA allows kinetic components present in the measured data to be readily identified after a priori definition of a range of basis functions covering the expected kinetic behavior of the PET tracer in the tissue. Its major advantage is that SA does not require the a priori definition of the number of numerically identifiable components present in the data. However, as shown by Cunningham *et al.* (1993b, 1998) this technique is sensitive to noise in the data and thus suffers by detecting spurious components of the system as well as shifting components away from their true values. For this reason often filtering techniques are required to remove the ambiguity. In addition, SA bears some nonnecessary assumptions, e.g., exponential amplitudes are assumed to be positive, which limit its generality. This was first discussed by Bertoldo *et al.* (1998) and recently by Kathleen Schmidt (1999). In particular, in Schmidt (1999) it is elegantly shown that this SA constraint allows the identification of only a class of nonoscillating linear systems; e.g., models containing feedback loops must be excluded.

Here we propose a new more general and statistically robust SA method and compare it with classical SA. Our approach relaxes the constraint on positive exponential amplitudes, uses parsimony criteria to identify the model order, i.e., the number of exponentials of the impulse response, and provides estimates of both exponential amplitudes and time constants with their precision. This generalized SA gives a more accurate and clearer model-independent picture of the system, thus allowing a more reliable structural specification of system model.

*Physiological Imaging
of the Brain with PET*

II. MATERIALS AND METHODS

A. Theory

SA, as introduced in Cunnigham and Jones (1993a), assumes that, if the system is linear, the impulse response can be written as

$$h(t) = \sum_{j=1}^{M} \alpha_j e^{-\beta_j t}, \tag{1}$$

with $\alpha_j, \beta_j \geq 0$ for every j, and the tissue tracer concentration $C_i(t)$ is simply the convolution of $h(t)$ with the plasma tracer concentration $C_p(t)$:

$$C_i(t) = \sum_{j=1}^{M} \alpha_j \int_0^t C_p(\tau) e^{-\beta_j(t-\tau)} d\tau. \tag{2}$$

If one fixes the β_j values and if $C'_j(t)$ denotes the known value of the input function with the β_j exponential term, the model becomes linear in the α_j:

$$C_i(t) = \sum_{j=1}^{M} \alpha_j C'_j(t). \tag{3}$$

Fixing the eigenvalues requires choosing an upper and a lower bound of the β_j range as well as their distribution within the chosen interval. Several candidate distributions can be considered including the linear, quadratic, logarithmic, and the one suggested by DiStefano (1981). In Turkheimer *et al.* (1994), this last distribution was chosen with lower limit $\beta_1 = 1/3T_f$ (T_f end time of the experiment), upper limit $\beta_M = 3/T_i$ (T_i duration of the first scan), and spacing given by

$$\beta_j = \frac{1}{\tau_j}, \quad \tau_j = \tau_{j-1} \left[\frac{T_f}{T_i} \right]^{[1/(M-1)]}, \quad j = 1, \ldots, M, \tag{4}$$

with $M = 100$.

The amplitude α corresponding to the highest eigenvalue ($\beta \to \infty$) gives a measure of the vasculature within the ROI since $\alpha \int_0^t C_p(\tau) e^{-\beta(t-\tau)} d\tau \to \alpha' C_p(t) = \alpha_\infty C_p(t)$. The number of amplitudes α_i corresponding to the intermediate β_i gives the number of reversible compartments which can be discriminated in the tissue. However, nothing can be said in terms of compartment connectivity; e.g., two amplitudes at the intermediate frequencies do not establish whether the corresponding reversible compartments are in parallel (heterogeneous tissue) or in cascade (homogeneous tissue) since these two structures are kinetically indistinguishable. Finally, the amplitude α corresponding to the lowest eigenvalue ($\beta \to 0$) reveals the presence of an irreversible process within the region since

$$\alpha \int_0^t C_p(\tau) e^{-\beta(t-\tau)} d\tau \to \alpha \int_0^t C_p(\tau) d\tau.$$

Thus, the intermediate and low-frequency components of the spectrum reflect the extravascular behavior of the tracer, i.e., the activity of the tracer within the tissue. The method estimates by linear least squares the number $N \leq M$ of nonzero values of α_i which best describe the data.

The SA method described in Cunningham and Jones (1993a) exhibits a number of problems. First it requires α_j to be positive. Second, it can give rise to α_j which show very poor precision and, even if one can improve the estimation by using a new β_j grid to avoid line doubling, some bias may occur since the "true" β_j is not necessarily the arithmetic mean of the doubled line eigenvalues. The constraint on α_j to be positive is not necessary and, in principle, can be misleading. From compartmental theory it is known that, while the impulse response of a generic compartmental model is always positive, its coefficients do not have to be positive, unless input and output are in the same compartment.

Further, if the aim of SA is to find the model order, i.e., the number of exponentials of the system impulse response without having to define a structure beforehand, the definition of a grid of eigenvalues β_j with estimation of the corresponding amplitudes α_j also is not necessary and may provide an unclear picture. The correct approach is to estimate the number of exponentials necessary to provide a good fit to the data by using models of increasing order. For instance, one can start first with a two-exponential model,

$$C_i(t) = \alpha_\infty C_p(t) + \alpha_1 \int_0^t C_p(\tau) e^{-\beta_1(t-\tau)} d\tau$$
$$+ \alpha_2 \int_0^t C_p(\tau) e^{-\beta_2(t-\tau)} d\tau, \tag{5}$$

and estimate by nonlinear weighted least squares $\alpha_\infty, \alpha_1, \beta_1, \alpha_2$, and β_2, then try a three exponential model,

$$C_i(t) = \alpha_\infty C_p(t) + \alpha_1 \int_0^t C_p(\tau) e^{-\beta_1(t-\tau)} d\tau$$
$$+ \alpha_2 \int_0^t C_p(\tau) e^{-\beta_2(t-\tau)} d\tau \tag{6}$$
$$+ \alpha_3 \int_0^t C_p(\tau) e^{-\beta_3(t-\tau)} d\tau,$$

and estimate $\alpha_\infty, \alpha_1, \beta_1, \alpha_2, \beta_2, \alpha_3$, and β_3, and so on. Then, one chooses the best model by using standard model parsimony criteria (e.g., the Akaike Information Criteria). This approach provides not only the precision of the α_j, but also of the β_j (fixing the β_j prevents the obtaining a measure of their precision). It is worth noting that the estimation of the β_j avoids the problem of line doubling.

B. Data Base

Classical SA and generalized are compared on a set of [18F]fluorodeoxyglucose ([18F]FDG) brain data (courtesy of Dr. D. Feng, University of Sidney, Sidney, Australia). Two normal subjects and one patient with epilepsy, each with gray (frontal, temporal, and visual cortex) and white ROIs (hemisphere), were studied. A bolus of [18F]FDG was injected intravenously and arterial blood samples were taken during the 90-min experimental period. A PET scanner with an eight-ring, 15-slice machine with a spatial resolution in the image plane of 6.5 mm FWHM was used. Twenty-one scans were collected and reconstructed on a 128×128 matrix.

III. RESULTS

Table 1 shows the results of the two methods on one representative white and gray ROI for the three subjects. As ex-

TABLE 1. Classical and Generalized SA Results of a Representative Grey and White ROI in the Three Subjects

	Classical SA			Generalized SA			
	β (min^{-1})	α (ml/ml/min)	CV	β (min^{-1})	CV	α (ml/ml/min)	CV
White 1	0	0.0072	(2)	0^a		0.0072	(2)
	0.1256	0.0159	(204)	0.1299	(10)	0.0286	(9)
	0.1366	0.0128	(270)	∞		0.0178^b	(32)
	∞	0.0156^b	(31)				
White 2	0	0.0042	(112)	0^a		0.0088	(3)
	0.0037	0.0055	(103)	0.1906	(14)	0.0493	(13)
	0.1911	0.0416	(27)	∞		0.0148^b	(60)
	0.4422	0.0090	(237)				
	7.0492	0.1086	(71)				
White3	0	0.0048	(8)	0^a		0.0051	(5)
	0.0459	0.0019	(1382)	0.0632	(13)	0.0145	(7)
	0.0499	0.0100	(272)	∞		0.0151^b	(26)
	0.5688	0.0099	(94)				
	4.2610	0.0066	(12255)				
	4.6339	0.0506	(1660)				
Gray1	0	0.0120	(3)	0^a		0.0120	(3)
	0.0826	0.0133	(732)	0.0853	(38)	0.0230	(62)
	0.0898	0.0098	(1140)	0.3663	(77)	0.0360	(38)
	0.3438	0.0062	(5293)	∞		0.0533^b	(23)
	0.3739	0.0299	(1083)				
	∞	0.0469^b	(22)				
Gray2	0	0.0138	(40)	0^a		0.0186	(1)
	0.0037	0.0059	(110)	0.1806	(13)	0.0486	(14)
	0.1911	0.0525	(13)	3.3172	(49)	0.1598	(35)
	2.0024	0.0111	(7005)	∞		0^a	
	2.1777	0.0672	(1208)				
Gray3	0.0048	0.0046	(6480)	0^a		0.0104	(14)
	0.0052	0.0123	(2526)	0.0310	(37)	0.0147	(10)
	0.0590	0.0075	(2514)	0.4650	(17)	0.0854	(12)
	0.0642	0.0022	(8233)	∞		0.0210^b	(37)
	0.4809	0.0846	(13)				
	∞	0.0209^b	(32)				

^{a}The value was constrained to zero by the estimation alghoritm.
^{b}Unitless.

pected from theory (Cunningham *et al.*, 1993b; Turkheimer *et al.*, 1994) noise causes a shift and spread of the distribution of the spectrum in classical SA. In fact, when applied to white ROIs, classical SA exhibits multiple lines together with a poor precision of the amplitude estimates and the presence of shifting of the higher spectrum components, thus making the inference on the minimum number of compartments difficult. In contrast, generalized SA unequivocally shows the presence of three components: an integral component, i.e., an irreversible process within the region, an intermediate one, i.e., a reversible process, and a blood component. This result is compatible with the tissue homogeneity assumption of the 3K model (Sokoloff *et al.*, 1977), i.e., there is a single reversible free [^{18}F]FDG tissue pool, and [^{18}F]fluorodeoxyglucose-6-phosphate ([^{18}F]FDG-6-P) is irreversibly trapped in tissue for the duration of the experiment.

As far as the gray ROIs are concerned, multiple lines surface again even if four distinct components in the spectrum emerge by lumping adjacent components: an integral component, a slowly equilibrating component, a more rapidly equilibrating component, and a blood component. Generalized SA results are less sensitive to noise and do not present the multiple-line problem, thus giving a clearer picture of the data kinetics. Of interest, classical SA of gray2 ROI does not permit the detection of the vascular component and the discrimination between rapidly and slowly equilibrating components, while in gray3 the detection of the integral component fails. On the contrary, generalized SA allows the estimation of the four components of the spectrum without shift, spread, and necessity of filters. The estimation of two equilibrating components plus an integral component may indicate either tissue heterogeneity or tissue homogeneity but with a model more complex than the 3K one.

IV. DISCUSSION

The SA original method (Cunningham and Jones, 1993a) is basically an input–output model which aims to identify the kinetic components of the tissue tracer activity without specific model assumptions, e.g., presence or absence of [^{18}F]FDG dephosphorylation or homogeneity in the tissue. SA is thus one important conceptual tool for the quantification of dynamic PET data. However, the method proved sensitive to data noise and was subsequently improved to reduce noise sensitivity for the very fast or very slow kinetic components by using some penalty functions (Cunningham *et al.*, 1993b). However, the positive constraint on exponential amplitudes and the fixed eigenvalues grid were maintained. Since a general approach to the derivation of suitable penalty functions has proven to be difficult, Cunningham *et al.* (1998) have recently proposed an alternative

approach to reduce the noise to sensitivity drawback. Our generalized SA overcomes both the structural, i.e., positive constraint amplitude and fixed grid eigenvalues, as well as the numerical limitations of classical SA by posing the estimation of spectral kinetic components as an input–output model identification problem. In particular, we relax the assumption of $\alpha_i \geq 0$ and estimate directly from the data the α_i and β_i necessary and sufficient to describe the tissue activity data. Generalized SA gives not only the precision of α_i but also of β_i estimates and avoid the problem of line doubling. Finally, generalized SA explicitly accounts for measurement error and thus the sensitivity to noise, from which classical SA suffers, is not an issue. In conclusion, generalized SA is more general than classical SA; i.e., it can handle a larger class of system models, and, by providing a statistically sound model-independent picture, can guide on more firm grounds the selection of the most appropriate among the potential candidate compartmental structures to describe the data.

References

Bertoldo, A., Vicini, P., Sambuceti, G., Lammertsma, A. A., Parodi, O., and Cobelli, C. (1998). Evaluation of compartmental and spectral analysis models of [18F]FDG kinetics for heart and brain studies with PET. *IEEE Trans. Biomed. Eng.* **45**: 1429–1448.

Cunningham, V. J., and Jones, T. (1993a). Spectral analysis of dynamic PET studies. *J. Cereb. Blood Flow Metab.* **13**: 15–23.

Cunnigham, V. J., Ashburner, J., Byrne, H., and Jones, T. (1993b). Use of spectral analysis to obtain parametric images from dynamic PET studies. *In* "Quantification of Brain Function. Tracer Kinetics and Image Analysis in Brain PET" (K. Uemura, N. A. Lassen, T. Jones, and I. Kanno, Eds.), pp. 101–112. Elsevier, Amsterdam and New York.

Cunningham, V. J., Gunn, R. N., Byrne, H., and Matthews, J. C. (1998). Suppression of noise artifacts in spectral analysis of dynamic PET data. *In* "Quantitative Functional Brain Imaging with Positron Emission Tomography" (R. E. Carson, M. E. Daube-Witherspoon, and P. Herscovitch, Eds.), pp. 329–334. Academic Press, San Diego.

DiStefano, J. J. (1981). Optimized blood sampling protocols and sequential design of kinetic experiments. *Am. J. Physiol.* **240**: R259–R265.

Meikle, S. R., Matthews, J. C., Brock, C. S., Wells, P., Harte, R. J., Cunningham, V. J., Jones T., and Price, P. (1998a). Pharmacokinetic assessment of novel anti-cancer drugs using spectral analysis and positron emission tomography: A feasibility study. *Cancer Chemother. Pharmacol.* **42**: 183–193.

Meikle, S. R., Matthews, J. C., Cunningham, V. J., Bailey, D. L., Livieratos, L., Jones, T., and Price, P. (1998b). Parametric image reconstruction using spectral analysis of PET projection data. *Phys. Med. Biol.* **43**: 651–666.

Richardson, M. P., Koepp, M. J., Brooks, D. J., Fish, D. R., and Duncan, J. S. (1996). Benzodiazepine receptors in focal epilepsy with cortical dysgenesis: An 11C-flumazenil PET study. *Ann. Neurol.* **40**: 188–198.

Schmidt, K. (1999). Which linear compartmental systems can be analyzed by spectral analysis of PET output data summed over all compartments? *J. Cereb. Blood Flow Metab.* **19**: 560–569.

Tadokoro, M., Jones, A. K. P., Cunningham, V. J., Sashin, D., Grootoonk, S., Ashburner, J., and Jones, T. (1993). Parametric images of 11C-diprenorphine binding using spectral analysis of dynamic PET images acquired in 3D. *In* "Quantification of Brain Function. Tracer Kinetics *and Image Analysis in Brain PET*" (K. Uemura, N. A. Lassen, T. Jones, and I. Kanno, Eds.), pp. 289–295. Elsevier, Amsterdam and New York.

Turkheimer, F., Moresco, R. M., Lucignani, G., Sokoloff, L., Fazio, F., and Schmidt, K. (1994). The use of spectral analysis to determine regional cerebral glucose utilization with positron emission tomography and [18F]fluorodeoxyglucose: Theory, implementation, and optimization procedures. *J. Cereb. Blood Flow Metab.* **14**: 406–422.

Sokoloff, L., Reivich, M., Kennedy, C., Des-Rosiers, M. H., Patlak, C. S., Pettigrew, K. D., Sakurada, O., and Shinohara, M. (1977). The [14C]deoxyglucose method for the measurement of local cerebral glucose utilization: Theory, procedure, and normal values in the conscious and anesthetized albino rat. *J. Neurochem.* **28**: 897–916.

27

Small Animal PET Enables Parametric Mapping of Saturation Kinetics at the 5-HT$_{1A}$ Receptor

ROGER N. GUNN, SUSAN P. HUME, ELLA HIRANI, IMTIAZ KHAN, and JOLANTA OPACKA-JUFFRY

MRC Cyclotron Unit, Hammersmith Hospital, London, UK

*A method is presented for the generation of parametric images of apparent half saturation dose, *ED$_{50}$, in small animals using position emission tomography. This involves the scanning of animals with a site selective radioligand at different levels of receptor occupancy. Images of binding potential (BP) are calculated using a simplified reference tissue model. Subsequently, parametric images of *BP$_{max}$ (apparent maximal binding potential) and *ED$_{50}$ are derived using a basis function implementation of a saturable binding model. The model can be extended to take into account the occupancy of the site by endogenous neurotransmitters.*

I. INTRODUCTION

Currently, there are a number of methods for generating parametric maps of ligand binding with positron emission tomography (PET). These methods attempt to quantify specific binding and are usually derived from a compartmental description of the ligands' behavior (Mintun *et al.*, 1984). Here, a method that uses a series of these binding maps, obtained in the presence of an increasing dose of competitor, is described to characterize saturation kinetics. Parametric maps of BP are generated, for each scan, using a basis function implementation of the simplified reference tissue model (Gunn *et al.*, 1997). These binding volumes are then used as the input to a basis function implementation of saturation kinetics, yielding parametric maps of the *ED$_{50}$ (half saturation dose) of the competitor. The models require no blood sampling and are therefore attractive for both human and animal studies.

The techniques are applied to a set of [^{11}C]WAY-100635 (a selective 5-HT$_{1A}$ receptor ligand) studies in rat brain. Here, the signal was blocked with (-)-pindolol, a β-adrenoceptor antagonist with some affinity for the 5-HT$_{1A}$ receptor as described in Hirani *et al.* (1999). The data led to biomathematical modeling of 5-HT receptor interactions and demonstrated that, in principle, it may be possible to obtain estimates of baseline receptor occupancy by endogenous neurotransmitter.

II. MATERIALS AND METHODS

A. Experimental Procedures and Data Acquisition

All work was carried out by licensed investigators in accordance with the Home Office's "Guidance in the Operation of Animals (Scientific Procedures) Act 1986." Isoflurane-anesthetized male Sprague–Dawley rats were scanned in 3D for 1 h (21 temporal frames) following intravenous administration of ~10 MBq of [^{11}C]WAY-100635 (sp. act. ~100 GBq/μmol) using a purpose built small animal scanner (spatial resolution at FWHM ~3 mm, Bloomfield *et al.* (1997)). The animals were maintained in the same position in the FOV of the camera by using a stereotaxic headholder. Four rats were scanned with [^{11}C]WAY-100635 alone, 4 were scanned after pretreatment with stable WAY-100635 (1 mg/kg *i.v.* at −10 min), and 13 were scanned after pretreatment with (-)-pindolol at doses ranging from 0.001 to 3 mg/kg i.v. at −10 min. The specific activity of [^{11}C]WAY-100635 used gave rise to an occupancy of ~7% in the tracer rats, as determined previously (Hume *et al.*, 1998). Three-dimensional reconstruction of the data was performed and a plane-to-plane calibration applied. All studies were normalized to a standard injection of 10 MBq.

B. Parametric Imaging–Binding

Parametric images of BP were generated for each scan using a basis function implementation of the simplified reference tissue model (Gunn *et al.*, 1997) with the reference tissue input (cerebellum) derived from rats pretreated with cold WAY-100635 at a dose sufficient to fully saturate the receptors. The reference tissue model is defined by

$$C_T(t) = R_I C_R(t) + \left\{ k_2 - \frac{R_I k_2}{1+BP} \right\} C_R(t) \otimes e^{-\left(\frac{k_2}{1+BP}+\lambda\right)t},$$

(1)

where $C_R(t)$ is the concentration time course in the reference region, $C_T(t)$ is the concentration time course in the target tissue, BP is the binding potential and is defined here as $((k_3/k_4) = B_{max} f_2/(K_{d_{Tracer}}(1 + \sum_i (F_i/K_{d_i}))))$, k_3 and k_4 refer to the rates of exchange between the free and the specifically bound compartments, B_{max} is the total concentration of specific binding sites, $K_{d_{Tracer}}$ is the equilibrium dissociation constant of the radioligand, f_2 is the "free fraction" of the unbound radioligand in the tissue, and the bracketed term takes account of any competing ligands (F_i and K_{d_i} are the free concentration and equilibrium disassociation constants of i competing ligands; cf. Mintun (1984)), k_2 is the efflux rate constant from the target tissue, R_I is the ratio of the delivery in the tissue region of interest compared to that in the reference region (ratio of influx), λ is the physical decay constant of the isotope, and $\otimes$ is the convolution operator.

C. Parametric Imaging—Saturation Kinetics

The following equation for saturable binding assuming a single-site competition model may be derived from Michaelis-Menten kinetics and the definition of binding potential.

$$BP_i(P_i) = \frac{{}^*BP_{max} {}^*ED_{50}}{P_i + {}^*ED_{50}} + NS,$$

(2)

where BP_i is the binding potential map in the rat pretreated with a dose of pindolol equal to P_i, ${}^*BP_{max} = f_2 B_{max}/(K_{d_{Tracer}})$ is the apparent maximum binding potential, ${}^*ED_{50}$ is the apparent dose of pindolol which would cause a 50% reduction in specific binding and NS is a term to account for any nonspecific binding. These image volumes were then fitted to a basis function implementation of Eq. (2). A set of $n(= 100)+1$ basis functions were generated as follows,

$$BF_j = \frac{{}^*ED_{50_j}}{P + {}^*ED_{50_j}}, \quad j = 1, \ldots, n(= 100),$$

$$NS = 1. \tag{3}$$

A discrete logarithmic spectrum of values for ${}^*ED_{50}$ is constructed (${}^*ED_{50_{min}}(= 0.01 \text{ mg/kg}) \leq {}^*ED_{50_j} \leq {}^*ED_{50_{max}} = (10 \text{ mg/kg})$). The single site model may then be solved for each basis function in turn using a NNLS algorithm which constrains ${}^*BP_{max} > 0$ and $NS > 0$.

For each j, solve

$$BP = \lfloor BF_j \ NS \rfloor \alpha, \quad \alpha \geq 0. \tag{4}$$

For each j record the residual sum of squares rss_j. Then the minimum rss_j is chosen along with its associated parameter values for ${}^*BP_{max}(= \alpha_1)$ and ${}^*ED_{50}$ and $NS(= \alpha_2)$. The problem is solved in a manner similar to that of the simplified reference region (Gunn *et al.*, 1997). Parametric images of ${}^*BP_{max}$, ${}^*ED_{50}$, and NS (nonspecific or nonsaturable binding) were generated.

D. Model for 5-HT Competition

As will be seen under Results it is possible that there are changes in endogenous levels of 5-HT. We now examine the consequences of this on the model. In brain, [^{11}C]WAY-100635 labels 5-HT$_{1A}$ receptors located both postsynaptically, in regions such as frontal cortex and hippocampus, and presynaptically in the midbrain raphe nuclei, where they act as autoreceptors. Agonist action presynaptically reduces synaptic levels of 5-HT, which could affect specific binding of the radioligand.

The single-site competition model is now extended to consider partial receptor occupancy with 5-HT. If a set of assumptions is met, then it is possible to obtain a function for the changing 5-HT levels. The major assumptions are as follows:

1. The values of ED_{50} for pindolol in pre- and postsynaptic regions are equal

$$ED_{50}^{pre} = ED_{50}^{post}. \tag{5}$$

2. The baseline 5-HT concentration, $[5HT]_{Base}$, and the affinity are assumed to be equal pre- and postsynaptically

$$[5HT]_{Base}^{pre} = [5HT]_{Base}^{post},$$

$$K_{d_{5HT}}^{pre} = K_{d_{5HT}}^{post}, \tag{6}$$

(i.e., the baseline occupancy is equal pre- and postsynaptically).

3. 5-HT remains at the same concentration in the presynaptic region for all doses of pindolol.

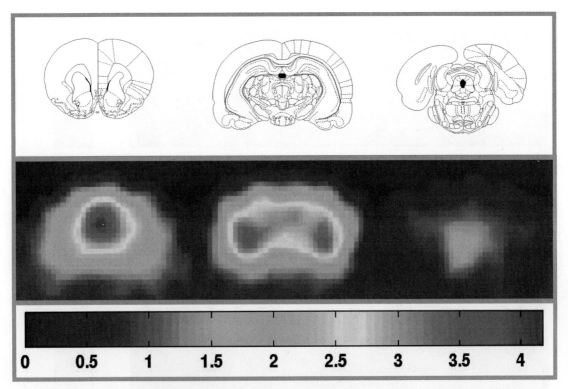

FIGURE 1. Rat brain slices incorporating, from left to right, frontal cortex, hippocampus, and midbrain raphe nuclei. Parametric images of BP for $[^{11}C]$WAY-100635 from a tracer alone study (bottom row). Illustrations from a rat brain atlas (top row, Paxinos and Watson, 1996).

Under these conditions and introducing 5-HT competition, the following equations may be derived,

$$BP^{Pre} = \frac{BP^{Pre}_{\max}}{\left(1 + \dfrac{P}{ED_{50}} + \dfrac{[5HT]_{Base}}{K_{d_{5HT}}}\right)} = \frac{{}^*BP^{Pre}_{\max}}{\left(1 + \dfrac{P}{{}^*ED_{50}}\right)}$$

$$BP^{Post} = \frac{BP^{Post}_{\max}}{\left(1 + \dfrac{P}{ED_{50}} + \dfrac{[5HT]_{Base}}{K_{d_{5HT}}} + \dfrac{\Delta[5HT](P)}{K_{d_{5HT}}}\right)}$$

$$= \frac{{}^*BP^{Post}_{\max}}{\left(1 + \dfrac{P}{{}^*ED_{50}} + \dfrac{\Delta[5HT](P)}{K_{d_{5HT}}\left(1 + \dfrac{[5HT]_{Base}}{K_{d_{5HT}}}\right)}\right)}, (7)$$

where $\Delta[5HT](P)$ is the change in 5-HT concentration from baseline as a function of pindolol (i.e., $\Delta[5HT](0) = 0$) in the postsynaptic region, $K_{d_{5HT}}$ is the equilibrium disassociation constant for 5-HT. Equation (7) may be transformed to yield

$$\frac{\Delta[5HT](P)}{K_{d_{5HT}} + [5HT]_{Base}} = \frac{{}^*BP^{post}_{\max}}{BP^{post}} - \frac{{}^*BP^{pre}_{\max}}{BP^{pre}}. \quad (8)$$

This function may now be calculated from the dose response curves obtained in the pre- and postsynaptic regions.

III. RESULTS AND DISCUSSION

A. Parametric Imaging—Binding

Good fits were obtained to the time-activity data. Parametric images of $[^{11}C]$WAY-100635 binding in a single rat (tracer) are shown in Fig. 1. The images of binding potential are displayed adjacent to these illustrations from a rat brain atlas (Paxinos and Watson, 1996) which correspond to the mid-slice of the PET planes. This model has been previously applied to $[^{11}C]$WAY-100635 binding in humans (Gunn *et al.*, 1998).

B. Parametric Imaging—Saturation Kinetics

The ${}^*BP_{\max}$ map indicated high binding of $[^{11}C]$WAY-100635 in frontal cortex and hippocampus (postsynaptic), with a hot spot consistent with the location of the midbrain raphe nuclei (presynaptic), all regions rich in 5-HT$_{1A}$ receptors (Fig. 2a). The parametric map of pindolol ${}^*ED_{50}$ (Fig. 2b) showed regional differences (raphe ${}^*ED_{50}$ = 0.30 mg/kg, frontal cortex ${}^*ED_{50}$ = 0.67 mg/kg, hippocampus

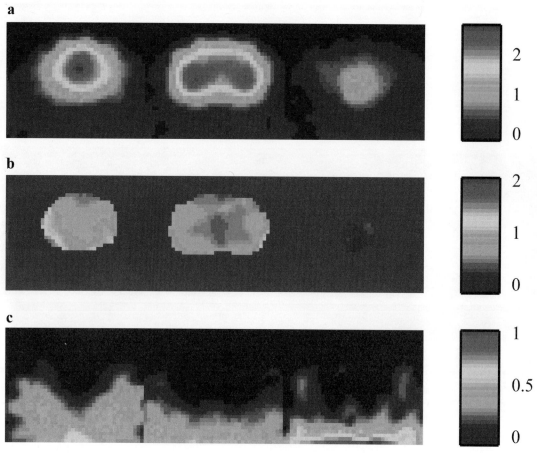

FIGURE 2. Parametric images of saturation kinetic parameters for pindolol blocking of $[^{11}C]$WAY-100635 binding at the 5-HT$_{1A}$ receptor site. (a) Apparent maximal binding, $^*BP_{max}$, (b) apparent half saturation dose, $^*ED_{50}$, and (c) nonspecific binding, *NS*. The slices correspond to those illustrated in Fig. 1.

$^*ED_{50} = 0.61$ mg/kg). Nonspecific binding outside the brain is shown in the NS images (Fig. 2c). Region of interest sampling showed a good fit to the single-site model in the raphe (Fig. 3a). However, in the postsynaptic regions there was a significant increase in binding above tracer studies for low doses of pindolol ($^*BP_{max} > 98\%$ confidence limits of tracer BP estimates) and the model fit was not as good (see hippocampus, Fig. 3b). There are several possibilities for this effect which center on factors affecting the $^*BP_{max}$, such as endogenous 5-HT levels, the free fraction of the radioligand in the tissue, the total concentration of receptor sites, and the affinity of the radioligand. Let us assume for the moment that the effect is due to changes in endogenous 5-HT levels.

C. Model for 5-HT Competition

The function for changing 5-HT levels calculated from Eq. (8) is shown in Fig. 4. It may be seen that this function drops to around -0.7 (although these three points may be susceptible to dividing by a small number in the raphe at this dose and a more conservative estimate of -0.4 may be more appropriate). The following inequalities allow us to derive limits on the baseline occupancy,

$$\frac{\Delta[5HT](P)}{K_{d_{5HT}} + [5HT]_{Base}} < -0.7,$$

and

$$[5HT]_{Base} + \Delta[5HT](P) > 0. \qquad (9)$$

These yield the following constraint for the occupancy,

$$Occ = \frac{[5HT]_{Base}}{K_{d_{5HT}} + [5HT]_{Base}} > 0.7. \qquad (10)$$

From this it is possible to deduce that the baseline occupancy must be greater than or equal to 70% (by taking -0.4 instead, see above, occupancy >40%).

A method for parametric imaging of binding and saturation kinetics in studies in small animals has been presented. Initial examination of the dose ranging data using the simple single-site competition model indicates a difference of potency for pindolol between pre- and postsynaptic 5-HT$_{1A}$ sites. This difference was a factor of two, and it was the

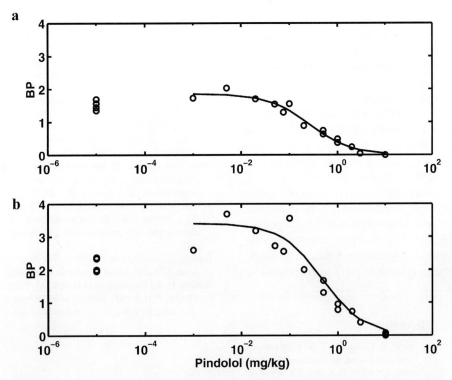

FIGURE 3. Dose–response data for competition of pindolol at the 5-HT$_{1A}$ site. (a) Midbrain raphe nuclei (presynaptic) and (b) hippocampus (postsynaptic). Also shown (at pindolol dose $= 10^{-5}$) are the control, tracer alone, group.

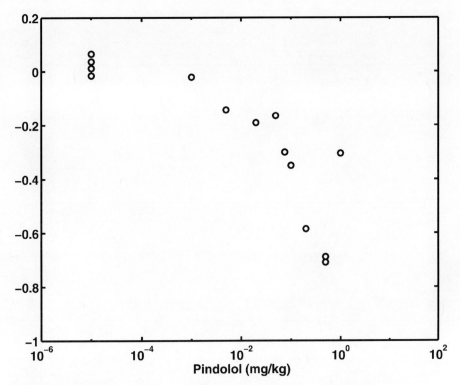

FIGURE 4. Function for changing 5-HT levels in the hippocampus as a function of pindolol as derived from Eq. (8): $(\Delta[5HT](P)/(K_{d_{5HT}} + [5HT]_{Base}))$.

presynaptic sites that indicated the greater sensitivity to pindolol. These results were in agreement with a more limited set of dose–response data obtained in humans which also indicated a greater sensitivity at presynaptic 5-HT$_{1A}$ sites for pindolol (Rabiner *et al.*, 1998).

On examination of the dose response data at the postsynaptic sites, it was evident that following a low dose of pindolol the binding of [^{11}C]WAY-100635 was higher than in tracer alone studies. If pindolol was acting as an agonist at the presynaptic 5-HT$_{1A}$ sites, 5-HT levels would drop postsynaptically (given the inhibitory role of the somatodendritic 5-HT$_{1A}$ autoreceptors). Partial receptor occupancy by 5-HT in the control rats could feasibly explain the data set more comprehensively. Manipulation of the competition model, now assuming that the affinity of pindolol was the same pre- and postsynaptically, indicated that, in principle, it was possible for changing levels in 5-HT to give rise to the observed data.

References

Bloomfield, P. M., Myers, R., Hume, S. P., Spinks, T. J., Lammertsma, A. A., and Jones, T. (1997). Three-dimensional performance of a small-diameter positron emission tomograph. *Phys. Med. Biol.* **42**: 389–400.

Gunn, R. N., Lammertsma, A. A., Hume, S. P., and Cunningham, V. J. (1997). Parametric imaging of ligand-receptor binding in PET using a simplified reference region model. *NeuroImage* **6**: 279–287.

Gunn, R. N., Sargent, P. A., Bench, C. J., Rabiner, E. A., Osman, S., Pike, V. W., Hume, S. P., Grasby, P. M., and Lammertsma, A. A. (1998). Tracer kinetic modelling of the 5-HT$_{1A}$ receptor ligand [*carbonyl*-^{11}C]WAY-100635 for PET. *NeuroImage* **8**: 426–440.

Hirani, E., Opacka-Juffry, J., Gunn, R. N., Khan, I., Sharp, T., and Hume, S. P. (2000). Pindolol occupancy of 5-HT$_{1A}$ receptors measured in vivo using small animal positron emission tomography with carbon-11 labelled WAY 100635. *Synapse*, **36**: 330–341.

Hume, S. P., Gunn, R. N., and Jones, T. (1998). Pharmacological constraints associated with positron emission tomographic scanning of small laboratory animals. *Eur. J. Nucl. Med.* **25**: 173–176.

Mintun, M. A., Raichle, M. E., Kilbourn, M. R., Wooten, G. F., and Welch, M. J. (1984). A quantitative model for the in vivo assessment of drug binding sites with positron emission tomography. *Ann. Neurol.* **15**: 217–227.

Paxinos, G., and Watson, C. (1996). *The Rat Brain in Stereotaxic Coordinates*, Compact 3rd ed. Academic Press, London.

Rabiner, E. A., Sargent, P. A., Gunn, R. N., Bench, C. J., Cowen, P. J., and Grasby, P. M. (1998). Imaging pindolol binding to 5-HT$_{1A}$ receptors in man using PET. *Int. J. Neuropsychopharmacol.* **1**(1): S65.

SECTION

V

DOPAMINE ENZYME, RECEPTOR, TRANSPORTER, and RELEASE IMAGING

Determination of 6-[18F]Fluoro-L-dopa Metabolites: Importance in Interpretation of PET Results

H. M. RUOTTINEN,* J. R. BERGMAN,[†] V. J. OIKONEN,[‡] M. T. HAAPARANTA,[†] O. H. SOLIN,[†] and J. O. RINNE*,[‡]

*Department of Neurology, University of Turku
[†]Radiochemistry Laboratory, Turku PET Centre, Turku, Finland
[‡]Turku PET Centre, Turku, Finland

The peripheral metabolism of 6-[18F]fluoro-L-dopa (FDOPA) is complex, with formation of numerous labeled metabolites. In metabolite analysis there may be overlapping between FDOPA and its metabolite peaks in the chromatogram. We investigated to what extent variation in FDOPA input influences the FDOPA uptake constant ($K_{i(p)}$) and the loss rate constant (k_{loss}). We compared the concentration of FDOPA metabolites in arterial plasma from six Parkinson's disease patients and six controls, without and with peripheral catechol-O-methyltransferase inhibition during a 210-min PET scan. The metabolite analysis used yields FDOPA and 3-O-methylfluoro-L-dopa (3-OMFD) peaks which are not "contaminated" by other metabolites, e.g., sulfates. As input in the graphical analysis we used FDOPA from which sulfates were separated, FDOPA + FDOPA-sulfate, and FDOPA + FDOPA-sulfate + 6-[18F]fluorodopamine-sulfate. FDOPA input containing sulfates leads to a significant underestimation of the $K_{i(p)}$ values and overestimation of the k_{loss} values compared with input where FDOPA is separated from sulfates and other FDOPA metabolites. Also changes in FDOPA metabolites other than 3-OMFD, which do not cross the blood–brain barrier, influence the interpretation of FDOPA uptake expressed as striatal region-to-occipital ratios and the FDOPA uptake constant using occipital input ($K_{i(o)}$).

I. INTRODUCTION

The peripheral metabolism of 6-[18F]fluoro-L-dopa (FDOPA) is complex, with formation of numerous labeled metabolites, as shown in Fig. 1. In metabolite analysis, there may be overlapping between FDOPA and its metabolite peaks in the chromatogram. The determination of FDOPA and its metabolites may influence the estimates of the FDOPA influx constant (Patlak *et al.*, 1983), the loss rate constant (Holden *et al.*, 1997), and results obtained by compartmental modeling, e.g., decarboxylation coefficient (Kuwabara *et al.*, 1993). Therefore, we investigated the extent to which a "contamination" of FDOPA input with metabolites influences the FDOPA influx constant using plasma input ($K_{i(p)}$) and the loss rate constant (k_{loss}).

II. MATERIALS AND METHODS

A. Subjects

We compared the concentrations or fractions of FDOPA metabolites in plasma from 12 subjects without and with peripheral catechol-O-methyltransferase (COMT) inhibition. Six Parkinson's disease (PD) patients (60.3 ± 10.3 years, mean ± SD, range 47–74 years) and six healthy controls (59.8 ± 3.6 years; range 53–63 years) were investigated twice with positron emission tomography (PET) using FDOPA as a tracer, without and with peripheral

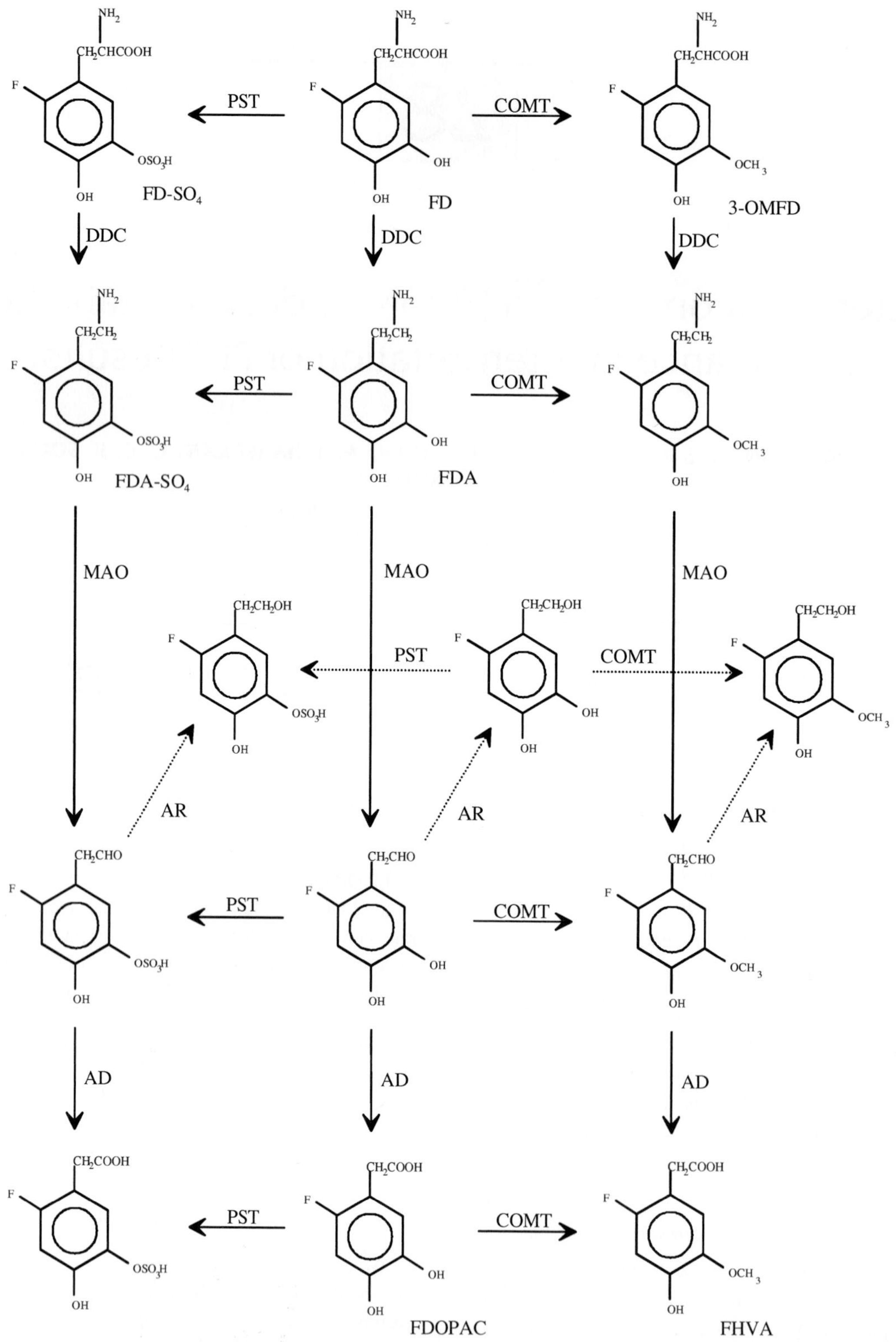

FIGURE 1.   The main metabolic pathways of 6-[^{18}F]fluoro-L-dopa (FDOPA). FDA, 6-[^{18}F]fluorodopamine; 3-OMFD, 3-*O*-methylfluoro-L-dopa; FDOPAC, l-3,4-dihydroxy-6-[^{18}F]fluorophenylacetic acid; FHVA, 6-[^{18}F]fluorohomova-nillic acid; AD, aldehyde dehydrogenase; AR, aldehyde reductase; COMT, catechol-*O*-methyltransferase; DDC, dopa decarboxylase; MAO, monoamine oxidase; PST, phenolsulphotransferase.

COMT inhibitor, entacapone (Orion Pharma, Espoo, Finland). The study followed a double-blind, randomized, placebo-controlled design.

B. PET Imaging

In PD patients, all antiparkinsonian medication was discontinued 10 h before the PET imaging. For 1 day before the PET investigation the subjects were on a low-protein diet and they fasted for 6 h before the PET imaging in order to minimize the competition for transport across blood–brain barrier between FDOPA and large neutral amino acids in plasma. Entacapone (400 mg) was administered orally 60 min before the injection of FDOPA. The peripheral decarboxylation of FDOPA was inhibited by carbidopa, 100 mg orally 60 min before and 50 mg 5 min before the FDOPA injection.

FDOPA was synthesized according to the method of Namavari *et al.* (1992) and Bergman *et al.* (1994). The specific radioactivity of FDOPA was 3500 ± 1000 MBq/μmol at the time of injection, and the mass injected was 14 ± 4 μg. The radiochemical purity was >98%.

In the PET investigations (GE Advance, 3D mode with scatter correction), on average, 209 ± 11 MBq of FDOPA was injected intravenously over 80 s, using an automated perfusor (Perfusor fm syringe pump, B. Braun, Melsungen AG, Germany). Data were collected for 210 min, with a 30-60-min break, 90–120 min after the FDOPA injection.

C. Plasma Metabolite Analysis

To obtain the input function, arterial blood was continuously drawn with a pump and the radioactivity concentration in whole blood during the first 5 min was measured using a two-channel on-line detector system (GE Medical Systems, Model KAL 200-31001, GEMS PET Systems AB, Uppsala, Sweden) cross-calibrated with an automatic gamma-counter. From the whole blood data the plasma radioactivity was calculated using individual hematocrit (Ruottinen *et al.*, 1997). Otherwise, 21 discrete arterial blood samples were taken manually at increasing intervals, maximally at 15-min intervals, up to 210 min after injection. Arterial samples were placed immediately on ice to minimize further metabolism of FDOPA and transport to erythrocytes. After centrifuging, the total plasma radioactivity concentration was measured with an automatic gamma counter (Wizard 1480 3″, Wallac/EG&G, Turku, Finland) cross-calibrated with the tomograph.

FDOPA and its metabolites were analyzed from 21 blood samples taken into gel filtration tubes between 2 and 210 min after injection. Aliquots of 250 μL of arterial serum were analyzed after removal of proteins by ultracentrifugation (cutoff 30 kDa), as it was found that acid precipitation of proteins may affect the recovery of sulfated metabolites.

The analysis of labeled metabolites was done using high-pressure liquid radiochromatography (radioHPLC). The radiochromatograms were analyzed with a peak fitting routine (Origin Peak Fitting) that allowed the determination of the relative amounts of labeled metabolites in each sample, even if the chromatographic separation did not reach baseline. The relative amounts of FDOPA, 3-OMFD and other metabolites, e.g., FDOPA-sulfate (FD-SO$_4$), 6-[^{18}F]fluorodopamine-sulfate (FDA-SO$_4$), and L-3,4-dihydroxy-6-[^{18}F]fluorophenylacetic acid (FDOPAC), were determined as a function of time for each individual.

The area under the plasma concentration–time curve for FDOPA, 3-OMFD, the combination of FDOPA and 3-OMFD, metabolites other than 3-OMFD, and total radioactivity concentration was calculated from injection time to 210 min (plasma AUC$_{0-210}$). Similarly, the AUC$_{0-210}$ of fractions for FD-SO$_4$, FDA-SO$_4$, and for FDOPAC was calculated. Because the concentration of these metabolites was occasionally very low, instead of concentrations, their fractions were used for calculation of AUCs.

D. Image Analysis

Regions of interest (ROIs) were delineated on magnetic resonance images, registered with the PET images, and copied on the dynamic PET images for calculation of time–activity curves (TACs).

In the brain, the area under the radioactivity concentration–time curve for the reference region (occipital region) from 0 to 210 min (tissue AUC$_{0-210}$) was calculated and was corrected to each individual's weight and injected radioactive FDOPA dose. The graphical analysis method with metabolite-corrected arterial plasma input was used to calculate the $K_{i(p)}$ (Gjedde *et al.*, 1981; Patlak *et al.*, 1983), and occipital input was used to calculate the FDOPA uptake constant ($K_{i(o)}$) (Patlak and Blasberg, 1985; Hoshi *et al.*, 1993). $K_{i(p)}$ and $K_{i(o)}$ were calculated with linear regression between 15 and 90 min after injection.

We estimated the k_{loss} by means of extended graphical analysis (Holden *et al.*, 1997). The k_{loss} reflects the metabolism of 6-[^{18}F]fluorodopamine (FDA) comprising both the rate constant for conversion of FDA to FDOPAC and to 6-[^{18}F]fluorohomovanillic acid (FHVA), and the rate constant for the rate of loss of the diffusible metabolites, FDOPAC and FHVA, from the brain. The k_{loss} was calculated in the time range from 30 to 210 min after injection.

The metabolite analysis used in this study yields FD and 3-OMFD peaks which are not contaminated by other metabolites, e.g., sulfates or FDA. As input in the graphical methods we used the metabolite-corrected FDOPA, from which the sulfates and other metabolites were subtracted. In addition, we used as input either FDOPA added with measured FD-SO$_4$ concentration, or FDOPA added with both measured FD-SO$_4$ and FDA-SO$_4$ concentration. Thereby, the effect of incorrect FDOPA input, contaminated with sulfates, on the $K_{i(p)}$ values was assessed.

E. Statistical Analysis

The difference in FDOPA metabolites, and in tissue AUC$_{0-210}$ values for occipital region without and with COMT inhibition was analyzed applying a one-sample t test (= paired t test). Similarly, the difference in FDOPA uptake values between correct and incorrect input function was analyzed with a one-sample t test. A probability value of <0.05 was considered statistically significant.

III. RESULTS

If the input consisting of FDOPA + FD-SO$_4$ was applied in the studies without peripheral COMT inhibitor, the $K_{i(p)}$ values were significantly underestimated by 9.6 ± 6.7% in controls ($p = 0.033$) and by 9.5 ± 8.2% in patients ($p = 0.037$) (Table 1). If both sulfates, FD-SO$_4$ and FDA-SO$_4$,

TABLE 1. The Influence of Variation in 6-[^{18}F]fluoro-L-dopa (FDOPA) Input on the FDOPA Influx Constant ($K_{i(p)}$): Comparison between FDOPA Input, from which Sulfates and Other Metabolites Were Subtracted, and Input with FDOPA + FD-Sulfates (FD-SO$_4$)

	Input FDOPA	Input FD + SO$_4$	Change %
$K_{i(p)}$, Putamen [×10^{-3} min^{-1}]			
Controls			
1	13.65	11.15	−18.3
2	16.40	13.90	−15.2
3	NA	NA	NA
4	16.30	15.60	−4.3
5	13.95	13.45	−3.6
6	14.10	13.15	−6.7
Mean	14.88	13.45	−9.6*
SD	1.35	1.60	6.7
$K_{i(p)}$, Putamen [×10^{-3} min^{-1}]			
PD patients			
1	6.95	5.60	−19.5
2	4.74	3.76	−20.6
3	5.67	5.35	−5.7
4	6.98	6.68	−4.4
5	7.50	7.30	−2.7
6	8.72	8.34	−4.3
Mean	6.76	6.17	−9.5*
SD	1.40	1.62	8.2

*$p < 0.05$.

were included in the FDOPA input, the underestimation in $K_{i(p)}$ values was 11.2 ± 6.4% in controls ($p = 0.017$) and 12.0 ± 7.3% in patients ($p = 0.010$).

The k_{loss} values increased significantly from 0.0022 ± 0.0011 to 0.0047 ± 0.0016 in controls ($p < 0.0001$), and from 0.0047 ± 0.0021 to 0.0068 ± 0.0021 in patients ($p = 0.019$), using false input function of FDOPA + FD-SO$_4$ + FDA-SO$_4$. Similarly, the k_{loss} increased in controls ($p = 0.0004$) and in patients ($p = 0.071$) when using FDOPA + FD-SO$_4$ as input.

COMT inhibition did not influence the total radioactivity concentration in plasma significantly either in controls ($p = 0.07$) or patients ($p = 0.54$), indicating that COMT inhibition induced only relative changes between FDOPA and its metabolite fractions (Fig. 3). Peripheral COMT inhibition increased the AUC$_{0-210}$ for FD by 53 ± 17% in controls ($p = 0.0023$) and by 61 ± 11% in patients ($p < 0.0001$) (Figs. 2 and 3). The AUC$_{0-210}$ for 3-OMFD decreased by 55 ± 4% in controls ($p < 0.0001$) and by 57 ± 4% in patients ($p < 0.0001$), as expected. However, the AUC$_{0-210}$ of metabolites other than 3-OMFD remained unchanged both in controls ($p = 0.94$) and in patients ($p = 0.30$) (Fig. 3). Correspondingly, the combined AUC$_{0-210}$ of FDOPA and 3-OMFD, the two fractions that penetrate the blood–brain barrier, remained unchanged in controls ($p = 0.91$) and in patients ($p = 0.97$). After COMT inhibition, the AUC$_{0-210}$ of fractions for FD-SO$_4$ increased significantly in controls by 46.8 ± 20.6% ($p = 0.007$), but remained unchanged ($p = 0.41$) in PD patients. The AUC$_{0-210}$ of the fraction for FDA-SO$_4$ and FDOPAC remained unchanged both in controls ($p = 0.25$) and in PD patients ($p = 0.82$).

In the brain, the AUC$_{0-210}$ of the tissue time–activity curve in the reference region did not change significantly in PD patients ($p = 0.59$), but decreased slightly in controls by 8.6 ± 8.1% ($p = 0.048$). In PD patients and in controls, the AUC of occipital activity remained unchanged when calculated from 0 to 90 min ($p = 0.12$ controls; $p = 0.74$ patients).

IV. DISCUSSION

This study showed that contamination of FDOPA input by FD-SO$_4$ or FD-SO$_4$ and FDA-SO$_4$ leads to significant underestimation of the $K_{i(p)}$ values by 10–12% both in controls and in patients, compared with FDOPA input from which the sulfates and other metabolites were subtracted. On the other hand, FDOPA input containing FD-SO$_4$ or FD-SO$_4$ and FDA-SO$_4$ caused a significant overestimation of the k_{loss} values both in controls and in patients.

FD-SO$_4$ or other sulfates may overlap with the FDOPA peak in chromatograms. If the FD-SO$_4$ or other sulfates cannot be separated from the FDOPA peak, this probably leads

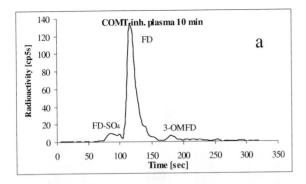
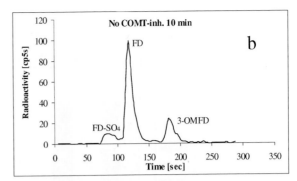
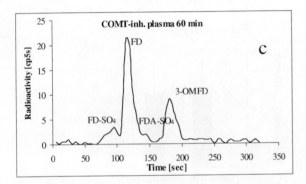
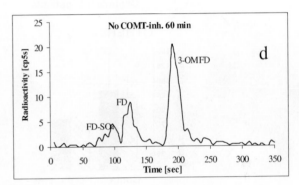
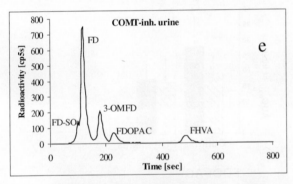
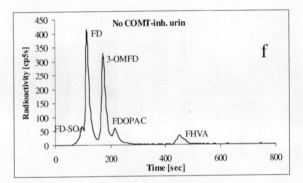

FIGURE 2. HPLC chromatograms of arterial plasma samples taken 10 min after 6-[^{18}F]fluoro-L-dopa (FDOPA) injection with (a) and without (b) COMT inhibition; 60 min after FDOPA injection with (c) and without (d) COMT inhibition; and from urine collected during PET investigation with (e) and without (f) COMT inhibition. Note that the FD-SO$_4$ peak is seen just before the FDOPA peak. In the radioHPLC chromatograms analyzed from urine collected during the PET investigation, the various peaks are more clearly visualized. The chromatograms are neither time-corrected nor corrected with injected FDOPA dose, and hence only the relative sizes of the peaks in the separate chromatograms are comparable. (cp5s = counts per 5 s).

to slight but significant underestimation of the values obtained by the graphical analysis method with plasma input. We have found that, compared with the present method, acid precipitation of the proteins seems to lower relative amounts of sulfated metabolites (unpublished results). This leads to an overestimation of the FDOPA amount and further to a decrease in the $K_{i(p)}$ values.

In the present study during every PET scan the dopa decarboxylase (DDC) was inhibited by carbidopa. The

combined COMT and DDC inhibition did not affect the AUC$_{0-210}$ for metabolites other than 3-OMFD, either in controls or in PD patients, compared with DDC inhibition alone. Among the other metabolites, however, the AUC$_{0-210}$ of fractions for FD-SO$_4$ increased significantly after COMT inhibition in controls, but remained unchanged in patients. This increase in the FD-SO$_4$ level can be explained by the combined DDC and COMT inhibition; DDC inhibition minimizes decarboxylation of FDOPA to FDA, and COMT inhi-

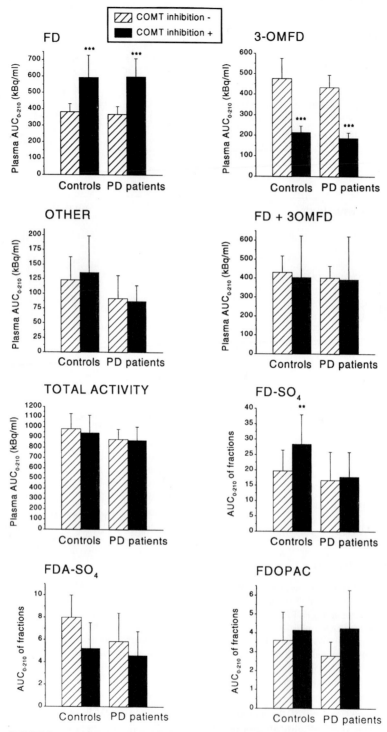

FIGURE 3. The effect of peripheral COMT inhibition with entacapone on the plasma concentrations of 6-[^{18}F]fluoro-L-dopa (FDOPA), and 3-O-methylfluoro-L-dopa (3-OMFD), on the sum of the area under the radioactivity concentration–time curves (AUCs$_{0-210}$) for both FDOPA and 3-OMFD, and on the total plasma radioactivity concentration. Also the effect of entacapone on the AUCs$_{0-210}$ of fractions for FDOPA sulfates (FD-SO$_4$), FDA sulfates (FDA-SO$_4$), and L-3,4-dihydroxy-6-[^{18}F]fluorophenylacetic acid (FDOPAC) was assessed. ** $= p < 0.01$, *** $p < 0.0001$, placebo vs entacapone.

bition minimizes its *O*-methylation to 3-OMFD. Thus, there may be some increase in the sulfation of FDOPA by phenol-sulphotransferase to FD-SO$_4$ (Fig. 1). Further decarboxylation of FD-SO$_4$ to FDA-SO$_4$ by DDC is inhibited by DDC-inhibitor. Thus, there may be some increase in the FD-SO$_4$ level after combined DDC and COMT inhibition compared with DDC inhibition alone, even though the sulfates are probably excreted into urine more vigorously than FDOPA and 3-OMFD. Otherwise, this study confirmed the previous finding of significantly decreased AUC$_{0-210}$ for 3-OMFD and a corresponding significant increase in the AUC$_{0-210}$ for FDOPA in plasma after peripheral COMT inhibition (Ruottinen *et al.*, 1997).

The present findings differ to some extent from those in a previous study using COMT inhibiton and FDOPA PET (Ishikawa *et al.*, 1996). A significant increase of 20 percentage units has been reported (Ishikawa *et al.*, 1996) in the fraction of an unknown metabolite after COMT inhibition with entacapone. This unknown metabolite was a small peak in the chromatogram overlapping the FDOPA peak. In the light of the present study, the unknown metabolite could have been FD-SO$_4$, which is known to be closely related and partly overlapping the FD peak in the chromatogram. A higher dose of carbidopa (200 mg) than in the present study was administered 90 min before the PET investigation. This higher dose of carbidopa may have induced a more effective DDC inhibition that more effectively prevented both decarboxylation of FDOPA to FDA and decarboxylation of FD-SO$_4$ to FDA-SO$_4$, thus leading to a higher increase in the FD-SO$_4$ fraction after DDC and COMT inhibition.

In the present study, the sum of plasma AUCs$_{0-210}$ for FDOPA and 3-OMFD, the fractions that penetrate the blood-brain barrier, remained unchanged after COMT inhibition. This is in line with the present finding of unchanged total radioactivity concentration in the reference region; i.e., the tissue AUC$_{0-210}$ for the occipital region was not affected by COMT inhibition in patients. Moreover, the unchanged AUC$_{0-210}$ of metabolites other than 3-OMFD and the unchanged total activity in plasma are in line with the unchanged sum of AUCs$_{0-210}$ for FDOPA and 3-OMFD after COMT inhibition. In controls, there was a slight but significant decrease in the tissue AUC of the occipital region calculated from 0 to 210 min after COMT inhibition. Calculated from 0 to 90 min, the tissue AUC of the occipital region remained unchanged also in controls.

An increase in the plasma concentration of the sulfated metabolite with COMT inhibitor pretreatment in one earlier study (Ishikawa *et al.*, 1996) led to a concomitant reduction in the combined FDOPA and 3-OMFD plasma fraction. This lowered the total activity in the occipital region. Such a reduction in occipital activity serves to increase the $K_{i(o)}$ without necessarily affecting striatal FDOPA uptake. Thus, the increased FD-SO$_4$ concentration in plasma and the use of FDOPA input including sulfates influence the interpreta-

tion of the results obtained with the ratio method or with the graphical analysis method using occipital input. The varying degree of peripheral DDC inhibition and the possibly different method used for metabolite analysis may, to some extent, explain the differing results. In addition, in the earlier study (Ishikawa *et al.*, 1996), instead of concentrations and AUCs, the fractions of metabolites were used, and these are not directly comparable with the AUCs based on concentrations used in the present study.

In conclusion, FDOPA input contaminated with sulfates leads to a significant underestimation of $K_{i(p)}$ values and overestimation of k_{loss} values, compared with input where FDOPA is separated from sulfates and other FDOPA metabolites. Also changes in FDOPA metabolites other than 3-OMFD, which do not cross the blood-brain barrier, influence the interpretation of FDOPA uptake expressed with striatal region-to-occipital ratios and $K_{i(o)}$ values. Thus, the accuracy of FDOPA metabolite analysis influences the comparison of FDOPA PET results between different PET laboratories as well as interpretation of mechanism of action of some drugs, e.g., enzyme inhibitors.

Acknowledgments

The assistance from the staff of the Turku PET Centre is gratefully acknowledged.

References

Bergman, J., Haaparanta, M., Lehikoinen, P., and Solin, O. (1994). Electrophilic synthesis of 6-[18F]fluoro-L-dopa, starting from aqueous [18F]-fluoride. *J. Label Comp.* **35**: 476–477. [Abstract]

Gjedde, A. (1981). High- and low-affinity transport of D-glucose from blood to brain. *J Neurochem.* **36**: 1463–1471.

Holden, J. E., Doudet, D., Endres, C. J., Chan, G. L.-Y., Morrison, K. S., Vingerhoets, F. J. G., Snow, B. J., Pate, B. D., Sossi, V., Buckley, K. R., and Ruth, T. J. (1997). Graphical analysis of 6-fluoro-L-dopa trapping: Effect of inhibition of catechol-*O*-methyltransferase. *J. Nucl. Med.* **38**: 1568–1574.

Hoshi, H., Kuwabara, H., Léger, G., Cumming, P., Guttman, M., and Gjedde, A. (1993). 6-[18F]fluoro-L-DOPA metabolism in living human brain: A comparison of six analytical methods. *J. Cereb. Blood Flow Metab.* **13**: 57–69.

Ishikawa, T., Dhawan, V., Chaly, T., Robeson, W., Belakhlef, A., Mandel, F., Dahl, R., Margouleff, C., and Eidelberg, D. (1996). Fluorodopa positron emission tomography with an inhibitor of catechol-*O*-methyltransferase: Effect of the plasma 3-*O*-methyldopa fraction on data analysis. *J. Cereb. Blood Flow Metab.* **16**: 854–863.

Kuwabara, H., Cumming, P., Reith, J., Léger, G., Diksic, M., Evans, A. C., and Gjedde, A. (1993). Human striatal L-DOPA decarboxylase activity estimated in vivo using 6-[18F]fluoro-DOPA and positron emission tomography: error analysis and application to normal subjects. *J. Cereb. Blood Flow Metab.* **13**: 43–56.

Namavari, M., Bishop, A., Satyamurthy, N., Bida, G., and Barrio, J. R. (1992). Regioselective radiofluorodestannylation with [18F]F$_2$ and [18F]CH$_3$COOF: A high yield synthesis of 6-[18F]-fluoro-L-dopa. *Appl. Radioact. Isot.* **43**: 989–996.

Patlak, C. S., and Blasberg, R. G. (1985). Graphical evaluation of blood-to-brain transfer constants from multiple-time uptake data. Generalizations. *J. Cereb. Blood Flow Metab.* **5**: 584–590.

Patlak, C. S., Blasberg, R. G., and Fenstermacher, J. D. (1983). Graphical evaluation of blood-to-brain transfer constants from multiple-time uptake data. *J. Cereb. Blood Flow Metab.* **3**: 1–7.

Ruottinen, H. M., Rinne, J. O., Oikonen, V. J., Bergman, J. R., Haaparanta, M. T., Solin, O. H., Ruotsalainen, U. H., and Rinne, U. K. (1997). Striatal 6-[^{18}F]fluorodopa accumulation after combined inhibition of peripheral catechol-O-methyltransferase and monoamine oxidase type B: Differing response in relation to presynaptic dopaminergic dysfunction. *Synapse* **27**: 336–346.

Special Characteristics of 6-Fluorodopa Can Cause Biased Estimates of DOPA Decarboxylation and Dopamine Loss Rates

VESA JUHANI OIKONEN,* HANNA MARIA RUOTTINEN,† and ULLA RUOTSALAINEN*

Turku PET Centre, University of Turku, Finland
†Department of Neurology, University of Turku, Finland

We studied five healthy subjects and five Parkinson's disease patients with 6-fluorodopa (FDOPA) PET for 3 1/2 h, with a 1/2- to 1-h break, to assess the validity of estimates of dopa decarboxylase (DDC) activity and dopamine loss rates provided by extended graphical analysis and nonlinear fitting of a FDOPA compartment model. Simulations show that the concentration of FDOPA in the reference region is higher than in striatum, which causes underestimation of the graphical analysis uptake constants. In the graphical analysis with the reference region as input function, all these errors result into a negative curvature of the plot, making the slope time-dependent and preventing the estimation of loss rate. With plasma input, a negative curvature is a result of the loss of FDOPA metabolites alone. Partial volume effect leads to underestimation of DDC activity with the extended graphical analysis and nonlinear fitting; the loss rates are overestimated with nonlinear fitting, but with the graphical analysis the bias is small. The uptake constants and loss rates were higher when determined with the nonlinear fitting than with the extended graphical analysis.

I. INTRODUCTION

DOPA decarboxylation and dopamine elimination rates in striatum have been estimated with 6-fluorodopa (FDOPA) PET studies using compartmental models (Huang *et al.*, 1991; Kuwabara *et al.*, 1993; Cumming and Gjedde, 1998; Danielsen *et al.*, 1999) and extended graphical analysis (Holden *et al.*, 1997) based on the Gjedde–Patlak plot (Patlak and Blasberg, 1985). We assessed the validity of these

methods with FDOPA PET studies and simulations based on the compartmental model.

II. MATERIALS AND METHODS

A. Subjects

We studied five healthy volunteers and five Parkinson's disease (PD) patients. Written informed consent, according to the Declaration of Helsinki, was obtained from all subjects. These studies belonged to a larger study which was approved by the Joint Ethical Committee of the Turku University and Turku University Central Hospital.

B. PET Studies

All subjects were premedicated with carbidopa and were studied for 3 1/2 h using PET (GE Advance, 3D mode with scatter correction), with a 1/2- to 1-h break 1 1/2–2 h after a FDOPA injection (210 MBq). Arterial plasma samples were taken and metabolites were analyzed throughout the study. Regions of interest (ROIs) were drawn on MR images on putamen and on occipital and frontal cortex, and dynamic time–activity curves (TACs) were calculated from PET images.

C. Data Analysis

The extended graphical analysis (Patlak and Blasberg, 1985; Holden *et al.*, 1997) was applied to calculate the putaminal uptake ($K_{i(o)}$, $K_{i(p)}$) together with the loss (k_{loss})

Physiological Imaging
of the Brain with PET

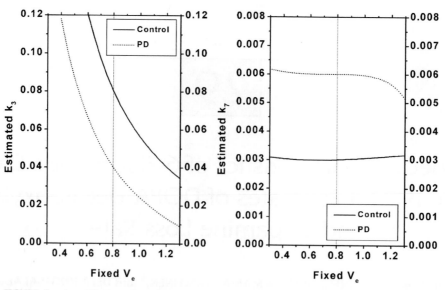

FIGURE 1.   (a and b) Simulation of the bias in putamen k_3 and k_7 caused by fixing the V_e ($= K_1/k_2$) to a wrong value in nonlinear fitting of the FDOPA compartment model.

rate constants using occipital cortex input or metabolite-corrected plasma input. The blood volume and the rate constants, including the unidirectional transfer rates across the blood–brain barrier (K_1 and k_2), DOPA decarboxylase (DDC) rate (k_3), and loss rate (k_7) were estimated applying the FDOPA compartment model (Huang *et al.*, 1991; Kuwabara *et al.*, 1993; Danielsen *et al.*, 1999), assuming that $k_5 = 0$, and $q = K_1^{OMFD}/K_1^{FDOPA} = 1.1$, and fixing $V_e = K_1/k_2$ to the estimate of V_e in frontal cortex. We assumed that fluorodopamine (FDA) and its acidic metabolites are eliminated from the brain as a single compartment at the rate k_7 (Huang *et al.*, 1991; Danielsen *et al.*, 1999), contrary to Cumming and Gjedde (1998), who estimated both k_7 and k_9 and set them to be equal. We compared these models and found that in 9 of 10 cases the k_7 model gave better but statistically nonsignificant (*F* test) fit than the $k_7 = k_9$ model.

D. Simulations

Average curves of plasma radioactivity and metabolite fractions were calculated and were used to simulate the brain TACs for estimation of bias in analysis. Using the method of Huang *et al.* (1991), we studied the effects of partial volume effect (PVE) and tissue heterogeneity by mixing the TACs of putamen and frontal cortex in various amounts, assuming that the TAC in tissue adjacent to striatum is similar to that of frontal cortex.

III. RESULTS AND DISCUSSION

Permeability ratio $q = K_1^{OMFD}/K_1^{FDOPA}$ has been measured in human brain and three different results (1.1, 1.7,

1.9) are available (Cumming and Gjedde, 1998). We estimated the parameters of the FDOPA model in frontal cortex using these values, and found that in 9 of 10 cases the 90- and 210-min fits were better when q was set to 1.1. From the 90-min fits the estimated V_e was 10 or 11% lower when q was set to 1.7 or 1.9; with $q = 1.1$ the V_e was closer to the expected value of 0.8 (Cumming and Gjedde, 1998). Our simulations also showed that if q would be considerably higher than 1, TACs of occipital cortex after peripheral COMT inhibition would be lower, which is not supported by our earlier results (Ruottinen *et al.*, 1997).

In 90-min fits for frontal cortex, the estimate of V_e increased from 0.72 ± 0.09 to 0.78 ± 0.09, when k_3 was set to 0. Fitting k_3 gave better frontal cortex fits in 8 of 10 cases, which was significant in 3 cases (*F* test, $p < 0.05$). In 210-min fits, fitting k_3 had been significantly better than fixing it to zero in 7 cases. The traditional way of fitting the blood volume, K_1, V_e, and k_3 in frontal cortex was selected. However, the fits of putamen would have been slightly better in all cases if the V_e had been fixed to values estimated by setting frontal cortex k_3 to 0. In our simulations, the estimated k_3 values of putamen are highly affected by the V_e from frontal cortex (Fig. 1a): if 0.8 is correct, but 0.7 is used, k_3 will be overestimated by 22% in controls and 29% in PD patients. Instead, estimates of the k_7 are almost independent of V_e (Fig. 1b).

In the graphical analysis with metabolite corrected plasma input, the reference region TAC is subtracted from the striatal TAC to correct for the fraction of 3-*O*-methyl-FDOPA (OMFD) in striatum. In effect, the unmetabolized FDOPA in tissue is also subtracted; its concentration is lower in putamen (Figs. 2a and 2b) due to the high DDC activity in striatum, which causes continuous drainage of FDOPA to FDA.

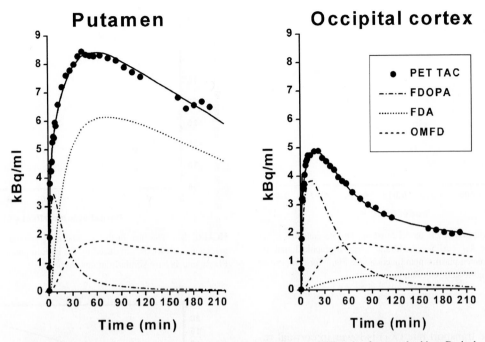

FIGURE 2. (a and b) The TACs of putamen and occipital cortex in one PET study of a control subject. Dashed lines show the constituents of PET TACs, based on the nonlinear fits for the FDOPA compartment model.

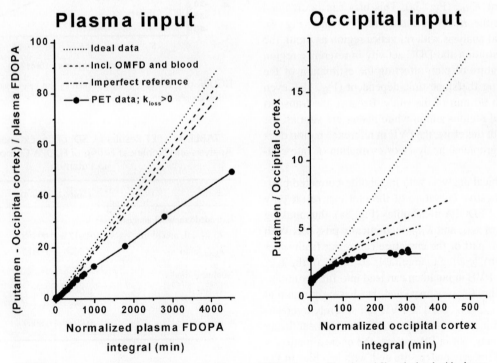

FIGURE 3. (a and b) Gjedde–Patlak plots from the same PET study as in Figs. 2a and 2b, calculated with plasma and occipital cortex input function. Dashed lines show the plots when errors are introduced to the ideal data, based on the nonlinear fitting, in three stages: (1) the vascular fraction and OMFD concentration in tissue, (2) the difference in transport rates between putamen and occipital cortex and the DDC activity in occipital cortex, and (3) the elimination of FDA from putamen, corresponding to the measured PET data.

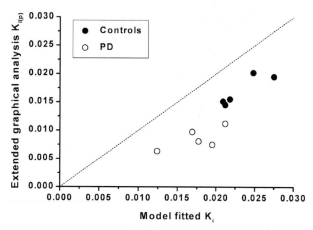

FIGURE 4. Comparison of the model fitted uptake rate constant $K_i = K_1 k_3/(k_2 + k_3)$ and $K_{i(p)}$ estimated with the extended graphical analysis with metabolite-corrected plasma input function. See Fig. 7 for comparison of the loss rates from the same calculations.

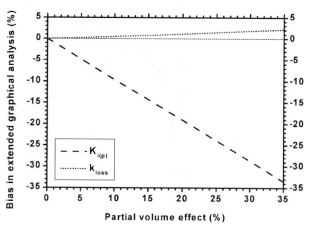

FIGURE 5. The bias in the K_i and k_{loss} estimated with the extended graphical analysis, as introduced by the PVE in putamen. The controls and PD patients behave similarly in this case.

Furthermore, the transport rate (K_1) is lower in cortical regions than in putamen (0.028 ± 0.004 mL min^{-1} mL^{-1} in frontal cortex vs 0.040 ± 0.007 mL min^{-1} mL^{-1} in putamen, $p < 0.001$), causing slower equilibration of OMFD concentrations between plasma and reference regions. Together with the low yet important DDC activity in reference region this results in obvious underestimation of the uptake constant $K_{i(p)}$ (Fig. 3a). This bias can be avoided by calculating the K_i using fitted rate constants (Fig. 4). In the graphical analysis with reference region as input, the OMFD concentration and DDC activity in reference region lead into curvature of plot, affecting the estimation of the $K_{i(o)}$ by making the slope time-dependent (Fig. 3b), even during the first 90 min of the study. Because the transport rates in cortical regions and in white matter are similar, the V_e is not biased; therefore, the PVE in reference region does not affect the graphical analyses or estimation of rate constants.

In the graphical analysis with metabolite-corrected plasma input, a negative curvature of the plot can result only from the loss of FDOPA metabolites (Fig. 3a), thus making the estimation of k_{loss} and $K_{i(p)}$ theoretically possible. With reference input, part of the curvature can result from other reasons (Fig. 3b), leading to unreliable estimates of the k_{loss} and $K_{i(o)}$. The PVE in putamen can lead into further underestimation of the $K_{i(p)}$, as expected, but has only minimal effect on the k_{loss} (Fig. 5). The PVE can severely overestimate the dopamine loss rates estimated with nonlinear fitting in control subjects, but in PD patients the overestimation is small; the k_3 is linearly underestimated, with less bias in PD patients (Fig. 6). The dopamine loss rates estimated with the nonlinear compartment model were on average two times higher than with the extended graphical analysis (Table 1; Fig. 7). Still, the difference between controls and patients

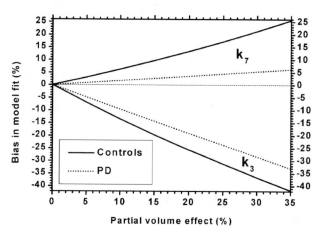

FIGURE 6. The bias in the putamen k_3 and k_7 estimated with nonlinear fitting, caused by the partial volume effect.

TABLE 1. PET Results ($\pm$ SD) of the Extended Graphical Analysis and Nonlinear Fitting of FDOPA Compartment Model in Putamen

	Controls ($n = 5$)	PD ($n = 5$)
Extended graphical analysis		
$K_{i(p)}$ (mL min^{-1} mL^{-1})	0.017 ± 0.003	0.009 ± 0.002
k_{loss} (min^{-1})	0.0020 ± 0.0012	0.0037 ± 0.0007
Nonlinear model		
k_3 (min^{-1})	0.086 ± 0.022	0.042 ± 0.006
K_i (mL min^{-1} mL^{-1})	0.023 ± 0.003	0.018 ± 0.003
k_7 (min^{-1})	0.0033 ± 0.0009	0.0072 ± 0.0015

is preserved, being more pronounced in the k_7 than in the k_{loss} (Table 1; Fig. 7), contrary to the K_i and $K_{i(p)}$ (Table 1; Fig. 4).

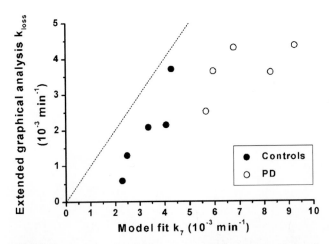

FIGURE 7. Comparison of the fitted k_7 and the extended graphical analysis k_{loss}.

In conclusion, the graphical analysis with reference region input is severely disturbed by possible DDC activity in the reference region and OMFD, and cannot be used to estimate the loss of FDOPA metabolites. The nonlinear fitting and the extended Gjedde–Patlak plot with metabolite-corrected plasma input can be used to estimate the DDC activity and dopamine loss rate.

References

Cumming, P., and Gjedde, A. (1998). Compartmental analysis of dopa decarboxylation in living brain from dynamic positron emission tomograms. *Synapse* **29**: 37–61.

Danielsen, E. H., Smith, D. F., Gee, A. D., Venkatachalam, T. K., Hansen, S. B., Hermansen, F., Gjedde, A., and Cumming, P. (1999). Cerebral 6-[18F]Fluoro-L-DOPA (FDOPA) metabolism in pig studied by positron emission tomography. *Synapse* **33**: 247–258.

Holden, J. E., Doudet, D., Endres, C. J., Chan, G. L. Y., Morrison, K. S., Vingerhoets, F. J. G., Snow, B. J., Pate, B. D., Sossi, V., Buckley, K. R., and Ruth, T. J. (1997). Graphical analysis of 6-fluoro-L-dopa trapping: effect of inhibition of catechol-O-methyltransferase. *J. Nucl. Med.* **38**: 1568–1574.

Huang, S.-C., Yu, D.-C., Barrio, J. R., Grafton, S., Melega, W. P., Hoffman, J. M., Satyamurthy, N., Mazziotta, J. C., and Phelps, M. E. (1991). Kinetics and modeling of L-6-[18F]fluoro-DOPA in human positron emission tomographic studies. *J. Cereb. Blood Flow Metab.* **11**: 898–913.

Kuwabara, H., Cumming, P., Reith, J., Léger, G., Diksic, M., Evans, A. C., and Gjedde, A. (1993). Human striatal L-DOPA decarboxylase activity estimated in vivo using 6-[18F]fluoro-DOPA and positron emission tomography: Error analysis and application to normal subjects. *J. Cereb. Blood Flow Metab.* **13**: 43–56.

Patlak, C. S., and Blasberg, R. G. (1985). Graphical evaluation of blood-to-brain transfer constants from multiple-time uptake data. Generalizations. *J. Cereb. Blood Flow Metab.* **5**: 584–590.

Ruottinen, H. M., Rinne, J. O., Oikonen, V. J., Bergman, J. R., Haaparanta, M. T., Solin, O. H., Ruotsalainen, U. H., and Rinne, U. K. (1997). Striatal 6-[18F]fluorodopa accumulation after combined inhibition of peripheral catechol-O-methyltransferase and monoamine oxidase type B: Differing response in relation to presynaptic dopaminergic dysfunction. *Synapse* **27**: 336–346.

6-[18F]Fluoro-*m*-tyrosine and 6-[18F]Fluoro-L-dopa Kinetic Estimate Correlations with Plasma LNAA Concentrations

DAVID B. STOUT,[*,†] **SUNG-CHENG HUANG,**[*,†,‡] **RANDA E. YEE,**[*,†] **MOHAMMAD NAMAVARI,**[†] **N. SATYAMURTHY,**[†]
KOORESH SHOGHI-JADID,[*,‡] **and JORGE R. BARRIO**[*,†]

[*]*Department of Molecular and Medical Pharmacology, UCLA School of Medicine, Los Angeles, California 90095*
[†]*Laboratory of Structural Biology and Molecular Medicine (DOE), UCLA School of Medicine, Los Angeles, California 90095*
[‡]*Department of Biomathematics, UCLA School of Medicine, Los Angeles, California 90095*

The effect of plasma large neutral amino acid (LNAA) concentrations on kinetic estimates of 6-[18F]fluoro-L-dopa (FDOPA) and 6-[18F]fluoro-m-tyrosine (6-FMT) uptake in squirrel monkeys was investigated. Similar to recent findings in vervet monkeys, an inverse relationship between plasma LNAA concentrations and FDOPA and 6-FMT kinetic estimates of blood–brain barrier transport (K_1) and uptake constant K_i was observed. This relationship was equally described by both K_1 and K_i, suggesting that carrier competition at the blood–brain barrier was sufficient to describe the relationship and that other parameters (k_2 and k_3) were not affected. Kinetic estimates of 6-FMT appeared to have a stronger correlation to LNAA concentrations than FDOPA estimates, resulting in lower kinetic estimate variability for 6-FMT than for FDOPA following adjustment for LNAA competition.

I. INTRODUCTION

Large neutral amino acids (LNAA) enter the brain by carrier-mediated transport, using the neutral amino acid transport system (Castagna *et al.*, 1997). Recently the effects of amino acid competition on 6-[18F]fluoro-L-dopa (FDOPA) transport kinetics in vervet monkeys using PET were described (Stout *et al.*, 1998). Similar, but not necessarily identical, relationships are expected in other animal species. The present study investigates the dependency of 6-[18F]fluoro-*m*-tyrosine (6-FMT) and FDOPA kinetic esti-

mates on LNAA competition using PET in squirrel monkeys (body wt ~0.8–1.0 kg, brain wt ~25 g).

Previous investigations have indicated that radiofluorinated meta-tyrosines can be useful for imaging the central dopaminergic system (Barrio *et al.*, 1996; Doudet *et al.*, 1999), primarily since they are not a substrate for catechol-*O*-methyl transferase (COMT). One drawback to 6-FMT, however, is the reported higher variability of the kinetic estimates, compared with FDOPA estimates, in the absence of adjustments for plasma LNAA concentrations (Doudet *et al.*, 1999). The following work investigates the effect of plasma LNAA competition on 6-FMT kinetic estimates and compares LNAA-adjusted estimates of both FDOPA and 6-FMT.

II. MATERIALS AND METHODS

Squirrel monkeys (*Saimiri sciureus*) were fasted overnight and examined following i.v. injections (2 mCi/kg) of either FDOPA (Luxen *et al.*, 1990) ($n = 12$) or 6-FMT (Namavari *et al.*, 1993) ($n = 16$) under ketamine/versed anesthesia (30/0.3 mg/kg/h). Arterial blood samples ($n = 20$) were acquired during the experiments to provide plasma time activity curves, to measure peripheral metabolites and to determine plasma LNAA concentrations. Dynamic uptake was assessed using regions of interest drawn around the cerebellum and left and right striatum. Patlak analysis estimates of blood–brain barrier (BBB) passage (K_1) and the uptake constant K_i ($K_i = (K_1 * k_3)/(k_2 + k_3)$) were determined us-

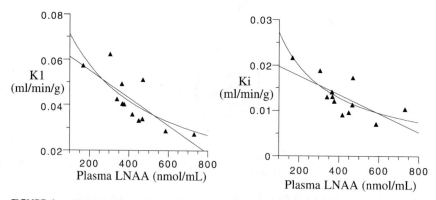

FIGURE 1. FDOPA kinetic estimates of K_1 and K_i (ml/min/g) versus plasma LNAA concentrations (nmol/mL, see Table 1 for fitting results).

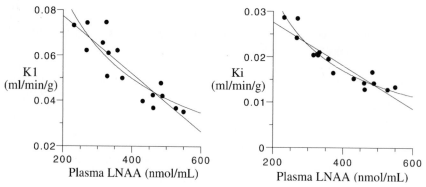

FIGURE 2. 6-FMT kinetic estimates of K_1 and K_i versus plasma LNAA concentrations (nmol/mL, see Table 1 for fitting results).

ing data from 0 to 7.5 min and 30 to 120 min postinjection, respectively.

Plasma samples taken at 7 min postinjection were used to measure LNAA concentrations by florescence detection (Keen *et al.*, 1989). In brief, plasma was deproteinated with an equal volume of 0.8 M perchloric acid, centrifuged, filtered, and stored for subsequent analysis. Samples were partially neutralized with an equal volume of 0.3 M potassium hydroxide, chilled for 5 min, and filtered. Equal volumes of sample and *o*-phthaldialdehyde (1 mg/mL, Sigma Chemical Co., St. Louis, MO) were mixed, incubated for 2 min, and injected onto a ^{18}C HPLC column (Beckman Instruments, Fullerton, CA) at 0.8 mL/min. The solvent was ramped from 70:30 0.1 M sodium phosphate:methanol to 35:65 at 30 min. The output was measured by fluorescence detection and peak areas were converted to micromolar values by comparison with known dilutions of LNAA standards. The total plasma LNAA concentration was calculated as the sum of tyrosine, phenylalanine, methionine, leucine, isoleucine, tryptophan, and valine concentrations.

Correlations between plasma LNAA concentrations and kinetic parameter estimates of K_1 and K_i for FDOPA and 6-FMT were determined using both Michaelis–Menten and linear fitting. Using these correlations, kinetic estimates were adjusted for LNAA competition (Stout *et al.*, 1998).

III. RESULTS AND DISCUSSION

Both FDOPA (Fig. 1) and 6-FMT rate constants (Fig. 2) show an inverse correlation with plasma LNAA concentrations in squirrel monkeys. Based on the difference in the correlation slopes (Table 1), 6-FMT kinetic estimates appear to be more affected by plasma LNAA concentrations than those of FDOPA.

The correlation of FDOPA and 6-FMT kinetics with plasma LNAA concentrations (Table 1) was essentially the same for K_1 and K_i. This implies that plasma LNAA competition affects only the forward BBB transport (K_1), leaving the reverse transport (k_2) and decarboxylation process (k_3) unaffected. This is consistent with the relatively stable tissue LNAA concentrations determined in rats (Glanville and Anderson, 1985; Voog and Eriksson, 1992; Stout *et al.*, 1999) and confirms the validity of the model that describes the distribution volume of radiolabeled amino acids in brain tissue (Huang *et al.*, 1998).

TABLE 1. Correlation Equations

	FDOPA		6-FMT	
	Equation	R^2	Equation	R^2
$K_{1_L}{}^a$	$0.067 - 6.1e^{-5}(\text{LNAA})$	0.60	$0.103 - 1.27e^{-4}(\text{LNAA})$	0.80
$K_{1_{MM}}{}^b$	$30.0/(321.3 + \text{LNAA})$	0.60	$23.2/(67.3 + \text{LNAA})$	0.79
$K_{i_L}{}^a$	$0.022 - 2.1e^{-5}(\text{LNAA})$	0.48	$0.037 - 4.8e^{-5}(\text{LNAA})$	0.85
$K_{i_{MM}}{}^b$	$7.3/(167.6 + \text{LNAA})$	0.61	$6.8/(3.2 + \text{LNAA})$	0.92

[a] Linear fit.
[b] Michaelis–Menten fit.

TABLE 2. Kinetic Estimates

	FDOPA		6-FMT	
	mL/min/g	%SD	mL/min/g	%SD
$K_1{}^a$	0.042	27	0.054	26
$K_{1_L}{}^b$	0.044	16	0.054	12
$K_{1_{MM}}{}^c$	0.040	17	0.051	12
$K_i{}^a$	0.013	32	0.019	27
$K_{i_L}{}^b$	0.014	22	0.019	11
$K_{i_{MM}}{}^c$	0.013	23	0.018	8

[a] Unadjusted data.
[b] Linear LNAA adjustment.
[c] Michaelis–Menten LNAA adjustment.

The relationship between plasma LNAA concentrations and kinetic estimates was used to adjust these estimates for LNAA competition (Table 2), resulting in substantially reduced percent standard deviations of the estimated parameters. The low variability of 6-FMT estimates (K_1, ±12%, K_i, ±8%) indicates that 6-FMT may provide more sensitive information related to changes in the central dopaminergic system compared with FDOPA estimates. The strong correlation of 6-FMT to plasma LNAA concentrations indicates that, without adjustment, 6-FMT estimates would be subject to variations in plasma LNAA concentrations and can explain previous reports of higher variability of 6-FMT results (Doudet et al., 1999).

Acknowledgments

The authors thank Joe Cook and the cyclotron staff for their contribution to the production of FDOPA and 6-FMT. We also thank Ron Sumida, Judy Edwards, Der-Jenn Liu, Zarina Kiziloglu, and Waldemar Ladno for their expertise with the PET procedures. This work was made possible by financial support from the National Institutes of Health (RO1 NS 33356) and the Department of Energy (DE FCO387-ER60615).

References

Barrio, J. R., Huang, S. C., Yu, D. C., Melega, W. P., Quintana, J., Cherry, S. R., Jacobson, A., Namavari, M., Satyamurthy, N., and Phelps, M. E. (1996). Radiofluorinated L-m-tyrosines: New in-vivo probes for central dopamine biochemistry. J. Cereb. Blood Flow Metab. 16: 667–678.

Castagna, M., Shayakul, C., Trotto D., Sacchi, V. F., Harvey, W. R., and Hediger, M. A. (1997). Molecular characteristics of mammalian and insect amino acid transporters: Implications for amino acid homeostasis. J. Exp. Biol. 200: 269–286.

Doudet, D. J., Chan, G. L., Jivan, S., DeJesus, O. T., McGeer, E. G., English, C., Ruth, T. J., and Holden, J. E. (1999). Evaluation of dopaminergic presynaptic integrity: 6-[18F]Fluoro-L-dopa versus 6-[18F]fluoro-L-m-tyrosine. J. Cereb. Blood Flow Metab. 19: 278–287.

Glanville, N. T., and Anderson, G. H. (1985). The effect of insulin deficiency, dietary protein intake, and plasma amino acid concentrations on brain amino acid levels in rats. Can. J. Phys. Pharm. 63: 487–494.

Huang, S. C., Stout, D. B., Yee, R. E., Satyamurthy, N., and Barrio, J. R. (1998). Distribution volume of radiolabeled large neutral amino acids in brain tissue. J. Cereb. Blood Flow Metab. 18: 1288–1293.

Keen, R. E., Barrio, J. R., Huang, S. C., Hawkins, R. A., and Phelps, M. E. (1989). In vivo cerebral protein synthesis rates with leucyl-transfer RNA used as a precursor pool: Determination of biochemical parameters to structure tracer kinetic models for positron emission tomography. J. Cereb. Blood Flow Metab. 9: 429–445.

Luxen, A., Perlmutter, M., Bida, G. T., Van Moffaert, G., Cook, J. S., Satyamurthy, N., Phelps, M. E., and Barrio, J. R. (1990). Remote, semiautomated production of 6-[18F]fluoro-L-dopa for human studies with PET. J. Appl. Radiat. Isot. 41: 275–281.

Namavari, M., Satyamurthy, N., Phelps, M. E., and Barrio, J. R. (1993). Synthesis of 6-[18F] and 4-[18F]fluoro-L-m-tyrosines via regioselective radiofluorodestannylation. J. Appl. Radiat. Isot. 44: 527–536.

Stout, D. B., Huang, S. C., Namavari, M., Satyamurthy, S., Yee, R. E., and Barrio, J. R. (1999). Direct measurement of striatum and cerebellum distribution volumes of 3-O-methyl-6-[18F]FDOPA in rats. J. Nucl. Med. 40: 145P.

Stout, D. B., Huang, S. C., Melega, W. P., Raleigh, M. J., Phelps, M. E., and Barrio, J. R. (1998). Effects of large neutral amino acid concentrations on 6-[18F]Fluoro-L-DOPA kinetics. J. Cereb. Blood Flow Metab. 18: 43–51.

Voog, L., and Eriksson, T. (1992). Relationship between plasma and brain large neutral amino acids in rats fed diets with different compositions at different times of the day. J. Neurochem. 59: 1868–1874.

31

Evaluation of Extrastriatal Specific Binding of [11C]Dihydrotetrabenazine Using Active and Inactive Enantiomers

R. A. KOEPPE, M. R. KILBOURN, S. F. TAYLOR, J. M. MEADOR-WOODRUFF, K. A. FREY, and D. E. KUHL

Division of Nuclear Medicine, University of Michigan, 3480 Kresge III Box 0552, Ann Arbor, Michigan 48109-0552

[11C]Dihydrotetrabenazine ([11C]DTBZ) has been demonstrated to be a suitable ligand for the in vivo imaging of the vesicular monoamine transporter (VMAT2) system. The (+)-isomer has a high affinity (~1 nM) for the VMAT2 binding site while the (−)-isomer has an extremely low affinity (~2 μM). Efforts to model dynamic (+)-DTBZ data demonstrated the difficulty of separating the specific binding component from the free + nonspecific component of the overall PET measure. While kinetic estimates as well as in vitro studies indicate that there is little specific binding in extrastriatal regions, the precise amount, however, was uncertain. Six normal control subjects were scanned once with (+)-DTBZ and once with (−)-DTBZ (order counterbalanced). A single patient with schizophrenia was scanned at baseline and 2 weeks following onset of treatment with a clinical dose of reserpine. Dynamic PET studies were acquired for 60 min and arterial samples were taken throughout, counted, and corrected for radiolabeled metabolites. Both compartmental and equilibrium analyses were performed to yield measures of the total tissue distribution volume of DTBZ for each scan. Results from the pairs of (+)- and (−)-DTBZ scans indicate very little extrastriatal binding. Midbrain structures, cerebellum, and thalamus have binding potentials for (+)-DTBZ of 0.15 to 0.4. Cortical regions exhibited binding potentials of less than 0.1. The single pair of pre- and postreserpine (+)-DTBZ scans also demonstrate the lack of significant specific binding in cortex. While the striatal binding was reduced by 60–65% postreserpine, cortical DV was almost unchanged. We conclude that the level of specific binding of (+)-DTBZ in cortical regions cannot be quantified accurately. However, this low level of binding indicates that reference region methods, such as are commonly used for [11C]raclopride and [11C]flumazenil, are applicable for [11C]DTBZ and can significantly improve precision in estimates of VMAT2 binding in the human basal ganglia.

I. INTRODUCTION

Previous studies have demonstrated the utility of [11C]dihydrotetrabenazine ([11C]DTBZ) as a ligand for *in vivo* imaging of the vesicular monoamine transporter (VMAT2) system (Henry and Scherman, 1989; DaSilva and Kilbourn, 1992; Vander Borght *et al.*, 1995; Koeppe *et al.*, 1996; Chan *et al.*, 1999). The (+)-isomer has a high affinity ($K_i = 0.97 \pm 0.45$ nM) for the VMAT2 binding site while the (−)-isomer has an extremely low affinity ($K_i = 2.2 \pm 0.3$ μM) (Kilbourn *et al.*, 1995). Efforts to model (+)-DTBZ time–activity curves (TAC) were not able to separate reliably the specific binding component from the free + nonspecific component of the overall PET measure. One conclusion from our previous work (Koeppe *et al.*, 1996) was that an estimate of the total volume of distribution, rather than one of specific binding alone, provided the most stable parameter estimates. Our kinetic analysis (on the basis of individual rate constant estimates) suggested that approximately 40% of the binding in cortical regions was specific, although with the high variability in the estimates of the individual compartment volumes, we were suspicious that these values were too high. In contrast, *in vitro* studies indicate that there is little specific binding in cortex, with only about 10% of the total uptake being due to specific VMAT2 binding. This inconsistency in results left two main questions unanswered. First, is specific (+)-DTBZ binding in cortex sufficiently high to be estimated reliably with PET? Second, is binding in the cortex sufficiently low that it could be used as a reference tissue considered to contain only free + nonspecific binding? We have sought answers to these questions using paired studies with both active (+) and inactive (−) isomers of DTBZ and with an additional pair of (+)-DTBZ studies performed on a single schizophrenic subject pre- and post-treatment with reserpine, a drug known to inhibit VMAT2 binding.

II. METHODS

Six normal control subjects, ages 35 ± 10 years, four male and two female, were scanned twice each, once following injection of 16 ± 1 mCi of (+)-DTBZ and once following injection of 16 ± 1 mCi of (−)-DTBZ. To eliminate order effects, three subjects received (+)-DTBZ first, while three subjects received (−)-DTBZ first. A single male schizophrenic patient, age 50, was scanned twice with (+)-DTBZ, once to establish baseline binding levels and a second time 2 weeks after the onset of treatment with 1.0 mg/day dose of reserpine.

No-carrier-added (+)- or (−)-DTBZ (300–1000 Ci/mmol at time of injection) was prepared as reported by Jewett *et al.* (1997). One half the dose was administered as a bolus, while the other half was continuously infused over the remainder of the 1-h study to achieve steady-state conditions. Data acquisition for the six (+)-DTBZ and (−)-DTBZ scans consisted of a 15-frame sequence of four 30-s scans, three 1-min scans, two 2.5-min scans, two 5-min scans, and four 10-min scans. The studies for the schizophrenic patient consisted of a single 4-min early scan and three 10-min scans between 30 and 60 min postadministration. Images were reconstructed using a Hann filter with 0.5 cutoff yielding images of ~9 mm FWHM. Arterial samples were taken throughout, counted, and corrected for radiolabeled metabolites. For the six dual isomer studies, a two-tissue compartment kinetic model was used to estimate $K_1–k_4$, and hence both the free + nonspecific distribution volume, $DV_{F+NS} = [K_1/k_2]$, the specific distribution volume, $DV_{SP} = [K_1/k_2] * [k_3/k_4]$, and the total distribution volume, $DV_{TOT} = [K_1/k_2] * [1 + (k_3/k_4)]$. A single-tissue compartment model was also used to estimate DV_{TOT}. In addition, we calculated DV_{TOT} by equilibrium analysis (C_{PET}/C_{AP}) using the data from 30–60 min post-injection. VMAT2 binding for the patient with schizophrenia was estimated only by equilibrium analysis.

All scans were registered and nonlinearly transformed into stereotactic coordinates based on the methods of Minoshima *et al.* (1993, 1994). Data from various brain regions for all scan pairs were analyzed using both absolute measures of binding and binding measures normalized to white matter ($DV_{TOT(VOI)}/DV_{TOT(WM)}$). Volumes of interest (VOIs) were created for a total of 22 brain regions including caudate nucleus, putamen, thalamus, ventral diencephalic nuclei, cerebellar hemispheres, mesencephalon, pons, hippocampus, frontal cortex, parietal cortex, temporal cortex, occipital cortex, and white matter. Left and right hemisphere regions were analyzed separately and then averaged within each subject.

III. RESULTS

Figure 1 summarizes the results of this study. The four images on the left are from the dual enantiomer scans. Im-

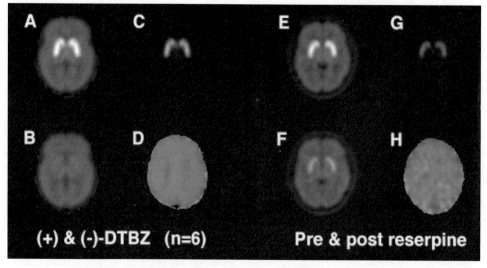

FIGURE 1. Assessment of VMAT2 binding with [^{11}C]DTBZ. See text for details.

ages A and B show the group average DV_{TOT} images of a slice through the basal ganglia for the (+)-DTBZ and (−)-DTBZ scans, respectively. Image C shows the difference between A and B and is suggestive that specific binding is present only in striatum. Image D is at a higher slice, through cortex and white matter, and shows the ratio of (+)-DTBZ to (−)-DTBZ scans. Note that this ratio image is nearly uniform, with only a slight elevation in the gray to white matter ratio for (+)-DTBZ, indicating very little specific binding in cortex. Images E through H are from the single schizophrenic subject. Image E and F show DV_{TOT} images of a slice through the basal ganglia for the pre- and postreserpine scans, respectively. The baseline scan has values very close to the mean of the group of six normal controls. The scan 2 weeks after initiation of reserpine treatment shows a marked, although not complete, decrease in (+)-DTBZ binding in the basal ganglia. The difference between the two scans is shown in image G and, as with the six controls, shows no cortical changes. At the higher level, the reserpine/baseline image, H, is nearly uniform. Again, this demonstrates that the cortical to white matter uptake of DTBZ did not change with reserpine treatment, suggesting minimal specific binding in cortex.

The kinetic estimates of the total DV and the fraction of the total uptake that is due to specific binding for the six subjects studied with both DTBZ isomers are shown in Table 1. The DV_{TOT} estimates in striatal regions demonstrate clearly the very low affinity of the (−) isomer. An interesting and unexpected finding was the slightly higher values of the (−) isomer observed in cortex and white matter. Our assumption was that the free + nonspecific binding would be the same for the two isomers, and thus DV values for (+)-DTBZ would always be as high or higher than for (−)-DTBZ. The time course of the authentic fraction in arterial plasma was similar for both tracers and thus did not explain the results. Based on *in vitro* human studies in our own laboratory showing specific (+)-DTBZ binding to be a factor of 4–5 lower in white matter than in cortex, we normalized each of the (−)-DTBZ scans by the ratio of the white matter DV_{TOT} values $(DV_{TOT(+)}/DV_{TOT(−)})$.

The inability to separate kinetically the free + nonspecific and specific compartments also is demonstrated in Table 1. Notice that the kinetically determined fraction of the total binding that is in the specific compartment, $\{(K_1/k_2) * (k_3/k_4)\}/\{(K_1/k_2) * [1 + k_3/k_4)]\} = (k_3/k_4)/[1 + k_3/k_4)]$, for (+)-DTBZ is inappropriately quite uniform across the regions whether total binding is high or low. Also note that a significant fraction of the total (−)-DTBZ uptake is associated with the second tissue compartment. Since we know for (−)-DTBZ that this is not specific VMAT2 binding, we can infer that the individual distribution volume estimates of the first and second tissue compartments for (+)-DTBZ do not accurately reflect free + nonspecific and specific binding, respectively.

Since the direct kinetic estimates of specific binding with (+)-DTBZ were not very accurate, we next calculated a specific binding index using the (−)-DTBZ DV_{TOT} estimates as a measure of free + nonspecific binding, assuming that $DV_{TOT(−)} = DV_{F+NS(−)} = DV_{F+NS(+)}$. Thus, the measure of specific DTBZ binding is given by $DV_{SP(+)} =$

TABLE 1. Estimates of Total DTBZ Uptake, DV_{TOT} (mL/g), and the Fraction of Total Uptake That Is Estimated to Be Specific Binding, DV_{SP}/DV_{TOT}, as Calculated from Compartment Model Estimates of the Individual Rate Parameters K_1, k_2, k_3, and k_4

Region	DV_{TOT} (mL/g) $= (K_1/k_2) * [1 + (k_3/k_4)]$		DV_{SP}/DV_{TOT} $= (k_3/k_4)/[1 + (k_3/k_4)]$	
	(+)-DTBZ	(−)-DTBZ	(+)-DTBZ	(−)-DTBZ
Putamen	13.6 ± 1.3	5.2 ± 0.6	0.61 ± 0.13	0.24 ± 0.08
Caudate nucleus	12.3 ± 1.8	4.9 ± 0.5	0.54 ± 0.14	0.24 ± 0.05
Thalamus	5.2 ± 0.4	5.0 ± 0.5	0.40 ± 0.10	0.16 ± 0.06
Ventral diencephalon	5.8 ± 0.8	4.6 ± 0.5	0.46 ± 0.10	0.26 ± 0.09
Mesencephalon	5.2 ± 0.5	4.4 ± 0.5	0.47 ± 0.10	0.28 ± 0.04
Pons	4.8 ± 0.3	4.7 ± 0.4	0.39 ± 0.08	0.24 ± 0.04
Cerebellum	5.2 ± 0.5	5.0 ± 0.5	0.37 ± 0.09	0.21 ± 0.09
Hippocampus	5.0 ± 0.5	4.7 ± 0.5	0.43 ± 0.10	0.24 ± 0.08
Frontal cortex	4.6 ± 0.6	5.1 ± 0.6	0.46 ± 0.06	0.38 ± 0.06
Parietal cortex	4.8 ± 0.5	5.2 ± 0.7	0.44 ± 0.06	0.33 ± 0.04
Temporal cortex	4.5 ± 0.5	4.9 ± 0.7	0.40 ± 0.06	0.31 ± 0.04
Occipital cortex	4.6 ± 0.5	4.9 ± 0.8	0.42 ± 0.07	0.32 ± 0.04
White matter	3.7 ± 0.3	4.2 ± 0.4	0.45 ± 0.14	0.42 ± 0.10

$DV_{TOT(+)} - DV_{TOT(-)}$. When $DV_{SP(+)}$ is normalized to the nonspecific uptake, $DV_{F+NS(+)}$, we get the familiar measure of binding potential, BP. Using the inactive isomer for the free + nonspecific binding, we get the BP $= [DV_{TOT(+)} - DV_{TOT(-)}]/DV_{TOT(-)}$. The second column of Table 2 shows binding potential estimates calculated using the (−)-isomer.

Binding potentials for (+)-DTBZ calculated using (−)-DTBZ are around 2.0 for striatal regions, in good agreement with previous results (Koeppe *et al.*, 1996; 1997). Extrastriatal regions show little specific binding, with the ventral diencephalon and mesencephalon having the highest BP values of around 0.35–0.45. Thalamus, cerebellum, and other brainstem structures had BP values between 0.15 and 0.25, while cortical BPs averaged just over 0.05. These results suggest that the answer to the first question, whether there is quantifiable specific (+)-DTBZ binding in extrastriatal regions, certainly is "no" for cortex, but "possibly" in certain midbrain structures.

Since cortical binding is very low, it does appear as though the cortex can be used as a reference tissue for free + nonspecific binding. Table 2 also gives binding potential estimates (column 3) calculated using the occipital cortex DV_{TOT} from the (+)-DTBZ as the free + nonspecific measure. Note that these values are similar but not identical to those calculated using the (−)-isomer (column 2). The differences are probably not important in the basal ganglia, however any attempt to quantify BP in the midbrain structures could be problematic. The nonlinearity in the relationship between the two methods of calculating BP arises from the fact that the regional DV values from the (−)-isomer scans are not identical to the occipital cortex DV from the (+)-isomer scans. The ratio of these DV_{F+NS} estimates is shown in the far right column of Table 2. Two effects are present. First is the small contribution of specific binding to the occipital cortex in the (+)-DTBZ scans, which causes underestimation of BP due to the overestimation of the free + nonspecific component. Second is the regional variation in DV in the (−)-DTBZ scans. This can cause further underestimation in BP if the free + nonspecific DV is smaller in the region be evaluated than in the reference region (occipital cortex in this instance), as is the case for ventral diencephalon, mesencephalon, and hippocampus, for example. On the other hand, if the free + nonspecific DV is larger in the region to be evaluated than in the reference region, as is the case in the putamen, cerebellum, and thalamus, the underestimation may be reduced or eliminated altogether.

Table 3 shows DV_{TOT} and BP values calculated using the occipital cortex as a reference region for both baseline and postreserpine (+)-DTBZ studies for the patient with schizophrenia. The values for the baseline study are in the normal range and comparable to the values from the (+) isomer scans of the six subjects shown in Table 1. The values for the postreserpine scan are much lower in basal ganglia, but almost unchanged in extrastriatal regions.

IV. DISCUSSION AND SUMMARY

The use of multiple stereoisomers of the same ligand has been employed previously by many groups for examining the degree of specific binding in a variety of neuropharmacological imaging studies using PET or SPECT. Qualitative imaging of multiple enantiomers showing differing regional uptake distribution has been used to demonstrate the presence of specific binding of a radiotracer. When one enantiomer has negligible affinity for the specific binding site

TABLE 2. Binding Potential (BP) Estimates Using Both the Inactive (−)-Isomer and an Occipital Cortex Reference Tissue Region as Measures of Free + Nonspecific Binding

Region	BP from inactive isomer $[DV_{TOT(+)} - DV_{TOT(-)}]/DV_{TOT(-)}$	BP from reference region $[DV_{TOT(+)} - DV_{OCC(+)}]/DV_{OCC(+)}$	Ratio $DV_{TOT(-)}/DV_{OCC(+)}$
Putamen	2.01 ± 0.20	1.97 ± 0.25	0.99 ± 0.04
Caudate nucleus	1.87 ± 0.21	1.71 ± 0.20	0.95 ± 0.04
Thalamus	0.18 ± 0.02	0.14 ± 0.04	0.97 ± 0.04
Ventral diencephalon	0.45 ± 0.12	0.29 ± 0.09	0.89 ± 0.04
Mesencephalon	0.36 ± 0.11	0.16 ± 0.06	0.86 ± 0.04
Pons	0.16 ± 0.05	0.07 ± 0.03	0.92 ± 0.04
Cerebellum	0.16 ± 0.03	0.14 ± 0.05	0.98 ± 0.04
Hippocampus	0.25 ± 0.08	0.12 ± 0.04	0.90 ± 0.03
Frontal cortex	0.05 ± 0.03	−0.01 ± 0.03	0.94 ± 0.04
Parietal cortex	0.06 ± 0.03	0.03 ± 0.01	0.97 ± 0.02
Temporal cortex	0.05 ± 0.02	−0.01 ± 0.02	0.94 ± 0.02
Occipital cortex	0.07 ± 0.03	0.00 ± 0.00	0.94 ± 0.03

TABLE 3.   Estimates of Total DTBZ Uptake, DV_{TOT} (mL/g), and Binding Potential (BP) at Baseline and Two Weeks Following Initiation of Reserpine Therapy

Region	DV_{TOT} (mL/g)		BP		
	Baseline	Reserpine	Baseline	Reserpine	% of Baseline
Putamen	12.15	7.42	1.98	0.84	42
Caudate nucleus	10.37	6.21	1.54	0.53	35
Thalamus	4.87	4.71	0.20	0.17	85
Ventral diencephalon	5.60	4.80	0.38	0.19	50
Mesencephalon	4.64	4.31	0.11	0.07	59
Pons	4.49	4.47	0.10	0.11	105
Cerebellum	4.62	4.84	0.13	0.20	148
Hippocampus	5.38	4.50	0.32	0.12	36
Frontal cortex	4.25	4.31	0.04	0.07	—
Parietal cortex	4.31	4.33	0.06	0.08	—
Temporal cortex	4.02	4.13	−0.01	0.02	—
Occipital cortex	4.07	4.04	0.00	0.00	—

relative to another, then quantitative imaging of the inactive enantiomer has been used as a direct measure of the free + nonspecific binding levels in the analysis of the active enantiomer images. In this study we used active (+) and inactive (−) isomers of DTBZ to determine the extent of extrastriatal specific binding of (+)-DTBZ to the vesicular monoamine transporter (VMAT2) binding site.

Our previous work with (+)-DTBZ indicated that two tissue compartments were necessary to describe time–activity curves in cortex and other extrastriatal regions, as well as in striatum (Koeppe *et al.*, 1996). This result suggested that quantifiable levels of specific VMAT2 binding might exist in cortex and extrastriatal regions. Kinetic estimates of the individual compartment DVs in cortex from the earlier work and the six subjects reported in Table 1 distribute ~40–45% of the equilibrium (+)-DTBZ uptake into the specific binding compartment. Note that the ratio DV_{SP}/DV_{TOT} (reported in Table 1) is equivalent to BP/(1+BP), and thus the kinetically estimated BP in cortical regions are around 0.65–0.85. In contrast, blocking studies in animals suggest lower VMAT2 levels in cortex (Vander Borght *et al.*, 1995). If the (−)-DTBZ scans could be fitted using a single-tissue compartment model, or when fitted using a two-tissue model, almost none of the (−)-DTBZ uptake was estimated as being in the second compartment, then we would have greater confidence that a quantifiable amount of VMAT2 binding uptake exists in cortex. However, a single-tissue model did not describe the kinetics of (−)-DTBZ. Not only was the goodness-of-fit significantly improved statistically by the inclusion of a second tissue compartment (data not shown), but as shown in Table 1, ~25% of the uptake was estimated kinetically to be in the second tissue compartment. From these results we concluded that the necessity of a second tissue

compartment for (+)-DTBZ, even in cortex, is due (at least partially) to something other than specific binding.

Since we have assumed that the uptake of (+) and (−) isomers is the same except that the (+) isomer has, in addition, specifically bound tracer, it was surprising that the total distribution volume for (−)-DTBZ in cortical regions was higher than for (+)-DTBZ. This increased uptake is difficult to explain. While the difference was small, it was consistent across subjects, ranging from 2 to 15%, averaging 8%. Order effects were ruled out due to counterbalancing and because all six subjects showed the effect. Differences in the rate of plasma metabolite formation or errors in our metabolite assays might explain this result, although we did not detect any differences in the arterial plasma curves between the two isomers. This "global" difference between isomers was removed by normalization to white matter as described above, but remains a confounding factor in this study.

When the scans of the inactive isomer are used to provide a measure of free + nonspecific binding, (+)-DTBZ binding potential estimates no longer conflict with those from our animal studies. Cortical BPs (Table 2, second column) averaged only 0.06±0.03, meaning that the specific binding signal is only 6% that of free + nonspecific signal. The highest binding levels outside the striatum were found in the midbrain structures but still were below 0.5. Thus, we conclude that there is no (+)-DTBZ binding in cortex to be quantified accurately with PET. There is measurable binding in some extrastriatal structures, however, the sensitivity for detecting changes in binding in these structures will be limited since less than one third of the PET signal is related to specific VMAT2 binding. It is important to point out that the specific binding measures have been calculated after normalizing the (−)-DTBZ scans so that distribution volumes in white matter were the same for both isomers. This, in effect, assumes white matter to be devoid of specific binding sites. Our *in vitro* studies using (+)-[³H]DTBZ showed white matter binding to be ~25% of cortical and 1–2% of striatal levels. If this is true *in vivo*, then we have underestimated cortical binding by about 25%, increasing BP values in cortex, but only to about 0.08, still not enough to quantify accurately.

Although cortical binding of (+)-DTBZ is not adequate for accurate quantification, it does appear appropriately low for use as a reference tissue region assumed to be devoid of specific binding sites. Such a region is used to provide an estimate of the level of free + nonspecific uptake of tracer. Reference regions have been used for a variety of tracers and include cerebellum for the dopamine D2 receptors (Farde *et al.*, 1986), occipital cortex for opiate receptors (Frost *et al.*, 1989), and brainstem or pons for benzodiazepine receptors (Blomqvist *et al.*, 1990). Biases in receptor density or the binding potential estimates occur if the reference region does not reflect accurately the level of free + nonspecific binding. If some specific binding does occur in the reference region, then binding site density will be underestimated (Litton *et al.*, 1995; Delforge *et al.*, 1995; Lassen *et al.*, 1995). Furthermore, if the free + nonspecific binding levels are not uniform across the brain, errors will occur whenever the free + nonspecific level of the reference region is not the same as the particular region being investigated.

Table 2 (third column) gives binding potentials using occipital cortex as a reference region. The differences between this and the column 2 BPs, calculated using the inactive isomer, arise from the combined effects of small levels of specific binding in the occipital cortex and nonuniformity in free + nonspecific distribution across the brain. The final column of Table 2 gives the ratio in the DV$_{F+NS}$ measures for the two methods. What little binding that does exist in the cortex is missed, but since cortical binding is too low to quantify anyway, this is not a concern. However, the binding potentials in the ventral diencephalic nuclei, mesencephalon, pons, and hippocampus are underestimated by around 50%, making accurate quantification even more difficult.

The effects of reserpine on (+)-DTBZ binding are seen clearly from the results presented in Table 3. Basal ganglia binding potentials were reduced by 60–65%. Effects are less clear elsewhere. DV values were nearly unchanged in cortex, again suggesting that specific VMAT2 binding of (+)-DTBZ in cortex is very low. There are some regions such as ventral diencephalon and hippocampus which have detectable amounts of binding at baseline that show inhibition by reserpine of roughly the same magnitude as seen in the basal ganglia. Other regions, such as pons and cerebellum, which had quantifiable amounts of VMAT2 binding in the six dual-isomer subjects, demonstrated no response to reserpine. However, these regions had baseline BPs <0.20. At these binding levels, where less than one-sixth of the total

uptake of (+)-DTBZ is due to specific binding, it is not surprising that precise quantitative measures of VMAT2 binding cannot be made.

In summary, we have performed studies on six normal volunteers using both active and inactive stereoisomers of [¹¹C]DTBZ and on a single schizophrenic subject at baseline and 2 weeks after initiation of reserpine treatment. Both demonstrate that quantification of VMAT2 binding density in cortex is not feasible with (+)-DTBZ. VMAT2 binding in other extrastriatal regions such as the ventral diencephalic nuclei and mesencephalon was detectable, but with limited sensitivity due to the low level of nonspecific binding and to the nonuniformity of the free + nonspecific distribution. While cortical binding levels are not quantifiable with (+)-DTBZ and PET, they may be used as a reference tissue region for estimating the level of free + nonspecific binding in the striatum.

Acknowledgments

This work was supported by the Department of Energy Grant DE-FG02-87ER60561 and the National Institutes of Health Grant P01 NS-15655.

References

Blomqvist, G., Pauli, S., Farde, L., Eriksson, L., Persson, A., and Halldin, C. (1990). Maps of receptor binding parameters in the human brain—A kinetic analysis of PET measurements. *Eur. J. Nucl. Med.* **16**: 257–265.

Chan, G. L. Y., Holden, J. E., Stoessl, A. J., Doudet, D. J., Dobko, T., Morrison, K. S., Adam, M., Schulzer, M., Calne, D. B., and Ruth, T. J. (1999). Reproducibility studies with ¹¹C-DTBZ, a monoamine vesicular transporter inhibitor in healthy human subjects. *J. Nucl. Med.* **40**: 283–289.

DaSilva, J. N., and Kilbourn, M. R. (1992). In vivo binding of [¹¹C]tetrabenazine to vesicular monoamine transporters in mouse brain. *Life Sci.* **51**: 593–600.

Delforge, J., Pappata, P., Millet, P., Samson, Y., Bendriem, B, Jobert, A., Crouzel, C., and Syrota, A. (1995). Quantification of benzodiazepine receptors inhuman brain using PET, [¹¹C]flumazenil, and a single-experiment protocol. *J. Cereb. Blood Flow Metab.* **15**: 284–300.

Farde L., Hakan, H., and Ehrin, E. (1986). Quantitative analysis of D2 dopamine receptor binding in the living human brain by PET. *Science* **231**: 258–261.

Henry, J.-P., and Scherman, D. (1989). Radioligands of the vesicular monoamine transporter and their use as markers of monoamine storage vesicles. *Biochem. Pharmacol.* **38**: 2395–2404.

Jewett, D. M., Kilbourn, M. R., and Lee, L. C. (1997). A simple synthesis of [¹¹C]dihydrotetrabenazine (DTBZ). *Nucl. Med. Biol.* **24**: 197–199.

Kilbourn, M. R., Lee L., Vander Borght, T., Jewett, D., and Frey, K. (1995). Binding of α-dihydrotetrabenazine to the vesicular monoamine transporter is stereospecific. *Eur. J. Pharmacol.* **278**: 249–252.

Koeppe, R. A., Frey, K. A., Kume, A., Albin, R., Kilbourn, M. R., and Kuhl, D. E. (1997). Equilibrium versus compartmental analysis for assessment of the vesicular monoamine transporter using (+)-α-[¹¹C]dihydrotetrabenazine (DTBZ) and PET. *J. Cereb. Blood Flow Metab.* **17**: 919–931.

Koeppe, R. A., Frey, K. A., Vander Borght, T. M., Karlamangla, A., Jewett, D. M., Lee, L. C., Kilbourn, M. R., and Kuhl, D. E. (1996). Kinetic evaluation of [¹¹C]dihydrotetrabenazine by dynamic PET: Measurement of the vesicular monoamine transporter. *J. Cereb. Blood Flow Metab.* **16**: 1288–1299.

Lassen, N. A., Bartenstein, P. A., Lammertsma, A. A., Prevett, M. C., Turton, D. R., Luthra, S. K., Osman, S., Bloomfield, P., Jones, T., Patsalos, P. N., O'Connell, M. T., Duncan, J. S., and Vanggaard Andersen, J. (1995). Benzodiazepine receptor quantification in vivo in humans using [11C]flumazenil and PET: application of the steady-state principle. *J. Cereb. Blood Flow Metab.* **15**: 152–165.

Litton, J. E., Hall, H., and Pauli, S. (1994). Saturation analysis in PET—Analysis of errors due to imperfect reference regions. *J. Cereb. Blood Flow Metab.* **14**: 358–361.

Minoshima, S., Koeppe, R. A., Fessler, J. A., Mintun, M. A., Berger, K. L., Taylor, S. F., and Kuhl, D. E. (1993). Integrated and automated data analysis method for neuronal activation studies using O-15 water PET.

In "*Quantification of Brain Function—Tracer Kinetics and Image Analysis in Brain PET*" (K. Uemura, N. A. Lassen, T. Jones, and I. Kanno, Eds.), International Congress Series 1030, pp. 409–418. Excepta Medica, Tokyo.

Minoshima, S., Koeppe, R. A., Frey, K. A., and Kuhl, D. E. (1994). Anatomical standardization: Linear scaling and nonlinear warping of functional brain images. *J. Nucl. Med.* **35**: 1528–1537.

Vander Borght, T. M., Sima, A. A. F., Kilbourn, M. R., Desmond, T. J., and Frey, K. A. (1995). [3H]Methoxytetrabenazine: A high specific activity ligand for estimating monoaminergic neuronal integrity. *Neuroscience* **68**: 955–962.

32

Amphetamine-Induced Dopamine Release: Duration of Action Assessed with [11C]Raclopride in Anesthetized Monkeys

RICHARD E. CARSON, MICHAEL A. CHANNING, BIK-KEE VUONG, HIROSHI WATABE,
PETER HERSCOVITCH, and WILLIAM C. ECKELMAN

W. G. Magnuson Clinical Center, Positron Emission Tomography Department, National Institutes of Health, Bethesda, Maryland 20892-1180

Amphetamine-induced dopamine release has been extensively studied with receptor-binding radiopharmaceuticals. Changes in specific binding have been interpreted as changes in synaptic dopamine, based on microdialysis experiments. Using data obtained with the bolus/infusion (B/I) method in combined PET-microdialysis experiments in monkeys, we previously extended the conventional compartment model for [11C]raclopride to account for the time-varying dopamine concentration. Model extrapolations predict that specific binding should return toward preamphetamine baseline values within 1–2 h postamphetamine. The goal of this study was to characterize the duration of action of amphetamine by performing multiple [11C]raclopride scans in the same animal on 1 day. Studies were performed in three isoflurane-anesthetized rhesus monkeys with the B/I method. Three [11C]raclopride scans were performed on each day, starting 2.5 h apart. Once equilibrium was reached during the first scan (50 min), 0.4 mg/kg of amphetamine was administered i.v. The binding potential (BP) was assessed from the concentration ratio of basal ganglia to cerebellum and by the reference region method. In the first scan, control BP values (30–50 min) were 2.51 ± 0.05, similar to values obtained previously. After amphetamine, the measured BP was reduced by 30 ± 3, 28 ± 6, and $30 \pm 10\%$ at 70, 220, and 330 min postamphetamine, respectively. Control studies consisting of 3 [11C]raclopride scans without amphetamine were performed in the same animals at a later date and showed no significant changes in BP. These data support the concept that the amphetamine-induced reduction in [11C]raclopride binding is produced by additional factors beyond the change in synaptic dopamine.

I. INTRODUCTION

Most PET neuroreceptor studies initially focused on determining changes in receptor concentration as a function of disease or measurement of receptor occupancy by drugs. A relatively new approach provides an estimate of changes in synaptic neurotransmitter concentration. This method determines the change in radiotracer binding levels after administration of pharmacologic agents that affect neurotransmitter levels. With appropriate modeling techniques, the change in radiotracer binding can be attributed to changes in the level of synaptic neurotransmitter that competes with the radiotracer for receptor binding.

Two experimental designs have been used to estimate receptor binding changes in such an intervention paradigm. In the first approach, two scans are performed, each with a bolus injection of high specific activity tracer. In the first scan, control levels of binding are measured, for example, by determining the distribution volume *V* by compartment modeling (Koeppe *et al.*, 1991) or graphical analysis (Logan *et al.*, 1990). Then, after the pharmacological intervention, a second measurement of binding is made with a second tracer injection. This approach has been used successfully with the D_2 ligand [11C]raclopride (Farde *et al.*, 1989), as well as with several other tracers. For example, Dewey *et al.* (1993, 1995) demonstrated the effects of changes in synaptic dopamine induced by direct manipulation of the dopamine system and by indirect pharmacological interventions.

The second study design consists of administering the tracer as a combined bolus plus continuous infusion (B/I).

Continuous infusion of tracers with reversible binding characteristics has been used for absolute quantification of receptor parameters (Carson *et al.*, 1993; Laruelle *et al.*, 1994). This approach can be extended to measure short-term changes in free receptor concentration. First, the B/I administration of tracer is performed to achieve constant radioactivity levels in all brain regions. Once equilibrium is achieved, control binding levels can be determined. Then, a stimulus is administered while the infusion of radiotracer continues, and the change in specific binding of the tracer can be monitored. Although a B/I scan is more complex technically than a bolus scan, this design has the advantage of measuring pre- and postintervention binding levels with a single administration of tracer. Data analysis is simpler, and the total study time for the patient is also reduced, although the statistical quality of the B/I results may be reduced as well, particularly for short-lived PET tracers. Note, however, that a B/I design is not possible for all tracers because the ability to reach equilibrium in an appropriate length of time depends upon the tracer kinetics in plasma and brain.

Our initial studies of amphetamine-induced dopamine release involved comparison of the dual bolus and B/I methods in monkeys (Carson *et al.*, 1997). With the dual bolus approach using a metabolite-corrected input function, the percentage decrease in specific binding, measured by various analytic schemes ranged from 22 to 42% with a dose of 0.4 mg/kg of amphetamine. In B/I scans, from the pre- and postamphetamine levels of specific binding determined directly from the tissue concentration values (basal ganglia/cerebellum −1), the decrease in specific binding after amphetamine was $19 \pm 16\%$ (no significant difference between bolus and B/I values). This B/I approach was then applied to compare amphetamine-induced changes in [^{11}C]raclopride in normal controls and patients with schizophrenia. A larger change was found in the patients (Breier *et al.*, 1997), replicating the finding obtained with single photon emission computed tomography (SPECT) and [^{123}I]IBZM (iodobenzamide) (Laruelle *et al.*, 1996) which also used the B/I approach.

Both the B/I and dual bolus methods provide an index of stimulus-induced changes in neurotransmitter concentration. To develop a more quantitative interpretation of these findings, we performed simultaneous PET and microdialysis experiments in rhesus monkeys (Breier *et al.*, 1997). Animals were studied twice on the same day using the B/I paradigm and two doses of amphetamine (0.2 and 0.4 mg/kg). Microdialysis measurements of striatal extracellular dopamine were made at 10-min intervals throughout each study. The time–activity data from PET were combined with the microdialysis data in an extension of the conventional compartment model for analysis of neuroreceptor ligands (Endres *et al.*, 1997). In this extended model (Fig. 1), the binding rate of [^{11}C]raclopride, $k_3(t)$, is a function of time because of the time-dependent occupancy of

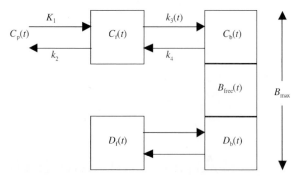

FIGURE 1. Extended receptor model for [^{11}C]raclopride that explicitly accounts for time-varying levels of free (D_f) and bound (D_b) dopamine. The total receptor concentration B_{max} is the sum of bound raclopride $C_b(t)$, bound dopamine $D_b(t)$, and free receptor $B_{free}(t)$. The parameters K_1 and k_2 represent transport across the blood–brain barrier. The binding rate of [^{11}C]raclopride, $k_3(t)$, which is proportional to $B_{free}(t)$, is time-dependent because of the time-varying dopamine concentration following amphetamine. k_4 represents the receptor dissociation rate.

the receptor by the changing dopamine concentration. The model assumes that bound and free dopamine are in rapid equilibrium and that dopamine binds to the receptor with a single affinity. This model was able to successfully fit the 90-min time–activity data from the paired scans in each animal with the addition of a single parameter, the K_d of dopamine.

Using this model, mathematical simulations of B/I experiments were performed to determine the relationship between the measured change in specific binding and the underlying release of dopamine. The simulations showed that the measured change in specific binding was proportional to the integral of the dopamine pulse (Endres *et al.*, 1997). In a subsequent analysis which examined the optimal characteristics of ligands for bolus or B/I scans, the relationship found for [^{11}C]raclopride was extended to other tracers and to the paired bolus method (Endres and Carson, 1998). Thus, based on the extended model, between-group differences in stimulus-induced changes in specific binding, like those found between schizophrenic patients and normal controls, can be interpreted as differences in the integral of the released neurotransmitter.

Analysis of the human data suggested that statistical noise in the measurements was substantial. Two approaches were taken to improve the quality of B/I results: (1) To maintain the simple clinical utility of the method, the time periods for pre- and poststimulus measurements were optimized; as a result, the statistical significance of the difference in specific binding between patients with schizophrenia and normal subjects was increased (Watabe *et al.*, 2000). (2) Several more complex model-based methods were also developed (Watabe *et al.*, 1998), including (a) fitting the preamphetamine data to the conventional receptor model and using that result to extrapolate the basal ganglia curve, and (b) a direct fit to the extended model.

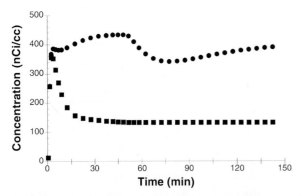

FIGURE 2. Predictions of the extended receptor model (Fig. 1) for the time course of basal ganglia (●) and cerebellum (■) uptake for a B/I scan with monkey microdialysis data used to predict the effect of amphetamine administration at 50 min. The model predicted that the basal ganglia [^{11}C]raclopride binding should show significant return toward baseline levels within 1–2 h postamphetamine.

Although the extended receptor model (Fig. 1) successfully described the 90-min time–activity data collected in the monkey B/I studies, further examination of the model suggested some inconsistencies. Because of the rapid clearance of synaptic dopamine after amphetamine administration (halftime of 10–15 min) (Endres *et al.*, 1997), model extrapolations (Fig. 2) beyond 90 min predict that specific binding should return towards preamphetamine baseline values within 1–2 h postamphetamine. However, SPECT studies with the D$_2$ antagonist [^{123}I]IBZM demonstrate long-lived reductions in receptor binding (Laruelle *et al.*, 1997), although this prolonged effect may be due at least in part to the kinetics of [^{123}I]IBZM, which are slower than [^{11}C]raclopride.

Therefore, we performed a study in rhesus monkey ($n = 3$) using multiple infusions of [^{11}C]raclopride in a single day to characterize the duration of action of amphetamine and to test the validity of the extended receptor model.

II. METHODS

[^{11}C]Raclopride studies were performed in three rhesus monkeys with the B/I method (Carson *et al.*, 1997). The animals were initially anesthetized with ketamine, endotracheal intubation was performed, and an intravenous line was inserted. The animals were transported to the PET suite, placed under isoflurane anesthesia (1–2%), and positioned in a stereotactic headholder on the scanning table. Blood pressure, EKG, temperature, and end-tidal pCO$_2$ were monitored continuously. All studies were performed under a protocol approved by the NIH Clinical Center Animal Care and Use Committee.

Three raclopride scans were performed on each day, starting 150 min apart. Data acquisition was performed contin-

uously with the third scan having a 90-min duration. The starting radioactivity dose was 4–6 mCi with specific activity at injection of 200–600 Ci/mmol. For all administrations, the total dose was delivered as a bolus plus continuous infusion with the bolus dose equal to that delivered in 60 min of infusion; i.e., $K_{bol} = 60$ min (Carson *et al.*, 1997). During the first scan only, once equilibrium was reached (50 min), 0.4 mg/kg of amphetamine was administered i.v. over 1 min. In previous microdialysis experiments in anesthetized rhesus monkeys (Endres *et al.*, 1997), this produced ∼1000% increases in extracellular dopamine. Two to 4 months later, control [^{11}C]raclopride studies were performed in the same animals following the same protocol.

Dynamic frames were acquired in 3D mode on the GE Advance scanner. Regions-of-interest were drawn manually on summed scans and used to produce time–activity curves (TACs) from basal ganglia (*B*) and cerebellum (*C*). Arterial blood samples drawn from an indwelling port were acquired throughout the study, although no metabolite correction was performed since ratio and reference region analyses were performed. Plasma free fraction was measured. In addition, the blood data were used to correct *B* and *C* for intravascular radioactivity in the TACs, resulting in changes of 5–15% in the cerebellum curves with smaller changes in the basal ganglia curves. Once equilibrium was achieved, baseline binding potential (*BP*) was assessed with a ratio measure ($BP = B/C - 1$) and the postamphetamine reduction in specific binding was calculated at later times. For the second and third [^{11}C]raclopride scans, *BP* was computed at later times once equilibrium was attained. In addition, *BP* was determined using the simplified reference tissue model which determines the input function from the cerebellum TAC (Lammertsma and Hume, 1996). This method was applied for various time intervals (0–50, 0–90, and 0–140 min) of the [^{11}C]raclopride scans, except for the amphetamine scan where data analysis was limited to 0–50 min.

III. RESULTS

The three administrations of [^{11}C]raclopride produced very similar cerebellum radioactivity levels (normalized to injected dose) and plasma free fraction values (mean of all studies $9.3 \pm 0.9\%$). In the first scan, baseline *BP* values by the ratio method (40–50 min) were 2.51 ± 0.05 and 2.44 ± 0.31 on the two scanning days (amphetamine and control). These values were very similar to those obtained previously (Carson *et al.*, 1997). The ratio *BP* values were 6–10% higher than the reference region values (fit over 0–50 min) of 2.26 ± 0.11 and 2.29 ± 0.34, perhaps because the B/I protocol did not achieve ideal equilibrium or because of bias in the simplified reference region model due to the use

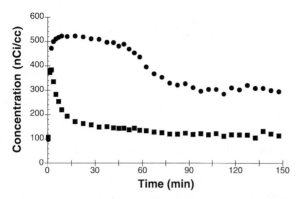

FIGURE 3. Time–activity curves in basal ganglia (●) cerebellum (■) for a [^{11}C]raclopride B/I scan acquired on the GE Advance with 3D acquisition. A dose of 0.4 mg/kg amphetamine was administered intravenously at 50 min postinjection which produced a reduction in basal ganglia specific binding.

TABLE 1. Percentage Reduction in Binding Potential (Tissue Ratio Method)

Scan	Ratio period	Time post-amphetamine (min)	Amphetamine Mean	Amphetamine SD	Control Mean	Control SD
1	40–50	0	0	0	0	0
	70–90	20–40	18.2	9.2	4.7	10.6
	100–140	50–90	30.2	2.9	5.6	11.6
2	40–50	140–150	25.7	4.2	4.2	4.5
	70–90	170–190	29.7	3.1	10.7	3.2
	100–140	200–240	28.0	5.5	12.9	5.1
3	40–50	290–300	21.8	3.8	3.2	3.7
	70–90	320–340	30.4	9.7	1.3	4.1

Note. Mean and standard deviation ($n = 3$) of the percentage reduction in binding potential (basal ganglia/cerebellum −1) measured at the specified times and calculated compared to the baseline value obtained at 40–50 min of the first B/I scan. In the amphetamine scans, 0.4 mg/kg was administered at 50 min.

of a single tissue compartment model. There was a slight (2–3%) rise in the reference region *BP* values when the fitting period was extended to 140 min. Similar patterns of *BP* values were found in scans 2 and 3.

Figure 3 shows an example of TACs from basal ganglia and cerebellum in a first scan with amphetamine administered at 50 min. The B/I method was designed to produce rapid equilibrium in basal ganglia and cerebellum, so that *BP* could be measured by 50 min. Following a dose of 0.4 mg/kg of amphetamine, specific binding of [^{11}C]raclopride was reduced by ~30% and showed no indication of returning toward baseline values during the 150-min scan. This result differs from the predictions of the extended receptor model (Fig. 2).

Table 1 shows the percentage reductions in *BP* calculated by the ratio method compared to the baseline value (40–50 min of the first scan). After amphetamine, the measured *BP* was reduced by 30±3% (100–140 min). In the second and third scans, *BP* at equilibrium did not increase. For 70–90 min postinjection of scans 2 and 3 (~3 and 5 h postamphetamine, respectively), *BP* was reduced by 30±3 and 30±10%, respectively, compared to the preamphetamine baseline. Thus, there was no evidence that [^{11}C]raclopride binding was returning to baseline levels over this 5.5-h period. Averaging the postamphetamine data together gives an estimated reduction in specific binding of 30.1±3.8%. When the same animals were studied with the same protocol but without amphetamine, the measured reduction in specific binding was 5.9±2.4%. Using the reference region method, the magnitude of the percentage reductions was smaller, but the pattern was unchanged: reductions of 20.6±3.4 and 3.5±3.1% over the course of the three scans for the amphetamine and control scans, respectively.

IV. DISCUSSION

This study shows that the reduction in [^{11}C]raclopride binding after amphetamine-induced dopamine release is long-lived, disagreeing with the prediction of the extended receptor model and consistent with previous studies with [^{123}I]IBZM. Thus, this prolonged reduction is not simply due to slower kinetics of [^{123}I]IBZM, and the simple model of competition between neurotransmitter and radioligand is not adequate to explain these data. Results of other recent studies also conflict with a simple competition model. Abi-Dargham *et al.* (1999) evaluated the effect of amphetamine on D$_1$ receptor binding with [^{11}C]NNC 756 and unexpectedly found no changes in binding. Tsukada *et al.* (1999) used several agents to modulate synaptic dopamine (including amphetamine) and measured changes in [^{11}C]raclopride simultaneously with dopamine microdialysis in monkeys. They found that the relationship between raclopride binding changes and extracellular dopamine was not consistent, suggesting that multiple mechanisms were involved.

There are many factors to consider in interpreting these data. The failure of the extended model to accurately predict these data could be due to a limitation in its assumptions. For example, the multiple affinity states of the dopamine receptor were ignored. Also, the extended receptor model assumes that the microdialysis data accurately reflect the shape of the concentration curve of free dopamine available to bind to the receptor, i.e., that extracellular and synaptic concentrations parallel each other. Furthermore, following the amphetamine-induced peak, extracellular dopamine levels eventually reach a new equilibrium level which is higher than the original baseline value (Endres *et al.*, 1997). This prolonged increase could contribute to long-lived [^{11}C]raclopride binding reductions, particularly if the additional dopamine occupies high affinity sites.

Clearly, these long-lived changes in specific binding may be due to prolonged alterations in the dopamine system induced by amphetamine (Laruelle *et al.*, 1997), including changes in the balance between high and low affinity sites, or changes in free receptor number, e.g., due to internalization. The [^{11}C]raclopride displacement (Fig. 3) may also represent a combination of the dopamine concentration changes and one or more of these factors. Additional studies are necessary to clarify the mechanism of the prolonged amphetamine effect on D$_2$ antagonist binding and to improve the accuracy of the models for receptor-binding ligands.

Acknowledgments

The authors acknowledge the excellent technical support of the NIH PET staff, particularly Ms. Wendy Greenley, and the veterinary assistance of John Bacher DVM and his staff.

References

Abi-Dargham, A., Simpson, N., Kegeles, L., Parsey, R., Hwang, D. R., Anjilvel, S., Zea-Ponce, Y., Lombardo, I., Van Heertum, R., Mann, J. J., Foged, C., Halldin, C., and Laruelle, M. (1999). PET studies of binding competition between endogenous dopamine and the D1 radiotracer [^{11}C]NNC 756. *Synapse* **32**: 93–109.

Breier, A., Su, T.-P., Saunders, R., Carson, R. E., Kolachana, B. S., de Bartolomeis, A., Weinberger, D. R., Weisenfeld, N., Malhotra, A. K., Eckelman, W. C., and Pickar, D. (1997). Schizophrenia is associated with elevated amphetamine-induced synaptic dopamine concentrations: Evidence from a novel positron emission tomography method. *Proc. Natl. Acad. Sci. USA* **94**: 2569–2574.

Carson, R. E., Breier, A., de Bartolomeis, A., Saunders, R. C., Su, T. P., Schmall, B., Der, M. G., Pickar, D., and Eckelman, W. C. (1997). Quantification of amphetamine-induced changes in [^{11}C]raclopride binding with continuous infusion. *J. Cereb. Blood Flow Metab.* **17**: 437–447.

Carson, R. E., Channing, M. A., Blasberg, R. G., Dunn, B. B., Cohen, R. M., Rice, K. C., and Herscovitch, P. (1993). Comparison of bolus and infusion methods for receptor quantitation: Application to [^{18}F]-cyclofoxy and positron emission tomography. *J. Cereb. Blood Flow Metab.* **13**: 24–42.

Dewey, S. L., Smith, G. S., Logan, J., Alexoff, D., Ding, Y. S., King, P., Pappas, N., Brodie, J. D., and Ashby, C. R. (1995). Serotonergic modulation of striatal dopamine measured with positron emission tomography (PET) and in-vivo microdialysis. *J. Neurosci.* **15**: 821–829.

Dewey, S. L., Smith, G. S., Logan, J., Brodie, J. D., Fowler, J. S., and Wolf, A. P. (1993). Striatal binding of the PET ligand ^{11}C-raclopride is altered by drugs that modify synaptic dopamine levels. *Synapse* **13**: 350–356.

Endres, C. J., and Carson, R. E. (1998). Assessment of dynamic neurotransmitter changes with bolus or infusion delivery of neuroreceptor ligands. *J. Cereb. Blood Flow Metab.* **18**: 1196–1210.

Endres, C. J., Kolachana, B. S., Saunders, R. C., Su, T., Weinberger, D., Breier, A., Eckelman, W. C., and Carson, R. E. (1997). Kinetic modeling of [C-11]raclopride: Combined PET-microdialysis studies. *J. Cereb. Blood Flow Metab.* **17**: 932–942.

Farde, L., Eriksson, L., Blomquist, G., and Halldin, C. (1989). Kinetic analysis of central [^{11}C]raclopride binding to D$_2$-dopamine receptors studied by PET: A comparison to the equilibrium analysis. *J. Cereb. Blood Flow Metab.* **9**: 696–708.

Koeppe, R. A., Holthoff, V. A., Frey, K. A., Kilbourn, M. R., and Kuhl, D. E. (1991). Compartmental analysis of [^{11}C]Flumazenil kinetic for the estimation of ligand transport rate and receptor distribution using positron emission tomography. *J. Cereb. Blood Flow Metab.* **11**: 735–744.

Lammertsma, A. A., and Hume, S. P. (1996). Simplified reference tissue model for PET receptor studies. *NeuroImage* **4**: 153–158.

Laruelle, M., Abi-Dargham, A., al-Tikriti, M. S., Baldwin, R. M., Zea-Ponce, Y., Zoghbi, S. S., Charney, D. S., Hoffer, P. B., and Innis, R. B. (1994). SPECT quantification of [^{123}I]iomazenil binding to benzodiazepine receptors in nonhuman primates. II. Equilibrium analysis of constant infusion experiments and correlation with in vitro parameters. *J. Cereb. Blood Flow Metab.* **14**: 453–465.

Laruelle, M., Abi-Dargham, A., van Dyck, C. H., Gil, R., D'Souza, C. D., Erdos, J., McCance, E., Rosenblatt, W., Fingado, C. L., Zoghbi, S. S., Baldwin, R. M., Seibyl, J. P., Zoghbi, S. S., Krystal, J. H., Charney, D. S., and Innis, R. B. (1996). Single photon emission computerized tomography imaging of amphetamine-induced dopamine release in drug-free schizophrenic subjects. *Proc. Natl. Acad. Sci. USA* **93**: 9235–9240.

Laruelle, M., Iyer, R. N., Al-Tikriti, M. S., Zea-Ponce, Y., Malison, R., Zoghbi, S. S., Baldwin, R. M., Kung, H. F., Charney, D. S., Hoffer, P. B., Innis, R. B., and Bradberry, C. W. (1997). Microdialysis and SPECT measurements of amphetamine-induced dopamine release in nonhuman-primates. *Synapse* **24**: 1–14.

Logan, J., Fowler, J. S., Volkow, N. D., Wolf, A. P., Dewey, S. L., Schyler, D. J., MacGregor, R. R., Hitzemann, R., Bendriem, B., Gatley, S. J., and Christman, D. (1990). Graphical analysis of reversible radioligand binding from time–activity measurements applied to [N-^{11}C-methyl]-(-)-Cocaine: PET studies in human subjects. *J. Cereb. Blood Flow Metab.* **10**: 740–747.

Tsukada, H., Nishiyama, S., Kakiuchi, T., Ohba, H., Sato, K., and Harada, N. (1999). Is synaptic dopamine concentration the exclusive factor which alters the in vivo binding of [C-11]raclopride?: PET studies combined with microdialysis in conscious monkeys. *Brain Res.* **841**: 160–169.

Watabe, H., Endres, C. J., Breier, A., Schmall, B., Eckelman, W. C., and Carson, R. E. (2000). Measurement of dopamine release with continuous infusion of [C-11]raclopride: Optimization and signal-to-noise considerations. *J. Nucl. Med.* **41**: 522–530.

Watabe, H., Endres, C. J., and Carson, R. E. (1998). Modeling methods for the determination of dopamine release with [^{11}C]raclopride and constant infusion. *Neuroimage* **7**: A57.

33

Amphetamine-Induced Dopamine Release Greater in Ventral Than in Dorsal Striatum

J. C. PRICE, W. C. DREVETS,* P. E. KINAHAN, B. J. LOPRESTI, and C. A. MATHIS

*Departments of Radiology and *Psychiatry, University of Pittsburgh School of Medicine, Pittsburgh, Pennsylvania 15213*

Amphetamine-induced dopamine release in the baboon striatum was assessed using positron emission tomographic (PET) measures of [^{11}C]raclopride specific binding to dopamine D_2/D_3 receptors at baseline and after d-amphetamine (AMPH) administration. The magnitude of the reduction in [^{11}C]raclopride binding after AMPH administration was twofold greater in the anteroventral striatum (ventral caudate, anteroventral putamen, and nucleus accumbens) than the dorsal striatum (dorsal caudate). Computer simulations were performed to examine aspects of the experiment that might falsely lead to the detection of a twofold greater AMPH response in the anteroventral striatum relative to the dorsal caudate. The simulations examined potential biases associated with the limited resolution of PET imaging (partial volume) and alignment errors in the coregistration of the PET and magnetic resonance image data. The simulation studies yielded biases that were significantly less in magnitude than the observed results. The observed regional differences (across the range tested) in the sensitivity to AMPH are consistent with microdialysis studies in rats that showed that the magnitude of dopamine release in response to AMPH concentration is greater in ventral than in dorsal striatal regions. The results of the present study strengthen the validity of PET imaging measures of mesolimbic endogenous dopamine release for investigations of dopaminergic function in relation to disorders such as depression, psychosis, and substance abuse.

I. INTRODUCTION

Extracellular dopamine increases several-fold in the striatum within 20–40 min after d-amphetamine (AMPH) administration (Zetterstrom *et al.*, 1983; Kuczenski and Segal, 1989). During the administration of drugs that inhibit dopamine reuptake (cocaine) or stimulate its release (AMPH), higher dopamine levels were found in the accumbens shell (anteroventral striatum, AVS) than in the dorsal caudate (DCA) (Carboni *et al.*, 1989; Kuczenski and Segal, 1992; Di Chiara *et al.*, 1993).

Dopaminergic function can be assessed using positron emission tomography (PET), [^{11}C]raclopride ([^{11}C]RAC), and AMPH. After AMPH administration, dopamine concentrations increase and this leads to a reduction in [^{11}C]RAC binding to dopamine D_2/D_3 receptors. Pre- and post-AMPH [^{11}C]RAC binding differences reflect dopaminergic function with post-AMPH (≤ 1 mg/kg) binding reductions ($\sim 20\%$) that exceed test-retest variability (Dewey *et al.*, 1993; Breier *et al.*, 1997; Carson *et al.*, 1997; Hartvig *et al.*, 1997; Price et al., 1998). Experiments showed that AMPH-induced [^{11}C]RAC binding reductions were well correlated with the post-AMPH microdialysis dopamine pulse integral (Endres *et al.*, 1997).

The aim of this study was to further examine pre- and post-AMPH [^{11}C]raclopride binding in baboons. This examination investigated the extent to which the twofold greater response observed in the AVS relative to the DCA could be due to partial volume averaging or image alignment errors.

*Physiological Imaging
of the Brain with PET*

211

II. METHODS

Isoflurane anesthetized baboons were studied with magnetic resonance (MR) imaging and PET. Ketamine and atropine were administered i.m. to anesthetize and control salivation and heart rate.

A. MR Methods

An MR scan was obtained for each baboon at the University of Pittsburgh MR Research Center on a GE Signa 1.5 Tesla scanner using a standard head coil. A T1-weighted sagittal series, an axial T1-weighted sequence, and a volumetric spoiled gradient recalled (SPGR) sequence with parameters for high contrast among gray matter, white matter, and cerebrospinal fluid (CSF) were acquired in the coronal plane (see Drevets et al., 1999, for imaging parameters). MR data were coregistered to the PET data using an automated image registration algorithm (Woods et al., 1993).

B. PET Methods

PET data were acquired using a Siemens-CTI 951R/31 PET scanner ($n = 10$) in 2D imaging mode (31 planes, axial slice width, 3.4 mm) and an ECAT HR$^+$ PET scanner ($n = 2$) in 3D imaging mode (63 planes, axial slice width, 2.4 mm). A transmission (10–15 min) scan was obtained before each emission scan ([^{68}Ge]/[^{68}Ga]) to correct for attenuation. Images were reconstructed with a Hanning filter (Nyquist frequency) and corrected for radioactive decay, dead time, and scatter (3D only).

High specific activity (>1500 Ci/mmol) [^{11}C]raclopride studies were performed at baseline and after d-amphetamine injection (0.3, 0.6, or 1.0 mg/kg). Twelve studies were performed in five baboons. The radiotracer was injected (10 mCi in 5–7 mL of saline) as a slow bolus (30 s); and a 90-min dynamic PET scan was performed. The AMPH was administered over 2 min, beginning 5 min before the second [^{11}C]RAC injection (Dewey et al., 1993). A minimum of 2 h separated the radioligand injections. The [^{11}C]RAC input function was determined from 35 hand-drawn arterial blood samples collected throughout the study (20 samples over the initial 2 min); plasma samples were assayed for radioactivity concentration using a gamma counter (Packard, Model 5003). Blood samples were collected for metabolite analyses (2, 10, 30, 45, 60, and 90 min) and these data yielded the fraction of unchanged [^{11}C]RAC over time. The [^{11}C]RAC input function was then determined from the total plasma radioactivity. Details of the analyses are described elsewhere (Drevets et al., 1999).

C. Data Analysis

Regions of interest (ROIs) were generated for the striatum and cerebellum as left/right averages for the dorsal caudate (DCA), middle caudate (MCA), dorsal putamen (DPU), anteroventral striatum (AVS), ventral putamen (VPU), and cerebellum (Cer). The mean striatal ROI volumes (mm^3) were as follows: AVS, 484 ± 145; DCA, 548 ± 67; MCA, 363 ± 131; DPU, 533 ± 191; VPU, 532 ± 157. ROIs were positioned for the 2D 951R images 0–17 mm (five planes) and for 3D HR$^+$ images 25–38 mm from the center of the axial field of view (Drevets et al., 1999). The kinetic data were analyzed using the Logan graphical method (Logan et al., 1990), which yields a slope that is directly related to the total radioligand distribution volume (DV). The Logan DV measures were used to determine the regional binding potential (BP) measures: BP = [DV Str/DV Cer] − 1, as described by Lammertsma et al. (1996).

D. Computer Simulations

Computer simulations were performed to examine the validity of the substriatal measures (e.g., AVS, DCA, and PUT:DPU or VPU) of dopamine activity (Drevets et al., 1999). These simulations were designed to examine potential biases that arise from the limited resolution of PET imaging (partial volume) and cross-modality image alignment errors, aspects of the experiment which might falsely lead to the detection of a two-fold greater AMPH response in the AVS relative to the DCA. For this work, the DV was the primary measure of interest and the BP value was then determined from the DV (see Methods). The DV parameter was chosen because it is equivalent to the equilibrium ratio of the radioactivity concentrations in tissue and plasma (Lassen and Perl, 1979); and a change in tissue radioactivity concentration would cause a linear change in DV.

1. Reference Baseline DV Image

A Logan distribution volume "reference image" was created at the "true" spatial resolution of the scanner. The image was generated from a measured DV image that was obtained after a baseline [^{11}C]RAC study in one baboon. The image was manually segmented into two parts: specific (striatum) and nonspecific (extrastriatal) binding. The segmented striatum was then manually divided into AVS, DCA, MCA, and putamen, across planes using MR-guided definitions.

The segmented DV image was then blurred using a FWHM value that was based on the measured intrinsic scanner resolution and resolution degradation associated with scatter, image smoothing, and image acquisition. For the 951R scanner (2D mode), the resolution was estimated to be 7.8 mm transverse and 5.3 mm axially, based on measures of the FWHM of an unsmoothed point source in air (transverse, 6.0 ± 0.5 mm and axial, 5 ± 1 mm, Spinks et al., 1992),

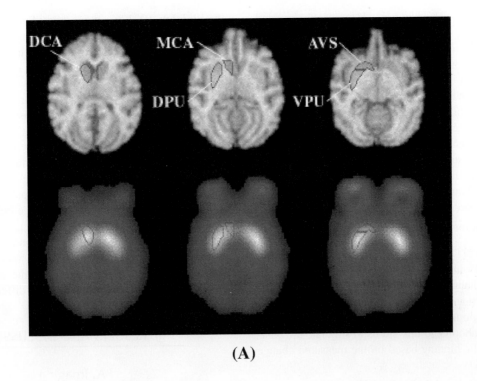

(A)

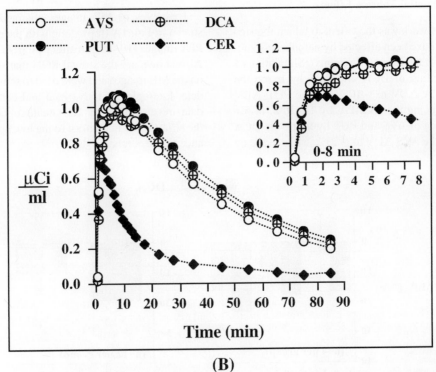

(B)

FIGURE 1.   (A) Examples of the striatal regions-of-interest (ROIs) that were created on MR images that were coregistered to the dynamic PET data (top row) (DCA, dorsal caudate, MCA, middle caudate; DPU, dorsal putamen; AVS, anteroventral striatum; VPU, ventral putamen). Also shown (bottom row), are these same ROIs on the coregistered summed PET images (0–15 min after [¹¹C]raclopride injection). (Reproduced from Drevets *et al.*, 1999.) (B) Examples of time–activity curves that were generated using the striatal and cerebellar regions of interest. (Inset) The early kinetics of the time–activity data (0–8 min after [¹¹C]raclopride injection). (Adapted from Drevets *et al.*, 1999.)

smoothing in the transverse plane by a Hanning window, and additional smoothing (~5 mm) associated with scatter inside the object (e.g., such as that observed for point sources in a water-filled phantom, Meltzer et al., 1999). For the HR$^+$ scanner (3D mode), Brix et al. (1997) reported a FWHM for a point source in air of about 5.0 ± 0.5 mm (transverse) and 4.5 ± 0.5 mm (axial). Additional smoothing due to the Hanning filter and scatter (unpublished data in our laboratory) led to a calculated effective resolution for the HR$^+$ of 7.1 mm transverse and 6.7 mm axially. As a conservative compromise between the two scanners, FWHM values of 8 mm (transverse) and 6 mm (axial) were assumed for the simulations and were applied to the segmented DV image.

2. Postamphetamine DV Image

The post-AMPH DV image was simulated using the baseline segmented DV image by setting its values to correspond to the observed changes in the actual measured post-AMPH DV image (striatal and extrastriatal). Differences in the pre- and post-AMPH nonspecific binding values were measured in the example baboon and this led us to simulate the post-AMPH nonspecific binding based on the measured change in the cerebellar DV.

3. Simulations (1). Partial Volume Effects

Of concern in this work was the extent to which the striatal measures might have been affected by neighboring subregions (see Fig. 1 for adjacency of regions). Simulations 1 examined: (A) Coupling of DCA and AVS by fixing the DCA ΔDV and MCA ΔDV to -10% and the PUT ΔDV to -20%, while reducing the AVS DV by 10–30%, in 5% steps and (B) coupling of AVS and PUT by fixing the DCA ΔDV to -10% and the AVS ΔDV to -20%, while reducing

the PUT DV 10–30%, in 5% steps. For all simulations, the change in DV was made in addition to changes in the free and nonspecific binding (e.g., beyond the Cer ΔDV).

4. Simulations (2). Image Alignment Errors

Another concern was how misalignment of the PET and MR might have affected the PET measures. These simulations were based on previous alignment validation studies in baboons (Black et al., 1997) that showed small subcortical errors (mean, 1.6 mm) and human studies in our laboratory that yielded an average error of <1 pixel based on visual assessments of internal landmarks (unpublished data). Simulations 2 examined misalignment of the PET image relative to MR image by shifting the PET image ± 1.5 mm in the left–right (x), anterior–posterior (y), and dorsal–ventral (z) directions. The impact of these errors was assessed by the overall ($\pm x$, $\pm y$, and $\pm z$) change in the shifted DV value relative to the unshifted value (percent difference).

III. RESULTS AND DISCUSSION

A. Imaging

Examples of MR images obtained in one baboon are shown in Fig. 1A (top) along with the coregistered summed PET images (10–15 min after [^{11}C]RAC injection) (bottom). Also shown are the striatal ROIs that were manually drawn on the MR images and were used to sample the dynamic PET data. Examples of the striatal and cerebellar time–activity data are shown in Fig. 1B. Greater uptake was observed for the striatal data in contrast to the low uptake and rapid clearance of the cerebellar data.

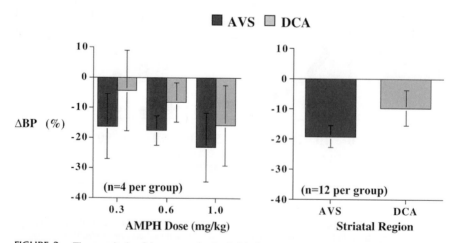

FIGURE 2. The magnitude of the mean reduction in binding potential (ΔBP) that was observed in the AVS was greater than that in the DCA and significant when the data were pooled (right) across doses (AVS reduced $99 \pm 80\%$ more than DCA, $P < 0.002$). At 0.6 mg/kg, the ΔBP values were $-17.7 \pm 4.9\%$ and $-8.4 \pm 6.5\%$ for the AVS and DCA, respectively. Changes in the neighboring putamen (ventral) and middle caudate were $-16.6 \pm 3.5\%$ and $-13.3 \pm 4.5\%$, respectively (0.6 mg/kg). These data demonstrated a similar change for the AVS and putamen. (Adapted from Drevets et al., 1999.)

B. Data Analysis

Application of the Logan method to the [^{11}C]RAC baboon data resulted in regional DV, BP, and ΔBP measures that were consistent with values obtained by other researchers in similar studies in nonhuman primates (Dewey *et al.*, 1993; Carson *et al.*, 1997; Laruelle *et al.*, 1997). The magnitude of the mean reduction in binding potential (ΔBP) that was observed in the AVS was greater than that in the DCA and significantly different when the data were pooled across AMPH doses (AVS reduced 99 ± 80% more than DCA, $t = 4.32$; $P < 0.002$), as shown in Fig. 2. At 0.6 mg/kg, the AVS and DCA ΔBP values were -17.7 ± 4.9

and $-8.4 \pm 6.5\%$, respectively; and changes in the adjacent VPU and MCA regions were -16.6 ± 3.5 and $-13.3 \pm 4.5\%$, respectively. A similar change was found for the AVS and PUT regions. The twofold ΔBP difference between the AVS and the DCA was not accounted for by changes in plasma clearance values (K_1) (data not shown). Similarly, the metabolism of [^{11}C]RAC in plasma did not significantly differ between the pre- and the post-AMPH conditions or across AMPH doses (data not shown) nor did the mean Cer DV (between conditions, 2.95 ± 11.0%, $t = 0.93$). Logan analyses yielded correlations (r^2) that were greater than 0.99 for all ROI and scans.

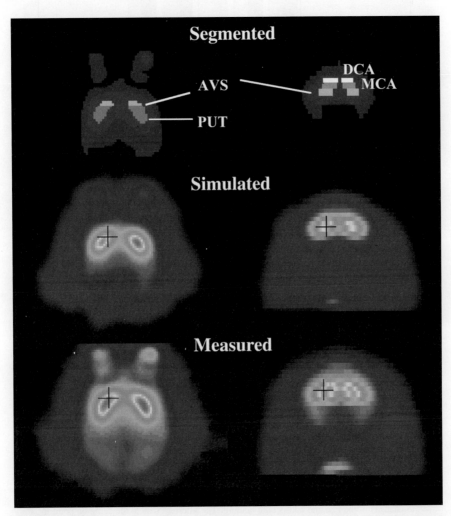

FIGURE 3. [^{11}C]Raclopride distribution volume images (transverse and coronal views) that were used in the computer simulation studies. The top row of images shows the segmented DV image that was generated from a measured baseline [^{11}C]raclopride distribution volume image that was obtained in one baboon (bottom row). The segmented striatal region was divided into subregions across multiple planes that corresponded to the anteroventral striatum (AVS), dorsal caudate (DCA), middle caudate (MCA), and putamen (PUT), based on MR-guided definitions (Drevets *et al.*, 1999). The segmented image was then blurred using a FWHM value that was based on the intrinsic scanner resolution and resolution degrading effects associated with scatter and image smoothing. In this manner, a simulated distribution volume "reference image" at the "true" spatial resolution of the scanner was generated (middle row) that compared well with the actual measured image (bottom row).

C. Computer Simulations

Figure 3 shows the segmented, simulated, and measured DV images used in the computer experiments. The simulated DV "reference image" at the "true" spatial resolution of the scanner (middle row) compared well with the original measured image (bottom row) with simulated (blurred segmented) DV image values that were within 3% of the measured values. Figure 4 shows the results of simulation 1 that were aimed at assessing the effects of regional measurement coupling. The DCA ΔBP was only slightly affected by changes in the AVS DV (Fig. 4A) and PUT DV (Fig. 4B) with an average variation in the DCA ΔBP of less than 1% in either case. As expected, the AVS ΔBP was more affected than the DCA by changes in the PUT ΔDV (re-

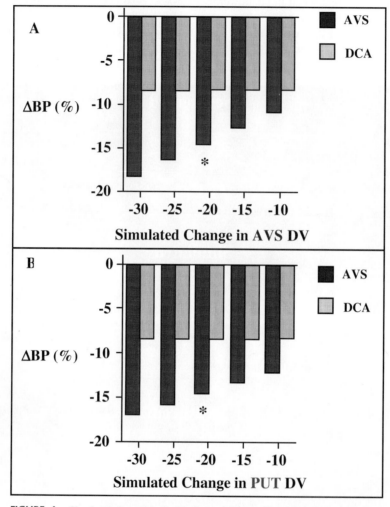

FIGURE 4.   Simulated changes in the binding potential (ΔBP) that arose from changes in the distribution volume (ΔDV) (simulation 1). (A) Result of reducing the AVS DV by 10–30%, while fixing the DCA and MCA ΔDVs to -10% and the PUT ΔDV to -20%. (B) Result of reducing the PUT DV by 10–30%, while fixing the DCA ΔDV to -10% and the AVS ΔDV to -20%. For all simulations, the change in DV was made in addition to changes in the free and nonspecific binding. The asterisks denote the common point in the two simulation data sets. The DCA ΔBP was only slightly affected by changes in the AVS DV (A) and PUT DV (B) with <1% variation in either case. The AVS ΔBP was more affected by changes in the PUT ΔDV due to regional adjacency. As expected, the AVS ΔBP was more affected by changes in the AVS DV (A) than by changes in the PUT DV (B), with linear regression slopes of 0.37 and 0.24, respectively. Changes in the PUT DV of -20 to -30% resulted in an absolute change in the AVS ΔBP of 2.5%. These studies showed that the PUT DV change would need to be much greater than the AVS DV change to account for the observed twofold difference between the AVS and DCA ΔBPs. In this work, the measured DV change was found to be similar between the AVS and the putamen regions. (See Drevets et al., 1999.)

gional proximity). Changes in the AVS BP were greater in response to changes in the AVS ΔDV than to changes in the PUT DV, as reflected by linear regression slopes of 0.37 and 0.24, respectively, for these data. A simulated change in the PUT DV of −20 to −30% resulted in an absolute change in the AVS ΔBP of 2.5%. These studies showed that the PUT ΔDV would have to be much greater than the AVS ΔDV to account for the twofold difference in the AVS and DCA ΔBPs. The average ($n = 12$) baseline DV and ΔBP values were similar for the AVS and PUT (Drevets et al., 1999). Simulation 2 examined the impact of errors in the PET/MR coregistration and yielded an overall ΔBP change of ~9% for the AVS and the DCA. The errors along the y-axis (3–48%, anterior–posterior) were greatest and smallest along the x-axis (1–7%). Nonetheless, the overall error was <10% and much lower than the 99% difference in the ΔBP that was measured between the AVS and the DCA. These results are consistent with previous studies in humans that showed that ROIs selected directly on the [11C]RAC PET images yielded similar results as those obtained from PET data that were co-registered to the MR (Wang et al., 1996).

IV. CONCLUSION

The AMPH-induced [11C]RAC-binding reductions in the striatum yielded a differential sensitivity for the anteroventral striatum (greater) relative to the dorsal caudate (lower). Potential biases due to scanner resolution effects and image alignment errors were significantly smaller than the magnitude of the observed differences between the anteroventral striatum and the dorsal caudate. These results serve to strengthen the validity of PET imaging measures of mesolimbic endogenous dopamine release for studies of dopaminergic function in relation to disorders such as depression, psychosis, and substance abuse.

Acknowledgments

This work was supported by NARSAD and also by MH52247 and the Whitaker Foundation. We thank James Ruszkiewicz and the PET staff, including Phil Greer, for assistance in this work.

References

Black, K. J., Gado, M. H., Videen, T. O., and Perlmutter, J. S. (1997). Baboon basal ganglia stereotaxy using internal MRI landmarks: Validation and application to PET imaging. *J. Comput. Assist. Tomogr.* **21**(6): 881–886.

Breier, A., Saunders, R., Carson, R. E., Kolachana, B. S., de Bartolomeis, A., Weinberger, D. R., Weisenfeld, N., Malhotra, A. K., Eckelman, W. C., and Pickar, D. (1997). Schizophrenia is associated with elevated amphetamine-induced synaptic dopamine concentrations: Evidence from a novel positron emission tomography method. *Proc. Natl. Acad. Sci. USA* **94**: 2569–2574.

Brix, G., Zaers, J., Adam, L.-E., Belleman, M. E., Ostertag, H., Trojan, H., Haberkorm, U., Doll, J., Oberdofer, F., and Lorenz, W. J. (1997). Performance evaluation of a whole-body PET scanner using the NEMA protocol. *J. Nucl. Med.* **38**: 1614–1623.

Carboni, E., Imperato, A., Perezzani, L., and Di Chiara, G. (1989). Amphetamine, cocaine, phencyclidine, and nomifensine increase extracellular DA concentrations preferentially in nucleus accumbens of freely moving rats. *Neuroscience* **28**: 653–661.

Carson, R. E., Breier, A., DeBartolomeis, A., Saunders, R. C., Su, T. P., Schmall, B., Der, M. G., Pickar, D., and Eckelman, W. C. (1997). Quantification of amphetamine-induced changes in [11C]raclopride binding with continuous infusion. *J. Cereb. Blood Flow Metab.* **17**: 437–447.

Dewey, S. L., Smith, G. S., Dewey, S. L., Brodie, J. D., Fowler, J. S., and Wolf, A. P. (1993). Striatal binding of the PET ligand 11C-raclopride is altered by drugs that modify synaptic dopamine levels. *Synapse* **13**: 350–356.

Di Chiara, G., Tanda, G., Frau, R., and Carboni, E. (1993). On the preferential release of dopamine in the nucleus accumbens by amphetamine: Further evidence obtained by vertically implanted concentric dialysis probes. *Psychopharmacology* **112**: 398–402.

Drevets, W. C., Price, J. C., Kupfer, D. J., Kinahan, P. E., Lopresti, B. J., Holt, D., and Mathis, C. A. (1999). PET measures of amphetamine-induced dopamine release in ventral versus dorsal striatum. *Neuropsychopharmacology* **21**: 694–709.

Endres, C. J., Kolachana, B. S., Saunders, R. C., Su, T., Weinberger, D., Breier, A., Eckelman, W. C., and Carson, R. E. (1997). Kinetic modeling of [11C]raclopride: Combined PET-microdialysis studies. *J. Cereb. Blood Flow Metab.* **17**: 932–942.

Hartvig, P., Torstenson, R., Tedroff, J., Watanabe, Y., Fasth, K. J., Bjurling, P., and Långström B. (1997). Amphetamine effects on dopamine release and synthesis rate studied in the Rhesus monkey brain by positron emission tomography. *J. Neural. Transm.* **104**: 329–339.

Kuczenski, R., and Segal, D. (1989). Concomitant characterization of behavioral and striatal neurotransmitter response to amphetamine using *in vivo* microdialysis. *J. Neurosci.* **9**(6): 2051–2065.

Kuczenski, R., and Segal, D. (1992). Differential effects of amphetamine and DA uptake blockers (cocaine, nomifensine) on caudate and accumbens dialysate DA and 3-methoxytyramine. *J. Pharmacol. Exp. Ther.* **239**: 219–228.

Lammertsma, A. A., Bench, C. J., Hume, S., Osman, S., Gunn, K., Brooks, D. J., and Frackowiak, R. S. J. (1996). Comparison of methods for analysis of clinical [11C]raclopride studies. *J. Cereb. Blood Flow Metab.* **16**: 42–52.

Laruelle, M., Iyer, RN., Al-Tikriti, M. S., Zea-Ponce, Y., Malison, R., Zoghbi, S. S., Baldwin, R. M., Kung, H. F., Charney, D. S., Hoffer, P. B., Innis, R. B., and Bradberry, C. W. (1997). Microdialysis and SPECT measurements of amphetamine-induced dopamine release in nonhuman primates. *Synapse* **25**: 1–14.

Lassen, N. A., and Perl, W. (1979). "*Tracer Kinetic Methods in Medical Physiology.*" Raven Press, New York.

Logan, J., Fowler, J. S., Volkow, N. D., Wolf, A. P., Dewey, S. L., Schlyer, D. J., MacGregor, R. R., Hitzemann, R., Bendriem, B., Gatley, S. J., and Christman, D. R. (1990). Graphical analysis of reversible radioligand binding from time–activity measurements applied to [N-11C-methyl]-(-)-cocaine PET studies in human subjects. *J. Cereb. Blood Flow Metab.* **10**: 740–747.

Meltzer, C. C., Kinahan, P. E., Greer, P. J., Nichols, T. E., Comtat, C., Cantwell, M. N., Lin, M. P., and Price, J. C. (1999). Comparative evaluation of MR-based partial volume correction schemes for PET. *J. Nucl. Med.* **40**: 2053–2065.

Price, J. C., Mason, N. S., Lopresti, B., Holt, D., Simpson, N. R., Drevets, W., Smith, G. S., and Mathis, C. A. (1998). PET measurement of endogenous neurotransmitter activity using high and low affinity radiotracers, *In "Quantitative Functional Brain Imaging with Positron Emission Tomography"* (R. E. Carson *et al.*, Eds.), pp. 441–448. Academic Press, San Diego.

Spinks, T. J., Jones, T., Bailey, D. L., Townsend, D. W., Grootoonk, S., Bloomfield, P. M., Gilardi, M. C., Casey, M. E., Sipe, B., and Reed, J. (1992). Physical performance of a positron tomograph for brain imaging with retractable septa. *Phys. Med. Biol.* **37**: 1637–1655.

Wang, G.-J., Volkow, N. D., Levy, A. V., Fowler, J. S., Logan, J., Alexoff, D., Hitzemann, R. J., and Schyler, D. J. (1996). MR-PET image coregistration for quantitation of striatal dopamine D_2 receptors. *J. Comput. Assist. Tomogr.* **20**(3): 423–428.

Woods, R. P., Mazziotta, J. C., and Cherry, S. R. (1993). MRI-PET registration with automated algorithm. *J. Comput. Assist. Tomogr.* **17**(4): 536–546.

Zetterstrom, T., Sharp, T., Marsden, C. A., and Ungerstedt, U. (1983). In vivo measurement of dopamine and its metabolites by intracerebral dialysis: Changes after *d*-amphetamine. *J. Neurochem.* **41**: 1769–1773.

34

Reproducibility of Radioligand Measurements of Pharmacologically Induced Striatal Dopamine Release in Humans

LAWRENCE S. KEGELES, YOLANDA ZEA-PONCE, ANISSA ABI-DARGHAM,
RONALD L. VAN HEERTUM, J. JOHN MANN, AND MARC LARUELLE

Departments of Psychiatry and Radiology, Columbia University College of Physicians and Surgeons, and Division of Brain Imaging, Department of Neuroscience, New York State Psychiatric Institute, New York, New York 10032

Binding competition between endogenous dopamine (DA) and D_2 receptor radiotracers permits the measurement of changes in synaptic DA following DA-releasing pharmacological challenges to be made with positron emission tomography (PET) or single photon emission computed tomography (SPECT) in the living human brain. Here we describe studies of the within-subject reproducibility and reliability of PET and SPECT measurements of decrease in binding potential (BP) as well as behavioral responses induced by these challenges. In a study by our group, six healthy male subjects, never previously exposed to psychostimulants, twice underwent measurement of striatal amphetamine-induced DA release (between-measurement interval 16 ± 10 days) using SPECT and the $[^{123}I]$iodobenzamide (IBZM) constant infusion technique. Results demonstrated a high within-subject reproducibility of amphetamine-induced DA release: decreases in IBZM BP stimulated by amphetamine were significant on each day, with an intraclass correlation coefficient of 0.89. Moreover, baseline BP values from the second experiment were not significantly different from the first, suggesting the absence of either sensitization or tolerance to the effect of amphetamine under these experimental conditions. Behavioral responses, as rated by the subjects on analog scales, were also highly reproducible. In a PET study using $[^{11}C]$raclopride and methylphenidate challenge, G. J. Wang et al. demonstrated comparable neurochemical and behavioral test–retest reproducibility. We conclude that these radioligand imaging techniques provide a reliable method of measuring pharmacologically induced DA release and permit the detection of between-subjects differences that appear stable over time.

I. INTRODUCTION

Neuroreceptor imaging with positron emission tomography (PET) and single photon emission computed tomography (SPECT) is commonly used to measure neuroreceptor distribution in the living human brain. More recently, several groups have demonstrated that *in vivo* neuroreceptor binding techniques can also be used to measure acute pharmacologically induced alterations in the concentration of endogenous transmitters in the vicinity of neuroreceptors (Seeman *et al.*, 1989; Dewey *et al.*, 1990; Bischoff *et al.*, 1991; Innis *et al.*, 1992; Carson *et al.*, 1997; Laruelle *et al.*, 1997). Competition between radiotracers and endogenous transmitters for binding to neuroreceptors is the principle underlying this technique. This new application of neuroreceptor imaging allows estimation of synaptic neurotransmitter levels in specific neurotransmitter systems in the brain, a correlation of these measurements with behaviors and symptoms, and an exploration of the role of neurochemical functional imbalances in the pathogenesis of psychiatric and neurological disorders.

Our group recently developed and validated a protocol to measure amphetamine-induced dopamine (DA) release with SPECT and $[^{123}I]$iodobenzamide ($[^{123}I]$IBZM) (Laruelle *et al.*, 1995, 1997). This radiotracer, an iodinated analog of raclopride, is a selective antagonist at the D_2 and D_3 receptors (Kung *et al.*, 1988). In human studies, amphetamine-induced DA release was detectable as a 5 to 15% decrease in the binding potential (BP) of [^{123}I]IBZM, measured under sustained equilibrium conditions (Laruelle *et al.*, 1995), in contrast to the absence of significant change resulting from injection of placebo or the absence of injection (Laruelle *et al.*, 1995; Booij *et al.*, 1997). Both [^{123}I]IBZM BP changes and subjective responses to amphetamine exhibited large between-subject variability. A significant but weak correlation between the BP change and subjective activation ($r^2 = 0.21$, $p = 0.0018$) was noted, suggesting the behavioral relevance of between-subject differences in DA release.

Determination of the relative magnitude of between- and within-subject variability in a test–retest paradigm provides data on reliability of this measure, as well as on sensitization or tolerance to i.v. injection of the DA-releasing agent. Here we describe two such studies: a SPECT study performed by our group (Kegeles *et al.*, 1999) is presented in detail, and a PET study is reviewed (Wang *et al.*, 1999) with a comparison of the findings.

II. MATERIALS AND METHODS

A. Subjects

Six healthy subjects (males, age 36 ± 10 years), never previously exposed to psychostimulants, were studied twice, at 16 ± 10 day interval (range from 7 to 35 days). Inclusion criteria were: (1) absence of past or present neurological or psychiatric illness; (2) no past or present cardiovascular conditions; (3) no present or prior exposure to psychostimulants; and (4) male gender, with the latter criterion intended to minimize the confounding effects of estrogens on DA function (Becker and Cha, 1989; Caster *et al.*, 1993; Xiao and Becker, 1994; Thompson and Moss, 1997). Administration of both [^{123}I]IBZM and amphetamine i.v. were approved by the U.S. Food and Drug Administration under an Investigational New Drug protocol and the entire protocol was approved by the Institutional Review Board of the hospital. All subjects gave written, informed consent for the study. A total of seven subjects were recruited into the study but because of a technical problem with the camera on the test day of one subject, six completed the protocol.

B. Radioligand Preparation

[^{123}I]IBZM was prepared by direct electrophilic radio-iodination of the phenolic precursor BZM [(S)(-)-N-[(1-ethyl-2-pyrrolidinyl)methyl]-2-hydroxy-6-methoxy-benzamide] with high-purity sodium [^{123}I]iodide solution, in the presence of a potassium biphthalate/sulfamic acid buffer, pH 2, and 3.2% aqueous peracetic acid, at ambient temperature for 10 min (Kung and Kung, 1989). Full details including purification and pyrogenicity testing are provided elsewhere (Zea-Ponce and Laruelle, 1999). Under these procedures, the labeling yield was $47 \pm 12\%$, the radiochemical yield $43 \pm 13\%$, and the radiochemical purity $97 \pm 2\%$ ($n = 12$).

C. SPECT Scan Protocol

The following protocol was used on each of the two scan days. Four fiducial markers were glued on the subject's head at the level of the cantho-meatal line. The radiotracer constant infusion technique was used to attain sustained equilibrium binding conditions (Laruelle *et al.*, 1994b, 1995). [^{123}I]IBZM was administered i.v. as a priming bolus (3.92 ± 0.02 mCi), immediately followed by i.v. infusion at a constant rate (0.85 ± 0.14 mCi/h) for 360 min, which corresponded to a total injected dose of 8.83 ± 1.56 mCi, decay corrected to the beginning of the experiment. The preamphetamine 60-min scan session was initiated at 180 min. SPECT data were obtained with the triple-head PRISM 3000 (Picker, Cleveland, OH), equipped with low-energy ultra-high-resolution (LEUHR) fan beam collimators (FWHM 8–10 mm), with the following acquisition parameters: continuous acquisition mode; matrix, $64 \times 64 \times 32$; angular range, 120; angular steps, 3; seconds per step, 18; frame duration, 12 min; number of frames, 5; radius of rotation, 13.5 cm. The second (postamphetamine) session was obtained using the same parameters from 300 to 360 min.

D. Behavioral Responses

Four items (happiness, restlessness, energy, and anxiety) from the Amphetamine Interview Rating Scale (van Kammen and Murphy, 1975) were self-rated during the 45 minutes following amphetamine injection. Each item was rated on a scale from 1 to 10, 1 being "not at all" and 10 being "most ever." Self-ratings were obtained at baseline and at 0, 5, 10, 20, 30, and 45 min post injection. These data were analyzed as previously described (Laruelle *et al.*, 1995).

E. Amphetamine Plasma Measurement

Amphetamine plasma concentration was measured (Analytical Psychopharmacology Laboratories, Nathan Kline Institute, Orangeburg, NY) in three venous samples obtained at 10, 20, and 40 min postamphetamine injection. The method was as described by Reimer *et al.* (1993) with the following modifications: A 30m DB-17 capillary column was substituted to improve separation and peak symmetry. Trideuterated amphetamine was used as the internal standard. Standard curves for both compounds were uniformly linear ($r > 0.99$) over the range tested (0.1–500 ng/mL) with negligible intercepts. Sensitivity was <0.1 ng/mL for each when 1 mL plasma was extracted. Interassay RSD% was 5.2% at 5 ng/mL.

F. Image Analysis

Image analysis was performed as previously described (Laruelle *et al.*, 1995). Briefly, data were reconstructed using a Butterworth filter (cutoff = 1 cm, power factor = 10), transferred into the MEDx software system (Sensor Systems, Sterling, VA), and attenuation corrected (uniform attenuation coefficient, $\mu = 0.10$ cm^2/g). Frames were aligned to each other using automated image registration (Woods *et al.*, 1992). All studies were analyzed with standard regions of interest of fixed size (striatal, 15,820 mm^3; frontal, 57,656 mm^3; occipital 58,497 mm^3), with right and left striatal regions averaged. Specific binding was calculated as striatal minus nonspecific activity. Nonspecific activity was calculated as the average of the frontal and occipital regions (Kegeles *et al.*, 1999). For each scanning session, the specific to nonspecific equilibrium ratio (V_3'') was calculated as the ratio of specific binding to nonspecific activity (Laruelle *et al.*, 1994a). Under steady-state conditions, assuming that amphetamine does not affect nonspecific binding, the percentage decrease in V_3'' is equal to that in BP (Laruelle *et al.*, 1997). Amphetamine-induced decrease in BP was expressed as a percentage of the preamphetamine value.

G. Statistical Analyses

All values are expressed as mean ±SD. The reproducibility of the measurements was assessed by the intraclass correlation coefficient, ICC (Kirk, 1982),

$$\frac{\text{MSBS} - \text{MSWS}}{\text{MSBS} + (n-1)\text{MSWS}},$$

where MSBS is the mean sum of squares between subjects, MSWS is the mean sum of squares within subjects, and n is the number of repetitions ($n = 2$ in this study). The ICC estimates the reliability of the measurement and can vary between -1 (no reliability) and $+1$ (maximum reliability, achieved in case of identity between test and retest; i.e., MSWS = 0). Repeated measures ANOVA was used to test for significant aggregate changes between test and retest experiments.

III. RESULTS

A. Amphetamine Effect on IBZM BP

Table 1 lists the values of specific to nonspecific equilibrium ratio ([^{123}I]IBZM V_3'') and the changes induced on the test and retest day by administration of amphetamine 0.3 mg/kg i.v. These changes were significant within each day ($p = 0.04$, $p = 0.03$, respectively, by repeated measures ANOVA). Table 2 summarizes the comparison between the test and the retest days of the amphetamine-induced changes in [^{123}I]IBZM BP. Amphetamine-induced change in [^{123}I]IBZM BP was $-8 \pm 8\%$ on test day and $-9 \pm 7\%$ on retest day, consistent with values previously reported in healthy controls of the same age range ($-8.7 \pm 7.2\%$, $n = 43$) (Laruelle *et al.*, 1996; Abi-Dargham *et al.*, 1998). No differences in [^{123}I]IBZM displacement were observed between test and retest conditions (repeated measures ANOVA, $p = 0.61$), suggesting the absence of either sensitization or tolerance in amphetamine-induced DA release under these conditions.

The reliability of the measurement of the amphetamine effect (change in BP) was high, with an ICC of 0.89. In other words, the magnitude of the amphetamine effect in a given subject on the test day was a very good predictor of the effect on the retest day ($r^2 = 0.76$, $p = 0.023$) (Fig. 1).

Baseline (preamphetamine) V_3'' was also reliably reproduced, with ICC of 0.74. Repeated measures ANOVA showed no differences between test and retest ($p = 0.79$),

TABLE 1.   Amphetamine-Induced Changes in Specific to Nonspecific Equilibrium Ratio, V_3''

	Test			Retest		
Subject No.	V_3'' baseline	V_3'' post-amphetamine	% Change	V_3'' baseline	V_3'' post-amphetamine	% Change
1	0.868	0.797	-8	0.822	0.779	-5
2	0.620	0.642	3	0.592	0.579	-2
3	0.682	0.624	-8	0.711	0.631	-11
4	0.823	0.700	-15	0.761	0.676	-11
5	0.628	0.517	-18	0.760	0.596	-22
6	0.659	0.635	-4	0.579	0.557	-4
Mean ± SD	0.713 ± 0.106	0.652 ± 0.092		0.704 ± 0.099	0.636 ± 0.081	
Repeated measures ANOVA	$F = 7.55$	$df = (1,5)$ $p = 0.04$		$F = 8.96$	$df = (1,5)$ $p = 0.03$	

TABLE 2.   Effect of Amphetamine (0.3 mg/kg) on IBZM BP on Test and Retest Days

Subject No.	Amphetamine effect, test day (%)	Amphetamine effect, retest day (%)	Relative difference (%)	Absolute difference (%)
1	−8	−5	−3	3
2	3	−2	6	6
3	−8	−11	3	3
4	−15	−11	−4	4
5	−18	−22	4	4
6	−4	−4	0	0
Mean ± SD	−8.2 ± 7.7	−9.2 ± 7.2	1.0 ± 3.8	3.2 ± 1.8
Reliability (ICC)				0.89

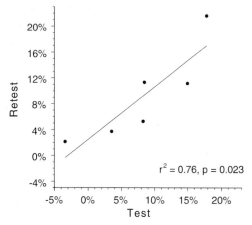

$r^2 = 0.76$, p = 0.023

FIGURE 1. Test/retest reproducibility of amphetamine-induced reduction in [^{123}I]IBZM BP in six healthy subjects. The regression is displayed of the test day (horizontal axis) against the retest day (vertical axis) of the percent decrease values in BP within each day induced by i.v. administration of amphetamine 0.3 mg/kg.

indicating the absence of effect on receptor density from a single injection of *d*-amphetamine approximately 2 weeks earlier.

B. Subjective Effects

The three-factor model previously derived from analysis of a larger group of healthy volunteers (Laruelle *et al.*, 1996) was used to analyze the subjective effects of amphetamine. The peak increase of each of these three factors over baseline following amphetamine is presented in Table 3 for each subject, under both test and retest conditions. Values were not statistically different between test and retest conditions (repeated measures ANOVA, time effect: euphoria, $p = 0.39$; restlessness, $p = 0.56$; anxiety, $p = 0.36$), failing to demonstrate either sensitization or tolerance to the subjective effects of amphetamine under these conditions. Euphoria and anxiety had high reliability (ICC of 0.79 and 0.77, respec-

tively), with restlessness less reproducible (ICC of 0.41). These data did not replicate the previously observed association between amphetamine-induced subjective activation (the sum of euphoria and restlessness factors) and decrease in BP (test day, $r^2 = 0.09$, $p = 0.54$; retest day, $r^2 = 0.01$, $p = 0.83$).

C. Amphetamine Plasma Levels

Amphetamine plasma levels were highly reproducible, with ICC of 0.71. We observed a trend toward an association between amphetamine plasma level and decrease in IBZM BP on both test ($r^2 = 0.54$, $p = 0.09$) and retest days ($r^2 = 0.60$, $p = 0.06$).

IV. DISCUSSION

A. IBZM SPECT Study

In this study, six participants twice underwent measurement of amphetamine–induced DA release and behavioral response at approximately a 2-week interval. The results indicated that both the biochemical and behavioral measures were remarkably similar under test and retest conditions. The high test–retest reliability of the amphetamine-induced DA release as measured with SPECT [^{123}I]IBZM (ICC = 0.89) indicates that, despite the relatively low magnitude of the effect, this method is sufficiently sensitive and reliable to permit measurement of between-subjects differences in DA release. The results suggest that the within-subject behavioral and biochemical responses to amphetamine are stable over time.

An interesting result from this study is the apparent absence of sensitization to the effect of amphetamine after single exposure, despite significant changes produced by amphetamine on each day. Long-term sensitization to psychostimulants is a phenomenon whereby exposure to these

TABLE 3.　Test/Retest Reproducibility of Subjective Effects of Amphetamine (0.3 mg/kg)

Subject No.	Euphoria factor			Restlessness factor			Anxiety factor		
	Test	Retest	Difference	Test	Retest	Difference	Test	Retest	Difference
1	6.0	6.8	−0.9	1.8	1.3	0.5	−0.2	0.6	−0.9
2	0.0	0.0	0.0	0.0	0.0	0.0	1.9	1.0	1.0
3	6.8	11.0	−4.2	2.2	5.1	−2.9	1.6	−0.3	1.9
4	0.0	3.4	−3.3	3.5	1.1	2.4	5.7	6.6	−0.9
5	2.6	0.0	2.6	−0.2	1.7	−1.9	−0.2	2.9	−3.1
6	0.9	0.8	0.0	−0.1	0.8	−0.9	4.7	7.6	−2.9
Mean ± SD	2.7 ± 3.0	3.7 ± 4.4	−1.0 ± 2.5	1.2 ± 1.5	1.7 ± 1.8	−0.5 ± 1.9	2.2 ± 2.5	3.1 ± 3.3	−0.8 ± 2.0
Reliability (ICC)			0.79			0.41			0.77

drugs results in an enhanced response at subsequent exposures, and in rodents this is very robust, reproducible, and detectable after only one injection of amphetamine (for reviews see Robinson and Becker, 1986; Kalivas and Stewart, 1991; Kalivas et al., 1993; Grace, 1995). In rodents, several studies have shown that sensitization is associated with increased stimulant-induced DA release in the axonal terminal fields (Robinson et al., 1988; Akimoto et al., 1990; Kalivas and Duffy, 1990; Patrick et al., 1991; Paulson and Robinson, 1995). Recent behavioral data in humans has suggested that behavioral sensitization is present 48 h after a single dose of amphetamine (Strakowski et al., 1996). In rodents, while behavioral sensitization resulting from multiple doses is stable, a single dose produces sensitization that may decrease with time (Kalivas et al., 1993). Thus it is possible that, in humans, a single amphetamine dose induces a transient sensitization, detectable at 48 h, but not after 2 weeks.

Amphetamine plasma levels appeared to account for some of the variability in BP change, since, on both days, an association was detected between amphetamine plasma levels and DA release. However, such a relationship was seen neither in two previously studied larger cohorts (Laruelle et al., 1996; Abi-Dargham et al., 1998)($n = 15$ each) nor in the [^{11}C]raclopride PET study by Breier et al. (1997) ($n = 10$). The present results indicate the importance of assaying plasma amphetamine levels in these studies.

Between-subjects variation in the amount of DA released following amphetamine and in the magnitude of the behavioral responses was similar to previous reports (Laruelle et al., 1995, 1996; Abi-Dargham et al., 1998). We previously reported an association between intensity of behavioral activation and magnitude of DA release in a larger cohort of healthy subjects ($r^2 = 0.21$, $p = 0.0018$, $n = 43$), but detected no such relation in the present study. The power to detect such a modest relationship with an n of 6 is less than 0.1. In addition, this relationship appears to be age dependent (more pronounced in young subjects), and the wide age range in the current study may have further limited the possibility of replication of this behavioral–neurochemical association.

B. [^{11}C]Raclopride PET Study

In a comparable study of striatal DA release induced by administration of methylphenidate (0.5 mg/kg) as measured by [^{11}C]raclopride PET, Wang et al. (1999) found high reproducibility under test–retest conditions. In their study, seven healthy subjects (two females and five males, ages $31.4 ± 7.9$ years) were studied at 1- to 2-week interval, with scans under both placebo and methylphenidate conditions, as well as behavioral measures, on each of the two evaluation days. Their subjects were found to have high reproducibility of baseline striatal BP (ICC = 0.82), consistent with earlier reports of test–retest stability of this measure (Nordström et al., 1992; Volkow et al., 1993). Methylphenidate-induced DA release resulted in BP percentage changes that were highly significant within each day but not significantly different between the two evaluation days by repeated measures ANOVA ($−18.4 ± 8.7\%$ on test day, $−13.4 ± 9.2\%$ on retest day, $p = 0.11$), with ICC of 0.58. Behavioral measures including subjective perception of high and restlessness were comparable under test and retest. Interestingly, in addition to the trend toward smaller values in BP percentage change on retest, two of the behavioral measures (rush and anxiety) were significantly lower on the retest day. The authors raised the possibility that the novelty component may explain these differences between the test and retest evaluations (Wang et al., 1999).

V. COMPARISON AND CONCLUSIONS

The high test–retest reproducibility shown by these studies of psychostimulant-induced DA release in striatum establishes the reliability of these measures and suggests an absence of neurochemical sensitization following single exposure to these agents at the 1- to 2-week time interval studied. The lower within-subject reproducibility of the methylphenidate effect on [^{11}C]raclopride BP (ICC of 0.58) than that for amphetamine on [^{123}I]IBZM BP (ICC of 0.89) might stem from differences in mechanisms underlying increased DA synaptic concentration between the two drugs. The increased DA synaptic concentration elicited by methylphenidate, for example, might be more impulse-dependent than that induced by amphetamine, and therefore more affected than the amphetamine effect by environmental factors and subjective state of mind (Johnson, 1983; Kuczenski *et al.*, 1990). Similarly, in response to methylphenidate, the decrease in some of the behavioral measures and the trend toward smaller DA release on retest raise the possibility of a greater role for novelty than in the amphetamine effect.

The high reproducibility found in these studies supports the use of measures of stimulant-induced striatal DA release in investigations of group differences in neuropsychiatric conditions. For example, a significantly larger effect of amphetamine on [^{123}I]IBZM BP was found in schizophrenic patients than in controls (Laruelle *et al.*, 1996; Abi-Dargham *et al.*, 1998) and independently confirmed by Breier *et al.* (1997) with [^{11}C]raclopride PET. Also, establishing reliability of these measures provides an important first step in developing a new line of investigation, the modulation *in vivo* of stimulant-induced DA release in humans by further pharmacological interventions, which might yield new insights into functioning of the neurochemical circuitry of the DA system in health and in neuropsychiatric conditions including schizophrenia and substance use disorders.

Acknowledgments

The authors thank Thomas Cooper for the amphetamine plasma assay, Satish Anjilvel, Ph.D., for data analysis support, and Ted Pozniakoff, Mali Pratap, Manuel DeLaNuez, Yolanda Rubino, Carl Campbell, and Suehee Chung, for excellent technical assistance. Supported by the Scottish-Rite Foundation and the Public Health Service (NIDA RO1-DA10219-01, NIMH 2 P30 MH46745, NIH M01 RR00645).

References

Abi-Dargham, A., Gil, R., Krystal, J., Baldwin, R. M., Seibyl, J. P., Bowers, M., van Dyck, C. H., Charney, D. S., Innis, R. B., and Laruelle, M. (1998). Increased striatal dopamine transmission in schizophrenia: confirmation in a second cohort. *Am. J. Psychiatry* **155**: 761–767.

Akimoto, K., Hamamura, T., Kazahaya, Y., Akiyama, K., and Otsuki, S. (1990). Enhanced extracellular dopamine level may be the fundamental neuropharmacological basis of cross-behavioral sensitization between methamphetamine and cocaine—An in vivo dialysis study in freely moving rats. *Brain Res.* **507**: 344–346.

Becker, J. B., and Cha, J. (1989). Estrous cycle-dependent variation in amphetamine-induced behaviors and striatal dopamine release assessed with microdialysis. *Behav. Brain Res.* **35**: 117–125.

Bischoff, S., Krauss, J., Grunenwald, C., Gunst, F., Heinrich, M., Schaub, M., Stocklin, K., Vassout, A., Waldmeier, P., and Maitre, L. (1991). Endogenous dopamine (DA) modulates [3H]spiperone binding in vivo in rat brain. *J. Recept. Res.* **11**: 163–175.

Booij, J., Korn, P., Linszen, D. H., and vanRoyen, E. A. (1997). Assessment of endogenous dopamine release by methylphenidate challenge using iodine-123 iodobenzamide single-photon emission tomography. *Eur. J. Nucl. Med.* **24**: 674–677.

Breier, A., Su, T. P., Saunders, R., Carson, R. E., Kolachana, B. S., deBartolomeis, A., Weinberger, D. R., Weisenfeld, N., Malhotra, A. K., Eckelman, W. C., and Pickar, D. (1997). Schizophrenia is associated with elevated amphetamine-induced synaptic dopamine concentrations: Evidence from a novel positron emission tomography method. *Proc. Natl. Acad. Sci. USA* **94**: 2569–2574.

Carson, R. E., Breier, A., deBartolomeis, A., Saunders R. C., Su, T. P., Schmall, B., Der, M. G., Pickar, D., and Eckelman, W. C. (1997). Quantification of amphetamine-induced changes in [C-11]raclopride binding with continuous infusion. *J. Cereb. Blood Flow Metab.* **17**: 437–447.

Caster, S. A., Xia, L., and Becker, J. B. (1993). Sex differences in striatal dopamine: In vivo microdialysis and behavioral studies. *Brain Res.* **610**: 127–134.

Dewey, S. L., Brodie, J. D., Fowler, J. S., MacGregor, R. R., Schyler, D. J., King, P. T., Alexoff, D. L., Volkow, N. D., Shiue, C. Y., Wolf, A. P., and Bendriem, B. (1990). Positron Emission Tomography (PET) studies of dopamine/cholinergic interaction in the baboon brain. *Synapse* **6**: 321–327.

Grace, A. A. (1995). The tonic/phasic model of dopamine system regulation: Its relevance for understanding how stimulant abuse can alter basal ganglia function. *Drug Alcohol Depend.* **37**: 111–129.

Innis, R. B., Malison, R. T., Al-Tikriti, M., Hoffer, P. B., Sybirska, E. H., Seibyl, J. P., Zoghbi, S. S., Baldwin, R. M., Laruelle, M. A., Smith, E., Charney, D. S., Heninger, G., Elsworth, J. D., and Roth, R. H. (1992). Amphetamine-stimulated dopamine release competes in vivo for [^{123}I]IBZM binding to the D2 receptor in non-human primates. *Synapse* **10**: 177–184.

Johnson, K. M. (1983). Phencyclidine: Behavioral and biochemical evidence supporting a role for dopamine. *Fed. Proc.* **42**: 2579–2583.

Kalivas, P., and Duffy, P. (1990). The effect of acute and daily cocaine treatment on extracellular dopamine in the nucleus accumbens. *Synapse* **5**: 48–58.

Kalivas, P. W., Sorg, B. A., and Hooks, M. S. (1993). The pharmacology and neural circuitry of sensitization to psychostimulants. *Behav. Pharmacol.* **4**: 315–334.

Kalivas, P. W., and Stewart, J. (1991). Dopamine transmission in the initiation and expression of drug and stress-induced sensitization of motor activity. *Brain Res. Rev.* **16**: 223–244.

Kegeles, L. S., Zea-Ponce, Y., Abi-Dargham, A., Rodenhiser, J., Wang, T., Weiss, R., Van Heertum, R. L., Mann, J. J., and Laruelle, M. (1999). Stability of [^{123}I]IBZM SPECT measurement of amphetamine-induced striatal dopamine release in humans. *Synapse* **31**: 302–308.

Kirk, R. E. (1982). "*Experimental Design: Procedures for the Behavioral Sciences.*" Brooks/Cole, Pacific Grove, CA.

Kuczenski, R., Segal, D. S., and Manley, L. D. (1990). Apomorphine does not alter amphetamine-induced dopamine release measured in striatal dialysates. *J. Neurochem.* **54**: 1492–1499.

Kung, H. F., Guo, Y.-Z., Billings, J., Xu, X., Mach, R. H., Blau, M., and Ackerhalt, R. E. (1988). Preparation and biodistribution of [^{125}I]IBZM: A potential CNS D-2 dopamine receptor imaging agent. *Nucl. Med. Biol.* **15**: 195–201.

Kung, M.-P., and Kung, H. (1989). Peracetic acid as a superior oxidant for preparation of [^{123}I]IBZM, a potential dopamine D-2 receptor imaging agent. *J. Labeled Compd. Radiopharm.* **27**: 691–700.

Laruelle, M., Abi-Dargham, A., van Dyck, C. H., Gil, R., De Souza, C. D., Erdos, J., McCance, E., Rosenblatt, W., Fingado, C., Zoghbi,

S. S., Baldwin, R. M., Seibyl, J. P., Krystal, J. H., Charney, D. S., and Innis, R. B. (1996). Single photon emission computerized tomography imaging of amphetamine-induced dopamine release in drug free schizophrenic subjects. *Proc. Natl. Acad. Sci. USA* **93**: 9235–9240.

Laruelle, M., Abi-Dargham, A., van Dyck, C. H., Rosenblatt, W., Zea-Ponce, Y., Zoghbi, S. S., Baldwin, R. M., Charney, D. S., Hoffer, P. B., Kung, H. F., and Innis, R. B. (1995). SPECT imaging of striatal dopamine release after amphetamine challenge. *J. Nucl. Med.* **36**: 1182–1190.

Laruelle, M., Baldwin, R. M., Rattner, Z., Al-Tikriti, M. S., Zea-Ponce, Y., Zoghbi, S. S., Charney, D. S., Price, J. C., Frost, J. J., Hoffer, P. B., and Innis, R. B. (1994a). SPECT quantification of [123I]iomazenil binding to benzodiazepine receptors in nonhuman primates. I. Kinetic modeling of single bolus experiments. *J. Cereb. Blood Flow Metab.* **14**: 439–452.

Laruelle, M., Iyer, R. N., Al-Tikriti, M. S., Zea-Ponce, Y., Malison, R., Zoghbi, S. S., Baldwin, R. M., Kung, H. F., Charney, D. S., Hoffer, P. B., Innis, R. B., and Bradberry, C. W. (1997). Microdialysis and SPECT measurements of amphetamine-induced dopamine release in nonhuman primates. *Synapse* **25**: 1–14.

Laruelle, M., van Dyck, C., Abi-Dargham, A., Zea-Ponce, Y., Zoghbi, S. S., Charney, D. S., Baldwin, R. M., Hoffer, P. B., Kung, H. F., and Innis, R. B. (1994b). Compartmental modeling of iodine-123-iodobenzofuran binding to dopamine D_2 receptors in healthy subjects. *J. Nucl. Med.* **35**: 743–754.

Nordström, A. L., Farde, L., Pauli, S., Litton, J. E., and Halldin, C. (1992). PET analysis of central [11C]-raclopride binding in healthy young adults and schizophrenic patients—Reliability and age effects. *Hum. Psychopharmacol.* **7**: 157–165.

Patrick, S. L., Thompson, T. L., Walker, J. M., and Patrik, R. L. (1991). Concomitant sensitization of amphetamine-induced behavioral stimulation and in vivo dopamine release from rat caudate nucleus. *Brain Res.* **538**: 343–346.

Paulson, P. E., and Robinson, T. E. (1995). Amphetamine-induced time-dependent sensitization of dopamine neurotransmission in the dorsal and ventral striatum: A microdialysis study in behaving rats. *Synapse* **19**: 56–65.

Reimer, M. L., Mamer, O. A., Zavitsanos, A. P., Siddiqui, A. W., and Dadgar, D. (1993). Determination of amphetamine, methamphetamine and desmethyldeprenyl in human plasma by gas chromatography/negative ion chemical ionization mass spectrometry. *Biol. Mass Spectrom.* **22**: 235–242.

Robinson, P. H., Jurson, P. A., Bennet, J. A., and Bentgen, K. M. (1988). Persistent sensitization of dopamine neurotransmission in ventral striatum (nucleus accumbens) produced by prior experience with (+)-amphetamine: A microdialysis study in freely moving rats. *Brain Res.* **462**: 211–222.

Robinson, T. E., and Becker, J. B. (1986). Enduring changes in brain and behavior produced by chronic amphetamine administration: A review and evaluation of animal models of amphetamine psychosis. *Brain Res. Review* **11**: 157–198.

Seeman, P., Guan, H.-C., and Niznik, H. B. (1989). Endogenous dopamine lowers the dopamine D_2 receptor density as measured by [^{3}H]raclopride: Implications for positron emission tomography of the human brain. *Synapse* **3**: 96–97.

Strakowski, S. M., Sax, K. W., Setters, M. J., and Keck, P. E., Jr. (1996). Enhanced response to repeated *d*-amphetamine challenge: Evidence for behavioral sensitization in humans. *Biol. Psychiatry* **40**: 872–880.

Thompson, T. L., and Moss, R. L. (1997). Modulation of mesolimbic dopaminergic activity over the rat estrous cycle. *Neurosci. Lett.* **229**: 145–148.

van Kammen, D. P., and Murphy, D. L. (1975). Attenuation of the euphoriant and activating effects of *d*- and l-amphetamine by lithium carbonate treatment. *Psychopharmacologia (Berl.)* **44**: 215–224.

Volkow, N. D., Fowler, J. S., Wang, G. J., Dewey, S. L., Schyler, D., MacGregor, R., Logan, J., Alexoff, D., Shea, C., Hitzemann, R., Angrist B., and Wolf, A. P. (1993). Reproducibility of repeated measures of carbon-11-raclopride binding in the human brain. *J. Nucl. Med.* **34**: 609–613.

Wang, G. J., Volkow, N. D., Fowler, J. S., Logan, J., Pappas, N. R., Wong, C. T., Hitzemann, R. J., and Netusil, N. (1999). Reproducibility of repeated measures of endogenous dopamine competition with [^{11}C]raclopride in the human brain in response to methylphenidate. *J. Nucl. Med.* **40**: 1285–1291.

Woods, R. P., Cherry, S. R., and Mazziotta, J. C. (1992). Rapid automated algorithm for aligning and reslicing PET images. *J. Comp. Assist. Tomogr.* **16**(4): 620–633.

Xiao, L., and Becker, J. B. (1994). Quantitative microdialysis determination of extracellular striatal dopamine concentration in male and female rats: Effects of estrous cycle and gonadectomy. *Neurosci. Lett.* **180**: 155–158.

Zea-Ponce, Y., and Laruelle, M. (1999). Synthesis of [123I]IBZM: A reliable procedure for routine clinical studies. *Nucl. Med. Biol.* **26**: 661–665.

SECTION

VI

OTHER TRANSMITTER ENZYME, RECEPTOR, and TRANSPORTER IMAGING

35

[α-11C]Methyl-L-tryptophan in Anesthetized Rhesus Monkeys: A Tracer for Serotonin Synthesis or Tryptophan Uptake?

SUSAN E. SHOAF,* RICHARD E. CARSON,[†] DANIEL HOMMER,* WENDOL A. WILLIAMS,*
J. DEE HIGLEY,* BERNARD SCHMALL,[†] PETER HERSCOVITCH,[†] and WILLIAM C. ECKELMAN[†]
Markku Linnoila,* in memorium

*Laboratory of Clinical Studies, DICBR, National Institute on Alcohol Abuse and Alcoholism, Bethesda, Maryland
[†]PET Department, National Institutes of Health Clinical Center, Bethesda, Maryland

The tracer [11C]-α-methyl-L-tryptophan (αMTP) has been used in an attempt to measure brain serotonin synthesis rates with positron emission tomography (PET). To address questions about the accuracy of the kinetic model, [14C]αMTP was used to directly measure conversion to [14C]-α-methyl-serotonin (αM5HT) in monkeys that were also studied with PET and [11C]αMTP. Four male, fasted, isoflurane-anesthetized rhesus monkeys were studied with [11C]αMTP and PET. Immediately following the initial 3-h scan, a second dose of [11C]αMTP was coinjected with 1 mCi of [14C]αMTP and additional PET data were collected. Approximately 90 min after the second αMTP administration, the animals were sacrificed with an overdose of phenobarbital, and brain samples from 21 regions were taken and analyzed by HPLC. Minimal conversion of [14C]αMTP to [14C]αM5HT occurred; HPLC analysis of 14C radioactivity showed that greater than 96% of the total counts were in fractions corresponding to the αMTP peak. Brain concentrations of serotonin, tryptophan, 5-hydroxyindole-3-acetic acid, and αMTP were also determined fluorometrically using external quantification; tryptophan and αMTP brain concentrations were highly correlated, with a range of r = 0.607–0.837, P < 0.001. No other correlations were statistically significant. Patlak plots generated from PET images acquired over 3 h showed no time period of linear increase and final slopes were not significantly different from zero, consistent with the finding of minimal conversion to [14C]αM5HT. These data indicate that in the 3-h period following injection, [11C]αMTP is acting predominantly as a tracer of tryptophan uptake, not serotonin synthesis.

I. INTRODUCTION

In the early 1990s, it was proposed by Diksic *et al.* (1990, 1991) and Nagahiro *et al.* (1990) that serotonin synthesis rate could be measured by using [11C]-α-methyl-L-tryptophan (αMTP) as a tracer of tryptophan, a precursor of serotonin. In theory, αMTP first acts as a tracer of tryptophan uptake into the brain. Then, in serotonergic neurons, αMTP is converted to α-methyl-serotonin (αM5HT). Since αM5HT is not a substrate for monoamine oxidase, it accumulates in the brain. This model was based on that used for 2-deoxyglucose, a tracer of glucose metabolism (Sokoloff *et al.*, 1977). However, there are important kinetic differences between αMTP and 2-deoxyglucose both in transport and metabolism due to differences in the transporter, enzyme kinetics (V_{max} and K_m), and substrate concentrations.

Several groups, including ourselves, have attempted to use αMTP and PET imaging to determination brain serotonin synthesis rates (Chugani *et al.*, 1997, 1998a, b; Nishizawa *et al.*, 1997, 1999; Muzik *et al.*, 1997; Shoaf *et al.*, 1998). During our analysis of data from anesthetized rhesus monkeys (Shoaf *et al.*, 1998), we became aware of some possible shortcomings with the tracer, while others were highlighted by the work of Muzik *et al.* (1997). In particular, it appeared that the apparent steady-state between the precursor pool and plasma may not be achieved until approximately 80 min (5 half-lives). Therefore, Patlak analysis (to determine the uptake of αMTP, K^*) of data obtained before this time would be invalid because the assumption of steady state would not have been met (Muzik *et al.*, 1997). Second, estimates of the amount of activity in the trapped or metabolized pool (i.e., αM5HT, made by multiplying the estimated Patlak K^* by the integral of the plasma input function) were much higher than was originally reported by Diksic *et al.* (1990) and later found by Gharib *et al.* (1999) (Shoaf *et al.*, 1998). This inconsistency between measurements of metabolized tracer and the model predictions of serotonin synthesis rate again suggested that the model estimates were incorrect.

Therefore, we performed a PET study in order to further examine αMTP as a tracer for serotonin synthesis. Scanning for a longer period of time with a more sensitive instrument would permit maximum accumulation of αM5HT and determination of a truly linear portion of the Patlak plot. Four rhesus monkeys, anesthetized with isoflurane, were scanned for 3 h in a high-sensitivity three-dimensional (3D) PET scanner following the administration of [^{11}C]αMTP. Immediately following an initial [^{11}C]αMTP study, we performed a second PET scan for 1 h following the administration of a mixture of [^{11}C]αMTP and [^{14}C]αMTP. The animals were then sacrificed and brain tissues were sampled in order to directly measure [^{14}C]αM5HT and [^{14}C]αMTP concentrations. This paired study design also permitted the assessment of any effect of the additional dose of unlabeled αMTP in the second run due to the coinjection of [^{14}C]αMTP.

II. METHODS

A. Animals

Four male rhesus monkeys were individually housed according to the NIH Animal Care and Use Guidelines. All procedures were approved by the NIAAA Animal Care and Use Committee (protocol LCS-82). Animals weighed 6.9–8.2 kg and were 12–20 years old. Preparation for scanning was performed as described in Shoaf *et al.* (1998).

B. Imaging

A GE Advance scanner was used in 3D mode. The animals were positioned prone in a stereotaxic frame designed for use in the scanner. A 6-mm lead shield was placed around the chest of the animal to minimize scatter and random counts from this region.

1. Scan 1

A transmission scan was obtained for attenuation correction. Two injections of 15 mCi of [^{15}O]water were performed and 60-s scans were acquired. These images were averaged and used for image registration and ROI placement. A 3-mL sample of arterial blood was taken for the determination of free and total plasma tryptophan. [^{11}C]αMTP was prepared according to the method of Schmall *et al.* (1995). The dose of 11–15 mCi (sp. act. (Ci/mmol): mean, 187; range, 24–320) was injected intravenously over a 1-min period using an infusion pump. Scanning and arterial blood sampling continued for 3 h.

2. Scan 2

The second scan of 1 h duration was begun 20–30 min following the completion of the first scan. A mixture of [^{14}C]- and [^{11}C]αMTP was administered intravenously over a 2-min period using an infusion pump. The αMTP mixture contained 1 mCi of [^{14}C]αMTP (sp. act. 56 mCi/mmol, radiochemical purity >99%, 100% L form, Moravek Biochemicals, Brea, CA) and 10–17 mCi of [^{11}C]αMTP (sp. act. (Ci/mmol): mean, 152; range, 52–260). Twenty-four arterial blood samples were taken and analyzed as above.

C. Data Analysis

K^* was calculated using a variant of the Patlak analysis (Patlak *et al.*, 1983). The tissue concentration data ($C_i(t)$, nCi/mL tissue) from each pixel over the specified time range are fit to the following linear equation:

$$C_i(t) = K \int C_p dt + V_e C_b(t), \tag{1}$$

where $C_p(t)$ and $C_b(t)$ are the plasma and whole blood concentrations of αMTP, respectively, and V_e is the exchangeable tissue volume (mL blood/mL tissue). The fit was performed pixel-by-pixel to estimate K^* and V_e and images of K^* were obtained. In order to compare the data from this study to those obtained in our previous study (Shoaf *et al.*, 1998), we repeated the analysis where we selected 30–60 min as the linear range. Brains were outlined on the [^{15}O]water scans and coregistered to the brain of one monkey by the gray-scale matching technique (Thévenaz *et al.*, 1995). Regions of interest (ROIs) were placed on the average water scan and the K^* values for each ROI were determined using the registered scans.

D. Curve Fitting

Irregular ROIs were drawn on cortical regions on the [^{15}O]water images and time–activity curves (TACs) were generated from the dynamic sequence of [^{11}C]αMTP images. These tissue data were fit to a four-parameter model (Diksic *et al.*, 1991; Muzik *et al.*, 1997)

$$C_i(T) = \frac{K_1 k_3}{k_2 + k_3} \int_0^T C_p dt + \frac{K_1 k_2}{k_2 + k_3} C_p$$
$$\otimes \exp(-(k_2 + k_3)t) + V_b C_b(T), \tag{2}$$

where K_1 is the uptake rate constant (mL/min/mL), k_2 is the tissue-to-blood clearance constant, k_3 is the rate constant of αMTP trapping, and V_b is the tissue blood volume. Fits were performed for time periods 0–60 min and 0–180 min for the first scan in each animal.

E. Brain Sampling

At 60 min following the second αMTP injection, the animal was removed from the stereotactic head frame, 0.8 mL of ketamine (i.m.) was administered and the animal was transported to a pathology laboratory. At approximately 90 min post-αMTP administration, an overdose of phenobarbital was administered intravenously. Dissection of the brain began 10–15 min later, following its removal from the cranium, and was completed in 10–15 min. Samples from 21 different regions (see Table 1) were taken and frozen at −80°C.

F. Chromatographic Assays

Free and total plasma tryptophan concentrations were determined by HPLC with fluorometric detection according to Shoaf et al. (1998).

Brain samples were analyzed for serotonin (5-HT), α-methyl-5-hydroxytryptophan (αM5HTP), α-methylserotonin (αM5HT), tryptophan (TP), 5-hydroxyindole-3-acetic acid (5-HIAA) and αMTP concentrations using HPLC with fluorometric detection. Since detectable concentrations of 5-hydroxytryptophan (5-HTP) were not expected to be found, the HPLC separation was optimized for the other compounds and quantitation of 5-HTP was not done.

Brain tissues were homogenized in a volume of buffer [0.1 N perchloric acid, 1% (v/v) ethanol, 0.02% EDTA (w/v)] equal to the weight of the tissue rounded up to next highest 100 μL. One half of the cerebellum, pons, medulla, and midbrain was homogenized after being sliced down the midline. The homogenate was then centrifuged at 5000g for 10 min; the supernatant was removed and placed in a microfuge tube which was then centrifuged at 12,000g for 3 min. One hundred and fifty microliters of supernatant was injected onto a Nova-Pak C-18 column, 300 × 4.6 mm (Waters, Milford MA). The mobile phase consisted of 100 mM citrate, 100 mM sodium acetate, and 0.03 mM EDTA buffer, methanol, and 4 N perchloric acid (ratio 94/5.5/0.5, v/v/v) run at 0.9 mL/min. External quantification was used.

The eluent from the HPLC was collected in 1-min fractions. Ten milliliters of Bio-Safe II (Research Products International, Corp., Mt. Prospect IL) was added to each fraction and counted four different times in a Scintillation counter (Beckman, model LS 6000) for ^{14}C activity. Background counts were subtracted from reported DPM values. Recovery of ^{14}C from brain tissue was determined by spiking samples of finely diced cold monkey brain tissue

with [^{14}C]αMTP stock solution, homogenizing as above, and counting the fractions of HPLC eluant. Recovery of [^{14}C]αMTP was >96% (data not shown). Recovery of total ^{14}C radioactivity was determined by comparing the total counts in the HPLC fractions to the total counts in the original brain homogenate supernatant and found to be ∼95%; the tissue pellet was found to contain less than 2% of the supernatant counts.

III. RESULTS AND DISCUSSION

Tryptophan concentrations during the scans were at steady state; total plasma tryptophan concentrations were 25.9 ± 1.2 (mean ± SD) and 24.3 ± 1.6 nmol/mL, respectively, in samples taken prior to and immediately after the PET scans. Free concentrations were 7.9 ± 3.5 and 7.5 ± 2.7 nmol/mL, respectively.

Mean concentrations of 5-HT, 5-HIAA, tryptophan, and αMTP (pmol/mg wet tissue) for 21 brain regions are presented in Table 1. Due to the low specific activity of the [^{14}C]αMTP (56 mCi/mmol), 17.9 μmol of αMTP was administered to each animal and this produced fluorometrically measurable brain tissue concentrations. However, αM5HTP or αM5HT concentrations in all tissues were below the level of detection, 0.2 and 10 pmol/mL, respectively. Tryptophan and αMTP brain concentrations were highly correlated, with a range of $r = 0.607–0.837$, $P < 0.001$. No other correlations were statistically significant.

Scintillation counting of 1-min fractions of HPLC eluant revealed that >96% of all counts were present in those fractions (20–23) belonging to αMTP (Table 2). An average of 0.61 ± 0.07% of the counts was present in the solvent front (fractions 3 and 4). Counts in these fractions were also present following injections of the standard and amounted to less than 1% of the total counts (data not shown). An average of 2.02 ± 0.19% of the counts eluted in fractions 9 and 10. Counts in these fractions were also present following injections of the standard and amounted to less than 1% of the total counts (data not shown). Counts in fractions 11 and 12 were assigned to αM5HT.

In cortical regions, αM5HT accounted for an average of 0.3% of the total ^{14}C radioactivity. The tissues with the highest radioactivity in fractions corresponding to the αM5HT peak were the midbrain and the pons. The dorsal raphe nucleus (an area reported to have a high serotonin synthesis rate, Diksic et al. (1990)) is a structure centered on the midline approximately 2 mm wide in the midbrain of the rhesus monkey, so determining the percentage of total DPM associated with αM5HT (range, 0.47–0.96%) in half of the midbrain would have diluted DPMs found in the raphe nucleus. In a second analysis, the remaining half of two of the midbrains were dissected into three sections, approximately

2 mm wide, by cuts parallel to the midline. The percentage of total DPMs associated with αM5HT in the innermost section was greater than that found in the entire half of the midbrain by a factor of 2–4 (2.1 vs 0.52 and 0.95 vs 0.47). Since the raphe is approximately 2 mm wide centered on the midline, it is possible that the percentage of measured counts due to αM5HT is still 50% lower than what would be measured if the raphe nucleus had been analyzed free of other tissue.

Our values of 0.95–2% of measured counts due to αM5HT in the dorsal raphe is considerably lower than the 31% (Diksic *et al.*, 1990) or 20% (Gharib *et al.*, 1999) reported for the awake rat 60 min after [^{14}C]αMTP administration. It is doubtful that isoflurane anesthesia would suppress serotonin synthesis rates to such a great extent. If serotonin synthesis rate is tightly coupled to αMTP uptake and anesthesia does suppress serotonin synthesis rate, then K^* should be lower in anesthetized subjects. However, Chugani *et al.* (1997) reported no differences in the uptake rates of αMTP (K^*) in normal adults and children that were scanned without anesthesia or following the administration of either midazolam or Nembutal. It is more likely that serotonin synthesis rates in larger animals are slower than that in rats and may be suppressed somewhat by anesthesia. Low oxygen tension, an inhibitor of tryptophan hydroxylase activity, would not be a problem in our monkeys as the isoflurane is carried in 100% oxygen and pO$_2$ concentrations are typically near 400 mm Hg.

A. Patlak Analysis

Figure 1 shows Patlak plots for 3 h of scanning data (corresponding to 400–450 min of transformed time) for each of the four animals (dashed lines) and the average (solid line).

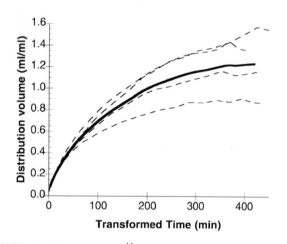

FIGURE 1. Patlak analysis of [^{11}C]αMTP study in four rhesus monkeys. The dashed lines represent the mean cortical Patlak curve for each animal and the solid line is the averaged curve after interpolation of the four individual curves to a common time base. Following the Patlak transformation, 180 min of true time corresponds to 400–450 min of transformed time.

Previous analyses (Shoaf *et al.*, 1998) used the time period of 30 to 60 min postinjection, which corresponds to approximately 60–120 min of transformed time. K^* values for this time period for the first and second PET scans were not significantly different, thus there was no measurable effect of the additional unlabeled αMTP on the tracer uptake, as expected, or an effect of prolonged anesthesia. None of the regional K^* values were significantly different from those given by Shoaf *et al.* (1998).

Examination of the 30- to 60-min portion (60–120 min of transformed time) of the plot (without benefit of the remainder of the curve) could permit the assertion that this is a linear portion of the curve and suitable for estimation of the uptake rate constant, K^*. However, as time proceeds, the Patlak curve continues to roll off, so slopes taken from 30 to 60 min substantially overestimate the true tracer uptake rate. Estimation of the final slope using values from 300 min of transformed time on, leads to estimates of K^* of 0.5 ± 0.7 μL/min/mL, which are not significantly different from zero. This final slope value is much lower than the value obtained from 30 to 60 min of ~4 μL/min/mL.

Unlike Chugani *et al.* (1998b), we found no correlation between serotonin concentration and K^* values (Fig. 2). We found that the pattern of distribution of serotonin and 5-HIAA concentrations (Table 1) matches that found by Brown *et al.* (1979) for young adult rhesus monkeys; our

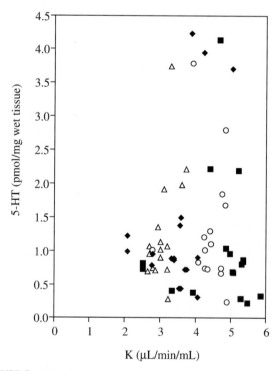

FIGURE 2. Plot of serotonin concentration (pmol/mg wet tissue) versus K^* value (μL/min/mL, determined from 30 to 60 min of PET data) for four rhesus monkeys; each symbol represents a different animal. See Methods for details.

TABLE 1.   Indoleamine Concentrations (pmol/mg Wet Tissue)[a] in Rhesus Monkey Brain

Brain region	Serotonin	5-Hydroxyindole-3-acetic acid	Tryptophan	α-Methyl-L-tryptophan
Cerebellum	0.26 ± 0.04	0.88 ± 0.31	21.7 ± 7.4	0.28 ± 0.05
Pons	2.93 ± 1.16	11.69 ± 3.52	17.9 ± 7.9	0.25 ± 0.08
Midbrain	3.49 ± 0.88	9.98 ± 3.75	20.1 ± 5.2	0.24 ± 0.03
Medulla	2.72 ± 1.26	7.47 ± 5.42	21.3 ± 10.6	0.22 ± 0.04
R. Medial orbital	0.85 ± 0.39	0.94 ± 0.21	20.0 ± 7.9	0.20 ± 0.05
L. Medial orbital	1.21 ± 0.47	0.88 ± 0.35	17.6 ± 7.1	0.25 ± 0.03
R. Lateral prefrontal	0.68 ± 0.19	0.73 ± 0.16	16.8 ± 5.5	0.25 ± 0.05
L. Lateral prefrontal	0.64 ± 0.17	0.88 ± 0.28	16.3 ± 3.8	0.25 ± 0.04
R. Motor cortex	0.68 ± 0.39	0.63 ± 0.11	15.1 ± 3.5	0.22 ± 0.04
L. Motor cortex	0.54 ± 0.16	0.73 ± 0.20	15.3 ± 7.2	0.22 ± 0.04
R. Cingulate	0.98 ± 0.21	1.17 ± 0.32	15.9 ± 4.6	0.21 ± 0.01
L. Cingulate	0.90 ± 0.15	1.21 ± 0.46	17.0 ± 8.7	0.22 ± 0.04
R. Lateral parietal	0.65 ± 0.24	1.04 ± 0.29	16.9 ± 5.5	0.21 ± 0.04
L. Lateral parietal	0.62 ± 0.23	0.87 ± 0.29	15.8 ± 3.6	0.23 ± 0.05
R. Occipital	1.00 ± 0.21	1.73 ± 0.49	23.7 ± 9.5	0.32 ± 0.05
L. Occipital	0.98 ± 0.20	1.67 ± 0.52	21.7 ± 10.9	0.30 ± 0.08
R. Superior temporal	1.10 ± 0.29	1.12 ± 0.22	16.9 ± 7.6	0.23 ± 0.01
L. Superior temporal	1.13 ± 0.35	1.17 ± 0.23	20.8 ± 8.1	0.28 ± 0.03
Hippocampus	1.17 ± 0.35	1.51 ± 0.19	17.4 ± 5.5	0.22 ± 0.05
Caudate	2.21 ± 0.58	3.17 ± 0.73	22.9 ± 7.1	0.32 ± 0.05
Thalamus	2.79 ± 0.82	6.10 ± 1.53	18.7 ± 5.6	0.23 ± 0.06

[a] Values are the mean $\pm$ SD for four monkeys at 90 min following administration of 17.9 μmol of αMTP.

values are slightly higher due to the older age of our monkeys. Brown *et al.* (1979) also report that for rhesus monkeys treated with NSD-1015 (an amino acid decarboxylase inhibitor), that "the accumulation of 5-HTP does not parallel serotonin levels." Therefore, it may be the case that serotonin concentration is not a good indicator of serotonin synthesis rate.

B. Curve Fitting

Fits to the four-parameter model over a three-parameter model (k_3 fixed to 0) showed a statistically significant improvement by the F test ($P < 0.01$). However, both models showed lack of fit as evidenced by nonrandom residuals. This lack of fit remained, even when only 60 min of data were fitted. From the fits of the average cortex curves from 0 to 60 min, the parameter values (mean $\pm$ SD) were K_1, 0.012 ± 0.003 mL/min/mL; k_2, 0.036 ± 0.008 min^{-1}; k_3, 0.020 ± 0.010 min^{-1}; V_b, 0.048 ± 0.018 mL/mL; and K^*, 4.3 ± 1.7 μL/min/mL. These fitted K^* values agreed well with the Patlak fitted values (30–60 min) of 4.1 ± 1.1 μL/min/mL. These values are very similar to those obtained by Muzik *et al.* (1997) in awake human subjects.

Ideally, the fitted parameters would be independent of the duration of the data used for fitting. However, upon extend-

ing the fitting procedure to 180 min, there was a statistically significant change ($P < 0.05$, two-tailed t test) in all parameters: K_1, 0.010 ± 0.002 mL/min/mL; k_2, 0.018 ± 0.003 min^{-1}; k_3, 0.003 ± 0.001 min^{-1}; V_b, 0.053 ± 0.017 mL/mL; and K^*, 1.7 ± 0.7 μL/min/mL. This reflects decreases of 48, 83, and 61% in k_2, k_3, and K^*, respectively, suggesting that kinetic parameters including K^* values determined from only 60 min of data may be significantly overestimated.

Based on the fitted parameters, the model can predict the fraction of the total radioactivity that will be in the trapped pool at any time. From the parameters of the 60-min fit, the predicted fraction of metabolized tracer (i.e., αM5HTP and αM5HT) at 90 min was $76 \pm 24\%$. From the parameters of the 180-min fit, the predicted metabolized fraction at 90 min was much lower, $23 \pm 6\%$. However, both predicted values were much higher than the ^{14}C measured value of 0.3%. This indicates that none of the estimated model parameters (60- or 180-min values) are valid.

C. Conclusion

After extending data collection to 180 min, we were unable to find a linear portion of the Patlak plot (Fig. 1). The final slope of this curve was very small and was not statistically different from zero. These data suggest that the net

TABLE 2.   Percentage of Total DPMs for Peaks in HPLC Eluate of Rhesus Monkey Brain Regions[a]

Brain region	Unknown 1	Unknown 2	α-Methyl-serotonin	α-Methyl-L-tryptophan
Cerebellum	0.51 ± 0.19	1.73 ± 0.56	0.22 ± 0.22	97.54 ± 0.77
Pons	0.60 ± 0.25	2.44 ± 0.51	0.91 ± 0.27	96.04 ± 0.76
Midbrain	0.62 ± 0.20	2.19 ± 0.67	0.70 ± 0.23	96.48 ± 0.83
Medulla	0.46 ± 0.24	2.26 ± 0.66	0.50 ± 0.11	96.78 ± 0.89
R. Medial orbital	0.53 ± 0.19	1.90 ± 0.71	0.28 ± 0.14	97.29 ± 0.94
L. Medial orbital	0.62 ± 0.22	2.01 ± 0.78	0.28 ± 0.08	97.09 ± 1.02
R. Lateral prefrontal	0.66 ± 0.28	2.10 ± 0.78	0.24 ± 0.08	97.00 ± 1.11
L. Lateral prefrontal	0.62 ± 0.22	2.05 ± 0.71	0.21 ± 0.14	97.12 ± 1.06
R. Motor Cortex	0.68 ± 0.35	2.21 ± 0.97	0.25 ± 0.17	96.86 ± 1.27
L. Motor Cortex	0.66 ± 0.24	1.99 ± 0.84	0.13 ± 0.06	97.22 ± 1.14
R. Cingulate	0.57 ± 0.13	2.10 ± 0.82	0.29 ± 0.13	97.05 ± 0.88
L. Cingulate	0.59 ± 0.16	2.09 ± 0.76	0.22 ± 0.09	97.10 ± 0.84
R. Lateral parietal	0.64 ± 0.21	2.03 ± 0.66	0.18 ± 0.14	97.15 ± 0.89
L. Lateral parietal	0.65 ± 0.24	2.07 ± 0.81	0.36 ± 0.19	96.92 ± 1.12
R. Occipital	0.71 ± 0.31	1.85 ± 0.62	0.39 ± 0.09	97.05 ± 0.90
L. Occipital	0.73 ± 0.28	1.81 ± 0.67	0.29 ± 0.14	97.18 ± 1.07
R. Superior temporal	0.63 ± 0.30	1.99 ± 0.66	0.32 ± 0.20	97.44 ± 0.79
L. Superior temporal	0.62 ± 0.19	2.06 ± 0.75	0.28 ± 0.11	97.03 ± 0.97
Hippocampus	0.62 ± 0.21	2.14 ± 0.77	0.30 ± 0.12	96.94 ± 1.04
Caudate	0.51 ± 0.11	1.63 ± 0.62	0.28 ± 0.10	97.58 ± 0.80
Thalamus	0.51 ± 0.25	1.86 ± 0.62	0.41 ± 0.20	97.22 ± 0.79

[a]Values are the mean ± SD for four monkeys 90 min following administration of 1 mCi [^{14}C]αMTP.

synthesis rate of αM5HT is very small, an interpretation that is validated by the finding that less than 1% of cortical radioactivity was present as [^{14}C]α M5HT. Thus, we conclude that the synthesis rate of αM5HT is negligible during the 180-min period of these studies.

In light of the small or nonexistent final uptake rate, the high correlation of αMTP to tryptophan concentrations, and the negligible presence of αM5HT in the tissue, these data demonstrate that [^{11}C]αMTP is simply tracing the uptake of a large neutral amino acid (tryptophan). In this case, the Patlak plot shows an initial uptake rate dependent upon the capacity and saturation of the LNAA transporter, and the ultimate equilibrium level (the asymptote in the Patlak plot) is controlled by the relative influx and efflux rates of the transporter, driven by the plasma and tissue concentrations of large neutral amino acids. We conclude that during the time frame of a typical ^{11}C study, [^{11}C]αMTP cannot be used to measure serotonin synthesis rates.

Acknowledgments

This study was supported by NIAAA Intramural Research funds. We thank the staff of the NIH PET Department, the surgical group of the NCRR Veterinary Resources Program, the staff of NIH Animal Transportation, and Justin L. Lockman for excellent technical assistance.

References

Brown, R. M., Crane, A. M., and Goldman, P. S. (1979). Regional distribution of monoamines in the cerebral cortex and subcortical structures of the rhesus monkey: Concentrations and in vivo synthesis rates. Brain Res. **168**: 133–150.

Chugani, D. C., Chugani, H. T., Muzik, O., Shah, J. R., Shah, A. K., Canady, A., Mangner,T., and Chakraborty, P. K. (1998a). Imaging epileptogenic tubers in children with tuberous sclerosis complex using α-[^{11}C]methyl-L-tryptophan positron emission tomography. Ann. Neurol. **44**: 858–866.

Chugani, D. C., Muzik, O., Chakraborty, P., Mangner, T., and Chugani, H. T. (1998b). Human brain serotonin synthesis capacity measured in vivo with α-[C-11]methyl-L-tryptophan. Synapse **28**: 33–44.

Chugani, D. C., Muzik, O., Rothermel, R., Behen, M., Chakraborty, P., Mangner T., da Silva, E. A., and Chugani, H. T. (1997). Altered serotonin synthesis in the dentatothalamocortical pathway in autistic boys. Ann. Neurol. **42**: 666–669.

Diksic, M., Nagahiro, S., Chaly, T., Sourkes, T. L., Yamamoto, Y. L., and Feindel, W. (1991). Serotonin synthesis rate measured in living dog brain by positron emission tomography. J. Neurochem. **56**: 153–162.

Diksic, M., Nagahiro, S., Sourkes, T. L., and Yamamoto, Y. L. (1990). A new method to measure brain serotonin synthesis in vivo. I. Theory and basic data for a biological model. J. Cereb. Blood Flow Metab. **10**: 1–12.

Gharib, A., Balende, C., Sarda, N., Weissmann, D., Plenevaux, A., Luxen, A., Bobillier, P., and Pujol, J.-F. (1999). Biochemical and autoradiographic measurements of brain serotonin synthesis rate in the freely moving rat: A reexamination of the α-methyl-L-tryptophan method. J. Neurochem. **72**: 2593–2600.

Missala, K., and Sourkes, T. L. (1988). Functional cerebral activity of an analogue of serotonin formed in situ. *Neurochem. Int.* **12**: 209–214.

Muzik, O., Chugani, D. C., Chakraborty, P., Mangner, T., and Chugani, H. T. (1997). Analysis of [C-11]alpha-methyl-tryptophan kinetics for the estimation of serotonin synthesis rate in vivo. *J. Cereb. Blood Flow Metab.* **17**: 659–669.

Nagahiro, S., Takada, A., Diksic, M., Sourkes, T., Missala, K., and Yamamoto, Y. (1990). In vivo measurement of the rate of synthesis of brain serotonin by autoradiography. II. A practical method tested in normal and lithium-treated rats. *J. Cereb. Blood Flow Metab.* **10**: 13–21.

Nishizawa, S., Benkelfat, C., Young, S. N., Leyton, M., Mzengeza, S., de Montigny, C., Blier, P., and Diksic, M. (1997). Differences between males and females in rates of serotonin synthesis in human brain. *Proc. Natl. Acad. Sci. USA* **94**: 5308–5313.

Nishizawa, S., Mzengeza, S., and Diksic, M. (1999). Acute effects of 3,4-methylendioxy-methamphetamine on brain serotonin synthesis in the dog studied by positron emission tomography. *Neurochem. Int.* **34**: 33–40.

Patlak, C. S., Blasberg, R. G., and Fenstermacher, J. D. (1983). Graphical evaluation of blood-to-brain transfer constants from multiple-time uptake data. *J. Cereb. Blood Flow Metab.* **3**: 1–7.

Schmall, B., Shoaf, S. E., Carson, R. E., Herscovitch, P., Sha, L., and Eckelman, W. C. (1995). Separation of racemic α-methyl-tryptophan into its L- and D-isomers and synthesis of α-(^{11}C-methyl)-L-tryptophan. *J. Labelled Compd. Radiopharm.* **37**: 689–691.

Shoaf, S. E., and Schmall, B. (1996). Pharmacokinetics of α-methyl-L-tryptophan in rhesus monkeys and calculation of the lumped constant for estimating the rate of serotonin synthesis. *J. Pharmacol. Exp. Ther.* **277**: 219–224.

Shoaf, S. E., Carson, R., Hommer, D., Williams, W., Higley, J. D., Schmall, B., Herscovitch, P., Eckelman, W., and Linnoila, M. (1998). Brain serotonin synthesis rates in rhesus monkeys determined by [^{11}C]α-methyl-L-tryptophan and positron emission tomography compared to CSF 5-hydroxyindole-3-acetic acid concentrations. *Neuropsychopharmacology* **19**: 345–353.

Sokoloff, L., Reivich, M., Kennedy, C., Des Rosiers, M. H., Patlak, C. S., Pettigrew, K. D., Sakurada, O., and Shinohara, M. (1977). The [14C]deoxyglucose method for the measurement of local cerebral glucose utilization: Theory, procedure, and normal values in the conscious and anesthetized albino rat. *J. Neurochem.* **28**: 897–916.

Thévanaz, P., Ruttimann, U. E., and Unser, M. (1995). Iterative multi-scale registration without landmarks. *In* "Proceedings International Conference on Image Processing. Los Alamitos, CA," pp. 228–231. IEEE Computer Society Press.

36

Serotonin Release and Reuptake Studied by PET Neuroimaging Using Fenfluramine, [15O]Water, [15O]Oxygen, [18F]Fluorodeoxyglucose, [11C]NS2381, and its Enantiomers in Living Porcine Brain

DONALD F. SMITH,[*,†] ANTONY GEE,[†] SØREN B. HANSEN,[†] PETER MOLDT,[‡]
ELSEBET ØSTERGAARD NIELSEN,[‡] JØRGEN SCHEEL-KRÜGER,[‡] KOICHI ISHIZU,[†] MASAHARU SAKOH,[†]
LEIF ØSTERGAARD,[†,§] PETER HØST POULSEN,[†] DIRK BENDER,[†] and ALBERT GJEDDE[†]

[*]Department of Biological Psychiatry, Institute for Basic Research in Psychiatry, Psychiatric Hospital, Risskov, Denmark
[†]PET Center, Aarhus University General Hospital, Aarhus, Denmark
[‡]NeuroSearch, Glostrup, Denmark
[§]Department of Neuroradiology, Aarhus University Hospitals, Aarhus, Denmark

Serotonin is a key neurotransmitter in a variety of neuropsychiatric disorders. It is therefore of interest to determine the functional status of serotonergic processes in the living brain. We have carried out positron emission tomography (PET) research on two strategies for doing so, namely the fenfluramine activation test and the pharmacokinetics of antidepressant drugs. We used fenfluramine and obtained quantitative measurements of the effects of serotonergic stimulation on regional cerebral blood flow (rCBF) and on regional cerebral metabolism of glucose and oxygen ($rCMR_{glc}$ and $rCMR_{O_2}$). Fenfluramine (25 mg/h i.v.) caused a significant rise in rCBF and, to a lesser extent, in $rCMR_{O_2}$, but it failed to affect $rCMR_{glc}$. The findings indicate that repeated quantitative estimation of rCBF was more sensitive than either $rCMR_{O_2}$ or $rCMR_{glc}$ for detecting effects of fenfluramine on serotonin neurotransmission in living porcine brain. Next, we used new selective serotonin reuptake inhibitors, NS2381 and its enantiomers, radiolabeled with [11C], for quantifying the serotonin uptake site in the living brain. [11C]NS2381 accumulated readily in brain and bonded reversibly in regions rich in serotonin uptake sites. In addition, [11C]NS2381 was displaced from brain tissue by citalopram, a potent inhibitor of serotonin uptake. However, [11C]NS2381 has a relatively high degree of nonspecific binding in brain regions, and the enantiomers of [11C]NS2381 were, in general, similar to the racemate in terms of their uptake and distribution in living pig brain.

Consequently, the search goes on for an ideal positron emission tomography radioligand for quantifying serotonin reuptake in the living brain.

I. INTRODUCTION

The invention of positron emission tomography (PET) for studying the function of the nervous system *in vivo* ushered in two decades of discovery that have been focused primarily on factors affecting either cerebral blood flow or the metabolism of oxygen and glucose in brain. Nowadays, however, PET is being applied more and more in experimental settings aimed at studying additional aspects of neuroreceptor function (Staley *et al.*, 1998). Such studies have the possibility of revealing disturbances in neuroreceptors that underlie neuropsychiatric disorders. As in the past, experiments using laboratory animals will most probably be necessary for finding the most appropriate PET radiotracers as well as the best procedures before testing hypotheses in humans.

Of particular interest for our work is the notion that disturbances in cerebral serotonergic processes play a key role in the occurrence and successful treatment of affective disorders. One strategy that has been used for examining cerebral serotonergic functions in affective disorders is the so-called fenfluramine activation test (Kapur *et al.*, 1994; Mann *et al.*,

1996; Meyer *et al.*, 1996). In it, the indirect serotonin agonist fenfluramine is given to subjects and PET studies of rCBF or rCMR$_{glc}$ are carried out. Fenfluramine releases serotonin into the synaptic cleft from the vesicles of the presynaptic terminal, and changes in rCBF or rCMR$_{glc}$ measured by PET after administration of fenfluramine are assumed to reflect the degree to which fenfluramine activates serotonergic neurotransmission. It is to be noted that fenfluramine was recently withdrawn from the market because its long-term usage as diet pill was linked to a disturbance in heart function. Nevertheless, the fenfluramine activation test could still be of value for investigating cerebral serotonergic functions under well-controlled conditions in humans and in laboratory animals, provided that the most appropriate conditions for the test are found. Here, we present our findings on whether quantification of regional cerebral of blood flow (rCBF), regional cerebral oxygen metabolism (rCMR$_{O_2}$), or regional cerebral glucose metabolism (rCMR$_{glc}$) provides the most reliable means of studying serotonergic processes by PET in the living brain.

Another strategy for examining cerebral serotonergic functions by PET is the use of radiolabeled compounds that bind to the neuronal serotonin reuptake site. It is noteworthy that almost all compounds that are effective antidepressant drugs have inhibitory effects on serotonin reuptake (Hyttel, 1982). At present, however, no radiotracer has been found to be ideal for examining serotonin uptake sites in the living brain (Sedvall *et al.*, 1986; Fletcher *et al.*, 1995; Pike, 1995, 1997; Smith *et al.*, 1996, 1997). Recently, a new compound with inhibitory actions at the serotonin reuptake site and having an appropriate structure for radiolabeling for PET became available (NS2381; NeuroSearch, 1997). We studied the properties of this compound as PET radioligand and describe our findings in this report.

II. MATERIALS AND METHODS

A. Animal Preparation

The experiments were approved by the Danish National Committee for ethics in animal research. Female pigs (Hampshire × Yorkshire × Duroc × Landrace crossbred) weighing 39–44 kg were used. They were housed and prepared for PET scanning as described in detail elsewhere (Poulsen *et al.*, 1997; Smith *et al.*, 1997). Briefly, we sedated them initially with i.m. injection of midazolam and ketamine HCl (approx. 1 and 20 mg/kg, respectively). Then, we intubated the animals and anesthetized them with isoflurane in O_2/N_2O (1:2). Catheters were installed in the femoral artery for blood sampling and in the femoral vein for injection of substances. Care was taken, using a thermostatically controlled heating blanket, to assure that the body temperature of the pig was in the normal range (39.0–39.4°C) during

scanning. Brain scanning typically began at least 2 h after the pig arrived at the PET Center so as to minimize possible aftereffects of transport to the PET Center and of midazolam and ketamine sedation (Åkeson *et al.*, 1993; Tao and Auerbach, 1994). Values of pH, pCO$_2$, and pO$_2$ were measured at 2-h intervals, and disturbances in body fluid balance were corrected by appropriate procedures (e.g., forced ventilation and/or changes in saline or glucose infusion rates) in order to maintain physiological parameters within the normal range.

B. PET Radiochemistry

[^{15}O]Oxygen, [^{15}O]water, and [^{18}F]fluorodeoxyglucose ([^{18}F]FDG) were produced by adaptation of standard procedures using commercially available devices (GE Medical Systems) (Clark *et al.*, 1987; Tooronigian *et al.*, 1990). For typical productions, 5 to 7 GBq [^{15}O]water and [^{18}F]FDG were obtained at end of synthesis as a sterile and pyrogen-free solution. Up to 12 GBq [^{15}O]oxygen gas was produced and at end of bombardment delivered on-line directly to the scanner room. [^{11}C]NS2381 and its enantiomers were prepared as follows: 0.5 mg of NS2435 free base was dissolved in 300 μL dimethylsulfoxide (DMSO), and alkylated with [^{11}C]methyl iodide by heating at 130°C for 5 min. Purification was performed by semipreparative HPLC (Nucleosil 5 CN column, 250 × 10 mm) eluted with acetonitrile:25 mM NaH$_2$PO$_4$ (1:2, v/v) at a flow rate of 10 ml/min and a UV detector wavelength of 225 nM. The fraction corresponding to the labeled product was evaporated to dryness, formulated in saline, and passed over a 0.22-μm filter into a sterile vial.

C. Brain PET

1. General Procedures

The pigs were placed prone in the scanner and were held by a custom-made head-holding device. PET was carried out using a Siemens ECAT Exact HR47 tomograph and began at the time of administration of radiotracer. The brain image data were reconstructed using a Hanning filter with a cutoff frequency giving a spatial resolution of 4.6 mm (FWHM) (Wienhard *et al.*, 1994). Correction for attenuation was made on the basis of a transmission scan acquired before injection. A standard set of brain regions of interest (ROIs) (cerebellum, basal ganglia, brainstem, occipital cortex, temporal cortex, thalamus, frontal cortex, and subthalamus) of approximately equal size was used for each pig based on a neuroanatomical atlas of the porcine brain (Yoshikawa, 1968; Smith *et al.*, 1999).

2. Fenfluramine Activation Test

rCBF, rCMR$_{O_2}$, and rCMR$_{glc}$ were measured before (i.e., baseline period) and during intravenous infusion of (±)-fenfluramine HCl (Sigma-Aldrich, 0.625 mg/mL,

25 mg/h, ca. 0.63 mg/kg/h) (i.e., fenfluramine period). Racemic fenfluramine was used because it is the most widely used serotonin challenge agent in psychiatric research (Mann *et al.*, 1996) and is far less expensive than the (+)-enantiomer. rCBF was determined by administering an i.v. bolus of [^{15}O]water (ca. 800 MBq) followed immediately by an i.v. injection of a heparin solution to flush the catheter. Pancuronium bromide (1 mg i.v.) was injected prior to each measurement of CMR$_{O_2}$ to ensure that the animal did not breathe voluntarily during administration of the [^{15}O]oxygen bolus via the lungs (Poulsen *et al.*, 1997). rCMR$_{O_2}$ was measured by single-breath inhalation of [^{15}O]oxygen (ca. 800 MBq) followed by 10 s breath holding (Poulsen *et al.*, 1997). rCBF and rCMR$_{O_2}$ were each measured six times in alternating order in two animals, thrice under baseline conditions and thrice during fenfluramine infusion. Glucose utilization was measured using the dual [^{18}F]FDG injection procedure described in detail elsewhere (Poulsen *et al.*, 1997). Briefly, an i.v. bolus of ca. 150 MBq was given at the start of the first period which lasted 45 min. Approximately 225 MBq was given at the start of the second period that lasted 30 min. The second injection of [^{18}F]FDG was given either immediately (early period) or ca. 2–3 h after (late period) the start of a continuous i.v. infusion of fenfluramine. The early period was studied in two animals and the late period was studied in two other animals.

3. [^{11}C]NS2381 and Its Enantiomers

Three other pigs received two intravenous bolus injections of racemic [^{11}C]NS2381 (760–1025 MBq), once under baseline conditions and once during continuous i.v. infusion of citalopram HBr (racemate), a potent SSRI (Hyttel, 1982; Hyttel *et al.*, 1984). Each scan lasted 90 min and consisted of 30 frames (6 × 10 s, 4 × 30 s, 7 × 1 min, 5 × 2 min, 4 × 5 min, 2 × 10 min, 2 × 15 min). Infusion of citalopram began upon completion of the baseline scan. Citalopram was dissolved in isotonic saline (1 mg/mL) and was given first as a 10-mL i.v. bolus and thereafter as continuous i.v. infusion (1 mL/min; ca. 60 nmol/kg/min). Arterial samples were obtained for HPLC estimation of the plasma concentration of citalopram which is expressed in μmol/L. Although this experimental design has some shortcomings that could be resolved by using either a counterbalanced design or an additional group of animals tested twice under baseline conditions (Ingle, 1972), it was used in the present study due mainly to the extremely slow washout of citalopram from brain tissue (Hume *et al.*, 1992) and the relatively great expense of additional days of scanning.

Another pig underwent four scans. First it received each enantiomer of ^{11}C-labeled NS2381 (baseline condition). They were given, one at a time, as an intravenous bolus (480–500 MBq) at the start of scans using the frame sequence and blood sampling protocol described above. As the absolute stereochemical configuration of the NS2381 enantiomers obtained by our procedure is presently unknown, they are arbitrarily named based on their order of elution from a DIACEL chiralcel OD-H column using *n*-hexane and ethanol (95:5) at a flow rate of 1 mL/min (first off, enantiomer A; second off, enantiomer B). The first scan under baseline conditions began immediately after injection of enantiomer A, and the second scan began ca. 90 min later, immediately after injection of enantiomer B. Then, citalopram was administered as described above and, 45 min after starting the citalopram infusion, the scans with enantiomers A and B were repeated (citalopram condition).

D. Blood Chemistry

Arterial blood samples (1–2 mL) were drawn for determination of total plasma radioactivity at times after injection of the radiotracer. Metabolite correction for [^{11}C]NS2381 and its enantiomers was done using 500 μL plasma samples drawn at 30 s and 2, 10, 30, and 60 min. Acetonitrile (0.5 mL) was added to plasma samples, they were centrifuged, and the supernatant was loaded into a 1-mL injection loop and chromatographed using the following analytical HPLC conditions: Nucleosil 5CN column (250 × 4.6 mm, 5 μm); eluent, 50 mM NaH$_2$PO$_4$/acetonitrile (2/1); flow, 2 cm^3/min. The amount of unchanged [^{11}C]NS2381, as racemate or enantiomer, in plasma was determined by two methods: integration of the peak corresponding to [^{11}C]NS2381 in relation to the sum of all radioanalytes and collection of fractions and counting of radioactivity. Biexponential fitting of the data to the total plasma radioactivity concentrations was performed to generate a metabolite-corrected input function.

E. Pharmacokinetic Analysis of PET Findings

Radioactivity concentrations were calculated for each ROI for the sequence of frames, were corrected for radioactive decay to the start of scan, and were plotted versus time. Estimation of pharmacokinetic parameters was carried out by graphical analysis using multilinear regression and a one-compartment model (Poulsen *et al.*, 1997). The resulting value of K_1 expresses the unidirectional clearance of the tracer from the circulation to a single-tissue compartment. CBF was expressed as mL/100 cm^3/min. rCMR$_{O_2}$ was calculated by the equation $K_{1O_2} \times Ca_{O_2}$, where K_{1O_2} is the unidirectional clearance of oxygen from blood into brain and Ca_{O_2} is the arterial concentration of oxygen (mM) estimated as described elsewhere (Poulsen *et al.*, 1997). rCMR$_{O_2}$ was expressed as μmol/100 cm^3/min. The analysis of the data for [^{11}C]NS2381 and its enantiomers was done assuming pools of free and bound radioligand, yielding the unidirectional clearance of the tracer from the circulation to a single tissue compartment (K_1'), the apparent rate constant for clearance of the tracer from the single tissue compartment (k'_2), the initial plasma volume of distribution (V_0), and the apparent volume of partition of the tracer between the circulation compartment and the single tissue compartment ($V'_e = K'_1/k'_2$), relative to the total concentration of the tracer in arterial blood (Logan *et al.*, 1990; Gjedde and Wong, 1991).

F. Synaptosomal Monoamine Uptake

Four male Wistar rats (150–200 g) were used. Brain tissue was prepared at 0–4°C. Dissected brain regions (hippocampus for [3H]serotonin, hypothalamus for [3H]noradrenaline, and striatum for [3H]dopamine) were homogenized in ice-cold 0.32 M sucrose containing 1 mM pargyline. Preparation of synaptosomes and incubation of samples were carried out according to conventional procedures (Low *et al.*, 1984; Bolden-Watson and Richelson, 1993). Incubations used concentrations from 0.00003 to 0.1 μM for serotonin uptake, from 0.1 to 30 μM for dopamine uptake, and from 0.01 to 30 μM for noradrenaline uptake. After incubation the samples were poured directly onto Whatman GF/C glass fiber filters under suction, the filters were then washed three times with 5 mL of ice-cold 0.9% (w/v) NaCl solution, and the amount of radioactivity on the filters was determined by conventional liquid scintillation counting. Specific uptake was calculated as the difference between total uptake and uptake measured in the presence of a selective uptake inhibitor (citalopram, benztropine, or desipramine at final concentration of 1 μM for inhibition of serotonin, dopamine, or noradrenaline uptake, respectively).

Six pigs were anesthetized and then killed by intravenous injection of saturated KCl. The brain was removed rapidly (within 5 min), was placed in ice-cold isotonic glucose, and was stored overnight at 4°C. Brain regions were dissected (ca. 200- to 300-mg pieces) using an anatomical atlas of the domestic pig brain (Yoshikawa, 1968). Synaptosomes were prepared as described in detail elsewhere (Smith *et al.*, 1992). Briefly, uptake was measured, on a blinded basis, in a modified Krebs–Ringer bicarbonate buffer (pH 7.4). [14C]Serotonin was added to the synaptosomal suspension at a final concentration of 2×10^{-8} M. The incubation was carried out for 15 min at 37 or 4°C. Uptake was stopped by rapid filtration under vacuum onto Millipore filters and radioactivity was measured after addition of scintillation fluid.

G. Statistics

Statistical analyses were carried out by ANOVA using SPSS PC$^+$ (Norusis, 1990).

III. RESULTS

A. Synaptosomal Uptake

Serotonin was actively transported into synaptosomes prepared from regions of the porcine brain (Fig. 1). The regions differed markedly in their uptake rates ($p < 0.001$). Inspection of the data shows that the highest rates of active uptake of serotonin occurred in subthalamic, raphé, and thalamic nuclei, whereas the lowest rates were in cortical and cerebellar regions.

The inhibitory effects of NS2381 and of citalopram were significantly greater for serotonin uptake than for uptake of dopamine and noradrenaline ($p < 0.02$) (Table 1). Moreover, NS2381 and citalopram differed significantly in their inhibitory actions on monoamine uptake *in vitro* ($p < 0.01$); NS2381 was 180 times more potent as an inhibitor of serotonin than of noradrenaline and 670 times more potent as an inhibitor of serotonin than of dopamine in rat brain synaptosomes.

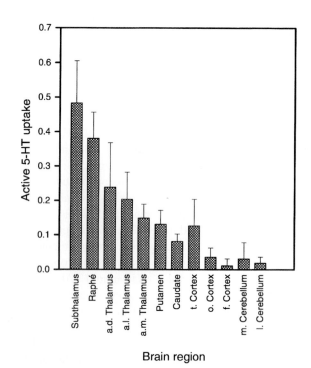

FIGURE 1. Regional rates of active uptake of serotonin (pmol/mg protein/min) measured *in vitro* in synaptosomes prepared from porcine brain. Values shown are means ± SE for six pigs. Active uptake was calculated by subtracting the rate of serotonin uptake obtained in cerebellar tissue incubated at 4°C (i.e., passive uptake and nonspecific binding of serotonin, average rate = 0.23 pmol/mg protein/min) from the total rate of serotonin uptake measured for each region. a.d., anterodorsal; a.l., anterolateral; a.m., anteromedial; t., temporal; o., occipital; f., frontal; m., medial; l., lateral.

TABLE 1. Inhibitory Effect (IC_{50}, μM) of NS2381 and of Citalopram, Both Racemic Compounds, on the Uptake of Serotonin, Dopamine, and Noradrenaline in Synaptosomes Prepared from Rat Brain Regions

Compound	[^{3}H]Serotonin	[^{3}H]Noradrenaline	[^{3}H]Dopamine
NS2381	0.0067 ± 0.0011	1.2 ± 0.2	4.5 ± 0.7
Citalopram	0.0029 ± 0.0003	5.1 ± 0.3	85.0 ± 11.0

B. Fenfluramine Activation Test

Figure 2 shows the levels of rCBF, $rCMR_{O_2}$, and $rCMR_{glc}$ obtained under the various conditions of the fenfluramine activation test. Under baseline conditions, differences in CBF were observed between the brain regions ($p < 0.001$), due mainly to the relatively low levels in the cerebellum and high level in the brainstem. Infusion of fenfluramine significantly elevated rCBF ($p < 0.001$).

Under baseline condition, $rCMR_{O_2}$ failed to differ significantly between brain regions. Fenfluramine infusion significantly elevated $rCMR_{O_2}$ ($p < 0.05$), but apparently to a lesser extent than rCBF.

No significant differences in $rCMR_{glc}$ were observed between brain regions either under baseline conditions or during fenfluramine infusion, and fenfluramine failed to have a significant effect on $rCMR_{glc}$, regardless of the time of testing.

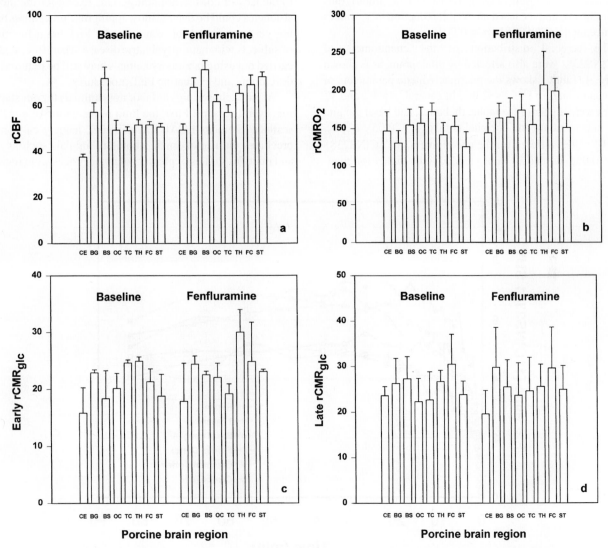

FIGURE 2. Fenfluramine's effects in the living porcine brain (a) on regional cerebral blood flow (rCBF) (mL/100 cm^3/min), (b) regional oxygen consumption ($rCMR_{O_2}$) (μmol/100 cm^3/min) and regional glucose consumption ($rCMR_{glc}$) (μmol/100 cm^3/min) either (c) immediately (early $rCMR_{glc}$) or (d) ca. 2–3 h after (late $rCMR_{glc}$). CE, cerebellum; BG, basal ganglia; BS, brainstem; OC, occipital cortex; TC, temporal cortex; TH, thalamus; FC, frontal cortex; ST, subthalamus. values shown are means $\pm$ SE.

C. [^{11}C]NS2381 and Its Enantiomers

Racemic [^{11}C]NS2381 entered the porcine brain readily and reached peak values between 20 and 40 min after injection (Fig. 3). Higher K_1' values were obtained under baseline conditions in the raphé region, thalamus, and basal ganglia than in cortical and cerebellar regions ($p < 0.01$). It is evident that radioactivity remained relatively constant in the thalamus and frontal cortex throughout the scan, whereas it tended to decline gradually in other regions.

Table 2 shows the pharmacokinetic parameters obtained for racemic [^{11}C]NS2381. There was a tendency under baseline conditions for V_e' values to differ between brain regions ($p = 0.06$), due primarily to the relatively high values obtained for the thalamus. Continuous infusion of citalopram produced plasma levels ranging from 1.2 to 2.0 μmol/L during the scans, and it caused an overall reduction in V_e' values for racemic [^{11}C]NS2381 ($p < 0.001$).

The cerebral distribution of the enantiomers of [^{11}C]NS2381 were also affected by citalopram, as is shown in Fig. 4. Table 3 shows the pharmacokinetic parameters of the enantiomers of [^{11}C]NS2381. In general, K_1', k_2', and V_e' values were higher in the raphé, thalamus, and basal ganglia than in cerebellum and occipital cortex ($p < 0.05$). K_1', k_2', and V_e' values differed for the enantiomers of [^{11}C]NS2381 ($p < 0.05$); K_1' and k_2' values tended to be higher for enantiomer A than for enantiomer B, whereas V_e' values tended to be lower for enantiomer A than for B. Continuous infusion of citalopram increased K_1' and k_2' values of the enantiomers ($ps < 0.005$), but k_2' more than K_1', and therefore markedly decreased their V_e' values ($p < 0.001$).

IV. DISCUSSION

PET is widely used for studying brain functions in a variety of neuropsychiatric disorders. These studies have indicated that cerebral malfunctions could be associated with certain disorders (Bonne and Krausz, 1997; Staley *et al.*, 1998), but there are still many unresolved issues in part due to the lack of consistent findings. One reason for the inconsistencies could be that the most appropriate procedures have not yet been found for examining by PET brain functions of subjects with neuropsychiatric disease. Therefore, studies carried out using laboratory animals may still be required for developing and evaluating PET procedures.

We have chosen pigs as laboratory animals for our studies for several reasons. First, the limit of resolution of our PET scanner (Siemens ECAT EXACT HR), namely ca. 4 mm, precludes the use of small laboratory animals such as rats and mice for studies of pharmacokinetic processes in regions

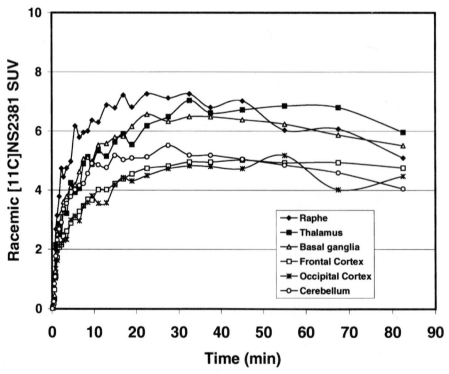

FIGURE 3. Time course of distribution of racemic [^{11}C]NS2381 in regions of the living porcine brain. SUV, standard uptake value, radioactivity measured in brain region (MBq/g tissue (i.e., MBq/cm^{-1})/dosage of racemic [^{11}C]NS2381 administered (MBq)/body weight of pig (kg)). SUV represents the percentage of the administered dosage of radioactivity that is present in the brain region.

of the living brain. On the other hand, the brain of the pig is large enough for studying regional cerebral pharmacokinetics, hemodynamics, and metabolism by PET. Second, the use of nonhuman primates for PET would have been expensive and potentially dangerous due to bite injuries and contagious disease. In contrast, pigs are relatively inexpensive and bear few health hazards for humans (Tumbleson, 1986). Third, pig brain shows sufficient affinity for drugs with specific actions on cerebral receptor sites (e.g., Brust *et al.*, 1996; Smith *et al.*, 1997; Danielsen, 1998). Fourth, current developments in genetic research and neuronal growth factors indicate that the pig may be suitable as a tissue donor for humans suffering from brain disorders (O'Brien, 1996; Deacon *et al.*, 1997), which may have implications for the use of PET.

Our findings show that, under baseline conditions, the rates of regional cerebral blood flow as well as the rates of cerebral metabolism of oxygen and glucose in pigs are almost identical to values obtained in humans (Madsen *et al.*, 1990; Åkeson *et al.*, 1992; Frackowiak *et al.*, 1980; Mintun *et al.*, 1984; Murase *et al.*, 1995; Ohta *et al.*, 1992; Poulsen *et al.*, 1997). Thus, the general properties of cerebral hemodynamics and metabolic rates in pigs resemble closely those of humans. One of the main findings obtained in our studies is the cerebral distribution of serotonin uptake sites in the porcine brain. We found that regions in the thalamus, subthalamus, and brainstem show a relatively high concentration of serotonin uptake sites in the porcine brain as in the brain of humans and nonhuman primates (Cortés *et al.*, 1988; Laruelle *et al.*, 1988; Szabo *et al.*, 1995, 1996a; Lavoie and Parent, 1991; Jagust *et al.*, 1993; Smith *et al.*, in press). It is also noteworthy that the human brain resembles the porcine brain more closely than the brain of rodents in terms of antidepressant binding to serotonin uptake sites (Erreboe *et al.*, 1995). We believe, therefore, that the outcomes of PET studies car-

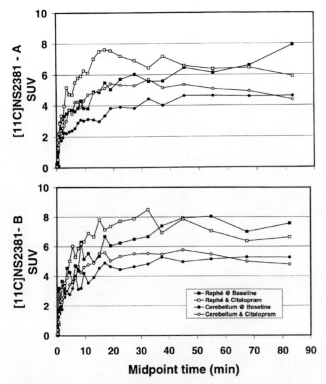

FIGURE 4. Time course of distribution of each enantiomer of [^{11}C]NS2381 (A and B) in regions of the living porcine brain. See legend to Fig. 3 for an explanation of the units.

TABLE 2. Kinetic Parameters of Racemic [^{11}C]NS2381 in Selected Regions of Living Porcine Brain under Baseline Conditions and during Infusion of Citalopram, a Potent SSRI

Brain region	Treatment	K_1' (mL/cm^3/min)	k_2' (min^{-1})	V_e' (mL/cm^3)
Cerebellum	Baseline	2.13 ± 0.33	0.024 ± 0.007	95.0 ± 14.9
	Citalopram	1.32 ± 0.02	0.024 ± 0.002	56.5 ± 4.5
Thalamus	Baseline	2.12 ± 0.07	0.014 ± 0.002	160.0 ± 17.3
	Citalopram	1.37 ± 0.09	0.013 ± 0.003	110.9 ± 21.6
Raphé	Baseline	2.70 ± 0.12	0.021 ± 0.002	128.5 ± 6.9
	Citalopram	1.50 ± 0.02	0.019 ± 0.003	79.9 ± 9.4
Basal ganglia	Baseline	2.32 ± 0.26	0.019 ± 0.005	129.7 ± 23.8
	Citalopram	1.42 ± 0.03	0.018 ± 0.003	80.8 ± 11.2
Frontal cortex	Baseline	1.50 ± 0.09	0.013 ± 0.003	127.0 ± 27.8
	Citalopram	1.13 ± 0.04	0.015 ± 0.004	80.1 ± 18.6
Occipital cortex	Baseline	1.76 ± 0.14	0.020 ± 0.005	93.1 ± 14.0
	Citalopram	1.10 ± 0.07	0.016 ± 0.004	73.3 ± 12.3

Note. Values shown are means $\pm$ SE for three animals.
[a] See Materials and Methods and Discussion for definitions of kinetic parameters.

TABLE 3.   Kinetic Parameters[a] (K_1, k_2, V_0 Model) of the Enantiomers of $[^{11}C]NS2381$ in Living Porcine Brain under Baseline Conditions and during Infusion of Citalopram

Brain region	Enantiomer	Inhibitor	K_1'	k_2'	V_0	V_e'
Cerebellum	A	None	1.18	0.027	0.38	43.7
	A	Citalopram	1.73	0.042	0.27	41.2
	B	None	1.53	0.033	0.35	46.4
	B	Citalopram	1.32	0.032	0.32	41.3
Raphé	A	None	1.70	0.026	0.15	65.4
	A	Citalopram	2.52	0.048	0.22	52.5
	B	None	2.01	0.030	0.76	67.0
	B	Citalopram	1.90	0.034	0.08	55.9
Thalamus	A	None	1.50	0.024	0.09	62.5
	A	Citalopram	2.01	0.031	0.45	64.8
	B	None	1.73	0.024	0.51	72.1
	B	Citalopram	1.60	0.024	0.23	66.7
Basal ganglia	A	None	1.86	0.030	0.33	62.0
	A	Citalopram	2.36	0.041	0.59	57.6
	B	None	2.19	0.033	0.39	66.4
	B	Citalopram	1.79	0.030	0.29	59.7
Frontal cortex	A	None	1.37	0.025	0.20	54.8
	A	Citalopram	1.70	0.033	0.23	51.5
	B	None	1.57	0.027	0.35	58.2
	B	Citalopram	1.31	0.024	0.19	54.6
Occipital cortex	A	None	1.11	0.025	0.21	44.4
	A	Citalopram	1.52	0.034	0.24	44.7
	B	None	1.32	0.027	0.31	48.9
	B	Citalopram	1.18	0.025	0.20	47.2

Note. Values shown are for one animal.

[a] See legend to Table 2 for units and text of report for definitions of the kinetic parameters.

ried out in the living porcine brain provide findings that are likely to be generalizable to humans.

Fenfluramine has several actions on serotonergic processes in the brain: it stimulates serotonin release into the synapse, it inhibits serotonin reuptake into presynaptic neurons, and it directly stimulates postsynaptic serotonin receptors (for review, see Davis and Faulds, 1996). We found that quantitative estimation of rCBF by dynamic PET using repeated injections of $[^{15}O]$water was more sensitive than either $rCMR_{O_2}$ or $rCMR_{glc}$ for detecting effects of fenfluramine given by continuous intravenous infusion (ca. 0.6 mg/kg/h). Our findings agree, in part, with those obtained in awake humans in which $[^{15}O]$water was used for measuring changes in relative CBF in response to intravenous fenfluramine (Meyer *et al.*, 1996). In that study, fenfluramine caused an increase in relative CBF within the frontal cortex, in accordance with the present results. However, fenfluramine was found in that study to cause a decrease in CBF in temporal cortex and thalamus in awake humans, whereas in our study, fenfluramine caused an increase in CBF in both

of those regions. The dosage of fenfluramine used in our study was selected on the basis of preliminary tests which showed higher doses to affect cardiovascular functions in anesthetized pigs. It is also noteworthy that our dosage is of the same order of magnitude as those typically used in studies carried out in humans (0.3 mg/kg i.v. or 60 mg orally) (Meyer *et al.*, 1996; Mann *et al.*, 1996; Kapur *et al.*, 1994). Thus, it seems unlikely that either direct pharmacological effects of fenfluramine in the cardiovasculature or differences in the dosage of fenfluramine used in the studies can account adequately for the findings.

With regard to $rCMR_{O_2}$, our findings show that fenfluramine acted throughout the brain, enhancing the utilization of $[^{15}O]$oxygen, which is also a measure of neuronal activity (Hattazawa *et al.*, 1998). There are, however, no reports on $rCMR_{O_2}$ in humans given fenfluramine with which to compare our findings. Although $rCMR_{O_2}$ levels failed to differ significantly between porcine brain regions in response to fenfluramine, inspection of the data suggests that the thalamus and frontal cortex may have been primarily affected.

With regard to rCMR$_{glc}$, we found no reliable effects of fenfluramine. Our finding contrasts sharply with studies carried out in awake humans given fenfluramine. In one study, fenfluramine enhanced the accumulation of [^{18}F]FDG mainly in the right prefrontal cortex and reduced it in occipital–temporal regions (Kapur *et al.*, 1994). In another study, fenfluramine enhanced the relative rates of glucose metabolism in left prefrontal and temperoparietal regions, whereas it tended to decrease glucose metabolism contralaterally. The asymmetric effect of fenfluramine on the accumulation of [^{18}F]FDG in human brain regions was attributed to serotonin nerve terminal discharge as well as alleged effects of such activation on GABAergic neurons (Mann *et al.*, 1996). Whether such processes can be studied quantitatively in the anesthetized porcine brain is an open question.

NS2381 is a new SSRI with an antidepressant profile (NeuroSearch, 1997). We reasoned, on the basis of information available to us, that the compound, in principle, could be used for neuroimaging of serotonin uptake sites in living brain, provided that it fulfills certain criteria (Sedvall *et al.*, 1986; Fletcher *et al.*, 1995; Pike, 1995, 1997). One criterion is that the precursor of a PET radioligand is available and suitable for rapid labeling with a positron-emitting radionuclide, yielding a radiochemically pure product. This is the case for racemic NS2381 and its enantiomers (Smith *et al.*, 1999). Another criterion is that sufficient amounts of the unmetabolized radioligand remain in the bloodstream long enough for accurate measurements of the compound to be made throughout the scan. This criteria was fulfilled for [^{11}C]NS2381 and its enantiomers in that roughly half of the radioactivity in the bloodstream stemmed from unmetabolized compound at the end of the scanning period (Smith *et al.*, 1999). A third criterion is that the cerebral distribution of the radioligand corresponds to the anatomical location of the neuroreceptor in question, in this case, the serotonin uptake site. [^{11}C]NS2381 and its enantiomers fulfilled this criteria in that they accumulated primarily in the raphé, thalamus, and basal ganglia, which are major locations of serotonin uptake sites in the brain of humans, nonhuman primates, and pigs (Cortés *et al.*, 1988; Laruelle *et al.*, 1988; Szabo *et al.*, 1995, 1996a; Lavoie and Parent, 1991; Jagust *et al.*, 1993; Smith, 1999). A fourth criterion is that prior administration of a selective neuroreceptor blocking agent reduces the accumulation of the PET radioligand at receptor sites. [^{11}C]NS2381 and its enantiomers fulfilled this criteria since citalopram, a potent SSRI and antidepressant drug (Hyttel, 1982), markedly reduced their accumulation, as measured by V_e' values in living porcine brain. Thus, [^{11}C]NS2381 and its ^{11}C-labeled enantiomers have some properties that make them suited for use as PET radioligands for investigating serotonin uptake in living brain. However, a shortcoming of [^{11}C]NS2381 and its enantiomers as PET radioligands is their nonspecific binding, which has always been a problem for the use of SSRIs for *in vivo* neuroimaging (Fletcher *et al.*, 1995; Pike, 1995, 1997, Shiue *et al.*, 1995; Smith *et al.*, 1997; Zea-Ponce *et al.*, 1997). Several strategies have been used for reconciling problems arising from the apparently high nonspecific binding of SSRIs (Szabo *et al.*, 1996b; Smith *et al.*, 1999), but further studies are required for determining whether they are valid and reliable. Thus, the search goes on for finding an ideal PET radiotracer for quantifying serotonin reuptake in the living human brain.

Acknowledgments

We thank the laboratory technicians of the PET Center (Ilse, Vikie, Helle and Gloria), of the Psychiatric Hospital in Risskov (Jetta, Inga, Helle, Kirsten, Nanna, and Solvej), and NeuroSearch (Ulla and Jane) for their skillful assistance. Dr. John Hyttel of Lundbeck A/S generously supplied citalopram. Financial support to D. F. Smith was provided by The Institute for Experimental Clinical Research of Aarhus University, Hørslev's Fund, The Danish Medical Research Council, and HjerneSagen.

References

Åkeson, J., Björkman, S., Messeter, K., and Rosen, I. (1993). Low-dose midazolam antagonizes cerebral metabolic stimulation by ketamine in the pig. *Acta Anesthesiol. Scand.* **37**: 525–531.

Åkeson, J., Nilson, F., Ryding, E., and Messeter, K. (1992). A porcine model for sequential assessments of cerebral haemodynamics and metabolism. *Acta Anaesthesiol. Scand.* **36**: 419–426.

Bolden-Watson, C., and Richelson, E. (1993). Blockade by newly-developed antidepressants of biogenic amine uptake into rat brain synaptosomes. *Life Sci.* **52**: 1023–1029.

Bonne, O., and Krausz, Y. (1997). Pathophysiological significance of cerebral perfusion abnormalities in major depression—Trait or state marker? *Eur. Neuropsychopharmacol* **7**, 225–233.

Brust, P., Bergmann, R., and Johannsen, B. (1996). High-affinity binding of [^{3}H]paroxetine to caudate nucleus and microvessels from porcine brain. *NeuroReport* **7**: 1405–1408.

Clark, J. C., Crouzel, C., Meyer, G.-J., and Strijkmans, K. (1987). Current methodology for oxygen-15 production in clinical use. *Appl. Radioact. Isot.* **38**: 597–600.

Cortés, R., Soriano, E., Pazos, A., Probst, A., and Palacios, J. M. (1988). Autoradiography of antidepressant binding sites in the human brain: Localization using [^{3}H]imipramine and [^{3}H]paroxetine. *Neuroscience* **27**: 473–496.

Danielsen, E., Smith, D. F., Poulsen, P. H., Østergaard, L., Gee, A., Ishizu, K., Venkatachalam, T. K., Bender, D., Hansen, S., Gjedde, A., Scheel-Krüger, J., and Møller, A. (1998). Positron emission tomography of living brain in minipigs and domestic pigs. *Scand. J. Lab. Anim. Sci.* (Suppl. 1) **25**: 127–135.

Davis, R., and Faulds, D. (1996). Dexfenfluramine: An updated reviews of its therapeutic use in the management of obesity. *Drugs* **52**: 696–724.

Deacon, T., Schumacher, J., Dinsmore, J., Thomas, C., Palmer, P., Kott, P., Edge, A., Penney, D., Kassissieh, S., Dempsey, P., and Isacson, O. (1997). Histological evidence of fetal pig neural cell survival after transplantation into a patient with Parkinson's disease. *Nat. Med.* **3**: 350–353.

Erreboe, I., Plenge, P., and Mellerup, E. T. (1995). Differences in brain 5-HT transporter dissociation rates among animal species. *Pharmacol. Toxicol.* **76**: 376–379.

Fletcher, A., Pike, W. W., and Cliffe, I. A. (1995). Visualization and characterization of 5-HT receptors and transporters in vivo and in man. *Semin. Neurosci.* **7**: 421–431.

Frackowiak, R. S. J., Lenzi, G. L., Jones, T., and Heather, J. D. (1980). Quantitative measurement of regional cerebral blood flow and oxygen metabolism in man using $^{15}O_2$ and positron emission tomography. Theory, procedure and normal values. *J. Comput. Assist. Tomogr.* **4**: 727–736.

Gjedde, A., and Wong, D. F. (1991). Modeling neuroreceptor binding of radioligands in vivo. In "*Quantitative Imaging: Neuroreceptors, Neurotransmitters, and Enzymes*" (J. J. Frost, and H. N. Wagner, Jr., Eds.), pp. 51–79. Raven Press, New York.

Hattazawa, J., Ito, M., Matsuzawa, T., Ido, T., and Watanuki, S. (1988). Measurement of the ratio of cerebral oxygen consumption to glucose utilization by positron emission tomography: Its consistency with the values determined by the Kety–Schmidt method in normal volunteers. *J. Cereb. Blood Flow Metab.* **8**: 426–432.

Hume, S. P., Lammertsma, A. A., Bench, C. J., Pike, V. W., Pascali, C., Cremer, J. E., and Dolan, R. J. (1992). Evaluation of S-[^{11}C]citalopram as a radioligand for in vivo labelling of 5-hydroxytryptamine uptake sites. *Int. J. Radiat. Appl. Instrum. B.* **8**: 851–855.

Hyttel, J. (1982). Citalopram—Pharmacological profile of a specific serotonin uptake inhibitor with antidepressant activity. *Prog. Neuro-Psychopharmacol. Biol. Psychiatry* **6**: 277–295.

Hyttel, J., Overo, K. F., and Arnt, J. (1984). Biochemical effects and drug levels in rats after long-term treatment with the specific 5-HT-uptake inhibitor, citalopram. *Psychopharmacology (Berlin)* **83**: 20–27.

Ingle, D. J. (1972). Fallacies and errors in the wonderlands of biology, medicine, and Lewis Carroll. *Perspect. Biol. Med.* **15**: 254–281.

Jagust, W. J., Eberling, J. L., Roberts, J. A., Brennan, K. M., Hanrahan, S. M., VanBrocklin, H., Enas, J. D., Biegon, A., and Mathis, A. (1993). In vivo imaging of the 5-hydroxytryptamine reuptake site in primate brain using single photon emission computer tomography and [^{123}I]5-iodo-6-nitroquipazine. *Eur. J. Pharmacol.* **242**: 189–193.

Kapur, S., Meyer, J., Wilson, A. A., Houle, S., and Brown, G. M. (1994). Modulation of cortical neuronal activity by a serotonergic agent: A PET study in humans. *Brain Res.* **646**: 292–294.

Laruelle, M., Vanisberg, M. A., and Maloteaux, J. M. (1988). Regional and subcellular location in human brain of [^{3}H]paroxetine binding, a marker of serotonin uptake sites. *Biol. Psychiatry* **24**: 299–309.

Lavoie, B., and Parent, A. (1991). Serotoninergic innervation of the thalamus in the primate: An immunohistochemical study. *J. Comp. Neurology* **312**: 1–18.

Logan, J., Fowler, J. S., Volkow, N. D., Wolf, A. P., Dewey, S. L., Schlyer, D. J., MacGregor, R. R., Hitzemann, R., Bendriem, B., Gatley, S. J., and Christman, D. R. (1990). Graphical analysis of reversible radioligand binding from time–activity measurements applied to [N-^{11}C-methyl]-(-)-cocaine PET studies in human subjects. *J. Cereb. Blood Flow Metab.* **10**: 740–747.

Low, W. C., Whitehorm, D., and Hendley, E. D. (1984). Genetically related rats with differences in hippocampal uptake of norepinephrine and maze performance. *Brain Res. Bull.* **12**: 703–709.

Madsen, F. F., Jensen, F. T., Waeth, M., and Djurhuus, J. C. (1990). Regional cerebral blood flow in pigs estimated by microspheres. *Acta Neurochir. (Wien)* **103**: 139–147.

Mann, J. J., Malone, K. M., Diehl, D. J., Perel, J., Nichols, T. E., and Mintun, M. A. (1996). Positron emission tomographic imaging of serotonin activation effects on prefrontal cortex in healthy volunteers. *J. Cereb. Blood Flow Metab.* **16**: 418–426.

Meyer, J. H., Kapur, S., Wilson, A. A., DaSilva, J. N., Houle, S., and Brown, G. M. (1996). Neuromodulation of frontal and temporal cortex by intravenous *d*-fenfluramine: An [15O]H2O PET study in humans. *Neurosci. Lett.* **207**: 25–28.

Mintun, M. A., Raichle, M. E., Martin, W. R. W., and Herscovitch, P. (1984). Brain oxygen utilization measured with oxygen-15 radiotracers and positron emission tomography. *J. Nucl. Med.* **25**: 177–187.

Murase, K., Kuwabara, H., Meyer, E., and Gjedde, A. (1995). Mapping of change in cerebral glucose utilization using [^{18}F]FDG double injection and new graphical analysis. *NeuroImage* **2**: S81.

NeuroSearch, A. S. (Moldt, P., Scheelkruger, J., Olsen, G. M., and Nielsen, E. O.) (1997). "8-Azabicyclo[3.2.1]oct-2-ene Derivatives, Their Preparation and Use," WO-09713770.

Norusis, M. J. (1990). "SPSS/PC+ 4.0 Base Manual." SPSS Inc, Chicago IL.

O'Brien, C. (1996). Yellow light for pig-human transplants. *Science* **271**: 1357.

Ohta, S., Meyer, E., Thompson, C. J., and Gjedde, A. (1992). Oxygen consumption of the living human brain measured after a single inhalation of positron emitting oxygen. *J. Cereb. Blood Flow Metab.* **12**: 179–192.

Pike, V. W. (1995). Radioligands for PET studies of central 5-HT receptors and re-uptake sites—Current status. *Nucl. Med. Biol.* **8**: 1011–1018.

Pike, V. W. (1997). Radioligands for the study of serotonin transporters and receptors in living human brain. *Serotonin* **2**(4).

Poulsen, P. H., Smith, D. F., Østergaard, L., Danielsen, E. H., Gee, A., Hansen, S. B., Astrup, J., and Gjedde, A. (1997). In vivo estimation of cerebral blood flow, oxygen consumption and glucose metabolism in the pig by [^{15}O]water injection, [^{15}O]oxygen inhalation and dual injections of [^{18}F]fluorodeoxyglucose. *J. Neurosci. Methods* **77**: 199–209.

Sedvall, G., Farde, L., Persson, A., and Wiesel, F.-A. (1986). Imaging of neurotransmitter receptors in the living human brain. *Arch. Gen. Psychiatry* **43**: 995–1005.

Shiue, C. Y., Shiue, G. G., Cornish, K. G., and O'Rourke, M. F. (1995). PET study of the distribution of [^{11}C]fluoxetine in a monkey brain. *Nucl. Med. Biol.* **22**: 613–616.

Smith, D. F. (1999). Neuroimaging of serotonin uptake sites and antidepressant binding sites in the thalamus of humans and 'higher' animals. *Eur. Neuropsychopharmacol.* **9**: 537–544.

Smith, D. F., Gee, A. D., Hansen, S. B., Moldt, P., Østergaard Nielsen, F., Scheel-Krüger, J., and Gjedde, A. (1999). Cerebral distribution of a new SSRI, NS2381, studied by PET in living porcine brain. *Eur. Neuropsychopharmacol.* **9**: 351–359.

Smith, D. F., Glaser, R., Gee, A., and Gjedde, A. (1996). ^{11}C-Nefopam as a potential PET tracer of serotonin reuptake sites: Initial findings in living pig brain. In "*Quantification of Brain Function: PET*" (T. Jones, V. Cunningham, R. Myers, and D. Bailey, Eds.), pp. 38–41. Academic Press, San Diego.

Smith, D. F., Jensen, P. N., Gee, A. D., Hansen, S. B., Danielsen, E., Andersen, F., Saiz, P.-A., and Gjedde, A. (1997). PET neuroimaging with [^{11}C]venlafaxine: Serotonin uptake inhibition, biodistribution and binding in living pig brain. *Eur. Neuropsychopharmacol.* **7**: 195–200.

Smith, D. F., Jensen, P. N., Gelbcke, M., and Tytgat, D. (1992). Effects of 3,*N,N*'-trimethyl-2-phenyl-1,4-piperazine diastereomers on monoamine uptake and monoamine oxidase in rat brain. *J. Neural Transm.* **88**: 177–185.

Staley, J. K., Malison, R. T., and Innis, R. B. (1998). Imaging of the serotonergic system: Interactions of neuroanatomical and functional abnormalities of depression. *Biol. Psychiatry* **44**: 534–549.

Szabo, Z., Kao, P. F., Mathews, W. B., Ravert, H. T., Musachio, J. L., Scheffel, U., and Dannals, R. F. (1996a). Positron emission tomography imaging of serotonin transporters in the human brain using [^{11}C](+)McN5652. *Synapse* **20**: 37–43.

Szabo, Z., Kao, P. F., Mathews, W. B., Ravert, H. T., Musachio, J. L., Scheffel, U., and Dannals, R. F. (1996b). Positron emission tomography of 5-HT reuptake sites in the human brain with C-11 McN5652: Extraction of characteristic images by artificial neural network analysis. *Behav. Brain Res.* **73**: 221–224.

Szabo, Z., Scheffel, U., Suehiro, M., Dannals, R. F., Kim, S. E., Ravert, H. T., Ricaurte, G. A., and Wagner, H. N. Jr., (1995). Position emission tomography of 5-HT transporter sites in the baboon brain with [^{11}C]McN5652. *J. Cereb. Blood Flow Metab.* **15**: 798–805.

Tao, R., and Auerbach, S. B. (1994). Anesthetics block morphine-induced increases in serotonin release in rat CNS. *Synapse* **18**: 307–314.

Tooronigian, S. A., Mulholland, G. K., Jewett, D. M., Bachelor, M. A., and Kilbourn, M. R. (1990). Routine production of 2-deoxy-2-[^{18}F]fluoro-D-glucose by direct nucleophilic exchange on a quarternary 4-aminopyridinium resin. *Nucl. Med. Biol.* **17**: 273–279.

Tumbleson, M. E. (Ed.) (1986). "Swine in Biomedical Research," Vol. 1. Plenum, New York.

Wienhard, K., Dahlbom, M., Eriksson, L., Michel, C., Bruckbauer, T.,

Pietrzyk, U., and Heiss, W. D. (1994). The ECAT EXACT HR: performance of a new high resolution positron scanner. *Comput. Assist. Tomogr.* **18**: 110–118.

Yoshikawa, T. (1968). "Atlas of the Brains of Domestic Animals. The Brain of the Pig." Pennsylvania State Univ. Press, University Park, PA.

Zea-Ponce, Baldwin, R. M., Stratton, M. D., Al-Tikriti, M., Soufer, R., Schaus, J. M., and Innis, R. B. (1997). Radiosynthesis and PET imaging of [N-methyl-^{11}C]LY257327 as a tracer for 5-HT transporters. *Nucl. Med. Biol.* **24**: 251–254.

Comparison of Kinetic Modeling Methods for the *In Vivo* Quantification of 5-HT$_{1A}$ Receptors Using WAY 100635

RAMIN V. PARSEY, MARK SLIFSTEIN, DAH-REN HWANG, ANISSA ABI-DARGHAM,
NORMAN SIMPSON, NINGNING GUO, ANN SHINN, OSAMA MAWLAWI,
RONALD VAN HEERTUM, J. JOHN MANN, and MARC LARUELLE

Departments of Psychiatry and Radiology, Columbia University College of Physicians and Surgeons, and Division of Brain Imaging, Department of Neuroscience, New York State Psychiatric Institute

Serotonin (5-HT) 1A receptors have been implicated in numerous psychiatric disorders. We compared several compartmental kinetic modeling methods for quantification of the 5HT$_{1A}$ receptors with [^{11}C]N-(2-(4-(2-methoxyphenyl)-1-piperazinyl)ethyl)-N-(2-pyridinyl)cyclohexane carboxamide ([^{11}C]WAY 100635) and positron emission tomography in five healthy volunteers. The cerebellum is a region devoid of 5HT$_{1A}$ receptors and was used as a measure of nonspecific binding. Several kinetic methods were compared: a two-compartment model (method 1) and an unconstrained three-compartment model, with direct (method 2) and indirect (method 3) derivation of the binding potential (BP), and a three-compartment model with the nondisplaceable distribution volume constrained to the cerebellum distribution volume (method 4). In addition, graphical analysis was also performed (method 5). Overall, methods 3, 4, and 5 provided comparable estimates of BP. Method 4 required longer scan duration to yield stable estimates of BP, and method 5 was subject to a noise-dependent bias that resulted in underestimation of BP in regions with larger noise, such as the dorsal raphe. In conclusion, kinetic analysis based on a three compartment model with indirect derivation of BP (method 3) appears to be a method of choice for derivation of [^{11}C]WAY 100635 BP in the human brain.

I. INTRODUCTION

Serotonin (5-HT) is a key neurotransmitter in the regulation of many behaviors including sleep, appetite, locomotion, and sexual activity. Yet, direct *in vivo* studies of these receptors in neuropsychiatric conditions are lacking. Imaging 5-HT$_{1A}$ receptors with positron emission tomography (PET) provides a method of characterizing 5-HT$_{1A}$ receptors in patients.

WAY-100635 [*N*-(2-(4-(2-methoxyphenyl)1-piperazinyl) ethyl)-*N*-(2-pyridinyl) cyclohexane carboxamide] is a potent and selective 5-HT$_{1A}$ antagonist with high affinity for 5-HT$_{1A}$ receptors ($K_d = 0.1$–0.4 nM) (Forster *et al.*, 1995; Gozlan *et al.*, 1995). An appropriate quantitative procedure is required to derive 5-HT$_{1A}$ receptor binding potential (BP, i.e., the product of receptor density, B_{max}, and affinity, $1/K_d$) with [^{11}C]WAY 100635 (heoreforth, we use [^{11}C]WAY 100635 to designate [carbonyl-^{11}C]WAY-100635). The goal of this study was to compare several analytical strategies to derive [^{11}C]WAY 100635 BP in humans.

II. MATERIALS AND METHODS

A. Human Subjects

Five healthy male volunteers (age 29 ± 4 years, with these and subsequent values given as mean $\pm$ SD) participated. The study was approved by Columbia Presbyterian

Medical Center Institutional Review Board, and subjects provided written informed consent after receiving an explanation of the study. The absence of medical, neurological, and psychiatric history (including alcohol and drug abuse) was assessed by history, review of systems, physical examination, routine blood tests, urine toxicology, and EKG.

B. Radiochemistry

WAY 100635 was prepared as previously described (Hwang *et al.*, 1999). Specific radioactivity at the time of injection was 1019 ± 336 Ci/mmol. Injected dose was 9.8 ± 3.4 mCi. Injected mass was 4.2 ± 1.3 μg.

C. PET Protocol

Subject preparation included placement of arterial and venous catheters, fiducial markers, and a polyurethane head immobilizer (Soule Medical, FL). PET imaging was performed with the ECAT EXACT 47 (Siemens/CTI, Knoxville, TN) (Wienhard *et al.*, 1992) (47 slices covering an axial field of view of 16.2 cm, axial sampling of 3.375 mm, 3D mode in plane, and axial resolution of 6.0 and 4.6 mm full width half-maximum at the center of the field of view, respectively). A 10-min transmission scan was obtained prior to radiotracer injection. Emission data were collected in the 3D mode for 120 min as 21 successive frames of increasing duration. Images were reconstructed to a 128×128 matrix (pixel size of 2.5×2.5 mm^2). Reconstruction was performed with attenuation correction using the transmission data and a Sheppe 0.5 filter (cutoff 0.5 cycles/projection rays).

D. Input Function Measurement

Plasma radioactivity was determined in 31 samples with a Wallac 1480 Wizard 3M Automatic Gamma Counter. Six samples were processed by protein precipitation using acetonitrile followed by high-pressure liquid chromatography (HPLC) to measure the fraction of plasma activity representing unmetabolized parent compound. The sum of one exponential plus a constant, with the value at time zero constrained to one, was fitted to the measured fractions of the parent. The product of total counts and interpolated fraction parent calculated the input function at each time. The sums of three exponentials were fitted to the input function and the fitted values were used as input to the kinetic and graphical analyses.

E. MRI Acquisition and Segmentation Procedures

MRIs were acquired on a GE 1.5 T Signa Advantage. Following a sagittal scout (localizer), performed to identify the AC–PC plane (1 min), a transaxial T1-weighted sequence with 1.5-mm slice thickness was acquired in a coronal plane orthogonal to the AC–PC plane over the whole brain with the following parameters: three-dimensional SPGR (spoiled gradient recalled acquisition in the steady state); TR 34 ms; TE 5 ms; flip angle of 45°; slice thickness 1.5 mm and zero gap; 124 slices; FOV 22 × 16 cm; with 256 × 192 matrix, reformatted to 256 × 256, yielding a voxel size of $1.5 \times 0.9 \times 0.9$ mm; and time of acquisition 11 min. MRI segmentation between gray matter (GM), white matter (WM), and cerebrospinal fluid (CSF) pixels was performed as previously described (Abi-Dargham *et al.*, 1999).

F. Image Analysis

Image analysis was performed with MEDx (Sensor Systems, Inc., Sterling, Virginia) according to the following steps. To correct for head movement, frames were coregistered to the first frame of the study, using automated image registration (AIR) software (Woods *et al.*, 1992). The 21 frames were then summed and coregistered and resampled to the MRI. The parameters of the spatial transformation matrix of the summed PET data set were then applied to each individual frame. ROI boundaries were drawn on the MRI according to criteria based on brain altlases (Duvernoy, 1991; Talairach and Tournoux, 1988) and on published reports (Kates *et al.*, 1997; Killiany *et al.*, 1997; Pani *et al.*, 1990). Regions included dorsolateral prefrontal cortex (DLPFC), medial prefrontal cortex (MPFC), orbitofrontal cortex (OFC), subgenual prefrontal cortex (SGPFC), anterior cingulate cortex (ACC), parietal cortex (PC), lateral temporal cortex (TC), occipital cortex (OC), uncus (UNC), amygdala (AMY), hippocampus (HIP), parahippocampal gyrus (PHG), entorhinal cortex (ERC), dorsal raphe nuclei (DRN), and cerebellum (CER). Except for the DRN, regions were drawn on the MRI to delineate the boundaries of the ROIs. Within these regions, only the voxels classified as GM were used to measure activity distribution. The DRN ROI was traced directly on the PET scan (coregistered summed image), given the absence of MRI criteria to identify DRN boundaries.

G. Quantitative Analysis

Derivation of [^{11}C]WAY 100635 regional distribution volumes (V_T) was performed with kinetic analysis and graphical analysis using the arterial input function. The total regional distribution volume (V_T) was defined as the equilibrium ratio between the total regional activity and arterial parent activity. In a region with receptors, V_T is equal to the sum of the specific (BP) and nondisplaceable (V_2) distribution volumes. Each analytical method shared two assumptions. First, given the negligible concentration of 5-HT$_{1A}$ receptors in the cerebellum, cerebellar V_T was assumed to represent only free and nonspecific binding (V_2) and to provide a reasonable estimate of V_2 in the ROIs. Second, the contribution of plasma total activity to the regional activity was calculated assuming a 5% blood volume in the regions of interest and subtracted from the regional activity prior to analysis (Mintun *et al.*, 1984).

1. Outcome Measures

Two outcome measures were compared. The first outcome measure, termed BP, was the equilibrium distribution volume of the specific binding compartment relative to the arterial parent concentration. The second outcome measure, termed V_3'' or k_3/k_4, corresponded to the BP normalized to the cerebellum distribution volume (V_2), i.e., the BP/V_2 ratio.

2. Kinetic Analysis with Arterial Input Function

Kinetic analysis was performed as described previously (Abi-Dargham *et al.*, 1999). Four approaches were used to derive BP and V_3''. The first model was a two-compartment model (2CM, method 1). Regional V_T was calculated as the K_1/k_2 ratio, and BP and V_3'' were derived as

$$BP = V_{T\ ROI} - V_{T\ CER},$$

and

$$V_3'' = \frac{BP}{V_{T\ CER}},$$

where $V_{T\ ROI}$ and $V_{T\ CER}$ are the total distribution volumes in the region of interest and the cerebellum, respectively.

The three other kinetic approaches (methods 2, 3, and 4) were based on a three compartment model (3CM). In method 2 (3CM D, direct), an unconstrained three-compartment model was fitted to regions of interest and BP (BP = $K_1 k_3/k_2 k_4$) and $V_3''(V_3'' = k_3/k_4)$ were calculated directly from the rate constants. In method 3 (3CM ID, indirect), an unconstrained three-compartment model was fitted to region of interest data and regional V_T were calculated as

$$V_T = \frac{K_1}{k_2}\left(1 + \frac{k_3}{k_4}\right),$$

and BP and V_3'' were derived as for method 1. In method 4, a three-compartment model with the K_1/k_2 ratio constrained to the value of cerebellum V_T (3CM CST) was fitted to the data, and BP and k_3/k_4 were derived as in method 2 (3CM D).

3. Graphical Analysis with Arterial Input Function

Regional time–activity curves were also graphically analyzed (Logan *et al.*, 1990). This method allows the determination of regional V_T of reversible ligands as the slope of the regression line without assuming a particular compartmental configuration. Assuming, as in the kinetic analysis, the equivalence between the cerebellum distribution volume (V_{TCER}) and the nondisplaceable distribution volume in the ROI, outcome measures BP and V_3'' were calculated as for 3CM ID.

4. Evaluation Criteria

Various models were compared in terms of goodness of fit, parameter identifiability, and stability over scan duration. Goodness of fit of models with different levels of complexity was compared using the Akaike Information Criterion (AIC, Akaike, 1974) and the F test (Carson, 1986; Landlaw and DiStefano, 1984). The identifiability was assessed by the standard error of the parameter at convergence. The standard error of the parameters was given by the diagonal of the covariance matrix (Carson, 1986) and expressed as percentage of the parameters (coefficient of variation, %CV). Experimental data were collected for 120 min. The relationship between BP derivation and the duration of the scan was evaluated by fitting shorter duration data sets and comparing the results to the reference value obtained with the 120-min data set. For each region and each duration, the average $\pm$ SD of the results were expressed as percentage of the reference value to provide an estimate of the bias (average) or dispersion (SD) induced in the outcome measure by analyzing shorter data sets. The solution was considered stable after time t if all results derived from time t to the end of the experiment had a mean within 10% of the reference value and a SD that did not exceed 15%.

III. RESULTS AND DISCUSSION

A. Metabolism

WAY 100635 underwent rapid metabolism. At 12 min, only $11 \pm 4\%$ ($n = 5$) of the total activity corresponded to the parent [^{11}C]WAY 100635 compound (Fig. 1).

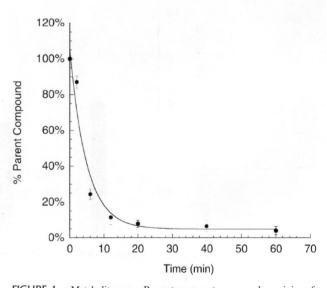

FIGURE 1. Metabolite curve. Percentage parent compound remaining after injection. Each point is the mean $\pm$ SD of five determinations. solid line is the fit of one exponential plus a constant. The first point is defined as 100% and included in the fit.

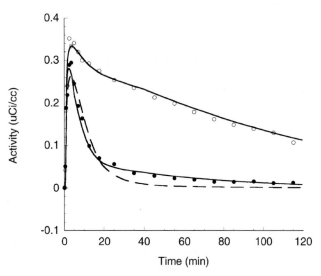

FIGURE 2. Time–activity curves. Decay-corrected activity in the cerebellum (solid circles) and cingulate (open circles). The solid lines are the result of a 3CM unconstrained fit. The dashed line is the result of a 2CM fit.

B. Brain Uptake

Activity concentrated in neocortical and limbic areas, with highest uptake in temporal cortex and medial temporal structures and lowest activity in the striatum, thalamus, and cerebellum (data not shown). An activity concentration was also visualized in the midbrain and pons, at the level of the floor of the fourth ventricle. This activity corresponded to the location of the dorsal raphe nuclei. In all regions, the uptake peaked early (within 10–15 min), and this peak was followed by appreciable washout.

C. Cerebellum

Cerebellum V_T was $44 \pm 25\%$ smaller as calculated by 2CM (0.58 ± 0.12 mL g^{-1}) than 3CM (0.85 ± 0.30 mL g^{-1}, paired t test, $p = 0.018$). In all subjects, the cerebellum was better fitted (Fig. 2) by the 3CM than the 2CM. This difference was statistically significant as measured by goodness of fit (residual sums of squares, AIC, and F test, all

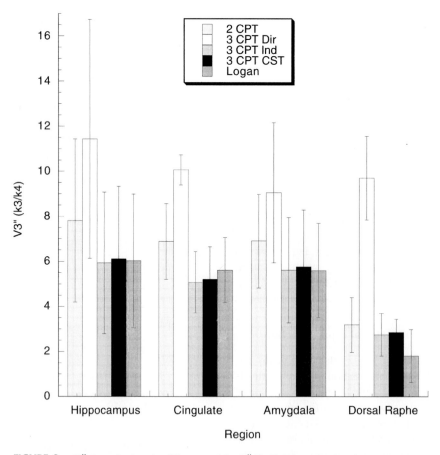

FIGURE 3. V_3'' determinations by different models. V_3'' (k_3/k_4) determinations in some representative regions as determined by five different methods. Data are mean and standard deviations of five independent experiments.

with $p < 0.05$). The identifiability of cerebellar V_T derived with a 3CM analysis (error of 8.43 ± 0.98%) was lower than 2CM analysis (5.40 ± 1.43%, paired t test, $p = <0.002$). All analysis performed on data sets longer than 30 min for 2CM returned V_T values within 10% of the reference value (V_T derived with 120 min of data) and with SD lower than 10%. In contrast, 3CM met these criteria after 70 min of data. Together, these data suggest that the 3CM is the model of choice to derive cerebellum V_T. Logan plots achieved linearity at times greater than 35 min, and a linear regression was performed on data from the 35- to 120-min interval. Cerebellum V_T by graphical analysis was 0.80 ± 0.22 mL g^{-1}, a value similar to cerebellum V_T derived by 3CM and larger than cerebellum V_T derived by 2CM. Since the graphical analysis does not assume any particular compartmental configuration, the similarity of graphical and the kinetic 3CM cerebellar V_T provided additional justification to use the 3CM rather than the 2CM model for kinetic cerebellar analyses.

D. Regions of Interest

1. V_3''

The different modeling methods displayed significantly different V_3'' determinations (several regions are shown in Fig. 3). V_3'' determinations from the 2CM and 3CM D overestimated V_3'' (repeated measures ANOVA, $p < 0.0001$ for all comparisons except for 3CM CST vs 3CM ID and 3CM CST vs Logan). This overestimation of V_3'' by 2CM or 3CM D was due to the fact that V_2 was underestimated by both methods.

2. BP

As compared to V_3'', BP determinations were in closer agreement between the methods (Fig. 4). All methods were highly correlated (in all comparisons the correlation coefficient was 0.980 or greater). However, significant between method differences were noted (repeated measures ANOVA,

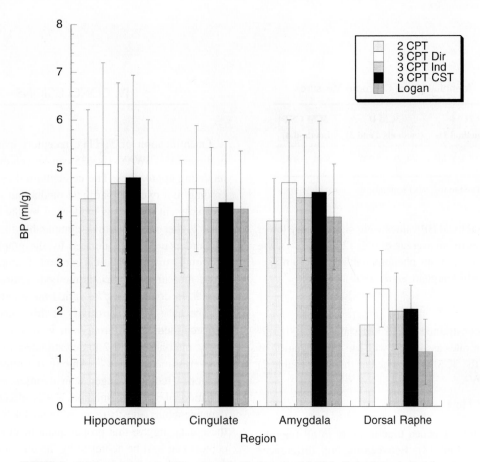

FIGURE 4. BP determinations by different models. Binding potential ($K_1 k_3 / k_2 k_4$) determinations in some representative regions as determined by five different methods. Data are mean and standard deviations of five independent experiments.

TABLE 1. Regional BP and V_3'' Determinations by Methods 3 (3CM D) and 5 (Graphical)

Region	BP (mL g-1)		V_3'' (unitless)	
	3CM IN	Graphical	3CM IN	Graphical
Entorhinal Ctx	6.06 ± 1.78	5.69 ± 1.45	7.31 ± 1.41	7.79 ± 1.98
Insula	5.77 ± 1.67	5.63 ± 1.55	7.02 ± 1.79	7.64 ± 1.80
Parahippocampal gyrus	5.17 ± 1.72	4.99 ± 1.45	6.13 ± 0.63	6.73 ± 1.43
Uncus	5.50 ± 1.16	5.24 ± 1.40	6.72 ± 1.15	7.24 ± 2.34
Temporal Ctx	5.09 ± 1.57	5.12 ± 1.60	6.08 ± 0.75	6.93 ± 1.66
Hippocampus	4.68 ± 2.11	4.25 ± 1.76	5.93 ± 3.14	6.02 ± 2.97
Cingulate Ctx	4.17 ± 1.25	4.14 ± 1.21	5.07 ± 1.36	5.61 ± 1.45
Amygdala	4.37 ± 1.31	3.97 ± 1.11	5.61 ± 2.33	5.60 ± 2.09
Subgenual prefrontal Ctx	3.94 ± 1.18	3.85 ± 1.08	4.88 ± 1.68	5.22 ± 1.32
Medial prefrontal Ctx	3.70 ± 1.04	3.76 ± 1.12	4.50 ± 1.08	5.08 ± 1.22
Dorsolateral prefrontal Ctx	3.16 ± 0.92	3.21 ± 0.99	3.84 ± 0.99	4.35 ± 1.16
Parietal Ctx	3.02 ± 0.79	3.03 ± 0.82	3.71 ± 0.99	4.20 ± 1.38
Orbitofrontal Ctx	2.86 ± 0.71	2.85 ± 0.77	3.47 ± 0.60	3.92 ± 1.12
Occipital Ctx	2.46 ± 0.53	2.49 ± 0.57	3.00 ± 0.48	3.42 ± 0.86
Dorsal raphe	2.00 ± 0.80	1.14 ± 0.68	2.75 ± 0.94	1.80 ± 1.17

Note. Values are mean $\pm$ SD, $n = 5$.

TABLE 2. Identifiability of Outcome Measures

Measure	2CM (method 1)	3CM D (methods 2 and 3)	3CM CST (method 4)
V_T	6.15 ± 1.76	$4.37 \pm 3.23^*$	5.91 ± 3.61

$^*p < 0.05$ compared to two other methods.

$p < 0.0001$). Graphical BP values were significantly lower than kinetic values by an average of $4 \pm 12\%$ (Table 1). We previously reported that this phenomenon is due to a noise-dependent bias (Abi-Dargham *et al.*, 1999).

3. Identifiability

The 3CM unconstrained model had statistically significantly lower error rates associated with the determination of V_T compared to the 3CM constrained or the 2CM (Table 2) ($p < 0.05$, ANOVA).

4. Stability over Time

BP determinations reached time stability in all regions, except the DRN (Table 3). Between-method differences were observed in scan duration needed to reach time stability (repeated measures ANOVA, $p = 0.010$). Post hoc analysis (Fisher PLSD) revealed that method 4 (3CM CST) required more time than the four other methods to reach time stability.

IV. CONCLUSIONS

Quantification of 5-HT$_{1A}$ receptors using the PET radioligand [^{11}C]WAY 100635 can be achieved with kinetic modeling approaches. The cerebellum does not behave according to a two-compartment model, but requires a three-compartment model, as evidenced by the better fit to the data, higher identifiability, comparable time stability, and concordance with V_T derived by the graphical method, a noncompartment analytical method. Using BP as the outcome measure, the various methods return overall similar results. In contrast, k_3/k_4 varied more between the methods, because of the sensitivity of this outcome measure to between-methods differences in V_2 determinations. Specifically, methods 1 and 2 underestimated V_2, which resulted in overestimation of V_3''. Among the remaining methods, method 4 requires longer scan duration to achieve determination of BP, and method 5 was affected by a noise-dependent bias (mostly detected in the DRN). In conclusion, these data suggest that proper quantification of 5-HT$_{1A}$ receptors can best be achieved by the three-compartment unconstrained model with indirect calculation of BP. Additional studies are required to compare the test/retest reproducibility of each methods, and to evaluate methods to derive V_3'' in the absence of plasma input function measurement.

TABLE 3. Minimal Scan Duration (Minutes) Needed for BP Determination

Region	2CM	3CM D	3CM ID	3CM CST	Logan
Entorhinal Ctx	80	90	110	110	110
Insula	80	60	70	90	100
Parahippocampal gyrus	90	100	110	110	110
Uncus	80	110	120	120	100
Temporal Ctx	80	80	80	80	90
Hippocampus	100	110	110	110	80
Cingulate Ctx	80	50	50	70	80
Amygdala	100	100	100	100	90
Subgenual prefrontal Ctx	80	90	90	90	90
Medial prefrontal Ctx	80	50	50	70	60
Dorsolateral prefrontal Ctx	80	50	50	80	60
Parietal Ctx	80	40	50	80	50
Orbitofrontal Ctx	70	70	80	90	80
Occipital Ctx	90	40	50	90	60
Dorsal raphe	110	120	120	120	110
Average ± SD	85 ± 10	77 ± 27	83 ± 27	94 ± 16	85 ± 5

Acknowledgments

We thank Rick Weiss, Justine Pidcock, Amy Anderson, Bryan Bergert, Julie Montoya, Analia Arevalo, Daniel Schneider, and Sue Chung for all of their hard work and dedication in helping acquire and analyze this data. Work supported by NARSAD.

References

Abi-Dargham, A., Martinez, D., Mawlawi, O., Simpson, N., Hwang, D. R., Slifstein, M., et al. (1999). Measurement of striatal and extrastriatal dopamine D1 receptor binding potential with [11C]NNC 112 in humans: Validation and reproducibility. J. Cereb. Blood Flow Metab., in press.

Akaike H. (1974). A new look at the statistical model identification. IEEE Trans. Automat. Contr. 19: 716–723.

Carson, R. E. (1986). Parameters estimation in positron emission tomography. In "Positron Emission Tomography. Principles and Applications for the Brain and the Heart." (M. E. Phelps, J. C. Mazziotta, and H. R. Schelbert, Eds.), pp. 347–390. Raven Press, New York.

Duvernoy, H. (1991). "The Human Brain. Surface, Three-Dimensional Sectional Anatomy and MRI." Sringer-Verlag Wien, New York.

Forster, E. A., Cliffe, I. A., Bill, D. J., Dover, G. M., Jones, D., Reilly, Y., et al. (1995). A pharmacological profile of the selective silent 5-HT1A receptor antagonist, WAY-100635. Eur. J. Pharmacol. 281: 81–88.

Gozlan, H., Thibault, S., Laporte, A. M., Lima, L., and Hamon, M. (1995). The selective 5-HT1A antagonist radioligand [3H]WAY 100635 labels both G-protein-coupled and free 5-HT1A receptors in rat brain membranes. Eur. J. Pharmacol. 288: 173–186.

Hwang, D. R., Simpson, N., Mann, J. J., and Laruelle, M. (1999). An improved one-pot procedure for the preparation of [11C-carbonyl]WAY 100635. Nucl. Med. Biol. 26: 815–819.

Kates, W. R., Abrams, M. T., Kaufmann, W. E., Breiter, S. N., and Reiss, A. L. (1997). Reliability and validity of MRI measurement of the amygdala and hippocampus in children with fragile X syndrome. Psychiatr. Res. Neuroimaging 75: 31–48.

Killiany, R. J., Moss, M. B., Nicholson, T., Jolez, F., and Sandor, T. (1997). An interactive procedure for extracting features of the brain from magnetic resonance images: The lobes. Hum. Brain Mapping 5: 355–363.

Landlaw, E. M., and DiStefano, J. J., III. (1984). Multiexponential, multicompartmental, and noncompartmental modeling. II. Data analysis and statistical considerations. Am. J. Physiol. 246: R665–R677.

Logan, J., Fowler, J., Volkow, N. D., Wolf, A. P., Dewey, S. L., Schlyer, D. J., et al. (1990). Graphical analysis of reversible radioligand binding from time–activity measurements applied to [N-11C-methyl]-(-)-cocaine PET studies in human subjects. J. Cereb. Blood Flow Metab. 10: 740–747.

Mintun, M. A., Raichle, M. E., Kilbourn, M. R., Wooten, G. F., and Welch, M. J. (1984). A quantitative model for the in vivo assessment of drug binding sites with positron emission tomography. Ann. Neurol. 15: 217–227.

Pani, L., Gessa, G. L., Carboni, S., Portas, C. M., and Rossetti, Z. L. (1990). Brain dialysis and dopamine: Does the extracellular concentration of dopamine reflect synaptic release? Eur. J. Pharmacol. 180: 85–90.

Talairach, J., and Tournoux, P. (1988). "Co-planar Stereotactic Atlas of the Human Brain. Three-Dimensional Proportional System: An Approach of Cerebral Imaging." Theime Medical, New York.

Wienhard, K., Eriksson, L., Grootoonk, S., Casey, M., Pietrzyk, U., and Heiss, W. D. (1992). Performance evaluation of the positron scanner ECAT EXACT. J. Comp. Assist. Tomgr. 16: 804–813.

Woods, R. P., Cherry, S. R., and Mazziotta, J. C. (1992). Rapid automated algorithm for aligning and reslicing PET images. J. Comp. Assist. Tomogr. 16(4): 620–633.

38

Partition Volume and Regional Binding Potentials of Serotonin Receptor and Transporter Ligands

DEAN F. WONG,*,† **GERALD NESTADT,**† **PAUL CUMMING,**‡ **FUJI YOKOI,*** and **ALBERT GJEDDE**‡

**Department of Radiology, Division of Nuclear Medicine, Johns Hopkins University, Baltimore, Maryland*
†*Department of Psychiatry, Johns Hopkins University, Baltimore, Maryland*
‡*Aarhus University PET Center, Aarhus, Denmark*

The calculation of the binding potential of several radioligands for serotonin receptors and uptake sites remains a challenge. The ubiquitous distribution of serotonin innervations makes difficult the isolation of the estimate of a radiotracer partition volume uncontaminated by specific binding components. In the present study we tested two methods of estimating the partition volume of the $5HT_{2A}$ receptor ligand [^{11}C]MDL100,907, from which we then calculated the binding potential at equilibrium. Both methods utilize a novel twice-integrated solution of the standard three-compartment model equations, the so-called "hyperplot." The first method involves the blocking of $5HT_{2A}$ binding in brain with unlabeled risperidone, a $5HT_{2A}$ antagonist. The second method involves the kinetic fitting of radioligand binding in brain regions having a wide range of receptor densities, without an additional blocking condition. In both methods, a linear regression then yields an estimate of the common partition volume, subsequently applied to all brain regions for calculation of regional binding potentials. The two methods gave results of similar magnitude. Using the hyperplot and the common partition volume, we then calculated the binding potentials of [^{11}C]MDL100,907 for serotonin $5HT_{2A}$ receptors in brain of a series of patients with schizophrenia (n = 9) and in brain of healthy volunteers (n = 9). Cortical binding potentials were in the range 2–4 and did not differ significantly between groups. We also tested the hyperplot for quantitative studies of the plasma membrane serotonin transporter labeled with [^{11}C]McN5652. The partition volume of this tracer was determined as above and used to calculate the binding potentials, which ranged from zero in cerebellum to unity in the diencephalon. Binding was low in the cerebral cortex. In the preliminary series of healthy volunteers and in patients with schizophrenia, there were no significant differences in the binding potential of [^{11}C]McN5652. The present technique is applicable to other radiotracers lacking a definitive reference region.

I. INTRODUCTION

The quantification of PET/SPECT studies of serotonin receptor and transporter radioligand binding in living brain requires the determination of the nonspecific binding at equilibrium, also known as the partition volume. In principle, the nonspecific binding can be calculated in an appropriate reference region, i.e., one in which the specific binding component is negligible. In corresponding studies of the dopamine system, the cerebellum is generally recognized as an appropriate reference region, but for the quantification of the serotonin $5HT_{2A}$ and $5HT_{1A}$ receptors and the plasma membrane serotonin transporters, the choice of the appropriate region has remained controversial.

Studies of the $5HT_{2A}$ radioligand [^{11}C]MDL100,907 at the Karolinska Institute (Ito *et al.*, 1998), as well as at the NIH (Watabe *et al.*, 1998), are believed to demonstrate significant specific binding in the cerebellum. Similarly, measurements of the brain uptake of $5HT_{1A}$ radioligand [^{11}C]WAY 100,602 failed to reveal an appropriate reference region. Calculation of regional binding potentials of these ligands requires a kinetic estimate of the partitioning of the radiotracer between brain and blood. It is often considered justified to assume that radiotracer partition has the same magnitude in all brain regions and is unaffected by disease. According to these assumptions, we now present a kinetic approach to the estimation of the partition volume for the subsequent calculation of the binding potential. This model is tested in the case of two radiotracers: [^{11}C]McN5652 for the plasma membrane serotonin transporters and [^{11}C]MDL100,907 for serotonin 5HT$_{2A}$ receptors (5HT$_{2A}$) in a group of normal volunteers and a group of patients with schizophrenia.

II. STUDY DESIGN AND METHODS

All subjects received at least one of the radioligands, [^{11}C]MDL100,907 and [^{11}C]McN5652, of high specific activity (>2000 Ci/mmol) as intravenous bolus injections. For those subjects receiving both radiotracers, the two studies were carried out on the same day, with a 2 to 3 h delay between radiotracer injections to allow for patient comfort and decay of residual radioactivity.

Dynamic emission data were recorded using the GE 4096+ PET scanner. A series of 50 frames, increasing in duration from 15 s to 6 min, were recorded over a total of 90 minutes. A series of 35 arterial blood samples were collected at intervals during the emission recording, and the radioactivity concentration measured with a gamma-counter cross-calibrated to the tomography. The plasma radioactivity plasma curve was corrected for the presence of metabolites by interpolation between HPLC fractionation of six selected plasma samples. For fractionations, raw plasma was first passed through a reverse phase precolumn for removal of proteins and then eluted to an analytical column equipped with an on line gamma detector. Recovery of total plasma radioactivity was in all cases close to 100%.

A volumetric MR image of the head of each subject was acquired using a "spoiled-grass" sequence (SPGR). The presence of significant brain pathology was first excluded by a neuroradiologist, and the images were subsequently used for registration of the emission images and the anatomical regions of interest.

A. Patient Population

Control subjects had no significant medical or psychiatric history, and were not using any illicit drugs or psychoactive medications. The schizophrenic subjects were diagnosed according to DSM IV criteria. All were either drug-naive, drug-free for several months, or on a stable regimen of a classical neuroleptic with minimal serotonin binding, i.e., fluphenazine or haloperidol (for numbers of subjects, see Table 1). There is little evidence that the binding site densities

TABLE 1.   Subject Population

Radioligand	Healthy controls	Patients with schizophrenia
MDL100,907	9	9
McN-5652	4	7

of serotonin 5HT$_{2A}$ antagonists or serotonin transporters in brain are altered by neuroleptics (Andree *et al.*, 1986).

B. Brain Uptake Analysis

The partition volume of [^{11}C]MDL100,907 was calculated in brain of four healthy volunteers, each of whom received a first tracer infusion in the baseline condition, and a second tracer infusion at 2 h after administration of a blocking dose (1–2 mg) of oral risperidone. We then tested a novel single-scan approach for the estimation of tracer partition volume (V_e) in the absence of a true reference region, and the subsequent calculation of regional binding potentials using a common estimate of V_e for brain. The new approach employs the "hyperplot," an extention of the original multitime graphical analysis published by Gjedde (1981, 1982). According to this analysis, a process of irreversible binding of a tracer to sites in brain tissue (i.e., $k_4 = 0$), approaches an asymptote as follows:

$$\frac{M(T)}{C_a(T)} = \left(\frac{K_1 k_3}{k_2 + k_3}\right) \frac{\int_0^T C_a \mathrm{d}t}{C_a(T)} + \frac{K_1 k_2}{(k_2 + k_3)}. \quad (1)$$

Blomqvist (1984) published a multilinear solution to the original differential equations underlying the 2-deoxyglucose method of Sokoloff *et al.* (1977),

$$M = K_1 \int_0^T C_a \, \mathrm{d}t + K_1 k_3 \int_0^T \int_0^u C_a \, \mathrm{d}t \mathrm{d}u$$
$$- (k_2' + k_3') \int_0^T M \, \mathrm{d}t, \quad (2)$$

which can be rearranged to yield the equation for an alternative multitime graphical analysis, the "hyperplot,"

$$\frac{\int_0^T M \, \mathrm{d}t}{\int_0^T C_a \, \mathrm{d}t} = \frac{(K_1 k_3)}{(k_2 + k_3)} \frac{\int_0^T \int_0^u C_a \, \mathrm{d}t \, \mathrm{d}u}{\int_0^T C_a \, \mathrm{d}t}$$
$$+ \left[\frac{K_1}{k_2' + k_3'}\right] \left[1 - \frac{M(T)}{K_1 \int_0^T C_a \, \mathrm{d}t}\right], \quad (3)$$

where $k_2' = k_2/(1+\rho)$, $k_3' = k_3/(1+\rho)$, and $\rho = k_5/k_6$, the latter equal to the partitioning between free radioligand and nonspecifically or rapidly and reversibly bound radioligand. Equation (3) can be rewritten in the form

$$V' = K\Theta' + V_f'\left(1 - \frac{E(T)}{E(0)}\right), \quad (4)$$

where

$$V' = \frac{\int_0^T M\,dt}{\int_0^T C_a\,dt}, \quad \Theta' = \frac{\int_0^T \int_0^u C_a\,dt\,du}{\int_0^T C_a\,dt}, \tag{5}$$

and

$$V'_f = \frac{K_1}{k_2 + k_3}, \quad E(T) = \frac{M(T)}{F \int_0^T C_a\,dt}, \quad \text{and } E(0) = \frac{K_1}{F}. \tag{6}$$

At steady state ($E(\infty)$ constant or zero), Eq. (4) describes a linear relationship with slope K and intercept $V'_f(1 - E(\infty)/(E(0)))$. For $k_3 = 0$, Eq. (4) describes an exponential approach to an asymptote of zero slope,

$$V' = V_e(1 + \rho)\left(1 - \exp\left[\frac{\Theta'(t_h)}{\tau}\right]\right), \tag{7}$$

where $V_e = K_1/k_2$ is the partition volume, and τ is an arbitrary time constant expressing the rate of approach to a total steady-state volume of distribution of magnitude,

$$V_h = V_e(1 + \rho), \tag{8}$$

which is the classical equation linking the total volume of distribution V_h to the specific binding potential ρ (Gjedde et al., 1986; Wong et al., 1986; Gjedde and Wong, 1990). Rearranging for the binding potential ρ,

$$\rho = \frac{V_h}{V_e} - 1. \tag{9}$$

C. Partition Volume Calculation

As the radioligands [^{11}C]MDL100,907 and [^{11}C]McN5652 were not known with certainty to have a reference region devoid of detectible specific binding, we used one or both of two methods to estimate the partition volume V_e for these ligands. For the inhibition method, we estimated the V_e of [^{11}C]MDL100,907 in brain from a linear plot of the pharmacologically blocked volume $V_h(I)$ versus the unblocked volume $V_h(0)$, measured in the same brain regions after treatment with risperidone. It can be shown that

$$V_h(I) = (1 - \sigma)V_h(0) + \sigma V_e, \tag{10}$$

where σ is the occupancy of the inhibition, and V_e was estimated from the ordinate intercept of the linear regression of measurements in a series of brain regions exhibiting a wide range of binding values.

For the regression method, a graph of V_h versus K' in a series of brain regions was used to calculate V_e, again from the ordinate intercept, plotted against

$$\Delta K^* = K^* - K^*_{\text{cerebellum}}, \tag{11}$$

where

$$K' = KK_1/(K_1 - K). \tag{12}$$

Here, K_1 is defined as the tangent to the hyperplot at $\tau = 0$, while K is the tangent at the onset of steady state $\tau = t_h$, which occured at about 90 min after tracer infusion in this study. In practice, the correct magnitude of the independent variable, $\Theta'(t_h)$, must be known to determine K. We first estimated the magnitudes of K_1 and K in each brain region, where K_1 is the unidirectional clearance, $K = K_1 k_3/(k_2 + k_3)$ is the net blood–brain clearance, and k_2 and k_3 are regional rate constants according to usual definitions. Numerical determination of the relevant slopes was accomplished by fitting a curve in the form,

$$R = A\Theta' + B\left(1 - \exp\left[\frac{\Theta'}{\tau}\right]\right), \tag{13}$$

where A, B and τ are the parameters of the fit.

Equation (13) is a variation of the hyperplot Eq. (4) where R and Θ' are defined in Eqs. (5) and (13). Thus,

$$K_1 = A + B/\tau, \tag{14}$$

and

$$K = A + \frac{B}{\tau}\exp\left(-\frac{\Theta'(t_h)}{\tau}\right), \tag{15}$$

at the time t_h. We next estimated each region's V_h by fitting the arterial hyperplot to the theoretically asymptotic form from Eq. (7),

$$R = V_h\left(1 - \exp\left[\frac{\Theta'}{\tau}\right]\right), \tag{16}$$

and then calculated the magnitude of K^* from regional estimates of K_1 and K, according to Eq. (12). Equations (13) and (16) are initially exclusive when $B \neq V_h$. For this reason, we plotted V_h versus $\Delta K^* = K^* - K^*(\text{cerebellum})$. The ordinate intercept of the linear regression is the estimate of V_e.

III. RESULTS

The analysis of [^{11}C]MDL100,907 distribution by the inhibition method yielded a $V_e\sigma \approx 18$, where σ is the (unknown) fractional receptor occupancy, and V_e is the partition volume for the normals. This result was in good agreement with the individual V_e values obtained by the two-scan regression method. Subsequently, the one-scan regression method alone was employed for the radioligand [^{11}C]McN5652, in which V_e was determined for each subject individually. Of the seven patients with schizophrenia studied with [^{11}C]McN5652, one was drug-free for several months, one was drug-naive and the other five were on stable doses of D_2 antagonists, four on fluphenazine, and one on thiothixene (1). Of the nine patients who received [^{11}C]MDL100,907, five were drug-free and the other four were on neuroleptics, three on fluphenazine, and one on thiothixene.

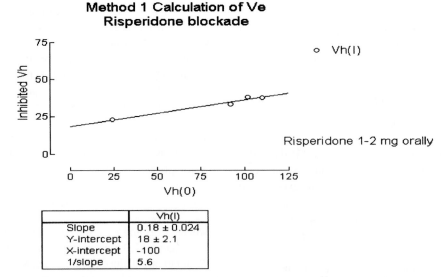

	Vh(I)
Slope	0.18 ± 0.024
Y-intercept	18 ± 2.1
X-intercept	–100
1/slope	5.6

FIGURE 1. The receptor blockade method for calculation of V_e of [^{11}C]MDL100907 in brain. Individual estimates of distribution volumes in the baseline unblocked condition ($V_h(0)$) and the after-risperidone treatment condition ($V_h(I)$) are plotted for several brain regions. Each point is an average obtained over four subjects. The ordinate intercept of the linear regression slope is V_e, the common partition volume.

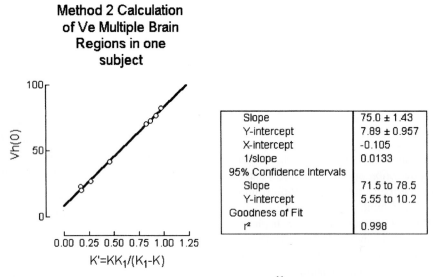

Slope	75.0 ± 1.43
Y-intercept	7.89 ± 0.957
X-intercept	–0.105
1/slope	0.0133
95% Confidence Intervals	
Slope	71.5 to 78.5
Y-intercept	5.55 to 10.2
Goodness of Fit	
r²	0.998

FIGURE 2. The regression method for calculation of V_e of [^{11}C]MDL100907 in brain. The data acquired in brain of a single subject are reported. The magnitude $V_h(0)$ in several brain regions is plotted as a function of the magnitude of K^* for several brain regions in the same subject. The ordinate intercept of the linear regression slope is V_e, the individual partition volume.

Figure 1 demonstrates the calculation of V_e in four normal subjects by the Inhibition Method, in which a blocking dose of risperidone was administered prior to the second [^{11}C]MDL100,907 emission scan.

Figure 2 illustrates the calculation by the regression method of the magnitude of V_e in a typical patient, in whom the magnitudes of V_h were first obtained from the hyperplot.

The linear relationship between V_h and $\Delta K'$ in a series of brain regions was used to calculate the tracer distribution.

Figure 3 summarizes the findings of tracer binding potentials in healthy volunteers and in patients with schizophrenia. Separate estimates of binding potentials of [^{11}C]MDL100,907 for 5HT$_{2A}$ receptors and [^{11}C]McN5652 for serotonin transporters density. Whereas the transporters

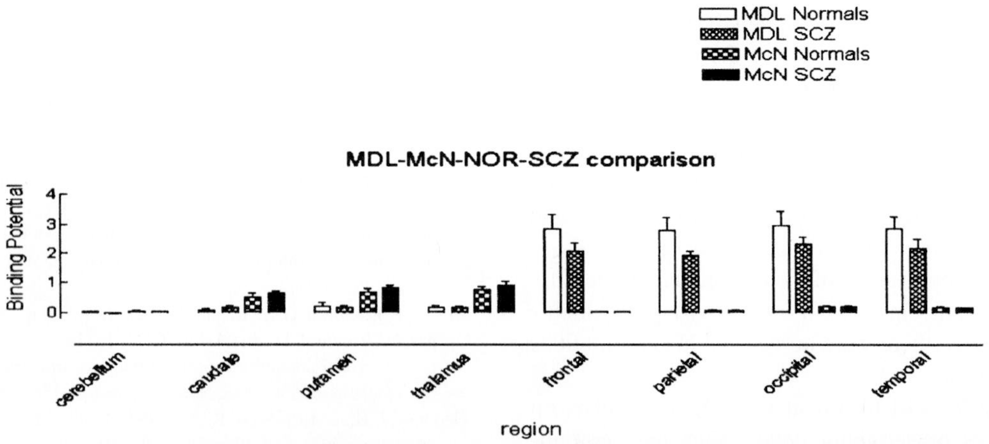

MDL-McN-NOR-SCZ comparison

FIGURE 3. Comparison of binding potentials in several brain regions for [^{11}C]MDL100907, a ligand for 5HT$_{2A}$ receptors, and [^{11}C]McN5652, a ligand for serotonin uptake sites, in a group of healthy volunteers and a group of patients with schizophrenia. Each estimate is the mean ($\pm$SD) of nine separate determinations.

were concentrated in the diencephalon, the 5HT$_{2A}$ receptors were concentrated in cortical regions (Fig. 3). There were no significant differences in tracer binding between the healthy volunteers and the patients with schizophrenia.

IV. DISCUSSION

We applied two novel methods of calculating binding potential based on the estimation of the partition volume (V_e) across several regions. Either method may be applicable with radioligands for which no region of very low specific binding can be idenitified. Consistent with the results of postmortem binding studies, we find that serotonin transporters are concentrated in the diencephalon while serotonin receptors of the 5HT$_{2A}$ type are concentrated in cerebral cortex. There is much evidence implicating serotonin in the pathophysiology of schizophrenia. Furthermore, serotoninergic effects play an important role in the mechanism of therapeutic action of some antipsychotic agents (Bleich *et al.*, 1988; Breier *et al.*, 1995). Thus, accurate methods for measuring markers of serotoninergic neurotransmission are needed.

The present lack of differences between the two study groups is in agreement with earlier reports of the binding of [^{18}F]setoperone to 5HT$_{2a}$ receptors in brain of drug-naive patients with schizophrenia (Trichard *et al.*, 1998; Lewis *et al.*, 1999).

Based on the observation that the serotonergic drug lysergic acid diethylamide (LSD) produces psychotic symptoms resembling in some aspects clinical features of schizophrenia, Gaddum and Hameed (1954) and Wooley and Shaw (1954) proposed the involvement of serotonin systems in the etiology of schizophrenia. More recently, it became clear that many "atypical" antipsychotic drugs are potent 5HT an-

tagonists, leading to renewed interest in the pathophysiology of serotonin in schizophrenia (Bleich *et al.*, 1988). Clozapine's "atypical" clinical profile was attributed to its relatively low occupancy of D$_2$-like dopamine receptors (20–30%) relative to its high occupancy of serotonin 5HT$_2$ receptors (80–90%) at very low, but clinically effective doses (Nordstrom *et al.*, 1993). Following upon clozapine, antipsychotics with combined dopamine and serotonin antagonism, such as risperidone or olanzapine, were developed to test the hypothesis that serotonin antagonism could provide the basis for antipsychotics lacking pronounced extrapyramidal side-effects. Large clinical studies of schizophrenia comparing the efficacy of risperidone with that of the typical neuroleptic haloperidol support the view that the strong antagonism of 5-HT$_2$ receptors, even in the presence of D2-antagonism, decreases the incidence of extrapyramidal side-effects (Chouinard *et al.*, 1993; Marder and Meibach, 1994).

Although serotonin antagonist may be clinically effective, postmortem binding studies with tritiated ligands have generally detected small declines in the density of cortical 5-HT$_2$ binding sites, especially in frontal regions of the brains of schizophrenic patients (Arora and Meltzer, 1991; Bennett *et al.*, 1979; Laruelle *et al.*, 1993; Mita *et al.*, 1986; Lewis *et al.*, 1999). The radiolgands employed in most of these studies ([^{3}H]spiperone, [^{3}H]LSD, and [^{3}H]ketanserine) lack selectivity for 5HT$_{2A}$ receptors. Indeed, LSD binds to at least six different types of dopamine and serotonin receptors, while spiperone and ketanserine both have a relatively high affinity for the α_1-receptor. In contrast, others report increased density of 5HT$_2$ receptors in the cortex of schizophrenic patients (Whitaker *et al.*, 1981; Joyce *et al.*, 1993). In previous PET studies (Lewis *et al.*, 1999; Trichard *et al.*, 1998) with the selective agent [^{18}F]setoperone, there were no large changes in specific binding, calculated using the cerebellum as a reference region. In the present study, we find no declines in cortical serotonin $5HT_{2A}$ binding site density, using a method insensitive to the presence of specific binding in a reference region.

Large increases in the density of serotonin uptake sites in the basal ganglia and decreases in the cortical binding have been reported for patients with schizophrenia in a postmortem study using a nonselective radioligand ($[^3H]$cyanoimipramine (Joyce *et al.*, 1993; Laruelle *et al.* 1993)). We now report results of the first attempt to measure serotonin uptake sites in brain of a small group of patients with schizophrenia; there were no detectable abnormalities in the clinical group.

In general, effects of treatment with D_2 antagonists on $5HT_{2A}$ receptors or serotonin uptake sites may complicate interpretation of these results. However, Andree *et al.* (1986) demonstrated no significant effect of haloperidol on $5HT_{2A}$ receptors in brain of rodents. It is thought that fluphenazine similarly lacks effects on serotonin receptors (Seeman, pers. commun.). Inferences from the present negative findings must be taken with great caution because of the small group size. The primary goal of this paper was to test two methods of determining V_e in brain for subsequent calculation of binding potential. Using these two methods, we estimate binding potentials for two tracers lacking a true reference region. The methods of inhibition with risperidone blockade of $[^{11}C]$MDL100,907 binding (Method 1) or from multiple brain regions with a range of non-specific binding (Method 2) have promise for application to $[^{11}C]$WAY100,635 studies of $5HT_{1A}$ receptors, and other tracer studies for which no reference region is available. The second method offers the distinct advantage of not requiring a second PET scan after administration of a blocking dose of medication.

Acknowledgments

This work was supported by the Scottish Rite Foundation, USPHS NIH Grants MH482821, DA11080, and DA09482, and MRC (Denmark) Grants 970055 and 9802563. Special thanks to M. Stephane, Semih Dogan, K. Neufeld, and J. Smith for PET and patient assistance. Special thanks to W. Bauer, R. Parker, C. Endres, and J. Hilton for technical advice and assistance.

References

Andree, T. H., Mikuni, M., Tong, C. Y., Koenig, J. I., and Meltzer H. Y. (1986). Differential effect of subchronic treatment with various neuroleptic agents on serotonin$_2$ receptors in rat cerebral cortex. *J. Neurochem.* **46**(1): 191–195.

Arora, R. C., and Meltzer, H. Y. (1991). Serotonin$_2$ (5HT$_2$) receptor binding in the frontal cortex of schizophrenic patients. *J. Neural Transm.* **85**: 19–29.

Bennett, J. P., Jr., Enna, S. J., Bylund, D. B., Gillin, J. C., Wyatt, R. J., and Snyder, S. H. (1979). Neurotransmitter receptors in frontal cortex of schizophrenics. *Arch. Gen. Psychiatry* **36**(9): 927–934.

Bleich, A., Brown, S. L., Kahn, R., and van Praag, H. M. (1988). The role of serotonin in schizophrenia. *Schizophr. Bull.* **14**(2): 297–315.

Blomqvist, G. (1984). On the construction of functional maps in positron emission tomography. *J. Cereb. Blood Flow Metab.* **4**: 629–632.

Breier, A. (1995). Serotonin, schizophrenia and antipsychotic drug action. *Schizophr. Res.* **14**: 187–202.

Chouinard, G., Jones, B., Remington, G., Bloom, d., Addington, D., MacEwan, G. W., Labelle, A., Beauclair, L., and Arnott, W. (1993). A Canadian multicenter placebo-controlled study of fixed doses of risperidone and haloperidol in the treatment of chronic schizophrenic patients. *J. Clin. Psychopharmacol.* **13**(1): 25–40.

Dewey, S. L., Smith, G. S., Logan, J., Alexoff, D., Ding, Y.-S., King, P., Pappas, N., Brodie, J. D., and Ashby, C. R., Jr. (1995). Serotonergic modulation of striatal dopamine measured with positron emission tomography (PET) and in vivo microdialysis. *J. Neurosci.* **15**(1): 821–829.

Gaddum, J. H., and Hameed, K. A. (1954). Drugs which antagonize 5-hydroxytryptamine. *Br. J. Pharmacol.* **9**: 240–248.

Gjedde, A. (1981). High- and low-affinity transport of D-glucose from blood to brain. *J. Neurochem.* **36**: 1463–1471.

Gjedde, A. (1982). Calculation of glucose phosphorylation from brain uptake of glucose analogs in vivo: A re-examination. *Brain Res. Rev.* **4**: 237–274.

Gjedde, A., Wong, D. F., and Wagner, H. N., Jr. (1986). Transient analysis of irreversible and reversible tracer binding in human brain in vivo. In *"PET and NMR: New Perspectives in Neuroimaging and Clinical Neurochemistry"* (L. Battistin, Ed.). A. R. Liss, New York.

Gjedde, A., and Wong, D. F. (1990). Modeling neuroreceptor binding of radioligands in vivo. In *"Quantitative Imaging: Neuroreceptors, Neurotransmitters, and Enzymes"* (J. J. Frost and H. N. Wagner, Jr., Eds.). Raven Press, Ltd.

Ito, H., Nyberg S., Halldin, C., Lundkvist, C., and Farde, L. (1998). PET imaging of central 5-HT2A receptors with carbon-11-MDL 100,907. *J. Nucl. Med.* **39**(1): 208–214.

Joyce, J. N., Shane, A., Lexow, N., Winokaur, A., Casanova, M. F., and Kleinman, J.E. (1993). Serotonin uptake sites and serotonin receptors are altered in the limbic system of schizophrenics. *Neuropsychopharmacology* **8**(4): 315–336.

Laruelle, M., Abi-Dargham, A., Casanova, M. F., Toti, R., Weinberger, D. R., and Kleinman, J. E. (1993). Selective abnormalities of prefrontal serotonergic receptors in schizophrenia. A postmortem study. *Arch. Gen. Psychiatry* **50**(10): 810–818.

Laruelle, M., Abi-Dargham, A., Van Dyck, C.H., Gil, R., De Souza, C. D., Erdos, J., McCance, E., Rosenblatt, W., Fingado, C., Zoghbi, S. S., Baldwin, R. M., Seibyl, J. P., Krystal, J. H., Charney, D. S., and Innis, R. B. (1996). Single photon emission computerized tomography imaging of amphetamine-induced dopamine release in drug free schizophrenic subjects. *Proc. Natl. Acad. Sci. USA* **93**: 9235–9240.

Lewis, R., Kapur, S., Jones, C., DeSilva, J., Brown, G. M., Wilson, A. A., Houle, S., and Zipursky, R. B. (1999). Serotonin 5-HT$_2$ receptors in schizophrenia: A PET study using $[^{18}F]$Setoperone in neuroleptic-naive patients and normal subjects. *Am. J. Psychiatry* **156**: 1.

Marder, S. R., and Meibach, R. C. (1994). Risperidone in the treatment of schizophrenia. *Am. J. Psychiatry* **151**: 825–835.

Mita, T., Hanada, S., Nishino, N., Kuno, T., Nakai, H., Yamadori, T., Mizoi, Y., and Tanaka, C. (1986). Decreased serotonin S2 and increased dopamine D2 receptors in chronic schizophrenics. *Biol. Psychiatry* **21**(14): 1407–1414.

Nordstrom, A. L., Farde, L., and Halldin, C. (1993). High 5-HT$_2$ receptor occupancy in clozapine treated patients demonstrated by PET. *Psychopharmacology* **110**: 365–367.

Sokoloff, L., Reivich, M., Kennedy, C., des Rosiers, M. H., Patlak, C. S., Pettigrew, K. D., Sakurada, O., and Shinohara, M. (1977). The $[^{14}C]$deoxyglucose method for the measurement of local cerebral glucose utilization: Theory, procedure, and normal values in the conscious and anesthetized albino rat. *J. Neurochem.* **28**: 897–916.

Trichard, C., Paillere-Martinot, M.-L., Attar-Levy, D., Blin J., Feline, A., and Martinot, J.-L. (1998). No serotonin 5-HT$_{2A}$ receptor density abnormality in the cortex of schizophrenic patients studied with PET. *Schizophr. Res.* **31**: 13–17.

Watabe, H., Channing, M. A., Der, M. G., Adams, H. R., Jagoda, E., Herscovitch, P., Eckelman, W. C., and Carson, R. E. (1998). Kinetic analysis of the 5-HT$_{2A}$ ligand [C-11]MDL 100,907. *J. Nucl. Med.* **39**: 5.

Whitaker, P. M., Crow, T. J., and Ferrer, I. N. (1981). Tritiated LSD binding in frontal cortex in schizophrenia. *Arch. Gen. Psychiatry* **38**: 278–280.

Wooley, D. W., and Shaw, E. (1954). A biochemical and pharmacological suggestion about certain mental disorders. *Proc. Natl. Acad. Sci. USA* **40**: 228–231.

Wong, D. F., Gjedde, A., and Wagner, H. N. (1986). Quantification of neuroreceptors in the living human brain. I. Irreversible binding of ligands. *J. Cereb. Blood Flow Metab.* **6**: 137–146.

39

Serotonin Transporter Binding *In Vivo*: Further Examination of [¹¹C]McN5652

BRIAN J. LOPRESTI,* CHESTER A. MATHIS,* JULIE C. PRICE,* VICTOR VILLEMAGNE,*
CAROLYN CIDIS MELTZER,*,† DANIEL P. HOLT,* GWENN S. SMITH,*,†
and ROBERT Y. MOORE‡

*PET Facility, Department of Radiology
†Department of Psychiatry
‡Department of Neurology, University of Pittsburgh Medical Center

Previous positron emission tomography imaging studies of the serotonin transporter (5-HTt) system indicated the need to utilize both the active ((+)-) and inactive ((−)-) enantiomers of carbon-11-labeled McN5652 to quantitate regional brain binding site densities. The goal of this study was to determine the simplest scan protocol and data analysis method that could reliably quantify 5-HTt densities in humans. Several specific binding parameters were compared, including an image-based subtraction of late-scan radioactivity concentrations, two- and three-compartment model distribution volumes (DV), and binding potentials (BP) derived from two-compartment DV estimates and a reference tissue model. Specific binding parameters were evaluated based upon: (1) the ease of implementation, (2) stability, (3) applicability across the physiologic range of brain 5-HTt densities, (4) correlation to the known distribution of 5-HTt as determined in vitro, and (5) the degree of intersubject variability. The BP measure derived from regional and cerebellar two-compartment DV estimates determined from a single 90-min scan of [¹¹C](+)-McN5652 best satisfied the aforementioned criteria in brain regions of high to medium 5-HTt densities. This method is not useful in regions of low 5-HTt density, and an additional scan using [¹¹C](−)-McN5652 is required for areas of low density such as cortex.

I. INTRODUCTION

A considerable body of clinical and basic science research implicates dysfunction of the serotonin (5-HT) neurotransmitter system in the pathophysiology of mood disorders such as depression (Meltzer *et al.*, 1991). Selective serotonin reuptake inhibitors (SSRIs) are an effective class of anti-depressant medications that act by blocking the 5-HT transporter (5-HTt). Studies of 5-HTt pathology have utilized autoradiographic and/or histologic analyses of postmortem brain tissues. However, these methods are not well-suited to examining dynamic changes in the 5-HTt system that result from disease progression, chronic pharmacotherapy, or drug abuse. The ability to study the 5-HTt *in vivo* using functional imaging techniques, such as positron emission tomography (PET), has motivated the development of selective radiotracers for use in human imaging studies.

Over the past 10 years, a number of ligands known to selectively bind to the serotonin transporter *in vitro* have been radiolabeled with the positron emitters carbon-11 or fluorine-18 for PET imaging studies. The most successful radioligand developed for this purpose is [¹¹C](+)-McN5652, the active enantiomer of 1,2,3,5,6β,10bβ-hexahydro-6α-[4-(methylthio)phenyl]pyrrolo[2,1-α]isoquinoline (Suehiro *et al.*, 1992, 1993a, b; Szabo *et al.*, 1995a, b). The *in vivo* binding of [¹¹C]McN5652 to the 5-HTt has been studied in baboons and humans using PET imaging (Szabo *et al.*, 1995a, b). These studies demonstrated the specificity and high affinity of $[^{11}C](+)$-McN5652 for the 5-HTt and determined regional measures of specific binding from the difference between the radioactivity concentration of $[^{11}C](+)$-McN5652 and that of the inactive enantiomer $[^{11}C](-)$-McN5652 over the period of 95–125 min postinjection. The rank order of specific binding was midbrain > caudate > thalamus > temporal cortex > frontal cortex > cerebellum. More recently, these investigators reported preliminary results of model-free and model-based kinetic analyses that utilized a single tissue compartment to describe the kinetics of $[^{11}C](+)$-McN5652 over 90 min (McCann *et al.*, 1998).

Previously, we reported the feasibility of obtaining routine estimates of 5-HTt binding using kinetic modeling of $[^{11}C]$McN5652 3D PET data in baboons (Price *et al.*, 1997). The studies involved the use of the (+)- and (−)-enantiomers analyzed with a conventional two-compartment (2C) model and yielded radioligand distribution volume (DV) estimates. In addition, regional specific binding was estimated from the late-scan (90–120 min) differences between the concentration of $[^{11}C](+)$-McN5652 and $[^{11}C](-)$-McN5652. High correlations ($r^2 \sim 0.9$) were found for the $[^{11}C](+)$-McN5652 2CDV values and the known distribution of the 5-HTt in human brain (Rosel *et al.*, 1997), and these values compared well with late scan difference measures obtained using $[^{11}C](-)$-McN5652 ($r^2 \sim 0.9$). The goal of the present work was to determine the optimum scan protocol and data analysis method for $[^{11}C]$McN5652 PET imaging. The ideal technique would: (1) allow specific binding to be determined from a single injection of $[^{11}C](+)$-McN5652, (2) demonstrate a degree of intersubject variability comparable to that reported for *in vitro* measures of 5-HTt densities (~ 10 to 15%), (3) be useful over the physiologic range of 5-HTt concentrations, (4) correlate well with the known distribution of the 5-HTt, and (5) be able to detect changes in 5-HTt occupancy by therapeutic doses of SSRIs. The analysis techniques were also applied to a series of baboon imaging experiments intended to determine their sensitivity to specific binding changes of $[^{11}C](+)$-McN5652 in the presence of a highly selective serotonin reuptake inhibitor, citalopram (Hyttel, 1982).

II. METHODS

A. Human PET Imaging

$[^{11}C]$ (+)- and -(−)-McN5652 were synthesized according to the method of Suehiro *et al.* (1992) and Huang *et al.* (1998). Single-day paired PET studies of $[^{11}C](+)$-McN5652 and $[^{11}C](-)$-McN5652 were performed in three normal human subjects (one female, two male; mean age, 25 ± 2 years). Dynamic PET scans were acquired over a duration of 120 min (23 frames of increasing length from 30 s to 10 min) following the injection of ~ 370 MBq (10 mCi) of either $[^{11}C](+)$- or $[^{11}C](-)$-McN5652 (sp. act. >56 GBq/μmol (1500 Ci/mmol) via an intravenous line placed in the antecubital vein. Images were acquired in 3D mode (septa retracted) on an ECAT HR$^+$ PET camera (CTI PET Systems, Inc., Knoxville, TN), which is capable of acquiring data over a 15.2 cm field-of-view. The images were reconstructed using a Hanning filter at 0.4 Nyquist. In 3D mode, the FWHM resolution of the camera for a point source in air is approximately 5.0 ± 0.5 mm transverse and 4.5 ± 0.5 mm axially (Brix *et al.*, 1997). Serial PET scans were aligned to each other and then to a single MR image (spoiled-gradient recalled sequence) acquired in the same subject using the automated image registration (AIR) algorithm (Woods *et al.*, 1992, 1993). Regions of interest (ROIs) were drawn manually on the MR image, resliced to the resolution and orientation of the PET image, and applied to the dynamic PET data sets to obtain decay-corrected tissue time–activity data. ROIs were defined for the dorsal raphe nucleus (DRN), thalamus (THL), striatum (STR), amygdala/hippocampus (AMH), lateral orbito-frontal (LOF), prefrontal cortex (PFC), and cerebellum (CER).

B. Baboon PET Imaging

Three baboons (papio anubis) were anesthetized with ketamine (10 mg/kg, i.m.) and isoflurane (0.5–1.5%) and paralyzed (pancuronium bromide) and injected with ~ 370 MBq (10 mCi) of $[^{11}C](+)$-McN5652 both at baseline and 30 min after treatment with 0.5 mg/kg of citalopram (injected intravenously over a period of 10 min). ROIs were defined for raphe nucleus (RAP), midbrain (MBN), thalamus (THL), striatum (STR), lateral temporal cortex (TEM), frontal cortex (FRT), and cerebellum (CER) on summed PET scans of the early (0–20 min for all regions but RAP) and late (80–120 min RAP only) frames of acquisition. As the baboons were anesthetized and immobilzed during the entire experiment, registration of the serial PET images was not required.

C. Plasma Analyses

In order to determine the radioactivity concentration of unmetabolized $[^{11}C](+)$-McN5652 and $[^{11}C](-)$-McN5652 in human and baboon plasma, arterial blood samples (0.5 mL) were drawn from the radial (human) or femoral (baboon) artery throughout the course of the study. Samples were drawn at a frequency of approximately every 6 s for the first 2 min of the study and with a reduced frequency thereafter (Fig. 1A). The samples were centrifuged at 13,000g for 2 min and 0.2 mL of plasma was separated and counted in a gamma radioactivity counter (Packard Cobra 5003, Meridan, CT). Additional blood samples were drawn

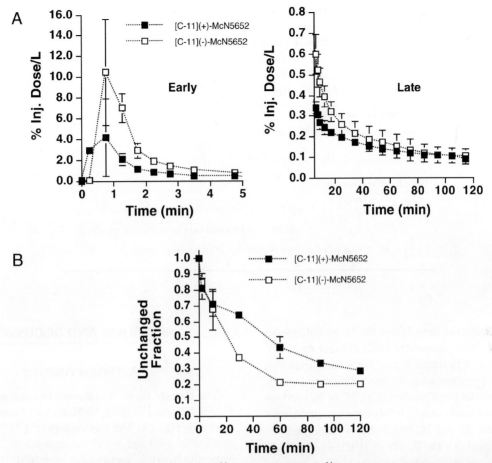

FIGURE 1. (A) Metabolite corrected average [^{11}C](+)-McN5652 and [^{11}C](−)-McN5652 plasma radioactivity for early (left) and late (right) samples in three human subjects. (B) Metabolism of [^{11}C](+)-McN5652 and [^{11}C](−)-McN5652 in humans ($n = 3$).

at 2, 10, 30, 60, 90, and 120 min postinjection and used to determine the fraction of unmetabolized [^{11}C](+)-McN5652 or [^{11}C](−)-McN5652 at each time point using a two-step extraction procedure (Mathis *et al.*, 1998).

D. Data Analyses

Regional brain specific binding of [^{11}C](+)-McN5652 was determined from the analysis of both the human and the baboon PET data in three different ways: (1) an image-based subtraction of late-scan radioactivity concentrations (ΔLS); (2) distribution volume (DV) determined by compartmental modeling; and (3) binding potential (BP) determined from two-compartment DV estimates and from a reference tissue model (Table 1).

Using the ΔLS method (Szabo *et al.*, 1995a,b), specific binding was determined by calculating the difference in the regional radioactivity concentrations between serial PET scans of [^{11}C](+)-McN5652 and [^{11}C](−)-McN5652 (ΔLS(+/−)) or between regional and cerebellar [^{11}C](+)-McN5652 radioactivity concentrations (ΔLS(+)). Both of

these specific binding measures were determined from summed late frames of acquisition (90–120 min). During this period it was assumed that maximal clearance of nonspecific binding was observed and steady-state had been achieved (Szabo *et al.*, 1995a,b). All measures of radioactivity concentration were normalized for injected dose and body mass.

Linear two-compartment (2C) and three-compartment (3C) models were applied to the dynamic [^{11}C](+)-McN5652 and [^{11}C](−)-McN5652 data to determine which provided more reliable kinetic parameter estimates. The 2C model estimated the kinetic variables K_1 and k_2, which describe tracer uptake and efflux from brain tissue. In the 3C configuration, these parameters represent ingress and egress from plasma to a tissue compartment representing the aggregate concentration of free and nonspecifically bound tracer. Two additional kinetic variables, k_3 and k_4, describe the on- and off-rates of the tracer to the 5-HTt. An additional parameter was added in both configurations to account for radioactivity contained in the vascular volume of brain tissue. In this work, 2CDV and 3CDV values were determined using

TABLE 1.   Formulation of Specific Binding Parameters

(ΔLS): Late-scan (90–120 min) difference measure

$$\Delta LS(+/-) = (+)NR_{ROI} - (-)NR_{ROI}$$

$$\Delta LS(+) = (+)NR_{ROI} - (+)NR_{CER}$$

$$NR = [\%ID/mL] \times \text{body mass (kg)}$$

(DV): Distribution volume

Two-compartment DV (2CDV) $= \dfrac{K_1}{k_2}$

Three-compartment DV (3CDV) $= \dfrac{K_1}{k_2}[1 + k_3/k_4]$

(BP): Binding potential

$$BP_{DV} = \left(\frac{DV_{ROI}}{DV_{CER}}\right) - 1 \quad \text{(2CBP: (+)CER) used the (+)-enantiomer for } DV_{CER}$$

$$\text{(2CBP: (-)CER) used the (-)-enantiomer for } DV_{CER}$$

$$\frac{k_3}{k_4} \propto BP = B_{max}/K_d \qquad \text{for a 3C model or reference tissue model}$$

the kinetic parameters obtained from the 2C and 3C model fits, respectively. The equations for the 2CDV and the 3CDV are shown in Table 1 in simple terms. We acknowledge that the physiological interpretation of these parameters depends on the specific model configuration (e.g., 2C or 3C) and refer the reader to Carson *et al.* (1993) for a comprehensive description of the 2C and 3C distribution volume parameters. It was assumed that the 2C DV of $[^{11}C](+)$-McN5652 in 5-HTt-rich tissue consisted of specific binding, nonspecific binding, and free tracer components, whereas the DV of $[^{11}C](-)$-McN5652 consisted only of nonspecific binding and free tracer.

Binding potentials (BP) derived from the 2CDV estimates were determined for $[^{11}C](+)$-McN5652 (Table 1), using both the $[^{11}C](+)$-McN5652 (2CBP:(+)CER) and $[^{11}C](-)$-McN5652 (2CBP:(-)CER) cerebellar DV values. In addition, a simplified reference tissue model was applied to the $[^{11}C](+)$-McN5652 data to determine the stability and reliability of BP without the need for arterial sampling or plasma metabolite corrections (Lammertsma *et al.*, 1996; Gunn *et al.*, 1997). Both the $[^{11}C](+)$-McN5652 and $[^{11}C](-)$-McN5652 cerebellar time–activity data were used as reference tissues.

To evaluate the sensitivity of the $[^{11}C](+)$-McN5652-specific binding parameters to changes in occupancy of the 5-HTt, regional comparisons were made between serial $[^{11}C](+)$-McN5652 baboon studies before and after treatment with 0.5 mg/kg of citalopram using the ΔLS(+) and the 2CBP:(+)CER methods. The analysis of these studies was limited to these two methods of specific binding determination, as it was not practical to conduct same day serial studies of both $[^{11}C](+)$-McN5652 and $[^{11}C](-)$-McN5652 before and after drug treatment in a baboon.

III. RESULTS AND DISCUSSION

A. Plasma Analyses

The average ($n = 3$) amounts of unchanged $[^{11}C](+)$-McN5652 and $[^{11}C](-)$-McN5652 in human plasma are shown in Fig. 1A. The metabolism of $[^{11}C](+)$-McN5652 resulted in ~80 and ~30% unchanged at 2 and 120 min, respectively, while metabolism of $[^{11}C](-)$-McN5652 resulted in ~85 and ~20% unchanged at 2 and 120 min. The metabolite corrected input functions for both $[^{11}C](+)$-McN5652 and $[^{11}C](-)$-McN5652 are shown in Fig. 1A. Of note are the substantially lower rate of metabolism (Fig. 1B) and lower metabolite-corrected plasma integral of $[^{11}C](+)$-McN5652 compared to $[^{11}C](-)$-McN5652. Together, these facts suggest that a substantial portion of the injected dose of $[^{11}C](+)$-McN5652 is not free within the plasma compartment immediately after injection. Stereoselective binding of the (+)-enantiomer to the platelet serotonin transporter or binding to sites in the lung (Hashimoto and Goromaru, 1991) or to other plasma proteins may explain this discrepancy.

B. Specific Binding Determination

Evaluation of the various approaches to determining the specific binding of $[^{11}C](+)$-McN5652 was based upon comparing the degree of intersubject variability (coefficient of variation, CV) in the specific binding parameter, the ability of the method to estimate specific binding over the complete range of tissue 5-HTt densities (Table 2), and the degree of correlation with *in vitro* 5-HTt densities.

The ΔLS(+) and ΔLS(+/−) methods of summing late-scan brain radioactivity concentrations exhibited a fairly high degree of intersubject variability, with the ΔLS(+)

TABLE 2.   Comparison of Specific Binding Measures for the Late-Scan Difference (ΔLS), Two-Compartment Distribution Volume (2CDV), and 2CDV Derived Binding Potential (2CBP)

	DRN	THL	STR	AMH	LOF	PFC	CER
ΔLS(+/−)							
Mean	0.28	0.17	0.17	0.09	0.07	0.08	NA
CV (%)	64	54	59	64	90	74	NA
(−)2CDV							
Mean	12	15	13	13	12	10	12
CV (%)	18	6	12	7	9	15	10
(+)2CBP:(−)CER							
Mean	5	2.2	1.9	1.5	0.8	0.6	NA
CV (%)	39	50	45	43	51	47	NA
ΔLS(+)							
Mean	0.24	0.16	0.13	0.06	0.03	—	NA
CV (%)	36	26	31	48	108	—	NA
(+)2CDV							
Mean	74	39	36	30	22	19	20
CV (%)	20	22	20	13	10	6	9
(+)2CBP:(+)CER							
Mean	2.8	1	0.8	0.6	0.1	—	NA
CV (%)	25	43	31	41	124	—	NA

Note. —, Negative specific binding measure.

method resulting in somewhat lower variability. However, the ΔLS(+) method often resulted in negative specific binding measures in regions of low 5-HTt density (e.g., PFC). This is a consequence of the low and nearly equivalent concentration of 5-HTt in frontal cortex and cerebellum (PFC B_{max} = 61 ± 10 fmol/mgP and CER B_{max} = 44 ± 9 fmol/mgP; Rosel *et al.*, 1997).

The 2CDV measures were more applicable for both [^{11}C](+)-MCN5652 and [^{11}C](−)-MCN5652 than the 3CDV measures (data not shown), as the latter exhibited large standard errors in parameter estimates (>100%), problems with model convergence, as reflected by higher Akaike Information Criteria (AIC) values for the 3C model fits (Akaike *et al.*, 1974), particularly in regions of high 5-HTt concentration using [^{11}C](+)-MCN5652. Compared to other methods, 2CDV measures showed the lowest intersubject variability and was stable across the entire range of 5-HTt tissue densities, although the [^{11}C](+)-MCN5652 cerebellar DV was not always lower than some of the cortical regions with comparable concentrations of the 5-HTt (e.g., PFC). No specific binding measure was able to resolve the small differences in 5-HTt binding between the cerebellum and most cortical regions. This made quantitation of the concentration of

5-HTt unreliable in regions of low 5-HTt density. Examples of 2C fits to [^{11}C](+)-MCN5652 and [^{11}C](−)-MCN5652 time–activity data for striatum and cerebellum are shown in Fig. 2.

The reference tissue model failed to provide stable measures of BP using either [^{11}C](+)-McN5652 or [^{11}C](−)-McN5652 cerebellar time–activity data as inputs (data not shown). This method had the greatest measure of intersubject variability and resulted in the greatest occurrence of negative BP values. As a result, this method did not exhibit a strong correlation with the known distribution of the 5-HTt as determined *in vitro* (Rosel *et al.*, 1997). For reasons that are unclear, the reference tissue model is not applicable to [^{11}C](+)-McN5652 data using either the [^{11}C](+)-McN5652 or [^{11}C](−)-McN5652 cerebellum as the reference tissue.

Binding potentials derived from the ratios of regional [^{11}C](+)-McN5652 to either cerebellar [^{11}C](+)-McN5652 (2CBP:(+)CER) or [^{11}C](−)-McN5652 (2CBP:(−)CER) DV estimates were more variable than the 2CDV measure alone. However, the 2CBP:(+)CER method often resulted in negative binding potentials in regions of low 5-HTt concentration, as would be expected when cerebellar DV exceeded that of other regions. BP calculated using the 2CBP:(−)CER method did not result in negative BP estimates for any region, as the cerebellar DV of [^{11}C](−)-McN5652 was about a factor of two lower than the cerebellar DV of [^{11}C](+)-McN5652. The higher DV value of the (+)-enantiomer can be attributed in part to a small amount of specific binding to 5-HTt in the cerebellum. Thus, the 2CBP:(−)CER may allow more accurate quantitation of 5-HTt concentration in regions of low 5-HTt density compared to the 2CBP:(+)CER method. However, these benefits are somewhat offset by the experimental complexity of an additional scan using [^{11}C](−)-McN5652.

Scan duration is a practical consideration in determining the optimum data analysis method. The ΔLS methods require at least 120 min of PET emission data to provide adequate specific binding measures, thereby permitting sufficient clearance of nonspecific binding. The 2CDV measure using 90 min of data exhibited the least error in the parameter estimates and the smallest degree of intersubject variability, as compared to 2CDV determination using either 65 or 120 min. A scan duration of 90 min was sufficiently long to observe reversibility of binding in 5-HTt-rich tissues but short enough that the quality of the model fits were not marred by noisy brain and plasma time-activity data as a result of the short half-life of carbon-11 (see Fig. 2).

The blocking experiments in the baboon demonstrated that decreases in [^{11}C](+)McN5652 specific binding can be reliably detected using the ΔLS(+) or the 2CBP:(+)CER methods. A comparable decrease in the specific binding of [^{11}C](+)McN5652 was observed with the 2CBP:(+)CER as compared to the ΔLS(+) method (see Table 3). How-

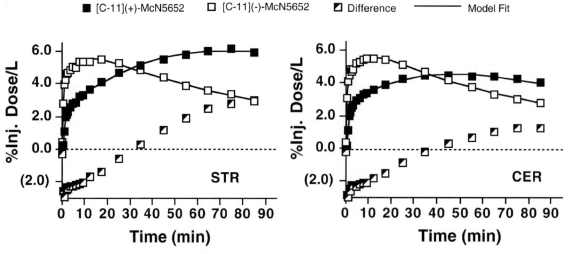

FIGURE 2. Time–activity curves of $[^{11}C](+)$-McN5652 and $[^{11}C](-)$-McN5652 in striatum (STR) and cerebellum (CER) in a representative human subject.

TABLE 3. Specific Binding Percentage Differences in Baboons ($n = 3$) Following Pretreatment with 0.5 mg/kg Citalopram in Raphe Nucleus (RAP), Midbrain (MBN), Thalamus (THL), Striatum (STR), Temporal Cortex (TEM); and Frontal Cortex (FRT)

	RAP	MBN	THL	STR	TEM	FRT
ΔLS(+)	−106	−74	−62	−56	−53	0
CV (%)	30	14	20	24	23	0
2CBP:(+)CER	−105	−87	−53	−46	−45	191
CV (%)	22	7	26	18	32	169

ever, the 2CBP:(+)CER method can be inaccurate in estimating changes in 5-HTt occupancy in regions containing concentrations of 5-HTt similar to the cerebellum (e.g., FRT: +191% with a CV of 161%).

IV. CONCLUSION

The method of choice for the analysis of human $[^{11}C]$McN5652 PET data in regions of mid- to high-5-HTt density is the 2CBP:(+)CER. This method does not require a second PET study using $[^{11}C](-)$-McN5652 and can be used to detect changes in 5-HTt occupancy. The optimum scan protocol for the 2CBP:(+)CER method is a single 90-min scan using $[^{11}C](+)$-McN5652. This method is easily implemented, displays acceptable intersubject variability, and correlates well with the known distribution of 5-HTt. To quantitate 5-HTt in brain regions of low density, the 2CBP:(−)CER method and a second scan of $[^{11}C](-)$-McN565 may be necessary.

Acknowledgments

This work was supported by grants from the National Institutes of Health (NS22899 and MH52247) and the Charles A. Dana Foundation.

References

Akaike, H. (1974). A new look at the statistical model identification. *IEEE Trans. Autom. Control* **19**: 716–723.

Brix, G., Zaers, J., Adam, L.-E., Belleman, M. E., Ostertag, H., Trojan, H., Haberkorn, U., Doll, J., Oberdorfer, F., and Lorenz, W. J. (1997). Performance and evaluation of a whole-body PET scanner using the NEMA protocol. *J. Nucl. Med.* **38**: 1614–1623.

Carson, R. E., Channing, M. A., Blasberg, R. G., Dunn, B. B., Cohen, R. M., Rice, K. C., and Herscovitch, P. (1993). Comparison of bolus and infusion methods for receptor quantitation: Application to [F-18]cyclofoxy and positron emission tomography. *J. Cereb. Blood Flow Metab.* **13**: 24–42.

Gunn, R. N., Lammertsma, A. A., Hume, S. P., and Cunningham, V. J. (1997). Parametric imaging of ligand–receptor binding in PET using a simplified reference region model. *Neuroimage* **6**(4): 2779–2787.

Hashimoto, K., and Goromaru, T. (1991). High-affinity [H-3]6-nitroquipazine binding to the 5-hydroxytryptamine transport system in rat lung. *Biochem. Pharmacol.* **41**(11): 1679–1682.

Huang, Y., Mahmood, K., Simpson, N. R., Mason, N. S., and Mathis, C. A. (1998). Stereoconservative synthesis of the enantiomerically pure precursors of [C-11](+)-McN5652 and [C-11](−)-McN5652. *J. Lab. Comp. Radiopharm.* **41**: 9–17.

Hyttel, J. (1982). Citalopram—Pharmacological profile of a specific serotonin uptake inhibitor with antidepressant activity. *Prog. Neuropsychopharmacol. Biol. Psychiatry* **6**(3): 277–295.

Lammertsma, A. A., and Hume, S. P. (1996). Simplified reference tissue model for PET receptor studies. *Neuroimage* **4**(3 Pt.1): 153–158.

Mathis, C. A., Mason, N. S., Holt, D. P., and Lopresti, B. J. (1998). A sensitive, rapid method to quantitate metabolites of (+)- and (−)-[C-11]McN5652. *J. Nucl. Med.* **39**: 239P.

McCann, U. D., Szabo, Z., Scheffel, U., Dannals, R. F., and Ricaurte, G. A. (1998). Positron emission tomographic evidence of toxic effect of MDMA ("Ecstasy") on brain serotonin neurons in human beings. *Lancet* **352**: 1433–1437.

Meltzer, H. (1991). Serotonergic dysfunction in depression. *Br. J. Psychiatry* **8**: 25–31.

Price, J., Lopresti, B., Huang, Y., Simpson, N., Mahmood, K., and Mathis, C. A. (1997). Kinetic analysis of serotonin transporter binding: [C-11]McN5652 PET studies. *Neuroimage* **5(4)3**: A19.

Rosel, P., Menchon, J. M., Oros, M., Vallejo, J., Cortadellas, T., Arranz, B., Alvarez, P., and Navarro, M. A. (1997). Regional distribution of specific high affinity binding sites for [H-3]-imipramine and [H-3]-paroxetine in human brain. *J. Neural Transmission.* **104**: 89–96.

Suehiro, M., Ravert, H. T., Dannals, R. F., Scheffel, U., and Wagner, H. N., Jr. (1992). Synthesis of a radiotracer for studying serotonin uptake sites with positron emission tomography: [C-11]McN5652-Z. *J. Lab. Comp. Radiopharm.* **31**: 341–343.

Suehiro, M., Scheffel, U., Dannals, R. F., Ravert, H. T., Ricaurte, G. A., and Wagner, H. N., Jr. (1993a). A PET radiotracer for studying serotonin uptake sites. Carbon-11 McN5652-Z. *J. Nucl. Med.* **34**: 120–127.

Suehiro, M., Scheffel, U., Ravert, H. T., Dannals, R. F., and Wagner, H. N., Jr. (1993b). [C-11](+)McN5652 as a radiotracer for imaging serotonin uptake sites with PET. *Life Sci.* **53**: 883–892.

Szabo, Z., Kao, P. F., Scheffel, U., Suehiro, M., Mathews, W. B., Ravert, H. T., Musachio, J. L., Marenco, S., Kim, S. E., Ricaurte, G. A., Wong, D. F., Wagner, H. N., Jr., and Dannals, R. F. (1995a). Positron emission tomography imaging of serotonin transporters in the human brain using [C-11](+)-McN5652. *Synapse* **20**: 37–43.

Szabo, Z., Scheffel, U., Suehiro, M., Dannals, R. F., Kim, S. E., Ravert, H. T., Ricaurte, G. A., and Wagner, H. N., Jr. (1995b). Positron emission tomography of 5-HT transporter sites in the baboon brain with [C-11]McN5652. *J. Cereb. Blood Flow Metab.* **15**: 798–805.

Woods, R. P., Cherry, S., and Mazziotta, J. (1992). Rapid automated algorithm for aligning and reslicing PET images. *J. Comput. Assist. Tomogr.* **16**: 620–633.

Woods, R. P., Mazziotta, J., and Cherry, S. R. (1993). MRI-PET registration with automated algorithm. *J. Comput. Assist. Tomogr.* **17**: 536–546.

40

A Tracer Kinetic Model for Measurement of Regional Acetylcholinesterase Activity in the Brain Using [11C]Physostigmine and PET

G. BLOMQVIST,*,† B. TAVITIAN,* S. PAPPATA,* C. CROUZEL,* A. JOBERT,* I. DOIGNON,*
and L. DI GIAMBERARDINO*

*INSERM U334, Service Hospitalier Frédéric Joliot, CEA/DSV, 4 Place du Général Leclerc, 91401 Orsay cedex, France
†Uppsala University PET Center, UAS, 75185 Uppsala, Sweden

Previous studies have shown that [11C]physostigmine, an acetylcholinesterase inhibitor, can be used in positron emission tomography for imaging of the cerebral concentration of the enzyme in animals and humans in vivo. The ratio between the uptakes in a target region and in white matter was found to become approximately constant after 20–30 min, indicating that with time the uptake mainly contains information about the distribution volume. Based on these previous results a quantitative and noninvasive method for measurement of the regional acetylcholinesterase concentration in the brain has been developed. A simplified reference tissue model with effectively one reversible tissue compartment and three parameters was found to give a good description of the data. White matter was used as reference tissue. One of the parameters in the model, the ratio between the total distribution volumes in the target and reference regions was found to correlate well with the regional acetylcholinesterase density measured postmortem in two monkeys and also found to give a considerably improved regional contrast compared to the relative magnitude of the regional uptake at late times (15–20 min), used in previous studies as an index of the enzyme activity. The method was applied to eight healthy male subjects. Also in this case the distribution volume ratio was found to correlate well with published values of regional acetylcholinesterase activity measured in vitro.

I. INTRODUCTION

Postmortem studies have shown large reductions in the activity of the enzyme acetylcholinesterase (EC 3.1.1.7, AChE) in the brains of patients suffering from neurodegenerative disorders such as Alzheimers disease (Geula and Mesulam, 1994), and Parkinson's disease (Ruberg et al., 1986). Different tracers for mapping of acetylcholinesterase activity have been proposed: [14C]- and [11C]-labeled N-methylpiperidin-4-yl acetate (AMP) (Iyo et al., 1997) and N-[11C]methylpiperidinyl proprionate ([11C]PMP) (Koeppe et al., 1999). AMP and PMP were designed to give a good measure in regions with low AChE density. These tracers have very high affinity for AChE, which implies that delivery to the brain, i.e., CBF, becomes a rate limiting factor in regions with high AChE density.

Recently, labeled physostigmine (PHY), an AChE inhibitor, has been proposed as a tracer for measurement of AChE concentration. PHY has lower affinity for AChE than AMP or PMP, which means that PHY should be a better estimator of the AChE density in regions rich of AChE, such as the putamen and the cerebellum. PHY labeled in the carboxylic carbon has been used for imaging of AChE density by autoradiography in rats (Planas et al., 1994) and by positron emission tomography (PET) in baboons (Tavitian et al., 1993) and in humans (Pappata et al., 1996). In the rat brain the [11C]PHY radioactivity was found to be essentially superimposable to AChE activity. In the primate brain it was observed that the early, blood-flow-dependent distribution of [^{11}C]PHY was followed by a redistribution to AChE-rich regions. The same pattern was observed in the human studies. In the monkey studies it was further observed that the uptake of [^{11}C]PHY was significantly reduced by competition with an excess of unlabeled PHY.

The aim of the present work was to develop a quantitative measure of regional AChE activity using [^{11}C]PHY and to validate the method by regional comparison of the measure with the corresponding value of the AChE density obtained postmortem.

II. MATERIALS AND METHODS

A. Measurement of Regional AChE Activity in Baboon's Brain Sections

For direct measurement of the AChE activity, brain regions from normal baboons were sampled immediately after sacrifice, homogenized in extraction buffer as previously described (Tavitian, 1985), and processed for the measurement of AChE activity by a standard procedure. The enzyme activity in absorbance units was converted to enzyme concentration (nmol/L) using the enzyme turnover values reported by Vigny et al. (1978).

B. PET and Magnetic Resonance Imaging

[^{11}C]PHY was synthesized as previously described (Bonnot-Lours et al., 1993). Five male adult baboons (*Papio papio*), body weight 15–20 kg, presedated and anaesthetized/ventilated with 1% isoflurane, 67% N$_2$O, and 33% O$_2$, were included in the study. The heads of the animals were placed in a solid headholder fixed to the examination bed so as to align their orbitomeatal line with the scanning plane of an ECAT953B/31 camera (31 slices; spatial resolution with Hanning 0.5 reconstruction filter $8.4 \times 8.4 \times 4.8$ mm at full-width half-maximum). After an i.v. bolus injection of [^{11}C]PHY 17–19 sequential PET scans were acquired for 80–120 min and 16–25 arterial blood samples collected for 50–80 min. Each animal was examined repeatedly. Eight such baseline experiments were performed. The dose range for the bolus injection of [^{11}C]PHY was 100–800 MBq (average $\pm$SD 422 ± 222), and specific radioactivity ranged between 2.0 and 41 (average $\pm$SD 9 ± 9) GBq/μmol. The method was applied to eight healthy male subjects (age 24–76 years). In these experiments data were acquired during 60 min following a bolus injection of [^{11}C]PHY (dose range 100–740 MBq, sp. act. 7.4 ± 5.6 GBq/μmol). Preliminary observations are given by Pappata et al. (1996).

C. Data Processing

Images of radioactivity concentrations, expressed as a percentage of the injected dose in 1 L of the tissue in question (%IDPL), were reconstructed with standard software. Based on the reconstructed images the time course of the radioactivity concentrations was obtained in a number of regions in the monkey brains. The definition of these regions of interest was based on magnetic resonance images (CGR 0.5 T, T1 images) of the animals' heads placed in the same headholder as used in the PET studies and on the anatomic atlas in the orbitomeatal plane of Riche et al. (1988). In the human studies the PET and MRI images were adjusted using an automatic three-dimensional coregistration method (Mangin et al., 1994). Irregular regions of interest were delineated in the MR images on the basis of sulcal anatomy and of the Talairach atlas and grouped to form anatomical functional regions.

D. Kinetic Analysis

PHY crosses the blood–brain barrier (BBB) by passive diffusion with high extraction rate (the early images after a bolus injection essentially show a blood flow distribution). In the brain tissue PHY can bind reversibly to the active site of AChE to form a PHY–AChE transition complex. The complex is rapidly carbamylated, inactivating the enzyme. By this process the label is transferred to methylcarbamic acid, which is unstable and rapidly cleaved into methylamine and [^{11}C]CO$_2$. The latter compound leaves the brain by diffusion into the blood stream. It should be stressed that, although the carbamylation and decarbamylation processes both are unidirectional, the overall kinetics of the label is reversible, and with time the radioactivity concentration in a region is predicted to reach equilibrium with the input function and with the radioactivity concentrations in other regions.

The rate of formation of the PHY–AChE transition complex is proportional to both the concentration of free PHY in the tissue and to the local AChE concentration. The rate of formation is expressed as $k_3 \cdot$ [free PHY] ([] denotes concentration) for unlabeled PHY and therefore the rate constant k_3 is predicted to be proportional to the AChE concentration. Consequently, k_3 is the parameter of primary interest in this model.

We have used a "reference tissue model" for the kinetic analysis. Common to such models is that the kinetics in a reference region is used in place of the input function and the transfer constant K_1 for influx across the blood–brain barrier. The distribution volume is replaced by the "distribution volume ratio" (DVR), i.e., the ratio between the distribution volume in the target and that in the reference region. Using a reference tissue model implies that the need for invasive arterial sampling and tedious and time consuming metabolite analysis of the blood/plasma is avoided. Systematic errors caused by an ill-defined input function are also eliminated. In this study a white matter region has been used as reference tissue.

If all tissue compartments except the one for free PHY is merged into one compartment and if the dissociation of the PHY–AChE complex is neglected, a simple configuration is obtained with an irreversible transfer between two tissue compartments and with loss ($[^{11}C]CO_2$) from the second one (rate constant k_L). For this compartmental configuration DVR becomes:

$$DVR = k_2(1 + k_3/k_L)/(k_2 + k_3). \qquad (1)$$

Expression (1) predicts that DVR (via k_3) depends nonlinearly on the AChE density.

The kinetics can be even more simplified provided the exchanges between the pools are so fast that the concentrations rapidly become proportional to each other and the time courses in the different compartments become difficult to resolve. In the "simplified reference tissue model" introduced by Lammertsma and Hume (1996) the kinetics of the tracer in the tissue is described by a single compartment and three parameters. The operational equation is:

$$
\begin{aligned}
C_{targ}(t) = {} & RC_{ref}(t) + k_2(1 - R/DVR) \\
& \times \int_0^t e^{[k_2(x-t)/DVR_{tot}]} C_{ref}(x)\, dx. \qquad (2)
\end{aligned}
$$

Provided that the time variation of the uptake in the reference tissue becomes small at late times, the model predicts that the ratio between the concentrations in a target and a reference region approaches a level close to the distribution volume ratio. Using expression (2), parameter estimates were obtained by standard regression analysis. Routines contained within the MATLAB software package were utilized.

III. RESULTS AND DISCUSSION

Figure 1 shows the ratios between the uptakes in two regions (putamen and thalamus) and the reference region (a white matter area) as a function of time obtained in the baseline monkey experiments. Averages and standard deviations of the control experiments have been calculated in each time frame. One observes that, after the initial flow-dependent phase, the ratio becomes constant within errors. The same pattern was observed in the human experiments. The displayed results show that under the late time period (after 20 min), when the regional net accumulation of tracer is observed to closely follow the known regional cerebral AChE activity (Tavitian *et al.*, 1993; Pappata *et al.*, 1996), near equilibrium conditions between the tracer in different regions and in white matter (the reference tissue) are reached. This feature is predicted by the simplified reference tissue model.

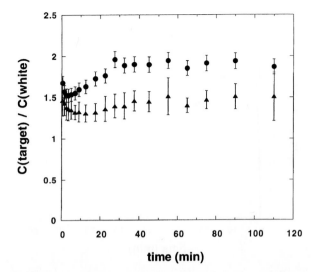

FIGURE 1. The ratio between the uptakes in two regions (putamen ●, thalamus ▲) and in a white matter area (the reference tissue region) as a function of time for monkeys. Averages of the control experiments ($n = 8$) are displayed. The bars indicate standard deviations. The time values are the midpoints of the frames chosen for the data acquisition.

Figure 2 shows examples of uptake curves from different regions measured in one of the monkeys. The results of fitting the simplified reference tissue model (Eq. (2)) to these uptake curves are also shown. The model curves are not smooth, but reflect the measured variations in the time–activity curve in the white matter region (Fig. 2d), which was used as reference tissue. Apart from these fluctuations, the simplified reference tissue model provides a good overall description of the uptake curves.

Figure 3a shows a comparison between the results obtained from the *in vitro* measurement of the regional AChE concentration in two monkeys and the estimated DVR values for the same regions in these monkeys. The points with the highest values are data from the putamen and the caudate nucleus. The data show that the DVR parameter obtained with the simplified reference tissue model correlates well with the *in vitro* AChE activity. The curve in Fig. 3a is obtained by fitting expression (1) to the data. The values 0.12 and 0.047 min^{-1} were obtained for k_2 and k_L, respectively. In addition, a scaling factor k_{on}, relating k_3 to the AChE density ($k_3 = k_{on} \cdot$ AChE-density) was fitted, and for this parameter the value 0.00118 (L min^{-1} nmol^{-1}) was obtained.

Figure 3b shows a comparison between the DVR values for some regions obtained in the human study and the corresponding AChE density obtained *in vitro*. The latter data are the latest found in the literature (Enz *et al.*, 1993). The plot indicates that also for humans the correlation between DVR and the AChE density is nonlinear. The curve is obtained by fitting the expression 1 to the data. Due to the small number of observations only the scaling factor k_{on} was allowed to vary in this case, whereas k_2 and k_L were fixed to the

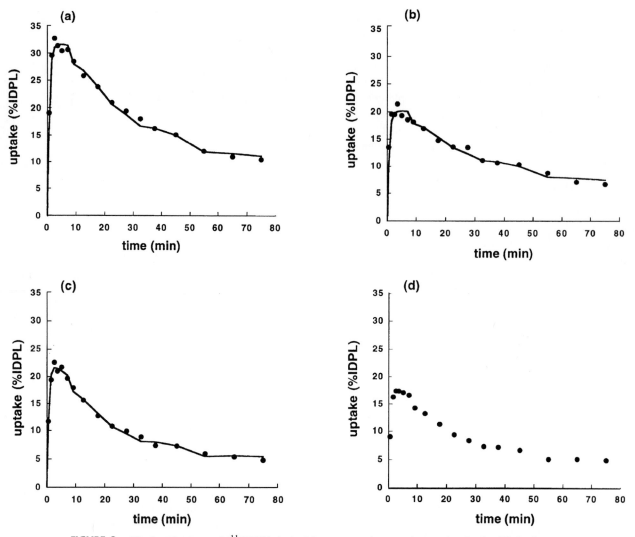

FIGURE 2. Fit of uptake curves of [^{11}C]PHY obtained from one monkey experiment using the simplified reference tissue model (full drawn lines). (a) Putamen, (b) pons, (c) temporal cortex, and (d) white matter. The last region is used as reference tissue and consequently no model fit it is provided for this region.

values obtained in the fit of the expression to the monkey data. The resulting value of k_{on}, 0.0013 L min^{-1} nmol^{-1}, is close to the value obtained with the monkey data. Table 1 gives regional results of the simplified reference tissue model applied to the human data. The values of DVR-1 (with no AChE actvity DVR = 1) were found to give a considerably enhanced contrast between regions, compared to the previouly used measure, based on uptake at late times (25–35 min after the injection; cf. Pappata *et al.*, 1996).

Koeppe *et al.* (1999) have used [^{11}C]PMP as tracer for AChE activity in the human brain. This tracer has lower affinity for AChE than AMP has (Iyo *et al.*, 1997), which means that regions with higher AChE density than in the cortex can be analyzed. The labeled metabolites are retained in the tissue. As index of regional AChE concentration the

rate constant for hydrolysis, k_3, was used. Very precise measurements of k_3 in regions with low AChE concentration were achieved, but the method performs progressively more poorly as the AChE concentration increases. The coefficient of variation COV for k_3 was, depending on the method, 12–13% in parietal cortex, 22–25% in hippocampus, 29–30% in thalamus, 16–21% in pons, 19–22% in cerebellum, 25–29% in putamen, and 40–42% in caudatus. Comparison with the corresponding values for DVR-1 (1 is the baseline value) obtained in this study (Table 1) shows that [^{11}C]PMP has much better precision in cortical areas where the AChE activity is low and that the two tracers have comparable precisions in regions with intermediate AChE activity, whereas [^{11}C]PHY has much better precision than [^{11}C]PMP in areas with high AChE activity.

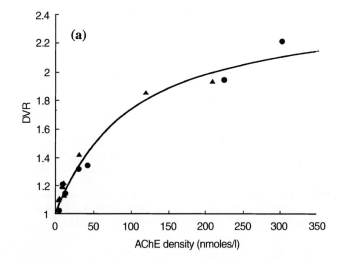

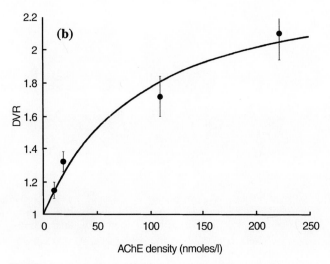

TABLE 1. Results Obtained by Applying the Simplified Reference Model on the Human Experiments ($n = 8$)

Region	Distribution volume ratio DVR			R-factor		k_2 (min^{-1})	
	Mean	SD	COV (%)	Mean	SD	Mean	SD
Corpus callosum	0.95	0.10	—	1.12	0.49	0.46	0.54
Thalamus	1.39	0.06	15	2.84	0.38	0.22	0.05
Caudate	1.95	0.11	12	2.44	0.40	0.64	0.27
Putamen	2.10	0.16	15	2.58	0.39	0.49	0.19
Ventral. striatum	1.86	0.16	19	2.33	0.35	0.42	0.15
Pons	1.54	0.09	17	2.39	0.41	0.38	0.11
Cerebellum	1.72	0.12	17	2.71	0.36	0.36	0.11
Amygdala	1.39	0.07	18	1.97	0.30	0.29	0.13
Hippocampus	1.32	0.06	19	2.06	0.29	0.23	0.08
Whole cortex	1.15	0.05	33	2.15	0.19	0.17	0.03
Calcarine	1.20	0.07	35	2.58	0.22	0.20	0.04
Prefrontal cortex	1.14	0.06	43	2.14	0.21	0.17	0.03
Parietal. assoc.cort.	1.15	0.05	33	2.21	0.22	0.17	0.03
Temporal assoc.cort.	1.19	0.06	32	2.17	0.23	0.16	0.04
Visual assoc. cortex	1.12	0.05	42	2.09	0.15	0.17	0.03

Note. The COV values are calculated from the values of DVR-1, the signal above the baseline value ($= 1$).

Acknowledgments

This study was supported by grants from Institut de la Santé et Recherche Medicale, INSERM (97/BPNP/CS/114), and from the Swedish Medical Research Counsil, MFR, Project K98-04F-12394-01.

FIGURE 3. (a) Estimated DVR versus the AChE concentration measured *in vitro* in some brain regions of two monkeys (● and ▲). The regions with low AchE density are white matter, pons, cerebellar cortex, hypothalamus, thalamus, and amygdala, and the two regions with the highest AChE densities are the caudate nucleus and the putamen. The curve is a fit of the expression (1) to the data with $k_3 =$ scaling factor · AChE density (see text). (b) Average DVR values obtained for the experiments on humans ($n = 8$) plotted versus the AChE concentration measured *in vitro* for some regions (average cortex, hippocampus, cerebellum, and striatum) of the human brain taken from Enz *et al.* (1993). In order of increasing AChE density the regions are cortex, hippocampus, cerebellum, and putamen. The curve is a fit of the expression to the data using the same values for k_2 and k_L as for the monkey data.

In summary, the proposed method using the simplified reference tissue method to analyze [^{11}C]PHY data is noninvasive and provides a robust index, DVR, of regional AChE concentration that correlates well with regional AChE density measured *in vitro*. The regions where [^{11}C]PHY and the alternative tracer [^{11}C]PMP, respectively, give the most precise estimates are complementary to each other.

References

Bonnot Lours, S., Crouzel, C., Prenant, C., and Hinnen, F. (1993). Carbon-11 labelling of an inhibitor of acetylcholinesterase: ^{11}C-physostigmine. *J. Label Compd. Radiopharm.* **33**: 277–284.

Davies, P. (1979). Neurotransmitter-related enzymes in senile dementia of the Alzheimer type. *Brain Res.* **171**: 319–327.

Ellman, G. L., Courtney, K. D., Andres, V., and Featherstone, R. M. (1961). A new and rapid colorimetric determination of acetylcholinesterase activity. *Biochem. Pharmacol.* **7**: 88–95.

Enz, A., Amstitz, R., Boddeke, H., Gemlin, G., and Malanowsky, J. (1993). Brain selective inhibition of acetylcholineeterase: a novel approach to therapy for Alzheimer's disease. *In* "Cholinergic Function and Dysfunction. Progress in Brain Research" (A. C. Cuello, Ed.), Vol. 98, pp. 431–438. Elsevier Science, Amsterdam.

Geula, C., and Mesulam, M. M. (1994). Cholinergic sytems and related neuropathological predilection patterns in Alzheimer disease. *In* "Alzheimer Disease" (R. D. Terry, R. Katzman, and K. L. Bick, Eds.), pp. 263–287. Raven Press, New York.

Iyo, M., Namba, H., Fukushi, K., Shinotoh, H., Nagatsuka, S., Suhara, T., Sudo, Y., Suzuki, K., and Irie, T. (1997). Measurement of acetylcholinesterase by positron emission tomography in the brains of healthy controls and with Alzheimers disease. *Lancet* **349**: 1805–1809.

Koeppe, R. A., Frey, K. A., Snyder, S. E., Meyer, P., Kilbourn, M. R., and Kuhl, D. E. (1999). Kinetic modelling of N-[^{11}C]methylpiperdin-4-yl proprionate: Alternatives for analysis of an irreversible positron emission tomography tracer for measurement of acetylcholinesterase activity in human brain. *J. Cereb. Blood Flow Metab.* **19**: 1150–1163.

Lammertsma, A. A., and Hume, S. P. (1996). Simplified reference tissue model for PET receptor studies. *Neuroimage* **4**: 153–158.

Mangin, J. F., Frouin, V., Bloch, I., Bendriem, B., and Lopez-Krahe, J. (1994). Fast nonsupervised 3D registration of PET and MR images of the brain. *J. Cereb. Blood Flow Metab.* **14**: 749–762.

Pappata, S., Tavitian, B., Traykov, L., Jobert, A., Dalger, A., Mangin, J. F., Crouzel, C., and DiGiamberardino, L. (1996). In vivo imaging of human cerebral acetylcholinesterase. *J. Neurochem.* **67**: 86–879.

Planas, A. M., Crouzel, C., Hinnen, F., Jobert, A., Né, F., Di Giamberardino, L., and Tavitian, B. (1994). Rat brain acetylcholinesterase vizualized with [^{11}C]physostigmine. *Neuroimage* **1**: 173–180.

Riche, D., Hantraye, P., and Guibert, B. (1988). *Brain Res. Bull* **20**: 283–300.

Ruberg, M., Rieger, F., Villageois, A., Bonnet, A. M., and Agid, T. Y. (1986). Acetylcholinesterase and butyrylcholinesterase in frontal cortex and cerebrospinal fluid of demented and nondemented patients with Parkinson's disease. *Brain Res.* **362**: 83–91.

Tavitian, B. (1985). "Transport axonal de l'acetylcholinesterase après inhibition par un organophosphate," Thesis 2818, René Descartes Académie de Paris.

Tavitian, B., Pappata, S., Planas, A., Jobert, A., Bonnot Lours, S., Crouzel, C., and DiGiamberardino, L. (1993). In vivo visualization of acetylcholinesterase with positron emission tomography. *Neuroreport* **4**: 535–538.

Vigny, M., Bon, S., Massoulié, J., and Leterrier, F. (1978). Active-site catalytic efficiency of acetylcholinesterase molecular forms in *Electrophorus, Torpedo,* rat and chicken. *Eur. J. Biochem* **85**: 317–323.

SECTION VII

MAPPING NEURONAL ACTIVATION

41

Noise Characteristic of 3D PET Scans from Single Subject Activation Studies: Effects of Focal Brain Pathology

ALEXANDER THIEL, KARL HERHOLZ, GUNTER PAWLIK, ALEXANDER SCHUSTER, UWE PIETRZYK,
KLAUS WIENHARD, and WOLF-DIETER HEISS

Max-Planck-Institut für neurologische Forschung und Neurologische Klinik der Universität zu Köln, Cologne, Germany

Estimating the variance of functional neuroimaging data from positron emission tomography (PET) activation studies obtained from a single subject is the first critical step in assessing significant signal changes between cerebral blood flow (CBF) images under a control and stimulation task. Variance estimates can either be derived locally, for each voxel across conditions with low degrees of freedom or globally across all voxels within the brain. The latter method yields a more stable estimator of variance but makes assumptions about the distribution of CBF changes and requires homoscedasticity of all voxels within the brain. We examined these assumptions for an [^{15}O]water-PET activation study on an ECAT EXACT HR scanner in 3D mode using a word-repetition task in 14 normal subjects and 60 patients with brain tumors. As an analytical derivation of noise distribution in PET images from activation studies in 3D are not feasable, empirical noise distributions were calculated by generating all of the 18 possible noise images (2 activation + 2 control − 2 remaining activation + 2 remaining control) from 8 scans. A variance image was computed from the 18 noise images and tested for homogeneity of variance at the level of resolution elements using the exceedence proportion test. The variance estimate from this image was used for generation of Z maps from the signal image (4 activation − 4 control). Signal intensities in all noise images were nearly normally distributed in all normal subjects and also in the vast majority of patients. In normal subjects no significant deviation from the null hypothesis of variance homogeneity was found at the 0.05 level. In only 8 of 60 patients with brain tumors could the hypothesis of variance homogeneity not be accepted. Significant local increases in vari-

ance were caused by strongly hyperperfused tumors or by patients who opened their eyes during the PET measurement leading to local variance increases within the primary visual cortex.

I. INTRODUCTION

For positron emission tomography (PET) activation studies in a clinical context, analysis of single subject data is of special interest. Due to the limited number of scans available, the power of conventional statistical approaches for the evaluation of PET activation studies is limited. The main problem in analyzing activation data from single subjects is the estimation of variance as a basis for subsequent significance testing. In general there are two possibilities for estimating variances from PET studies: local variance estimates on a pixel level across repetitive scans or global variance estimates across all pixels.

Conventional analysis (Friston *et al.*, 1995) of PET activation studies assumes a signal distribution under the Null hypothesis of no activation which is normally distributed with unit variance at every pixel and compares this assumed distribution to the distribution derived from the actual activation data on a pixel-by-pixel basis. A similar approach (Worsley *et al.*, 1992) assumes that the variance at all pixels within the brain is homogenous and a pooled variance across all pixels is used for calculation of Z scores. The first method has the advantage of taking possible variance inhomogeneities into account, whereras the latter yields a more stable variance estimator.

Other approaches to single-subject studies make use of an empirical noise distribution (Poline and Mazoyer, 1994; Chmielowska *et al.*, 1997), calculated from an equal number of activation and resting scans obtained on low-resolution 2D scanners. The variance of this empirical noise distribution is then used for significance assessment of the signal images. The advantage of this method lies in the higher sensitvity due to the lower variance of a noise distribution derived from all intracerebral voxels and the use of an empirical distribution derived from the actual data. Again, however, the assumption of homogeneity of noise variance is implicit.

Activation studies of single subjects are usually acquired on high-resolution PET scanners in 3D mode with septa retracted to increase scanner sensitivity and decrease the injected radioctivity dose to allow repeated measurments. An analytical derivation of noise propagation in 3D-reconstructed images, however, is not feasable (Defrise and Kinahan, 1998). Even if the radioactive decay can be modeled as a Poisson process, which tends toward a Gaussian process for large numbers of events, 3D data acquisition and reconstruction as well as image realignment and smoothing does not make the assumption of Gaussian noise in PET activation images evident. Another source of potential variance inhomogeneity may occur if PET activation studies are performed in patients with focal brain pathologies like brain tumors, where altered perfusion within the brain tumor can be a source of local variance change (Thiel *et al.*, 1998a). In these cases variance inhomogeneities may influence the sensitivity of the procedure to detect task-induced cerebral blood flow (CBF) changes close to brain tumors.

Due to the crucial nature of the assumptions implicit in the above statistical approaches for clinical application of functional brain imaging and the difficulties of modeling noise propagation in 3D reconstructed images analytically, we examined these issues applying a subsampling technique to create all possible different noise distributions which can be produced from a certain number of activation and resting scans to estimate noise variance empirically and investigate the spatial dependency of the pooled variance estimate. This noise variance was then used for calculating Z-statistic images.

II. PATIENTS AND METHODS

A. Normal Subjects and Patients

Normal subjects were 15 right-handed male volunteers (average age 49.4 ± 11.9 years) without previous history of neurological or psychiatric disorders. Informed consent to participate in the study was obtained from all subjects.

All 60 patients were right-handed also (41.7 ± 13.8 years), 24 female and 36 male. PET-activation studies were performed prior to surgical intervention. All brain tumors

were gliomas of different grading. Patients showed no aphasic symptoms which could prevent them from performing the word repetition task correctly.

B. Activation Paradigm

Patients and normal subjects performed a word repetition task (simple high frequency german nouns were read by the investigator and patients and subjects had to pronounce the words as fast as possible) and a baseline resting task (dark room, eyes closed, ears open with low ambient noise). Rest and activation condition were replicated four times each. A detailed description of the paradigm is given in Thiel *et al.* (1998b).

C. Data Acquisition and Processing

PET scans were performed on a CTI/Siemens ECAT EXACT HR scanner (Wienhard *et al.*, 1994) in 3D mode with septa retracted after a transmission scan of 15 min duration. Due to the linear relationship between cerebral blood flow and tissue activity of [^{15}O]water in units of nCi/ml, measures of relative cerebral blood flow were obtained. Data acquisition started with i.v. bolus injection of 370 MBq of [^{15}O]water and lasted for 90 s. Eight subsequent scans were performed on each patient with an interval of approximately 8 min between scans. Each scan was reconstructed to 47 slices (3.125 mm thickness and 2.16 mm pixel size within an 128×128 matrix) using filtered backprojection technique supplied by the manufacturer. A Hanning filter with cutoff frequency of 0.4 cycles/pixel was applied. Prior to reconstruction corrections for random coincidences, scatter (Watson *et al.*, 1997), and photon attenuation were applied.

Each subject had a T1-weighted MRI scan obtained on a Siemens Magnetom as 64 transaxial slices of 2.5 mm thickness acquired simultaneously with a 3D FLASH sequence. The anterior and posterior commissure (AC, PC) were identified on MRI, and scans were aligned to the AC–PC line according to the Talairach stereotactic atlas. The eight PET scans of each subject were automatically coregistered to the MRI in AC–PC postition. The multipurpose imaging tool (MPI Tool) was used for image realignment (Pietrzyk *et al.*, 1994).

For definition of brain tissue, an average image of all eight PET scans was calculated and a binary brain mask was created by applying a threshold of 40% maximum intensity to the average image, followed by slicewise processing to remove remaining extracerebral structures.

In order to account for differences in whole brain activity, all PET scans were normalized by dividing each intracerebral voxel by the average of all voxels within the brain mask multiplied by 100%, thus yielding normalized relative CBF values.

Reconstructed PET images were smoothed applying a spherical Gaussian filter with 12 mm FWHM. The filtered image data were corrected for edge artifacts according to a proposal of Maisog and Chmielowska (1998).

D. Generation of Signal and Noise Images

Images of task-associated CBF increase (further referred to as signal images) were generated by averaging the four normalized and smoothed scans under the activation condition and subtracting the average of the four normalized and smoothed resting scans.

For variance estimation, noise images were created by adding up 2 smoothed activation and 2 smoothed rest scans and subtracting the sum of the remaining 2 activation and 2 rest scans, thus yielding images which do not contain task-associated signal but have the same count statistic as the signal image. Having 4 activation images and 4 resting images, in total 18 unique noise images can be generated (exhaustive recombination yields 36 images with 18 images having identical absolute values, and only different sign):

$$\text{Noise image} = (2 \text{ rest} + 2 \text{ activation})$$
$$- (2 \text{ remaining rest}$$
$$+ 2 \text{ remaining activation}).$$

As each noise image is determined by the choice of the 2 activation and 2 rest scans for the augent (first summand)—the scans for the addend are then fixed—the maximum number of degrees of freedom for the noise images is 4 (or $2(n/2)$ with n the number of scans per condition).

E. Variance Estimation

For estiamtion of variance from noise images, an empirical intensity distribution of all voxels in all 18 noise images was obtained. This histogram was normalized so that the area under the curve became unity and was compared to a standard normal distribution. For quantification of deviation from standard normal distribution, r^2 was calculated based on the number of resolution elements (resels) (Table 2).

Next a variance image was calculated by computing the mean sum of squares (MSS) for each voxel across the 18 noise images. A global estimate of variance was obtained as average MSS across all voxels within the variance image and the signal image was Z-transformed using this global variance estimate.

In order to test the assumption of homogeneity of variance, Worsley's exceedence proportion test was applied to the variance images. This test included the transformation of the variance image into a χ^2-statistic image by multiplying the variance image with the degrees of freedom of the noise images ($df = 4$). The proportion of resels exceeding the χ^2-statistic at the 0.05 level was determined and the exceedence proprtion test (Worsley et al., 1995) was applied to

TABLE 1. Exceedence Proportion Test for Homogeneity of Variance in 15 Normal Subjects ($df = 4$, $P = 0.05$)

No.	Global variance	Resels above threshold	Resels total	Exceedence proportion	P for greater exceedence proportion	r-square for fit of normal distribution
1	111.4	50.7	677	0.0750	0.004[a]	0.94
2	45.1	34.9	599	0.0583	0.201	0.97
3	63.7	40.8	774	0.0527	0.379	0.97
4	27.4	27.9	491	0.0567	0.271	0.96
5	36.4	28.5	613	0.0465	0.638	0.98
6	84.4	33.4	772	0.0433	0.778	0.97
7	79.6	29.7	745	0.0399	0.874	0.99
8	47.8	25.1	515	0.0486	0.553	0.97
9	104.2	28.6	611	0.0452	0.689	0.96
10	76.4	35.2	767	0.0421	0.573	0.94
11	55.4	28.0	670	0.0417	0.810	0.91
12	41.5	27.8	592	0.0469	0.621	0.97
13	73.0	35.6	671	0.0531	0.372	0.96
14	102.7	34.4	613	0.0561	0.268	0.95
15	30.2	24.7	555	0.0445	0.705	0.97

[a] Significant deviation from homoscedasticity.

TABLE 2.   Exceedence Proportion Test for Homogeneity of Variance

No.	Global variance	Resels above threshold	Resels total	Exceedence proportion	P for greater exceedence proportion
1	93.04	40.0	582	0.0686	0.0318 Focal tumor
2	89.43	48.8	739	0.0660	0.0366 Focal tumor
3	154.64	48.4	722	0.0669	0.0304 Focal tumor
4	207.05	55.7	719	0.0775	0.0011 Focal tumor
5	86.04	73.1	994	0.0735	0.0011 Eyes open
6	79.57	45.0	571	0.0787	0.0023 Eyes open
7	54.12	46.3	574	0.0807	0.0012 Eyes open

Note. Data from 7 out of 60 patients with brain tumors and significant deviation from homoscedasticity of variance ($df = 4$, $P = 0.05$).

test the hypothesis that this proportion could have occurred by chance on a 0.05 level (Table 1).

III. RESULTS

A. Signal and Noise Distribution

The noise distribution for each normal subject, obtained from the resampling procedure, is plotted in Fig. 1, compared to the standard normal distribution. The plot illustrates the excellent reproducibility of noise estimation between subjects. Although there is a systematic skew in the empirical distribution, compared to standard normal distribution, this deviation is minimal as documemented by the r^2 values in Table 1. Especially the tails of the distribution follow the normal distribution closely, which is important for the assessment of quantiles for significance thresholds. A goodness-of-fit test based on the number of resolution elements did not show a significant deviation from standard normal distribution.

Although in patients with brain tumors, the noise distributions were not as closely normally distributed as for the normal controls (Fig. 2), noise distributions were symmetric, with a good fit of the distribution at the tails. Estimating variance at full-width half-maximum yields nearly identical values as for the normal distribution. Again a systematic skew in the empirical data is observed.

B. Homogeneity of Noise Images

The exceedence proportion test showed a significant deviation from the assumption of homogeneity for one subject. In all 14 other subjects this test was not significant, thus justifying the acceptance of the null hypothesis of variance homogeneity across voxels. Although this test does not prove

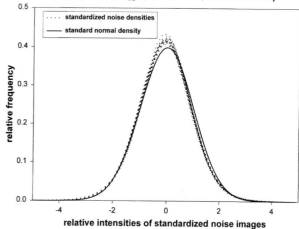

FIGURE 1. Intensity distributions of resampled noise images from 15 normal subjects.

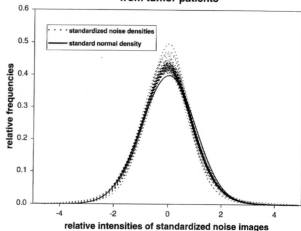

FIGURE 2. Intensity distributions of resampled noise images from 60 patients with brain tumors.

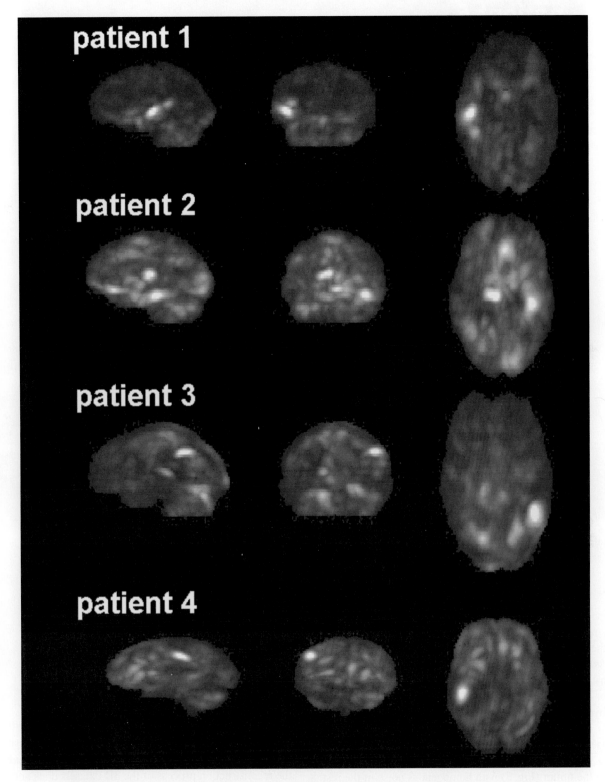

FIGURE 3.   Noise images from four tumor patients, maximum intensity projections (sagital, coronal, and transaxial views). The exceedence proportion test (Table 2) indicated significant deviation from homogenous noise distribution. Noise is increased focally within the hyperperfused brain tumors.

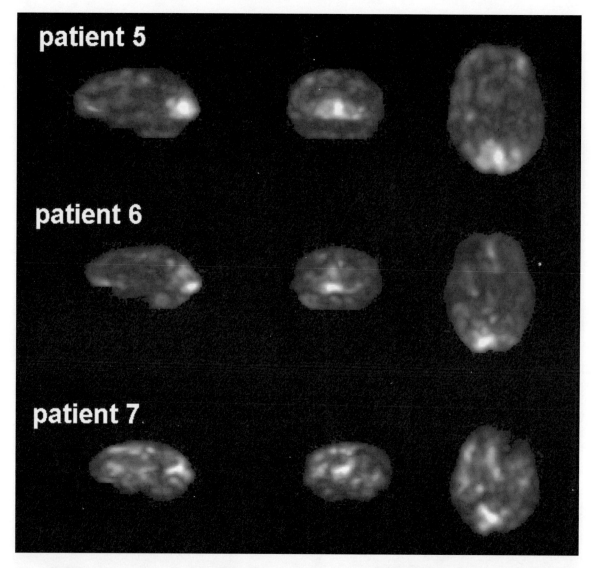

FIGURE 4. Noise images from three tumor patients, maximum intensity projections (sagital, coronal, and transaxial views). The exceedence proportion test (Table 2) indicated significant deviation from homogenous noise distribution. Noise is increased focally within the primary visual cortex because the subjects randomly opened their eyes during the stimulation procedure.

that the varince is indeed homogenously ditributed, it nevertheless shows that the assumption of homoscedasticity is not inconstistent with the data.

For only 7 subjects among the 60 tumor patients did the null hypothesis of homoscedasticity have to be rejected (Table 2) on the 5% level. Four of the 7 subjects had hyperperfused tumors (arteriovenous malformations or gliomas WHO grade IV) and showed a focal increase of variance within the tumor (Fig. 3) with homogenous noise in the rest of the brain. The other 3 patients had a focal increase of noise within the primary visual cortex (Fig. 4). These patients reportedly opened their eyes during some (but not all) activation scans.

IV. DISCUSSION

A. Gaussian Noise Distribution

The results of this study suggest that the assumption of homogenous Gaussian noise in 3D PET activations studies is justified. From these empirical noise data it is evident, that the reconstruction process of the 3D datasets does not influence the theoretically expected Gaussian noise distribution. Although an analytical explanation of the noise propagation in 3D PET images is difficult, it was shown that for the special case of a uniform sphere the propagation of Poisson noise is similar in 2D and 3D filtered backprojection (Defrise *et al.*, 1990).

A small but systematic deviation from normality, however, was found in all subjects. Empirical noise distributions were slightly skewed and leptokurtotic. This effect is probably caused by smoothing. The distribution of noise intensities derived from unfiltered images was even more perfectly Gaussian, but these distributions were also slightly skewed due to the reconstruction filter.

B. Homoscedasticity

The results of the exceedence proportion test are consistent with earlier data from low-resolution 2D PET scanners (Chmielowska *et al.*, 1997). In normal subjects the proportion of resels exceeding the proportion expected from stationary Gaussian noise is less than 5% of all resels. These resels are usually located at the top or bottom slices of the image, where a full 3D data acquisition is not possible and slices contain sections of the brainstem and vertex with a low number of voxels.

In patients with brain tumors, a significant departure from the assumption of homoscedasticity was found in 7 of 60 cases (12%). In three cases, the local increase in variance within the primary visual cortex was caused by randomly opening the eyes during the measurement. The remaining four cases (7%) exhibited a local variance increase within a hyperperfused tumor. These cases show that a careful inspection of variance images is indicated but the assumption of homogenous variances in the rest of the brain nevertheless is justified.

C. Global vs Local Variance Estimates

The decision for using a global or a local estimate of variance is a decision between sensitivity and specificity. In preopertaive brain mapping of tumor patients, the greater emphasis is probably put on sensitivity than on specificity because the avoidance of postoperative language deficits is of greater importance than the extent of resection. The results of this study suggest that the assumptions implicit in the use of globally pooled variance estimates are not inconsistent with the data from normal subjects and are also applicable in most cases of focal cerebral pathologies. The correctness of these assumptions, however, must be checked in each individual case. This can be done with the proposed methods.

V. CONCLUSION

This study demonstrates that the assumptions of normality and homoscedasticity of noise distribution in 3D-PET

activation studies are justified. The effects of the 3D reconstruction algorithm and scanner geometry are small and restricted to the top and bottom slices of the images. Thus the number of voxels ecxeeding that expected by chance in perfectly uniform Gaussian noise is less then 5%. In patients with brain tumors only 7% showed a significant departure from the homogeneity assumption caused by a local variance increase within a hyperperfused tumor.

References

Chmielowska, J., Maisog, J. M., and Hallet, M. (1997). Analysis of single subject data sets with low number of PET scans. *Hum. Brain Mapping* **5**: 445–453.

Defrise, M., and Kinahan, P. E. (1998). Data Acquisition and image reconstruction for 3D PET. *In* "*Theory and Practice of 3D PET*" (B. Bendriem, and D. W. Townsend, Eds.), pp. 11–53. Kluwer Academic, Dordrecht.

Defriese, M., Townsend, D. W., and Deconinck, F. (1990). Statistical noise in three-dimensional positron tomography. *Phys. Med. Biol.* **35**: 131–138.

Friston, K. J., Holmes, A. P., Worsley, K. J., Poline, J. B., Frith, C. D., and Frackowiack, R. S. J. (1995). Statistical parametric maps in functional imaging: A general approach. *Hum. Brain Mapping* **2**: 189–210.

Maisog, J. M., and Chmielowska, J. (1998). An efficient method for correcting the edge artifact due to smoothing. *Hum. Brain Mapping* **6**: 128–136.

Pietrzyk, U., Herholz, K., Fink, G., Jacobs, A., Mielke, R., Slansky, I., Wuerker, M., and Heiss, W. D. (1994). An interactive technique for three-dimensional image registration: Validation for PET, SPECTR, MRI and CT brain studies. *J. Nucl. Med.* **35**: 2011–2018.

Poline, J. B., and Mazoyer, B. M. (1994). Analysis of individual positron emission tomography activation maps by detection of high signal to noise ratio pixel clusters. *J. Cereb. Blood Flow Metab.* **13**: 425–437.

Thiel, A., Herholz, K., v. Stockhausen, H. M., Pawlik, G., and Heiss, W. D. (1998a). Suitability of O-15-water and PET to detect activation induced cerebral blood flow changes in brain tissue altered by brain tumors. *In* "*Quantitative Functional Brain Imaging with Positron Emission Tomography*" (R. E. Carson, M. E. Daube-Witherspoon, P. Herscovitch, Eds.), pp. 155–158. Academic Press, San Diego.

Thiel, A., Herholz, K., v. Stockhausen, H. M., van Leyen, K., Klug, N., and Heiss, W. D. (1998b). Language activation studies inpatients with gliomas using O-15-water and PET. *Neuroimage* **7**: 284–295.

Watson, C. C., Newport, D., Casey, M. E., deKemp, A., Beanlands, R. S., and Schmand, M. (1997). Evaluation of simulation-based scatter correction for 3D PET cardiac imaging. *IEEE Trans. Nucl. Sci.* **44**: 90–97.

Wiernhard, K., Eriksson, L., Grootoonk, S., Casey, S., Pietrzyk, U., and Heiss, W. D. (1994). The ECAT EXACT HR: Performance of a new high resolution positron scanner. *J. Comp. Assist. Tomogr.* **18**: 110–118.

Worsley, K. J., Evans, A. C., Marrett, S., and Neelin, P. (1992). A three dimensional statistical analysis for CBF activation studies in human brain. *J. Cereb. Blood Flow Metab.* **12**: 990–918.

Worsley, K. J., Poline, J. B., Vandal, A. C., and Friston, K. J. (1995). Tests for distributed, nonfocal brain activations. *Neuroimage* **2**: 183–194.

42

Reexamining the Correlation between $H_2^{15}O$ Delivery and Signal Size: Toward Shorter, More Efficient PET Sessions

J. J. MORENO-CANTÚ,[*,‡] **J.-S. LIOW,**[*,‡] **M. SAJJAD,**[‡] **D. A. ROTTENBERG,**[*,†,‡] **and S. C. STROTHER,**[*,†,‡]

[*]*Department of Radiology, University of Minnesota*
[†]*Department of Neurology, University of Minnesota*
[‡]*PET Imaging Service, Veterans Affairs Medical Center, Minneapolis, Minnesota*

We characterize the correlation between $H_2^{15}O$ PET activation signals and blood tracer concentration for three infusion techniques. We hypothesize that the length of PET sessions may be shortened if the duration of activation signals in single scans is extended by manipulating tracer infusion gradients. The temporal behavior of signals associated with a finger-movement task was studied in eight volunteers for the following infusion techniques: (1) standard fast bolus (FB), (2) continuous infusion (CI), and (3) multiple successive bolus (MB). In FB, tracer for 90-s scans is delivered as a single 6-mL bolus over 6 s; in CI, tracer is administered at a constant rate throughout 5-min scans; while in MB, tracer is delivered in three or five boluses, each bolus 1 min apart. FB, CI, and MB images were compared for activation-baseline signal/time and noise levels. Expected activation foci were detected by all techniques. The largest signal was obtained following the first injection of a bolus; however, the signal decreased rapidly as the tracer concentration decreased. CI generated the weakest signal but it was maintained for over 5 min. MB capitalized on the ability of a fast bolus to generate large signals while significantly increasing their temporal extent. Under MB, signals were the same as in FB after the first bolus, however, MB reduced significantly the signal drop. These data indicate that the signal resulting from differences in blood flow between two mental states can be extended for several minutes by manipulating the tracer infusion pattern.

I. INTRODUCTION

The signal-to-noise ratio (SNR) of $H_2^{15}O$ positron emission tomography (PET) activation images has been extensively studied as a function of tracer concentration, duration of task stimulation, and length of data acquisition (e.g., Silbersweig *et al.*, 1993; Volkow *et al.*, 1991; Cherry *et al.*, 1995; Moreno-Cantú *et al.*, 1998). When using bolus injections, the activation signal is enhanced if the stimulus-induced increases in regional cerebral blood flow are confined to the interval when tracer concentration in brain tissue is rising. Compartmental models for $H_2^{15}O$ suggest that the magnitude and duration of an activation signal are a function of the percentage change in blood activity concentration per unit time. These observations indicate that the duration of PET sessions may be shortened and the dose-efficiency improved by manipulation of tracer infusion gradients. The **objective of this study** was to characterize the correlation between three $H_2^{15}O$ intravenous infusion techniques and the magnitude and temporal extent of activation signals. We speculate that PET data acquisition sessions can be shortened if the drop in signal value commonly observed following a bolus injection is eliminated or reduced.

II. METHODS

The correlation between four blood tracer-concentration patterns and the magnitude/temporal-behavior of tracer-accumulation in brain tissue was investigated for $H_2^{15}O$ PET in eight healthy right-handed volunteers (five males/three females, ages 24–45 years). A simple motor task, extensively studied in our laboratories with $H_2^{15}O$ PET and fMRI (Strother *et al.*, 1995), was used to generate the PET activation signal. All the scans were performed using the CTI ECAT 953B PET tomograph (Mazoyer *et al.*, 1991) and were collected in accordance with the medicoethical guidelines for the study of human subjects in effect at our institution. Informed consent was obtained from each volunteer prior to their participation in this study.

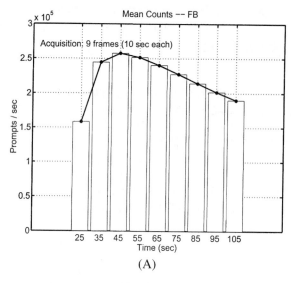

(A)

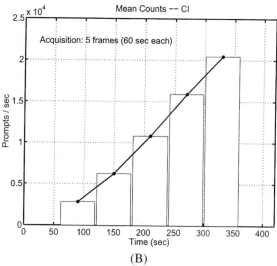

(B)

FIGURE 1 (A–D). Count levels for the infusion techniques tested. Graphs show values from sample sessions. Time $t = 0$ corresponds to the start of tracer infusion. The bars in each graph depict the time length of the acquisition frames (bar width) and the mean count rate estimated from the accumulated total counts detected during a given frame (bar height). Note that frame duration across infusion techniques is different. In protocols requiring the collection of data over several minutes, frame size had to be increased because of limitations in the acquisition hardware. The reader should consider the differences in frame size when comparing count rate values across techniques.

A. Tracer-Infusion Patterns

Figures 1A–1D illustrate the four tracer concentration patterns studied. These were: (1) a fast-bolus-injection pattern characterized by a sudden increase in brain counts lasting 20–30 s followed by a slow return to initial values over several minutes—this widely used pattern is generated by injecting intravenously a bolus of tracer (FB method); (2) an equilibrium pattern consisting of a slow increase in activity concentration that is maintained until a constant activity level is reached—this pattern is generated by continuously infusing tracer intravenously at a constant rate (CI method); (3) a multiple-bolus-injection pattern characterized by five sudden increases of activity concentration, each increase follows a bolus injection during a single scan—in this pattern, the relative changes in tracer concentration following injections 2 to 5 are smaller than the initial relative increase gen-

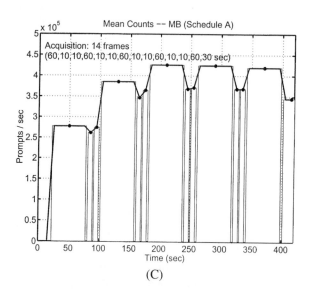

(C)

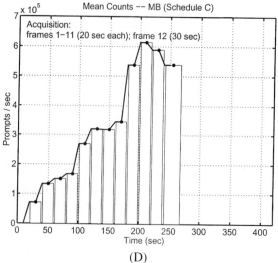

(D)

FIGURE 1.   (Continued).

erated after injection 1 (MB method, schedules A and B); and (4) a multiple-injection pattern similar to that seen in MB (schedules A and B); however, in this case, only three independent injections are used per scan—in this pattern the relative concentration of tracer is increased following every subsequent injection (MB method, schedule C).

B. Imaging Protocols

The schedules for tracer delivery, stimulus presentation and data acquisition studied in this chapter are shown in Diagram 1. The number of scans collected per protocol and other imaging parameters are described in Table 1. Images for each of the protocols were acquired as follows.

1. FB Protocol

A single bolus injection of 13 mCi is delivered in 6.2 mL of saline at a rate of 1 mL/s. In this protocol, the tracer was delivered using a computer-controlled syringe pump.

2. CI Protocol

Tracer was infused at a rate of 3 mL/min starting 1 min prior to stimulus presentation and was maintained throughout the scanning period (5 min). Two activity levels were tested (5 and 7.4 mCi/scan). The CI delivery system consists of a combination of two peristaltic pumps and a mobile furnace (Sajjad *et al.*, 2000). One pump withdraws the tracer from a reservoir and injects it into the subject while the other replenishes the reservoir with nonradioactive saline. Both pumps transfer the fluids at the same rate in order to maintain a constant reservoir volume. The saline in the reservoir is continuously mixed with the target gas transported from the cyclotron. The furnace is used to prevent condensation of the target gas during its transport from the cyclotron to the bubbling reservoir.

3. MB Protocol

The computer-controlled syringe pump from method FB was used to withdraw 27.7 or 18.7 mL of tracer (Diagram 1[III], schedules A, B and C, respectively). However, instead of injecting the entire withdrawn volume at once as in method FB, the tracer was injected as a series of five

(schedules A and B) or three (schedule C) independent boluses. Each bolus was 80 s apart and was delivered at a speed of 1 cc/s.

C. Tasks

Tracer infusion patterns were investigated by comparing the radioactivity concentration levels obtained for two mental states: (A) rest (baseline task) and (B) left-hand finger-to-thumb opposition (activation task). For the baseline task, the subjects lay quietly with their eyes covered and ears plugged. For the activation task, the subjects performed a left-hand movement consisting of touching the thumb with each digit successively at a rate of 1 Hz (Strother *et al.*, 1995). This movement was prompted by an auditory cue presented binaurally. As in the baseline condition, the subjects had their eyes covered.

D. Image Analysis

Dynamic 3D PET scans were reconstructed, aligned, transformed to Talaraich space and normalized by frame means. Images were grouped into sets according to the infusion scheme used during their acquisition. Images in each set were subtracted and averaged. The expected activation foci were then located in the averaged image from each set. The top 20 voxels from the foci under investigation were located and identified (signal). The average signal corresponding to each infusion scheme was characterized based on magnitude and temporal extent. Infusion schemes were also characterized based on image noise. Noise levels were estimated by measuring the variability across intra-cerebral voxels in five arbitrarily selected brain slices across subjects for each infusion scheme.

E. Expected Activation Foci

Based on evidence from previous PET and fMRI experiments, strong activation foci were expected in the left cerebellum and the right primary motor cortex. A weaker activation focus in supplementary motor cortex was also expected. The cerebellum signal was not used in this analysis. This brain region was not imaged because of limitations in the axial field of view (FOV) of our scanner.

III. RESULTS AND DISCUSSION

A. Whole Brain Tracer Levels

Figures 1A–1D depict global brain tracer concentration levels for each of the infusion techniques tested. Each data point shown represents the total prompt counts collected in a frame divided by the respective frame length in seconds.

TABLE 1.   Samples Analyzed

	FB	CI		MB		
Sessions	4	2	1	5		3
Scans/session	10	2	4	2	2	4
mCi/scan	13	5	7	26	38.5	31.2
Scan length (s)	90	300		410	410	250
Boluses/scan	1	Continuous		5	5	3

TABLE 2.   Activity Levels Used to Test MB Protocols

Schedule	Injection 1			Injection 2			Injection 3			Injection 4			Injection 5		
	t_i	t_j	t_t	t_i	t_j	t_t	t_i	t_j	t_t	t_i	t_j	t_t	t_i	t_j	t_t
A	0	12	12	8	8	16	10	6.5	17	11	6	17	11	6	17
B	0	8	8	5	5	11	6.8	4.4	11	7.1	4	11	7.2	4	11
C	0	6	6	4	9	13	8.1	17	25	—	—	—	—	—	—

Note. t_i, background activity in subject just prior to injection; t_j, injected activity; t_t, total activity in subject immediately after injection. All quantities shown are in mCi.

The figures indicate that a fast bolus infusion generates the sharpest tracer concentration gradient. The weakest gradient is yielded by the CI technique. Figures 1C and 1D illustrate the effect on whole brain counts of infusing multiple boluses during a single scan. The sudden count increase observed at 20, 100, 180, 260, and 340 s in Fig. 1C and at 20, 100, and 180 s in Fig. 1D correspond to the arrival of each single bolus in the brain. Table 2 describes the expected activity levels immediately before and after each bolus during MB scans. As shown in the table, all of the injections following the initial bolus for protocol MB schedules A and B were only fractions of the activity delivered by the first injection. Figure 1C shows the increases in global brain activity levels introduced by injections 2 to 5. In schedule C (Fig. 1D), the boluses that followed each initial injection infused aliquots of tracer substantially higher than the ones infused by the first injections. In that schedule, injections 2 and 3 were 167 and 300% larger, respectively, than the activity infused by the first bolus.

B. Signal Magnitude

As indicated before, three regions of increased blood flow are associated with the finger-opposition task used here. In this study, only two of the regions were monitored because of the restricted axial FOV of the scanner employed to collect the images. Figures 2A to 2E depict the behavior of the averaged subtracted activation signal for each of the protocols tested. Inspection of the figure suggests that the largest signal difference between activation and baseline states is obtained following the first injection of a bolus. However, the signal magnitude decreases very rapidly as the tracer concentration in blood decreases. This is in agreement with results reported by other groups (Silbersweig *et al.*, 1993; Volkow *et al.*, 1991; Cherry *et al.*, 1995; Moreno-Cantú *et al.*, 1998). The opposite behavior is obtained when using the CI protocol. The signal magnitude yielded by the CI protocol is only about 60% of the peak signal generated by the FB protocol. However, when the FB signal is averaged across the 90-s acquisition period, CI and FB signal magni-

tudes become similar. In the CI experiments, the signal was maintained for over 5 min.

MB protocols capitalize on the ability of fast boluses to generate large signals while significantly increasing the signal temporal extent. Note that the first group of measurements collected when using the MB protocol are, in fact, FB measurements. Thus, the data points shown in the 20- to 80-s time frame in Figs. 2C–2E are fast bolus measurements for those sessions. Figures 2C and 2D depict the subtracted signal magnitude patterns for schedules A and B, respectively. As anticipated, the signal magnitude did not change when using two protocols with equal relative activity-increases per injection even though the total activity levels were different (i.e., schedule A started with 300 MBq (8.1 mCi) injections, while schedule B started with 444 MBq (12 mCi) injections; however, the relative increase in activity injected was the same in both schedules) (Moreno-Cantú *et al.*, 1998b). Figure 2E depicts the signal behavior obtained when using schedule C. This figure shows that, although the subtracted signal was not kept at the same level as the signal yielded by the protocol's initial injection (perhaps because of a habituation effect (Raichle *et al.*, 1994; Chmielowska *et al.*, 1999)), the decrease appears to be substantially smaller. This effect is most likely explained by the increases in injected activity of 167 and 300% introduced in schedule C. It is important to point out that the apparent small rebound of the averaged subtracted signal in two of the subjects between 100 and 140 s (see Fig. 2E) is most likely caused by the selection of the sampling rate during acquisition. For this schedule, frames could not be made shorter than 20 s because of limitations on the scanner's data acquisition hardware. Given that once the first frame is started in a dynamic acquisition all of the subsequent frames are fixed in time, regardless of their temporal position with respect to the bolus injections, increases in signal magnitude following a bolus could be spread over two frames instead of one. Thus, the second peak value in two of the plotted curves, most likely, fell between the two 20-s frames collected by the dynamic scan.

As shown in Diagram 1, the volume of the initial bolus injection in protocols FB and MB (schedules A–C) was 6.2, 3.36, 3.36, and 1.7 mL, respectively. The results shown in

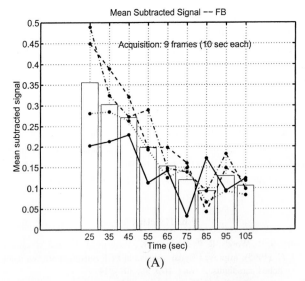

(A)

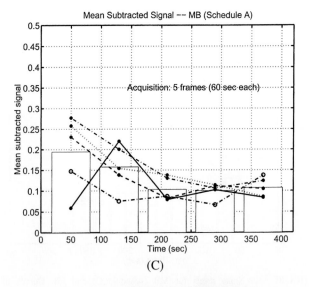

(C)

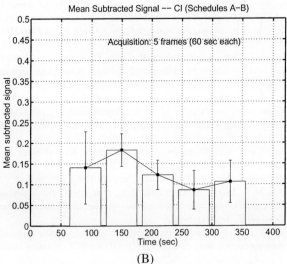

(B)

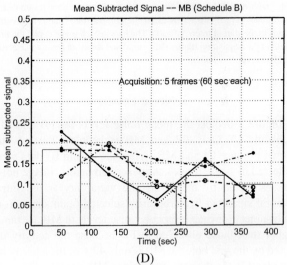

(D)

FIGURE 2 (A–E). Magnitude of the subtracted signal obtained for each of the infusion techniques tested. curves in each graph depict data from single subjects, except in B, where only one curve is shown. This curve describes mean values across subjects and schedules for the Cl technique. Time $t = 0$ corresponds to the start of tracer infusion. The bars in each graph depict the time length of the acquisition frames (bar width) and the mean signal value across subjects (bar height). The reader should consider the differences in frame size when comparing signal values across techniques.

Figs. 2A and 2C–2E suggest that the differences in bolus volume had little or no effect on the initial subtracted signal even though the injected volume varied over 300% between FB and MB (schedule C). As stated before, in the MB protocols, the group of measurements following the first bolus injection are comparable to FB measurements. However, for this MB protocol, the initial bolus injection consisted of 1.7 mL of tracer compared to the 6.2 mL used in the FB protocol.

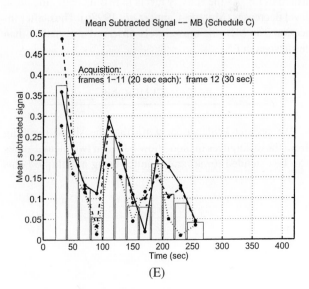

(E)

FIGURE 2. (Continued).

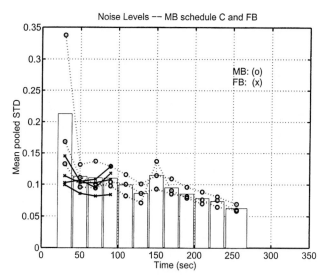

FIGURE 3. Noise levels for MB (schedule C) and FB. Curves in each graph depict data from single subjects. FB data were resampled to 20 s/frames to facilitate across technique comparison. Time $t = 0$ corresponds to the start of tracer infusion. The bars in each graph depict the time length of the acquisition frames (bar width) and the mean noise value across the MB curves (bar height).

C. Noise Levels

Figure 3 depicts the mean noise levels for FB and MB (schedule C). Only FB and MB (schedule C) values are shown because CI and MB (schedules A and B) data were collected in 1-min frames. As described before, image noise was estimated by measuring the variability across intracerebral voxels in five arbitrarily selected brain slices from each subject for each protocol. Noise levels in MB are much higher than those observed in FB because the activity injected by the first bolus in MB is only 42% of the activity injected in FB. As the activity levels in MB increase, the noise level decreases (Moreno-Cantú *et al.*, 1998b). Thus, after injection 3 in MB, the noise levels observed are smaller than those seen in FB.

IV. CONCLUSIONS

We present evidence showing that the signal resulting from the differences in blood flow between two mental states can be extended for several minutes using a multiple bolus infusion schedule. We speculate that this protocol, if implemented to maximize the particular count-rate of the scanner used, can be employed to decrease the length of $H_2{}^{15}O$ activation studies.

Acknowledgments

We are grateful to the technical staff from the PET Imaging Service of the Minneapolis VA Medical Center for their assistance during the acquisition and analysis of the PET data sets. This work was supported by the National Institutes of Health, Grants MH57180 and NS33721.

References

Cherry, S. R., Woods, R. P., Doshi, N. K., Banerjee, P. K., and Mazziotta, J. C. (1995). Improved signal-to-noise in PET activation studies using switched paradigms. *J. Nucl. Med.* **36:** 307–314.

Chmielowska, J., Coghill, R. C., Carson, R. E., *et al.* (1999). Comparison of PET [^{15}O]water studies with 6-minute and 10-minute interscan intervals: Single-subject and group analyses. *J. Cereb. Blood Flow Metab.* **19:** 570–582.

Mazoyer, B., Trebossen, R., Deutch, R., Casey, M., and Blohm, K. (1991). Physical characteristics of the ECAT 953B/31: A new high resolution brain positron tomograph. *IEEE Trans. Med. Imaging* **10:** 499–504.

Moreno-Cantú, J. J., Reutens, D. C., Thompson, C. J., Zatorre, R. J., Klein, D., Meyer, E., and Petrides, M. (1998a). Signal-enhancing switched protocols to study higher-order cognitive tasks with PET. *J. Nucl. Med.* **39:** 350–356.

Moreno-Cantú, J. J., Thompson, C. J., and Zatorre, R. J. (1998b). Evaluation of the ECAT Exact HR+ 3D PET scanner in ^{15}O-water brain activation studies. *IEEE Trans. Med. Imaging* **17**(6): 979–985.

Raichle, M. E., Fiez, V. A., and Videen, T. O., *et al.* (1994). Practice-related changes in human brain functional anatomy during nonmotor learning. *Cerebral Cortex* **4**(1): 8–26.

Sajjad, M., Liow, J.-S., and Moreno-Cantú, J. J. (2000). A system for continuous production and infusion of [^{15}O]H$_2$O for PET activation studies. *Appl. Radiat. Isot.* **52:** 205–210.

Silbersweig, D. A., Stern, E., Frith, C. D., *et al.* (1993). Detection of thirty-second cognitive activations in single subjects with positron emission tomography: A new low-dose H$_2{}^{15}$O regional cerebral blood flow 3D imaging technique. *J. Cereb. Blood Flow Metab.* **13:** 617–629.

Strother, S. C., Anderson, J. R., Schaper, K. A., *et al.* (1995). Principal component analysis and the scaled subprofile model compared to intersubject averaging and statistical parametric mapping. I. Functional connectivity of the human motor system studied with [^{15}O]water PET. *J. Cereb. Blood Flow Metab.* **15:** 738–753.

Volkow, N. D., Mullani, N., Gould, L. K., Adler, S. S., and Gatley, S. J. (1991). Sensitivity of measurements of regional brain activation with oxygen-15-water and PET to time of stimulation and period of image reconstruction. *J. Nucl. Med.* **32:** 58–61.

43

Precision of rCBF Measurement with [15O]Water—Whole Body Kinetics and Linear vs Nonlinear Estimation

R. P. MAGUIRE, K. L. LEENDERS, and N. M. SPYROU

Groningen NeuroImaging Program, University and University Hospital Groningen, The Netherlands
Department of Physics, University of Surrey, Guildford, United Kingdom

We have routinely used on-line arterial blood sampling during positron emission tomography studies with [15O]water. The measurements were carried out with a temporal resolution of 1 s and duration of 3.5 min. This has allowed us to examine the whole-body kinetics of [15O]water in man and to address the issue of an appropriate model. Seventy-one data sets were acquired by repeated measurement on 12 healthy human volunteers. Three models were tested; the simplest (SM) had a single tissue compartment and no further clearance from plasma. The next in complexity (CM) included a clearance from plasma to a second compartment, and the most complex (CM+) had two tissue compartments. Delay and dispersion were included in the models. In model building CM was found to be the most appropriate model (F tests: 89% favored vs SM; 83% vs CM+). Exchange with the tissue compartment yielded rate constants 3.0 and 0.73 min⁻¹ for influx and efflux, respectively. The clearance rate through other pathways was 0.82 min⁻¹. Linear and nonlinear parameter estimation methods were tested for the one tissue compartment model, using a noise model based on system parameters. Linear estimation of flow and partition coefficient was found to yield good estimates (although the flow estimate was biased by +5%) in sampling protocol and sensitivity variation simulations.

I. INTRODUCTION

Measurements using [15O]water to determine regional cerebral blood flow (rCBF) are routine at many PET cen-tres, and a number of models for the whole body distribution of water have been proposed (Smith *et al.*, 1994; Powers *et al.*, 1988; Bigler *et al.*, 1981) based on data from those and other measurements. As part of the routine blood sampling protocol for [15O]water studies, an on-line [15O]water detector (COIN-422, Magen Scientific, New York) was used to measure blood samples. This allowed the measurement of the kinetics of water in the human body with high temporal resolution (1 s). It was decided therefore to reexamine the minimum model necessary to explain these data, with the aim of being able to simulate typical [15O]water injection protocols, as input to a tissue model of uptake.

The Kety–Schmidt one tissue-compartment model (Kety and Schmidt, 1948) is universally used for the analysis of the tissue data from positron emission tomography studies. The nonlinear operational equation derived from this model can be used with nonlinear parameter estimation techniques to estimate flow and partition coefficient from PET data. However, a more appropriate model for application on a pixel-by-pixel basis has been described in the literature (Van den Hoff *et al.*, 1993). This method is attractive because of its computational efficiency, and since it is linear it could possibly be applied to projection space parametric image estimation (Maguire *et al.*, 1995).

Using the best whole body [15O]water model, derived from analysis of 71 data sets, time–activity courses of brain regions of interest were simulated, and analysed using linear and nonlinear methodology to validate its accuracy and precision.

Physiological Imaging
of the Brain with PET

II. METHODS

Healthy volunteers ($n = 12$) were injected with approx. 1 GBq of [^{15}O]water in a protocol designed to measure task-specific changes of rCBF. During the following 3.5 min, continuous sampling of the radial artery was achieved using a COIN-422 coincidence detector, based on two BGO blocks (efficiency, 21%; sensitive volume, 60 μL), with a time resolution of 1 s. The volunteers were measured repeatedly yielding 71 data sets.

The models in Fig. 1. were implemented in MATLAB 5.3 (Mathworks, Natick, USA) and used with the Simplex algorithm (Nelder and Mead, 1965) to determine optimum parameters for each data set. Early portions of the data curves before the initial bolus reached the detector were not included, although the timing was still related to time point 0 on the original time–activity course. The input bolus was modeled as an injection of a constant activity over 5 s. The most appropriate model was selected using an F test (Landaw and DiStefano, 1984) of the residual sum of squares from each fit individually. Determination was repeated on a reduced, 2-min, subset of the data.

The most appropriate model (CM) was then used with the one tissue compartment model to simulate measured plasma and tissue data, typical for region of interest analysis of data from an [^{15}O]water PET scan. Noise was added to the tissue and blood monitoring activity time courses using the following model:

$$C^*(t) = \frac{\mathbf{K}}{\varepsilon \Delta t}, \quad \langle \mathbf{K} \rangle = C(t)\varepsilon \Delta t, \quad (1)$$

where $\mathbf{K}$ is a Poisson random variable, $C(t)$ is the simulated data, ε ((Hz·ml)/Bq) is a calibration factor relating activity concentration to measured counts, and Δt is the duration of the measurement. For the blood measurements ε was determined directly, and for the tissue measurements it was modeled as:

$$\varepsilon = \left(\frac{S}{N}\right)\left(\frac{r^2}{R^2}\right) \left(\frac{1}{3.7 \times 10^4}\right), \quad (2)$$

where S is the sensitivity of the scanner (Hz/(μCi·mL)). (This is effectively also a calibration factor, measured using the specific geometry and attenuation of a 20 cm phantom.) N is the number of reconstructed planes, r and R are the region of interest radius and the phantom radius. Thus the noise characteristics of the PET time–activity curve could be varied depending on system parameters. Note that attenuation was not included in the analysis, thus the noise level is an intermediate level between the maximum at the center of an object and the minimum at the edge.

Two investigations were performed to determine the performance of the determination method under different conditions. First, the 50 simulations were performed at each of four sensitivities between 1×10^4 and 1×10^7 (Hz/(μCi·mL)). Second, the timing of the blood sampling was varied, and 50 simulations were performed at [1, 2.5, 5, 7, 10, and 12 s] measurement intervals. Parameter fitting was achieved using a nonlinear parameter estimation algorithm (Ralston and Jennrich, 1978).

III. RESULTS

Table 1 show the result of fitting the full 3.5 min data sets. The most appropriate model was determined to be CM, using the F test (89% of the studies showed CM to be superior to SM; in the analog comparison with CM+, 83% showed CM to be sufficient). Based on an initial distribution volume of 3500 ml and using an injection time of 5 s, the median activity injected was estimated to be 1.2×10^9 Bq, which is in good agreement with the target applied activity. The value for k_3 is much higher than that expected from urine production (0.005 min^{-1}). Using k_1 and k_2, the volume of

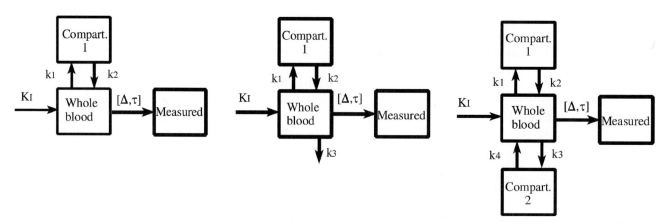

FIGURE 1. Model configurations considered in the whole body distribution determination.

distribution of the tissue compartment was estimated to be 14.3 L. The estimated volume of the compartments of the CM+ model was 3.2 and 11.3 L for the fast and slowly exchanging compartments, respectively.

Figure 2 compares fitted curves with a typical data set, all models fit the data well, but SM shows a marked deviation for longer times and does not fit the peak of the bolus well in the same way as CM and CM+.

A. Tissue Simulations

Table 3 presents the results for fitting using the linear method and the standard deviation of the estimates. The standard deviation was increased only at a very low sensitivity, which is lower than a coimmercial scanner operating in 2D acquisition mode. At higher sensitivities, a residual error of 1% was observed for both the linear and nonlinear methods,

TABLE 1. Medians of the Parameters Determined Using Each of the Models, Using the Whole (3 min) Time Set of Data

Model	$K_I \times 10^6$ (Bq/(mL·min))	k_1 (min^{-1})	k_2 (min^{-1})	k_3 (min^{-1})	k_4 (min^{-1})	Delay (Δt) (min)	Disp. (τ) (min)
CM+	4.02	2.59	0.799	1.04	0.0917	0.564	0.118
CM	**4.29**	**3.00**	**0.734**	**0.825**	—	**0.563**	**0.125**
SM	3.38	2.75	0.411	—	—	0.575	0.0836

TABLE 2. Medians of the Parameters Determined Using Each of the Models, for the Restricted (2-min) Time Set

Model	$K_I \times 10^6$ (Bq/(mL·min))	k_1 (min^{-1})	k_2 (min^{-1})	k_3 (min^{-1})	k_4 (min^{-1})	Delay (Δt) (min)	Disp. (τ) (min)
CM+	4.10	2.61	0.808	1.18	0.0925	0.564	0.119
CM	**4.45**	**3.00**	**0.836**	**0.934**	—	**0.565**	**0.125**
SM	3.46	2.92	0.462	—	—	0.575	0.0921

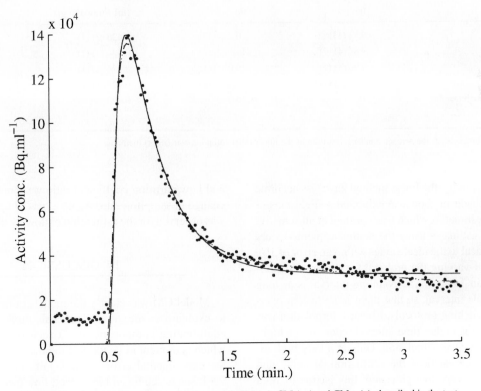

FIGURE 2. A comparison of fits using the models SM (−), CM (.−) and CM+ (..), described in the text.

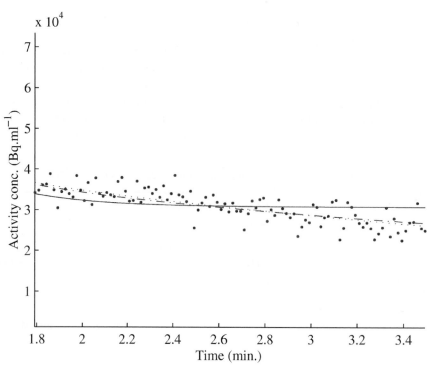

FIGURE 3. An expanded time axis view comparing the data in Fig. 1, over the time range 1.8 to 3.4 min.

TABLE 3. Table of Estimated Mean Parameter Values for Flow and Flow/Partition Coefficient Ratio for the Kety–Schmidt Compartment Model

Sensitivity log$_{10}$ (Hz/(μCi·mL))	Delay (Δt) (s)	Disp. (τ) (s)	Flow (mL/(min·100 g))	Flow/ (partition coeff.)
4	9.72 (13) (2)	10.2 (7) (3)	0.426 (2) (1)	0.506 (4) (1)
5 (2D ECAT 933)	9.84 (4) (2)	10.0 (2) (3)	0.424 (1) (1)	0.502 (1) (1)
6	9.90 (3) (1)	10.0 (1) (1)	0.424 (1) (1)	0.501 (1) (1)
7 (3D ECAT HR+)	10.0 (2) (1)	10.0 (0) (1)	0.423 (1) (1)	0.501 (1) (1)
Ideal values (and nonlinear fit results ±1%)	10	10.0	0.400	0.500

Note. Values in brackets are the percent standard deviation of the linear (left) and nonlinear (right) methods.

at the lowest sensitivity, the linear method gave less accurate estimates. The linear method also exhibited a slight bias of 5% on the flow estimates, which was constant at all sensitivities. The mean estimates using the nonlinear method were essentially identical to the ideal values within an error of 1%.

The results of varying the protocol for the arterial sampling are given in Table 4. The counting interval is as long as the intersample interval, so that there is no loss in overall counts by switching protocols. Linear interpolation was used to generate the higher time interval scans from a high-resolution plasma curve simulation. Only at a very high intersample time interval does the estimation of delay and dispersion show a very low precision, this is true of both the linear and the nonlinear methods. Estimations of flow

and flow/partition coefficient ratio were not affected by the change in sampling rate. Again a bias of approximately 5% was observed for the estimation of flow by the linear method.

IV. DISCUSSION

Model CM was clearly the model configuration required to explain our measured data. This model contains a fast equilibrating component, and a second lumped clearance, which is thought to be too fast to be urine excretion. Within the experimental errors, it is significantly lower than the 9.8 L suggested by all high-blood-flow tissues (flow volume >0.1 ml·min) (Smith *et al.*, 1994).

TABLE 4. Results of Varying the Time Between Blood Measurements

Time between blood measurements (s)	Delay (Δt) (s)	Disp. (τ) (s)	Flow (mL/(min·100 g))	Flow/ (partition coeff.)
1	9.93 (4) (1)	10.0 (2) (2)	0.424 (1) (2)	0.502 (1) (1)
2.5	9.96 (2) (1)	10.0 (0) (2)	0.424 (1) (0)	0.501 (1) (1)
5	9.98 (3) (1)	9.97 (1) (1)	0.424 (1) (1)	0.501 (1) (0)
7	11.3 (1) (3)	9.00 (0) (2)	0.419 (1) (0)	0.493 (1) (1)
10	11.9 (6) (1)	8.66 (5) (2)	0.421 (1) (0)	0.495 (2) (1)
12	13.2 (5) (8)	7.58 (6) (8)	0.420 (0) (0)	0.494 (2) (1)

Note. Values in brackets are the percent standard deviation of the linear (left) and nonlinear (right) methods.

If the results of fitting CM+ are considered, then they suggest two compartments, one with a volume of 3 L and the other 11 L. Clearly this is because CM+ is inappropriate for the explanation of a 3.5-min data set; thus the single volume of distribution of the correct model (CM) is distributed between the two apparent compartments of CM+. With longer acquisition times, a second compartment may be required to explain the data.

The noise model proposed in this paper has the advantage that it relates the time–activity data to measured system parameters, and can therefore be used to predict accuracy of parameter estimates in different instrumentational settings. The efficiency factor depends on the number of planes which are measured; this is important for more modern machines, which will exhibit lower statistical accuracy per plane as a result of the large number of thin tomographic planes which are reconstructed. In this study a constant 8-mm slice thickness was chosen for all the simulations.

Linear parameter estimation provided good estimates of the flow and flow/(partition coefficient) ratio in all the sensitivity situations. Although there is a small bias on the flow estimation of 5%, the precision of the estimate is comparable with the nonlinear method. However the computation speed of the linear method is much higher, and it can therefore be used for the estimation of parametric images on a pixel-by-pixel basis.

Simulations using different protocols for blood sampling were also favorable for the linear method. Again the same bias was observed on the flow parameter. Estimations of delay and dispersion were very poor for both the linear and the nonlinear methods, and there was an observed increase in the delay parameter and a matched decrease in the dispersion parameter as the intermeasurement interval increased. Since these two parameters are highly correlated, this effect did not affect the accuracy of measurement of flow or parti-

tion coefficient. It should be noted that the timing was considered exact, that is to say there was variation in the timing interval. An effect like this might be introduced if the blood sampling at greater intermeasurement intervals were carried out manually, and this should be investigated.

The results encourage the application of the linear model, which might also be applicable to projection data.

References

Bigler, E. B., Kostick, J. A., and Gillespie, J. R. (1981). Compartmenal analysis of the steady-state distribution of [O-15]O2 and [O-15]Water in total body. *J. Nucl. Med.* **22**: 959–965.

Kety, S. S., and Schmidt, C. F. (1948). The nitrous oxide method for the quantitative determination of cerebral blood flow in man: Theory, procedure and normal values. *J. Clin. Invest.* **27**: 476–483.

Landaw, E. E., and DiStefano, J. J., III (1984). Multiexponential, multicompartmental, and noncompartemental modeling. II. Data analysis and statistical considerations. *Am. J. Physiol.* **246**: R666.

Maguire, R. P., Calonder, C., and Leenders, K. L. (1995). Patlak analysis applied to sinogram data. *Quant. Brain Function Using PET.* (R. Myers, V. Cunningham, D. Bailey, and T. Jones, Eds.), pp. 307–311. Academic Press, London.

Nelder, J. A., and Mead, R. (1965). A simplex method for function minimization. *Comput J.* **7**: 308–313.

Powers, W. J., Stabin, M., Howse, D., Eichling, J. O., and Herscovitch, P. (1988). Radiation absorbed dose estimates for oxygen-15 radiopharmaceuticals (H2(15)O, C15O, O15O) in newborn infants. *J. Nucl. Med.* **29**: 1961–1970.

Ralston, M. L., and Jennrich, R. I. (1978). Dud, a derivative-free algorithm for nonlinear least squares. *Technometrics* **20**: 7–14.

Smith, T., Tong, C., Laimmertsma, A. A., Butler, K. R., Schnorr, L., Watson, J. D., Ramsay, S., Clark, J. C., and Jones, T. (1994). Dosimetry of intravenously administered oxygen-15 labelled water in man: A model based on experimental human data from 21 subjects. *Eur. J. Nucl. Med.* **21**: 1126–1134.

Van den Hoff, J., Burchert, W., Müller-Schauenburg, W., Meyer, G. J., and Hundeshagen, H. (1993). Accurate local blood flow measurements with dynamic PET: Fast determination of input function delay and dispersion by multilinear minimization. *J. Nucl. Med.* **34**: 1770–1777.

Visualization of Correlated Hemodynamic and Metabolic Functions in Cerebrovascular Disease by a Cluster Analysis with PET Study

H. TOYAMA,[*,†] **K. TAKAZAWA,**[‡] **T. NARIAI,**[§] **K. UEMURA**[‡] **and M. SENDA**[†]

[*]*National Institute of Radiological Sciences, Chiba, Japan*
[†]*Positron Medical Center, Tokyo Metropolitan Institute, Gerontology, Tokyo, Japan*
[‡]*Waseda University, Tokyo, Japan*
[§]*Department of Neurosurgery, Tokyo Medical and Dental University, Tokyo, Japan*

We developed a method for clustering brain pixels on the basis of stage of hemodynamic deficiency using three or four sets of PET images and applied it to evaluating the regional vasodilative and vasoconstrictive reactivity and oxygen metabolism before and after revascularization surgery in chronic occlusive cerebrovascular disease. In this study, we compared two types of clustering methods: agglomerative hierarchical and K-means. In comparison between the two methods, the K-means method divides into almost the same volume of clusters in the correlation map and the agglomerative hierarchical method creates the clustered region in a small volume at the extremely low or high values of variables in PET images. We also examined the effect of the number of variables and the number of clusters. Four anatomically and pathophysiologically different areas were delineated with four clusters in these two cases with cerebrovascular disease. Functional changes in the revascularized region are depicted in the clustered brain images. This clustering was considered to be useful for multivariate staging of hemodynamic deficiency in obstructive cerebrovascular disease and is suitable for objective representation of multiple PET parameters obtained in the activation study as well as in a study with ^{15}O-labeled CO_2, O_2, and CO gases.

I. INTRODUCTION

In nuclear medicine, several kinds of organ function can be measured simultaneously by means of imaging with various radio tracers and/or in the different conditions. The relationships among metabolism, blood flow, and hemodynamics give us useful information about the stage of disease progress. The purpose of this study is to develop a method of clustering the brain pixel by pixel by means of combinations of multiple functions.

II. MATERIALS AND METHODS

A. Subjects

The stage of hemodynamic deficiency was evaluated in a patient with occlusion of internal carotid artery (ICA) and in another with moyamoya disease before and after revascularized surgery.

1. Case 1

The subject in case 1 was a 60-year-old male with occlusion of the right cervical ICA. The territories of the right anterior cerebral artery (ACA) and middle cerebral artery (MCA) were fed only by a poorly developed anterior communicating artery from the left side. Since the right posterior cerebral artery (PCA) was not visualized by vertebral angiography, the occluded right ICA was thought to have had a fetal-type posterior communicating artery. The patient once presented transient left side weakness and was found to have multiple cerebral infarctions on MRI. The superficial temporal artery (STA)-MCA anastomosis was performed on the right side. PET studies were performed before and 3 and 11 months after surgery.

2. Case 2

The subject in case 2 was a 14-year-old girl with occlusion of bilateral ICA (Moyamoya disease). The left MCA had poorer antegrade perfusion than the right MCA. The territories of bilateral MCA and ACA were also perfused from PCA through leptomeningeal collateral flow. The patient complained about transient weakness of left or right extremities. MRI showed no abnormality. The patient was treated by indirect bypass surgery (encephalo-duro-arterio-synangiosis, EDAS), in which the anterior and posterior branches of left STA and the posterior branch of right STA were intracranially implanted. After the operation, she did not show transient weakness any more. Angiography 1 year after the operation demonstrated markedly improved collateral flow in the whole territory of the left anterior circulation through implanted vessels. It was demonstrated that the posterior part of the right anterior circulation was well fed by the external carotid artery. However, as the stage of the disease progressed during the follow-up period, the anterior part of right frontal lobe, where vessel was not surgically implanted, showed increasingly poor perfusion. An additional PET study was performed 14 months after the operation.

A **PET H$_2^{15}$O CBF study** was performed with HEAD-TOME IV (Shimadzu, Kyoto Japan) over 120 s, starting immediately after the intravenous bolus injection of ~1.0 GBq of H$_2^{15}$O. Arterial blood was continuously drawn with a peristaltic pump to obtain the time–activity curves using a beta detector equipped with plastic scintillator. The rCBF was calculated by the PET autoradiographic method (Herscovitch *et al.*, 1983) and 14 slices of CBF images were obtained with 6.5-mm intervals. The rCBF measurement was repeated under various conditions including resting state, hyperventilation, and acetazolamide loading.

Three sets of PET images were generated: "CBF," i.e., resting CBF, "AZ," representing the response to acetazolamide calculated as acetazolamide loading minus resting CBF, and "HV," representing the response to hyperventilation calculated as hyperventilation minus resting CBF divided by the decrease in PCO$_2$ by hyperventilation.

An ^{15}O gas study was performed within 2 months before or after the H$_2$O PET study. None of the patients presented any change in clinical symptoms between the two studies. The regional oxygen metabolic rate (CMRO$_2$) was measured using continuous and consecutive inhalations of C^{15}O$_2$ and ^{15}O$_2$ gas with continuous arterial blood sampling and using a table-lookup method (Senda *et al.*, 1988; Sadato *et al.*, 1993). The effect of regional CBV on the CMRO$_2$ was corrected by measured CBV with C^{15}O inhalation scan.

The four sets of PET images (pre- and postoperative H$_2^{15}$O scans and ^{15}O gas scans) were three-dimensionally registered to one another for each subject using an automatic image registration program (Ardekani *et al.*, 1995).

B. Analysis

A **three-variable correlation map** was generated, in which the pixel values of resting CBF, the hyperventilatory (HV) response, and the acetazolamide (AZ) response was plotted on *X*-, *Y*-, and *Z*-coordinates, respectively, as shown in Fig. 1. In four-variable correlation map, the fourth coor-

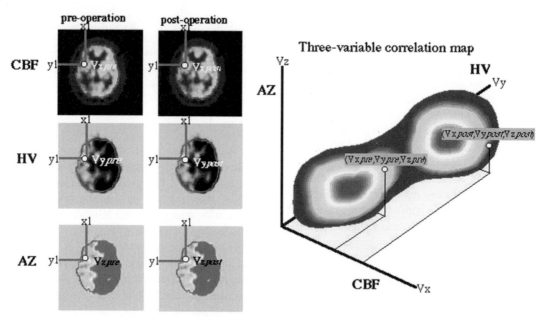

FIGURE 1. Three-variable correlation map, in which the pixel values of resting CBF, the hyperventilatory (HV) response, and the acetazolamide (AZ) response was plotted on *X*-, *Y*-, and *Z*-coordinates, respectively.

dinate (T) corresponding to CMRO$_2$ was added. The value of X-, Y-, Z,- and T-coordinates was normalized between minimum and maximum of each variable and transformed into an integer of 0–30 to facilitate computation and display.

C. Clustering

The number of clusters was empirically set to be four or six, which classified the tissues most appropriately. The pixels were clustered by the following two methods, and each pixel was assigned its cluster number. In the *agglomerative hierarchical method*, pixels are classified without preknowledge. First, each pixel is considered to constitute an independent cluster. Then we repeat merging the clusters until the number of clusters becomes the designated number (= 4 or 6). A classified correlation map is shown on the right side of Fig. 2. In the *K-means method*, the number of clusters (= 4 or 6) and the initial location of the center of each cluster are given at first. Each pixel is allocated to the cluster whose center is the nearest. The center of the cluster is recalculated until the center dose not change. A classified correlation map by this method is shown on the left side of Fig. 2.

A clustered brain image was generated by referring to the clustered three- or four-variable correlation map, where each pixel in the brain was labeled with a color representing its cluster number.

III. RESULTS

A. Clustering Method

The number of pixels within each cluster varied with the clustering method as shown in Fig. 2 (see color plate). The number of pixels tended to equalize with each other in the K-means method. In the agglomerative hierarchical method, a small region was assigned to be an independent cluster. The border of each cluster was straight in the K-means method but was curved in the agglomerative hierarchical method. The number of cluster was set to be four for three variable sets and to be six for four variable sets, respectively.

B. Case 1 (ICA Occlusion)

With four clusters for three-variable correlation map, four anatomically and pathophysiologically different areas were delineated as shown in Fig. 3 (see color plate). The left-hand side shows cross-sectional images of the clustered brain (top left) and 3D correlation map (bottom left) by K-means method: 1 (yellow), normal cortex; 2 (green), cortex with abnormal value in all three variables; 3 (red), cortex with higher blood flow and vasoreactivity; 4 (blue), white matter. The right temporo-occipital area represented by the green

(abnormal) cluster turned into the yellow cluster (normal) after the operation as shown in Fig. 3. The four-variable correlation map was clustered into six regions as shown in Fig. 4 (see color plate). In both clustering methods, the area of cluster 5 (cerulean blue) (impaired AZ and normal HV responses and higher CMRO$_2$) in the preoperative image turned into cluster 1 (normal) with revascularization in the postoperative image.

C. Case 2 (Moyamoya Disease)

The left-hand side in Fig. 5 (see color plate) shows cross-sectional images of the clustered brain map by the K-means method (top) and agglomerative hierarchical clustering (bottom). The right-hand side shows the original brain images. In the K-means method (top left), 1 (red), normal cortex; 2 (green), cortex with impaired acetazolamide response and normal hyperventilatory response; 3 (blue), cortex with abnormal value in all three variables; 4 (yellow), white matter and ventricles. In the agglomerative hierarchical clustering (bottom left), 1 (yellow), normal cortex; 2 (blue), cortex with impaired acetazolamide response and normal hyperventilatory response; 3 (green), cortex with abnormal value in all three variables. Pixels in the left frontal region belonging to cluster 3 before operation turned into the normal 1 cluster after operation. It represents the area with ameliorated hemodynamics following bypass surgery. Pixels in the right frontal region belonging to cluster 2 in the preoperative image turned into cluster 3. They represent the area with worsened hemodynamic due to disease progression. In the patient with moyamoya disease, regional abnormality of CMRO$_2$ was not found and the additional information was not obtained by the clustering for the four-variable correlation map.

IV. DISCUSSIONS

The number of cluster was empirically set to be four for three variable sets and to be six for four variable sets, respectively. In the determination of the number of clusters, it is important for each cluster to have an anatomically and pathophysiologically different meanings. Whereas a K-means method divides into almost same volume of clusters in the correlation map, an agglomerative hierarchical method creates the clustered region in small volume at the extremely low or high value of variables in PET images. The blue region in the clustered correlation map by an agglomerative hierarchical method is small and corresponds to a ventricle as shown on the bottom left of Fig. 4. This fact shows that an agglomerative hierarchical method might be better than a K-means method in medical use.

In clinical study, regional cerebral function was assigned by means of the combination of multiple functions such as all normals, normal CBF with less AZ response, all abnormals, and so on, which are depicted as a different cluster (color) in the clustered correlation map. in chronic occlusive cerebrovascular disease, the cortex with impaired vasodilative response to AZ and normal vasoconstrictive response to HV came to belong in the normal cluster after operation.

V. CONCLUSIONS

Four anatomically and pathophysiologically different areas were delineated with four clusters in two patients with cerebrovascular disease. Functional changes in the revascularized region are depicted on the clustered brain images.

We consider that this method is useful for multivariate staging of hemodynamic deficiency in obstructive cerebrovascular disease and that it is suitable for objective representation of multiple PET parameters obtained in activation studies as well as in studies with ^{15}O-labeled CO_2, O_2, and CO gases.

Acknowledgments

This work was supported by Grant-In-Aid for Scientific Research (C) 11670653 from the Ministry of Education, Science, Sports, and Culture, Japan.

References

Ardekani, B. A., Braun, M., Hutton, B. F., Kanno, I., and Iida, H. (1995). A fully automatic multimodality image registration algorithm. *J. Comput. Assist. Tomogr.* **19**: 615–623.

Herscovitch, P., Markham, J., and Raichle, M. E. (1983). Brain blood flow measured with intravenous $H_2(15)O$. I. Theory and error analysis. *J. Nucl. Med.* **24**: 782–789.

Sadato, N., Yonekura, Y., Senda, M., Iwasaki, Y., Matoba, N., Tamaki, N., Sasayama, S., Magata, Y., and Konishi, J. (1993). PET and the autoradiographic method with continuous inhalation of oxygen-15-gas: Theoretical analysis and comparison with conventional steady-state methods. *J. Nucl. Med.* **34**: 1672–1680.

Senda, M., Buxton, R. B., Alpert, N. M., Correia, J. A., Mackey, B. C., Weise, S. B., and Arckerman, R. H. (1988). The ^{15}O steady-state method: Correction for variation in arterial concentration. *J. Cereb. Blood Flow Metab.* **8**: 681–690.

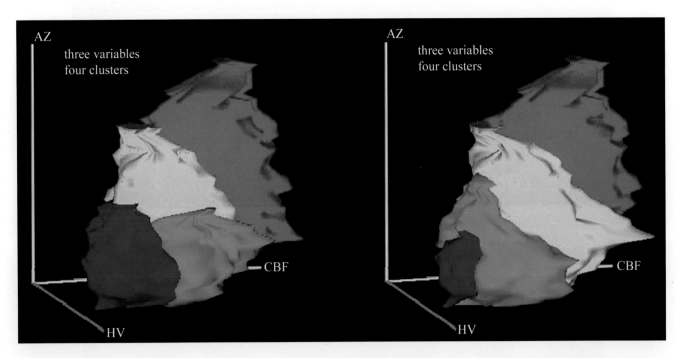

FIGURE 2. The clustered correlation maps of three variables by means of the *K*-means method (left) and agglomerative hierarchical method (right).

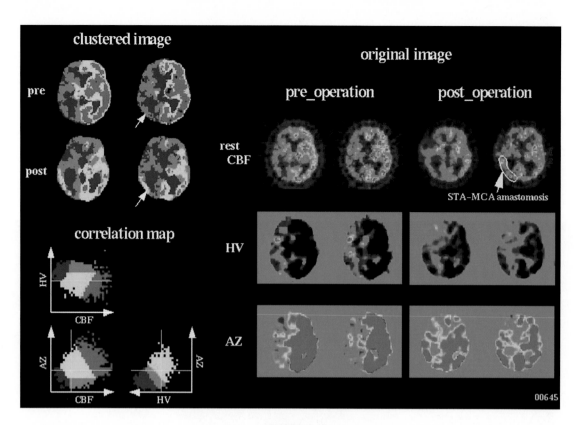

FIGURE 3

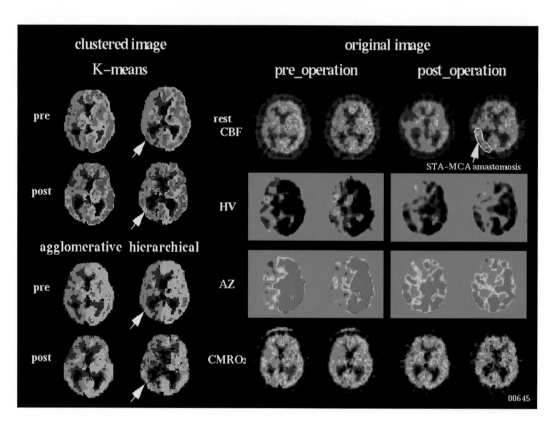

FIGURE 4

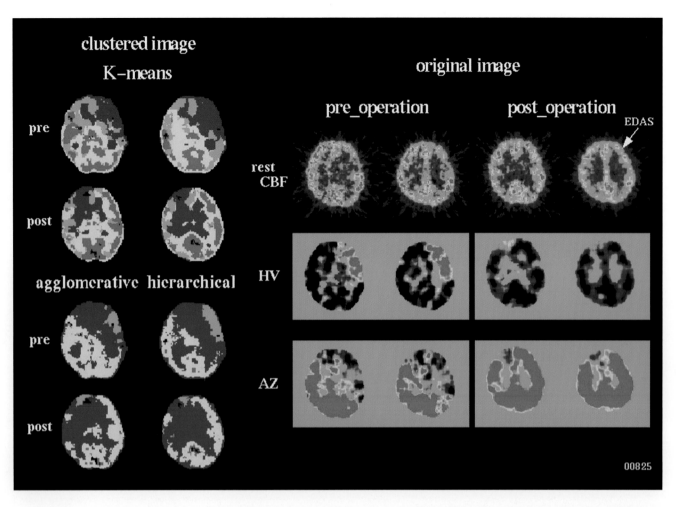

FIGURE 5. (Case 2, Moyamoya disease) Three-variable clustered images (left) and their original images (right) of the brain pre- and post-operation.

FIGURE 3. (Case 1, ICA occlusion) Clustered brain images (top left) cross-sectional images of 3D clustered correlation map (bottom left) and original PET images (right) of CBF, HV, and AZ pre- and post-operation. In the clustered images: yellow, normal cortex; green, cortex with abnormal values in all three variables; red, cortex with higher blood flow and vasoreactivity; blue, white matter and ventricles.

FIGURE 4. (Case 1, ICA occlusion) Clustered brain images (left) and original PET images (right) of CBF, HV, and CMRO2 pre- and post-operation. In the clustered images: yellow, normal cortex; green, cortex with abnormal values in all three variables; red, cortex with higher blood flow and vasoreactivity; cerulean blue, impaired AZ and normal HV response and higher CMRO2; blue, white matter and ventricles.

45

Establishing Behavioral Correlates of Functional Imaging Signals

JOHN J. SIDTIS, JON R. ANDERSON, STEPHEN C. STROTHER, and DAVID A. ROTTENBERG

Department of Neurology, University of Minnesota, and PET Imaging Service, Minneapolis Veterans Affairs Medical Center, Minneapolis, Minnesota

Functional imaging studies have sought to establish brain–behavior relationships in a variety of ways, many of which involve some form of image decomposition. The present study took a different approach, namely, decomposing the behavior rather than the image. A speech task produced during scanning was acoustically analyzed to produce a number of performance measures and multiple linear regression was used to examine the ability of image data to predict these performance measures. The results identified areas known to be important in speech production and further suggested that activity in multiple brain areas is needed to predict behavior in this task. This approach requires neither active nor passive control states, and avoids some of the difficulties associated with image decomposition.

I. INTRODUCTION

Many functional imaging studies seek to establish brain-behavior relationships by making two assumptions: subtraction of imaging data obtained during different behavioral states will yield meaningful data about relevant brain processes, and the magnitude of such differences in brain activity reflects the relative involvement of different areas in the behavior under study. Using a series of speech production tasks, it has been shown that the cerebral blood flow changes associated with syllable production are not accurately decomposed into articulatory and phonatory components using task subtraction (Sidtis *et al.*, 1999). Further, if the experimental results are restricted to a single task (syllable production) and a single baseline condition (quiet), neither the "activated" minus "baseline" subtraction or the covariance pro-

file that best separated the "activated" and "baseline" states has predictive value with respect to a primary performance measure, speech rate. In contrast, using data obtained during "activation" scans alone, speech rate was predicted by activity in two regions (Broca's area and the right caudate). It is important to note that neither area had the highest signal peak, and one area (right caudate) had a negative relationship with performance (Sidtis *et al.*, 1998). These results and others (e.g., Friston *et al.*, 1996; Jennings *et al.*, 1997) suggest that attempting to establish specific brain-behavior relationships based on the joint decomposition of images from different behavioral states may be problematic. The present study examined a different approach. In this approach, the behavior rather than the image is decomposed and brain-behavior relationships are examined using the relevant "activated" scan data without reference to either active or passive control states. The question is whether a range of performance characteristics of a single complex behavior can be meaningfully associated with specific patterns of brain activity.

II. METHODS

A. Subjects

A group of 13 right-handed normal volunteers (eight females and five males ages 43 ± 11 years) participated in this study after informed consent was obtained. All participants were native speakers of English.

---

*Physiological Imaging
of the Brain with PET*

B. Scanning Protocol

Bolus injection of [^{15}O]water was used as a marker of regional cerebral blood flow. Each study consisted of eight 90 s scans (four "baseline" scans alternating with four "activated" scans), separated by an interscan interval of approximately 9 min, acquired using a Siemens-ECAT 953B tomograph in 3D mode. During each "baseline" scan, subjects were required to remain awake and quiet. The speech task consisted of the repetition of the syllables *pa*, *ta*, and *ka*, produced as quickly as possible. Subjects were instructed to take a deep breath, then produce as many syllables as possible during expiration. Subjects repeated this process for 60 s. Syllable repetitions were recorded during the scans for subsequent acoustic analyses.

C. Image Analysis

A set of 22 regions of interest (ROI's) were generated based on evidence of a significant regional change during this task or either of two other speech tasks (sustained phonation, repetitive lip closure) completed by this group of subjects (Sidtis *et al.*, 1999). A threshold was applied to each ROI such that values represented the mean of the upper 25% of activity. Left (L) and right (R) side homologs of the following regions were examined: inferior portion of the cerebellar hemisphere, middle lobe of the cerebellar hemisphere, superior portion of the cerebellar hemisphere (anterior lobe), superior temporal gyrus, transverse temporal gyrus, putamen, caudate, thalamus, inferior frontal lobe, pre-/postcentral gyrus, supplementary motor area (SMA). A volume-mean normalized measure of cerebral blood flow was created by multiplying each subject's ROI value by the ratio of the highest volume mean in the dataset divided by that individual's volume mean (Arnt *et al.*, 1996).

D. Behavioral Decomposition

The recording of each subject's speech production during each scan was digitized at 20 kHz for acoustic analysis. Several measures were obtained: speech rates (syllables/second), mean duration for each syllable type and syllable duration variability across repetitions of the same syllable type, mean voice onset time (VOT, the time between initial articulatory movement and the onset of phonation) for each syllable type and VOT variability across repetitions of the same syllable type, and the number of breath groups during each 60-s production period.

E. Statistical Analysis

Multiple linear regression analysis (Norusis, 1988) was used to determine if there was some linear combination of regions that could predict each of the acoustic measures. Separate analyses with stepwise replacement were conducted for each measure using the following criteria: probability of F-to-enter (0.05), probability of F-to-remove (0.10), and tolerance (0.01).

III. RESULTS

Some examples of performance measures and brain regions where the functional imaging signal was significantly associated are presented in Table 1. None of the performance measures examined was significantly associated with activity in a single region. Across all syllable types, syllable production rate was predicted by a linear combination of activity in two regions such that increased speech rate was associated with increased signal in the left inferior frontal cortex and decreased signal in the right caudate ($F = 10.26$; $p < 0.0002$). The relative regression weights for these two regions are depicted in Fig. 1. Mean duration for the syllable /ka/, inversely related to speech rate, was also associated with activity in the left inferior frontal area and right caudate, but this relationship also included the left pre-/postcentral gyrus, the left and right putamen, and the left cerebellum ($F = 15.59$; $p < 0.0000$). The relative regression weights for this relationship are presented in Fig. 2. Duration variability for the syllable /ka/ was associated with activity in the left superior temporal gyrus, left thalamus, and right cerebellum ($F = 16.06$; $p < 0.0000$). Mean VOT for the syllable /ka/ was associated with activity in the left pre-/postcentral gyrus, the right transverse temporal gyrus, and the right putamen ($F = 10.82$; $p < 0.0000$). Voice onset time variability for the syllable /ka/ was associated with activity in the left inferior frontal area, the left supplementary motor area, the right caudate, and the left and right putamen ($F = 8.68$; $p < 0.0000$). Finally, the number of breath groups was associated with activity in the right pre-/postcentral gyrus,

TABLE 1. Selected Performance Measures Derived from an Acoustic Analysis of a Speech Production Task and the Regions Where the Linear Combinations of Activity Significantly Predicted Performance

Performance	Associated ROIs
Syllable rate	L inferior frontal area, R caudate
Syllable duration	L inferior frontal, L pre-/postcentral gyrus, R caudate, L,R putamen, L cerebellum
Duration variability	L superior temporal gyrus, L thalamus, R cerebellum
VOT mean	L pre-/postcentral gyrus, R transverse temporal gyrus, R putamen
VOT variability	L inferior frontal area, L SMA, R caudate, L,R putamen
Breath group	R pre-/postcentral gyrus, R thalamus, R cerebellum

Syllable rate

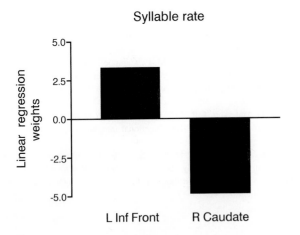

FIGURE 1. Linear regression weights for the left inferior frontal area and the right caudate in the linear combination that is significantly associated with syllable production rate.

Mean Duration/ka/

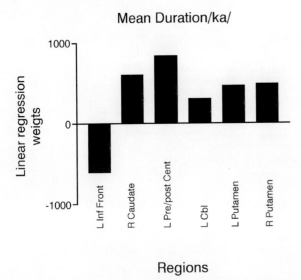

FIGURE 2. Linear regression weights for the left inferior frontal area, the right caudate, left pre-/postcentral gyrus, left cerebellum, and left and right putamen in the linear combination that is significantly associated with the mean duration of the syllable/ka/.

the right thalamus, and the right cerebellum ($F = 13.82$; $p < 0.0000$).

IV. DISCUSSION

Previous studies have suggested that in spite of their popularity, functional imaging methods that attempt to establish brain–behavior relationships by manipulating image data to isolate a single component of a behavioral task may have serious problems. At one level, the difficulties may be viewed

as a weakness in accounting for interactions between contrasting behavioral conditions (Friston *et al.*, 1996), at another, the problems may reflect an inadequate conceptualization of the relationship between imaging signals and underlying neurophysiological processes (Sidtis, 2000). The present study suggests a different approach to establishing brain–behavior relationships, namely, decomposing the behavior rather than the image.

Behavioral decomposition is based on the premise that even relatively simple behaviors involve multiple neurophysiologic processes at motor, sensory, perceptual, and cognitive levels. Some, if not all, of these processes will be expressed in a functional image to create the observed data in combinations that are not yet understood and not likely to be "controlled" for using the family of subtraction techniques. An important feature of behavioral decomposition is that while some performance measures may be correlated (e.g., syllable rate and syllable duration), the extraction of one set of performance measures does not alter the extraction of others, nor does it alter the behavioral record (e.g., the speech waveform). The same cannot be said of image manipulations, where the subtraction of either active or resting control states significantly alters the relationship between image data and behavior (Sidtis *et al.*, 1998, 1999).

The present approach is similar to studies that have sought to identify brain areas where activity is correlated with experimental parametric manipulation (e.g., Grafton *et al.*, 1992; Dettmers *et al.*, 1995; Blinkenberg *et al.*, 1996; Sadato *et al.*, 1996; Price *et al.*, 1996). Parametric manipulation is probably the most direct way to establish task-correlated changes in imaging data, but it should be noted that as in the present study, it is not necessary to parametrically vary the task to relate performance variability to variability in the functional signal. However, establishing a functional relationship between performance and imaging data depends on finding the relevant signals: relationships may involve multiple regions, and decreases as well as increases in activity may be important. There is no simple solution to this problem.

Given the current state of understanding regarding the connection between functional imaging signals and relevant neurophysiology, it may well be that more effort should be spent on decomposing behavior than on decomposing images and on directly establishing relationships between behavioral and image data.

References

Arnt, S., Cizadlo, T., O'Leary, D., Gold, S., and Andreasen, N. C. (1996). Normalizing counts and cerebral blood flow intensity in functional imaging studies of the human brain. *NeuroImage* **3**: 175–184.

Blinkenberg, M., Bonde, C., Holm, S., Svarer, C., Andersen, J., Paulson, O. B., and Law, I. (1996). Rate dependence of regional cerebral activation during performance of a repetitive motor task: A PET study. *J. Cereb. Blood Flow Metab.* **16**: 794–803.

Dettmers, C., Fink, G. R., Lemon, R. N., Stephan, K. M., Passingham, R. E., Silbersweig, D., Holmes, A., Ridding, M. C., Brooks, D. J., and Frackowiak, S. J. (1995). Relation between cerebral activity and force in the motor areas of the human brain. *J. Neurophysiol.* **74**: 802–815.

Friston, K. J., Price, C. J., Fletcher, P., Moore, C., Frackowiak, R. S. J., and Dolan, R. J. (1996). The trouble with cognitive subtraction. *NeuroImage* **4**: 97–104.

Grafton, S., Mazziotta, J., Presty, S., Friston, K. J., Frakowiak, R. S. J., and Phelps, M. (1992). Functional anatomy of human procedural learning determined with regional cerebral blood flow and PET. *J. Neurosci.* **12**: 2542–2548.

Jennings, J. M., McIntosh, A. R., Kapur, S., Tulving, E., and Houle, S. (1997). Cognitive subtractions may not add up: the interaction between semantic processing and response mode. *NeuroImage* **5**: 229–239.

Norusis, M. J. (1988). *"SPSS/PC + V2.0."* SPSS Inc., Chicago.

Price, C., Moore, C. J., and Frackowiak, R. S. J. (1996). The effect of varying stimulus rate and duration on brain activity during reading. *NeuroImage* **3**: 40–52.

Sadato, N., Ibanez, V., Deiber, M.-P., Campbell, G., Leonardo, M., and Hallett, M. (1996). Frequency-dependent changes of regional cerebral blood flow during finger movements. *J. Cereb. Blood Flow Metab.* **16**: 23–33.

Sidtis, J. J., Anderson, J. R., Strother, S. C., and Rottenberg, D. A. (1998). Predicting performance from functional imaging data. *NeuroImage* **7**: S749.

Sidtis, J. J., Anderson, J. R., Strother, S. C., and Rottenberg, D. A. (1999). Are brain functions really additive? *NeuroImage* **9**: 490–496.

Sidtis, J. J. (2000). From chronograph to functional image: What's Next? *Brain Cognition* **42**: 75–77.

46

Functional Neuroanatomy of Writing Revealed by 3D PET

NAOHIKO OKU, KAZUO HASHIKAWA, MASAYASU MATSUMOTO,*,† HARUKO YAMAMOTO,
YUJIRO SEIKE, MASARU NUKATA,‡ HIROSHI NISHIMURA,‡ TADANORI TERATANI,‡
MASATSUGU HORI,* and TSUNEHIKO NISHIMURA‡

Department of Nuclear Medicine
Department of Medicine and Therapeutics
†*Department of Neurology*
‡*Department of Tracer Kinetics, Osaka University Graduate School of Medicine, Osaka, Japan*

The purpose of this study was to evaluate the functional neuroanatomy of the Japanese writing system in the normal Japanese population. Ten paid young normal Japanese volunteers participated in the study. They all were right handed. Each underwent 13 serial rCBF measurements using 3D PET and ^{15}O-labeled water semibolus injection with autoradiographic method. During a scan, the volunteer was requested to write on a writing board to dictation in one of Kanji (morphogram), Kana (phonogram), or meaningless and soundless characters (Control). Statistical parametric maps of significant rCBF change were obtained by using SPM96 software. Writing in Kanji activated a part of the right superior frontal gyrus and anterior cingulate gyrus more than was seen in control. Writing in Kana activated the left inferior frontal gyrus, left cuneus, and left middle temporal gyrus more than was seen in control. Control task activated left supramarginal gyrus and a wide area extending from the right precuneus, supramarginal gyrus to the right inferior parietal lobule compared to that seen when writing in Kanji. The control task also activated the right supramarginal gyrus, a part of right middle frontal gyrus, and left supramarginal gyrus compared to that seen when writing in Kana. Writing in Kana activated the left calcarine-cuneus, medial part of lenticular nucleus, posterior part of cingulate gyrus, left primary somatosensory area for the

hand, left angular gyrus, and right orbital gyri more than did in Kanji. In contrast, writing in Kanji did not activate any sites significantly compared to that seen when writing in Kana. The results supported the dual neuropathway model in the Japanese writing system hypothesized by pathology-symptom analysis.

I. INTRODUCTION

From the extensive observation of the patients with alexia or agraphia, recent cognitive neurology has proposed a bidirectional functional neuronal pathway model in processing written words. Much interest has been focused on the Japanese writing system by cognitive neurologists because the Japanese writing system has two types of characters (Sugishita *et al.*, 1992). One is the phonogram, which has strict matching to the syllable (called Kana), and the another is the morphogram, which was derived from the Chinese character system and represents irregular sound (called Kanji). In the Indo-European language system, regular words correspond to Kana and irregular words correspond to Kanji. Some previous reports (Sakurai *et al.*, 1992, 1993) have shown that there were differential functional neuronal pathways of reading between Kana and Kanji in normal human brain. However, no study has shown the differential functional neuronal pathways of writing between Kana and Kanji in normal human brain. In the present study, we evaluate the functional neuronal pathways of writing in Japanese by using three-dimensional positron emission tomography (3D PET).

II. SUBJECTS AND METHODS

Ten paid young normal Japanese volunteers (mean age 21 years old) participated in the study. They all were native Japanese speakers and educated university students. All were right-handed confirmed by the Edinburg handedness inventory. Written informed consent was given by each volunteer.

The volunteers were rested supine on a scanner bed with their head immobilized and positioned in the gantry. Outside the gantry and above volunteers' chest, a writing board was set within arm's reach. A white paper (420 mm horizontally and 297 mm vertically in size) matrixed with black grid line was attached on the board. The span between lines was 3.5 cm vertically and horizontally. The volunteer filled the matrices with characters by using a felt pen in their right hand.

Regional cerebral blood flow (rCBF) was measured by using a 3D PET (SET-2400W, Shimadzu Co. Kyoto, Japan), whose specifications were 63 slices, 3.7 mm slice thickness, and spatial resolution of 4.1 mm FWHM. Before emission scans, 10 min of transmission scan was done by using Ge-68/Ga-68 line source. Each subject underwent 13 serial rCBF measurements by using ^{15}O-labeled water semibolus injection and the autoradiographic method. Arterial blood sampling was not performed. The dose of ^{15}O-labeled water and acquisition period were 296 MBq and 90 s, respectively. Time interval between scans was 8 min to allow the washout and decay of ^{15}O-labeled water in the brain. In a scan, the volunteer was requested to write on the writing board to dictation in either Kanji, Kana, or meaningless and soundless characters (control task). The words presented in the Kanji and Kana tasks were selected from daily use vocabulary (Fig. 1). The characters used in the control task were simple graphics of nonwords identified and named by their shape for convenience (Fig. 1). The volunteer was requested to draw a control character in quadruplet in single presentation. Words or control characters were presented orally in every 8 s by the examiner using a loud speaker. The volunteers practiced writing or drawing these characters for several minutes before being set in the PET gantry. Of the 13 scans, the first one was the control scan and its data was not used in further analysis. Three conditions noted above were presented in a random order in the remaining 12 scans. The time course of a scan is presented in Fig. 2.

Kana words: せつめい(se-tsu-me-i, explanation)
たくさん(ta-ku-sa-n, many)
こうどう(ko-u-do-u, behavior or hall)

Kanji words: 警察(ke-i-sa-tsu, police)
成人(se-i-ji-n, adult)
太陽(ta-i-yo-u, the sun)

non words: ≡≡≡≡ horizontal line
≋≋≋≋ horizontal wave
⊚⊚⊚⊚ spiral

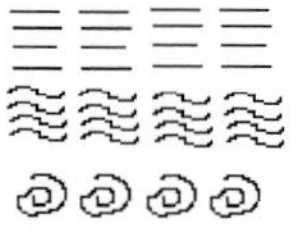

FIGURE 1. The examples of words or nonword characters used in this study. Words were selected from daily use vocabulary, which consisted of four syllables. Nonword characters consisted of simple graphics.

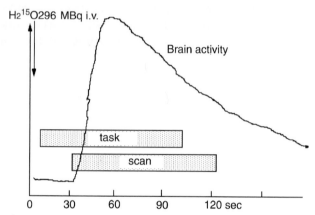

FIGURE 2. Schematic illustration for the protocol of a session.

Statistical parametric maps of significant rCBF differences between conditions were obtained by using SPM96 (The Wellcome Department of Cognitive Neurology, University Collage of London, UK) software on a workstation (Ultra-1, Sun microsystems, Palo Alto, CA). In the first step, SPM software checked the voxels independently whether they showed significant changes (statistical P value <0.001) among every voxels. In the second step, SPM software extracted the cluster of voxels with a significant change (statistical P value <0.05) according to the Gaussian field theory (i.e., the value of a voxel was not independent and influenced by surrounding voxels).

III. RESULTS

Writing in Kanji activated a part of the right superior frontal gyrus and the anterior cingulate gyrus ($z = 4.22$) more than did control (Fig. 3). Writing in Kana activated

SPM{Z}

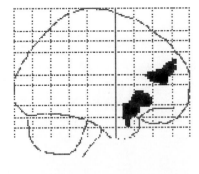
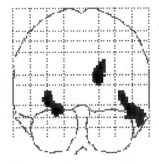
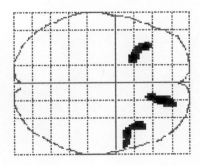

Cluster P level (k, Z)	Voxel P level (Z)	Coordinate (mm)		
		x	y	z
0.043 (244, 4.22)	0.259 (4.22)	18	46	14
0.067 (209, 3.87)	0.646 (3.87)	−24	18	−16
0.056 (223, 3.77)	0.757 (3.77)	48	12	−14

FIGURE 3.   SPM display of activated area by Kanji writing versus that seen under the control condition. The table presents the statistical results of subtraction analysis by SPM. Data of significant regions and *P* values are presented (in terms of cluster and voxel level inferences). *k* is the size of cluster by voxel and *Z* is the peak *Z* score transformed from *t* statistics in a cluster. Note that the last two regions were not significant because *P* values did not survive a correction for the cluster analysis.

the left inferior frontal gyrus ($z = 5.30$), the left cuneus ($z = 4.74$), and the left middle temporal gyrus ($z = 4.98$) more than did control (Fig. 4). The control task activated the left supramarginal gyrus ($z = 6.31$) and a wide area extending from the right precuneus, supramarginal gyrus, to the right inferior parietal lobule ($z = 5.02$), more than seen when writing in Kanji (Fig. 5). The control task also activated the right supramarginal gyrus ($z = 6.22$), a part of the right middle frontal gyrus ($z = 4.90$), and the left

supramarginal gyrus ($z = 4.41$), more then that seen when writing in Kana (Fig. 6). Writing in Kana activated the left calcarine-cuneus ($z = 6.36$), the medial part of the lenticular nucleus ($z = 5.13$), the posterior part of the cingulate gyrus ($z = 5.05$), the left primary somatosensory area for the hand ($z = 4.40$), the left angular gyrus ($z = 5.16$), and the right orbital gyri ($z = 4.22$) more than did in Kanji (Fig. 7). In contrast, compared to writing in Kana, writing in Kanji did not activate area significantly.

SPM{Z}

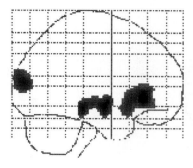
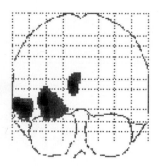
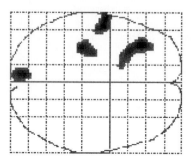

Cluster P level (k, Z)	Voxel P level (Z)	Coordinate (mm)		
		x	y	z
0.001 (777, 5.30)	0.003 (5.30)	−36	34	−2
0.019 (313, 4.98)	0.013 (4.98)	−62	−4	−6
0.027 (281, 4.79)	0.029 (4.79)	−32	−28	−12
0.021 (302, 4.74)	0.036 (4.74)	−8	−86	14

FIGURE 4. SPM display of activated area by Kana writing versus that seen under the control condition. The table presents the statistical results of subtraction analysis by SPM. Data of significant regions and P values are presented (in terms of cluster and voxel level inferences). k is the size of cluster by voxel and Z is the peak Z score transformed from t statistics in a cluster.

IV. DISCUSSION

In the present study, we evaluated the functional neuronal pathways of writing in Japanese by using 3D PET. The results suggested that the left occipital lobe, the posterior part of cingulate gyrus, and the posterior part of left superior temporal gyrus played more important roles in writing phonograms than in writing morphograms. These re-sults partially supported an observation of writing impairment in Alzheimer's disease in which a dissociation impairment of lexical spelling over phonological spelling occurred (Penniello, 1995). Comparisons of Kana or Kanji writing against the control condition simply suggested the differential functional neuronal pathways. Drawing graphical characters without meaning and sounds differed completely from writing words in morphograms or in phonograms.

SPM{Z}

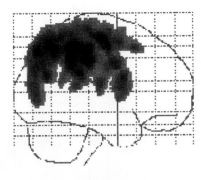 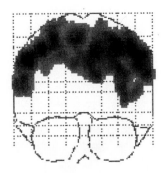

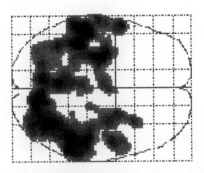

Cluster *P* level (*k, Z*)	Voxel *P* level (**Z**)	Coordinate (mm)		
		x	*y*	*z*
0.000 (19232, 6.31)	0.000 (6.31)	−54	−28	40
0.001 (777, 5.02)	0.010 (5.02)	52	−2	36

FIGURE 5. SPM display of activated area by control condition versus that observed in Kanji writing. The table presents the statistical results of subtraction analysis by SPM. Data of significant regions and *P* values are presented (in terms of cluster and voxel level inferences). *k* is the size of cluster by voxel and *Z* is the peak *Z* score transformed from *t* statistics in a cluster.

Writing in Kana (phonograms) may be a strict process that needs the procedures of matching syllables to characters one by one. In contrast, the main process in writing in Kanji (morphogram) may be a relatively simple process of retrieving the shape of the character from visual memory according to the meanings of presented word and sending it to the motor engram.

However, we did not detect any regions activated more in writing Kanji than in writing Kana. These results contradicted the hypothesis of Kanji writing neuroanatomy in which the posterior part of left inferior temporal gyrus played an important role (Iwata, 1984; Sakurai *et al.*, 1997). Further investigations are needed to explain this discrepancy.

Finally, It revealed that the 3D PET system was a powerful tool in investigating the generation of text in the brain.

Acknowledgments

We gave special thanks to Dr. Kinoshita H., Mr. Fujino K., Mr. Matsuzawa H, Mr. Nakamura S., Mr. Sugimoto K., and Mr. Suezawa K. for technical support.

SPM{Z}

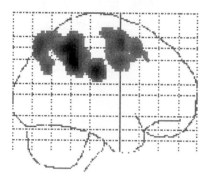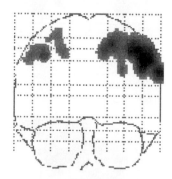

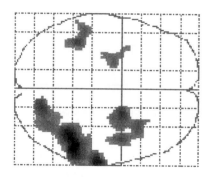

Cluster *P* level (*k, Z*)	Voxel *P* level (*Z*)	Coordinate (mm)		
		x	*y*	*z*
0.000 (3305, 6.22)	0.000 (6.62)	50	−44	40
0.000 (1046, 4.90)	0.018 (4.90)	22	0	56
0.015 (332, 4.41)	0.131 (4.41)	−42	−40	46
0.046 (238, 3.86)	0.650 (3.86)	−24	−8	48

FIGURE 6. SPM display of activated area by control condition versus that observed in Kana writing. The table presents the statistical results of subtraction analysis by SPM. Data of significant regions and *P* values are presented (in terms of cluster and voxel level inferences). *k* is the size of cluster by voxel and *Z* is the peak *Z* score transformed from *t* statistics in a cluster.

SPM{Z}

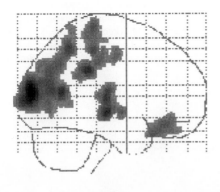
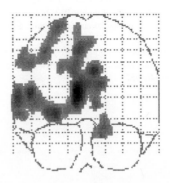

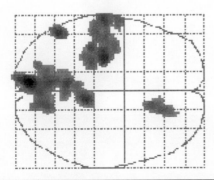

Cluster *P* level (*k, Z*)	Voxel *P* level (*Z*)	Coordinate (mm)		
		x	*y*	*z*
0.000 (3201, 6.36)	0.000 (6.36)	−8	−86	−14
0.013 (240, 5.16)	0.005 (5.16)	−52	−60	12
0.000 (1076, 5.13)	0.006 (5.13)	−28	−18	−6
0.010 (369, 5.05)	0.009 (5.05)	8	−34	32
0.001 (788, 4.40)	0.139 (4.40)	−38	−30	50
0.011 (365, 4.22)	0.254 (4.22)	14	26	−24

FIGURE 7.   SPM display of activated area by Kana writing versus that observed in Kanji writing. The table presents the statistical results of subtraction analysis by SPM. Data of significant regions and *P* values are presented (in terms of cluster and voxel level inferences). *k* is the size of cluster by voxel and *Z* is the peak *Z* score transformed from *t* statistics in a cluster.

References

Iwata, M. (1984). Kanji versus Kana: Neuropsychological correlates of the Japanese writing system. *Trends Neurosci.* **7**: 290–293.

Penniello, M. J., Lambert, J., Eustache, F., Petit-Taboue, M. C., Barre, L., Viader, F., Morin, P., Lechevalier, B., and Baron, J. C. (1995). A PET study of functional neuroanatomy of writing impairment in Alzheimer's disease: The role of the left supramarginal and left angular gyri. *Brain* **118**: 697–706.

Sakurai, Y., Matsumura, K., Iwatsubo, T., and Momose, T. (1997). Frontal pure agraphia for Kanji or Kana: Dissociation between morphology and phonology. *Neurology* **49**: 946–952.

Sakurai, Y., Momose, T., Iwata, M., Watanabe, T., Ishikawa, T., and Kanazawa, I. (1993). Semantic process in Kana word reading: Activation studies with positron emission tomography. *NeuroReport* **4**: 327–330.

Sakurai, Y., Momose, T., Iwata, M., Watanabe, T., Ishikawa, T., Takeda, K., and Kanazawa, I. (1992). Kanji word reading process analysed by positron emission tomography. *NeuroReport* **3**: 445–448.

Sugishita, M., Otomo, K., Kabe, S., and Yunoki, K. (1992). A critical appraisal of neuropsychological correlates of Japanese ideogram (Kanji) and phonogram (Kana) reading. *Brain* **115**: 1563–1585.

PATHOPHYSIOLOGY OF
ACQUIRED BRAIN DISORDERS

47

Brain Activation during Somatosensory and Auditory Stimulation in Acute Vegetative State of Anoxic Origin

S. LAUREYS,[*,†] **M. E. FAYMONVILLE,**[‡] **G. DEL FIORE,**[†] **N. JANSSENS,**[‡] **C. DEGUELDRE,**[†] **J. AERTS,**[†]
M. LAMY,[‡] **A. LUXEN,**[†] **G. MOONEN,**[*,†] **and P. MAQUET**[*,†]

[*]*Department of Neurology and* [‡]*Department of Anesthesiology and Intensive Care Medicine, CHU Sart Tilman, Liège, Belgium*
[†]*Cyclotron Research Center, University of Liège, Belgium*

Patients in vegetative state (VS) are thought to lack the capacity to experience the external world. Here, we investigate the brain reactivity to exteroceptive stimulation in two posthypoxic VS patients using positron emission tomography (PET). Cerebral blood flow was measured with the $H_2{}^{15}O$ technique during rest, somatosensory stimulation (electrical stimulation of the median nerve), and auditory stimulation (95-dB clicks). Results were compared to those of 10 age-matched controls. Quantified metabolic rates for glucose were measured, during rest, using the fluorodeoxyglucose method. In both patients, global brain metabolism was more than 70% below normal. Nevertheless, regional cerebral blood flow data showed that in both patients external stimulation resulted in an activation of primary cortices: primary somatosensory cortex during somatosensory stimulation and primary auditory cortex in response to auditory stimulation. In contrast, secondary somatosensory cortex and auditory association areas were significantly less activated during the respective stimulation modality in patients than that seen in controls. Although these results do not tell what VS patients actually perceive, the lack of activation beyond the primary cortex suggests that the observed residual cortical processing does not lead to the integrative processesing thought to be necessary to attain the normal level of awareness.

I. INTRODUCTION

Based on clinical, electrophysiological, positron emission tomography (PET), and neuropathologic findings, patients in a vegetative state (VS) are thought to lack the cerebral capacity to consciously experience inner and external world while having a preserved arousal (The Multi-Society Task Force on PVS, 1994). In clinical practice, however, it is difficult to evaluate the cerebral function of VS patients. If the absence of intentional behavior can readily be assessed at the bedside, the appreciation of their ability to consciously perceive external stimuli remains uncertain. Most PET studies of patients in VS have measured *resting* brain metabolism and demonstrated a global reduction in cerebral metabolic rate for glucose (CMRGlu) ranging from 50 to 70% below normal (see Laureys *et al.* Chapter 48 in this volume). We have previously shown that the metabolically most impaired regions were frontal and parietal associative cortices (Laureys *et al.*, 1999a, c) and that cortico-cortical (Laureys *et al.*, 1999b) and thalamo-cortical (Laureys *et al.*, 2000a, b) functional connectivity was altered in these patients. Measurement of residual cognitive processing in VS, using $H_2{}^{15}O$ PET (de Jong *et al.*, 1997; Menon *et al.*, 1998; Owen *et al.*, 1999) or magnetoencephalography (MEG) (Ribary *et al.*, 1998) are so far limited to anecdotal case reports. The present study is the first aimed at measuring regional CM-RGlu at rest and changes in regional cerebral blood flow (rCBF) in VS patients during somatosensory and auditory stimulations.

II. MATERIALS AND METHODS

The study was approved by the Ethics Committee of the Faculty of Medicine of the University of Liège. Written informed consent was obtained from the family of both patients and from all control subjects, in accordance with the declaration of Helsinki.

A. Patients

Two patients with the clinical diagnosis of VS according to established criteria (The Multi-Society Task Force on PVS, 1994) were included in the present study. In both cases, pupillary, corneal and vestibulo-ocular reflexes were preserved; T1 and T2 weighted magnetic resonance imaging (MRI) was normal; somatosensory evoked potentials of the median nerve (SEPs) showed the presence of P14 (medial lemniscus) and N20 (primary somatosensory cortex) potentials; and auditory evoked potentials (AEPs) showed preserved midbrain function.

Patient 1 was a 42-year-old right-handed woman who attempted suicide through hanging and was found unconscious and in respiratory arrest. She was reanimated but evolved to a VS. The patient died after having remained vegetative for 4.5 months. PET was performed 5 days after the clinical diagnosis of VS. Electroencephalograms (EEG) showed a nonreactive low-voltage 8-Hz basal activity.

Patient 2 was a 52 year-old-woman who was in a VS after prolonged respiratory insufficiency due to a severe pneumonia. She remained vegetative for 37 days and subsequently regained consciousness and partial autonomy. PET was performed when the patient had been in a VS for 6 days. The EEG showed a disorganized theta rhythm.

Both patients were already intubated prior to the study (but not ventilated) and both presented episodic spontaneous head and trunk movement necessitating muscular blockade during the PET study. In patient 1, exteroceptive stimulation resulted in important transpiration, tachycardia, and increased pulse rate. In order to obtain stable hemodynamic conditions during scanning, this patient was studied while receiving light intravenous sedation (midazolam 0.057 mg/kg/h and alfentanyl 4.4 μg/kg/h).

B. Controls

For the study of rCBF, data were obtained from 10 drug-free, healthy volunteers (mean age 33 years; range 19–45 years). For the study of CMRGlu, data were obtained from 53 subjects (mean age, 42 years; range, 18–76 years).

C. Data Acquisition

1. PET

PET data were obtained on a Siemens CTI 951 16/32 scanner (Siemens, Erlangen). A transmission scan was performed to allow a measured attenuation correction. Data were reconstructed using a Hanning filter (cutoff frequency: 0.5 cycle/pixel) and corrected for attenuation and background activity. CMRGlu was measured in 2D mode after injection of 5 to 10 mCi (185–370 MBq) [^{18}F]fluorodeoxyglucose ([^{18}F]FDG) and used arterial blood sampling (Phelps *et al.*, 1979). rCBF was measured after injection of 6 to 8 mCi (222–296 MBq) H$_2^{15}$O. Scans were acquired at 8-min intervals in 3D mode. Each scan consisted of 2 frames: a 30-s background frame and a 90-s frame. The intravenous water infusion was totally automated and begun just before the second frame in order to observe the head curve rising within the first 10 s of this frame.

2. MRI

For each patient a high resolution (voxel size, 0.96 × 0.96 × 1.35 mm) T_1-weighted structural MRI scan (1.5 T Magnetom imager, Siemens, Erlangen) was performed for coregistration with PET data.

3. Polygraphic Recording

EEG (C3-A2 and C4-A1), horizontal electrooculogram, and chin electromyogram were monitored throughout the scanning procedure.

4. Evoked Potentials

Stimulation and acquisition of evoked potentials was performed using a Nicolet Viking system 4.0 (Nicolet Biomedical, Madison Wisconsin). For SEPs, electrodes were positioned on Fz and 2 cm behind the 10–20 system C3 and C4 positions. Further leads explored the spinal potentials at the C6 level and the brachial plexus potentials at Erb's point. Somatosensory stimulation consisted of a 0.2-ms electrical square-wave pulse delivered at a frequency of 5.1 Hz to the median nerve at the wrist. Intensity was set above motor threshold for the abductor pollicis brevis and adapted until symmetrical amplitudes of the brachial plexus potential were obtained. Stimulation intensities were higher in the two patients (mean, 12 mA; range, 4–20 mA) than in the controls (mean, 6 mA; range, 2–15 mA). For AEPs, electrodes were positioned on each earlobe, Cz served as a reference. Auditory stimulation consisted of 95-dB rarefied clicks (0.1 ms duration) presented monaurally at a rate of 5.1 Hz with masking white 55-dB noise presented to the other ear.

D. Experimental Conditions

Scanning was performed under five conditions: resting state, median nerve stimulation (left then right), and auditory stimulation (left then right). Each condition was repeated three times. The presentation was pseudorandomized using a Latin square design. Stimulation started 10 s prior to the second scan frame. Ambient light and noise were kept to a minimum. Control subjects kept their eyes closed and patients' eyes were taped. Patients' vital parameters (temperature, electrocardiogram, blood pressure, O_2 saturation, respiratory rate, tidal volume, airway pressures, FiO_2, and P_{CO_2} capnography) were continuously monitored.

E. Data Analysis

The data were analyzed using the statistical parametric mapping software (SPM96 version; Welcome Department of Cognitive Neurology, Institute of Neurology, London, UK) implemented in MATLAB (Mathworks, Sherborn, MA). Data obtained during left-sided stimulation and during the rest condition were flipped. Scans from each subject were realigned (Friston et al., 1995a), transformed into a standard stereotaxic space (Talairach and Tournoux, 1988) using a symmetrical template (Montreal Neurological Institute; Ashburner et al., 1999), and smoothed (16 mm full-width half-maximum isotropic kernel; Friston, 1997). In patients, PET data were coregistered to their T1-weighted MRI (Friston et al., 1995b).

The condition and subject (block) effects were estimated according to the general linear model at each voxel (Friston et al., 1995c). Global flow normalization was performed by

TABLE 1. Cerebral Metabolic Rate for Glucose (in mg/100g · min) in Primary Sensory Cortex (SI; coordinates ±42 − 26 58) and Primary Auditory Cortex (AI; coordinates ±40 − 28 12) in Patients and Controls

	Area		
	SI	AI	Average gray matter
Patient 1	1.73	1.67	1.96
Patient 2	1.90	1.71	2.00
Normal controls	6.82 ± 1.31	6.3 ± 1.23	7.10 ± 1.32

Note. Stereotaxic coordinates (Talairach and Tournoux, 1988) were peak voxels activating during somatosensory and auditory stimulation in controls.

TABLE 2. Cerebral Areas that Showed Activation during Somatosensory (A) and Auditory (B) Stimulation in Normal Controls and in Both Vegetative Patients

Region	Coordinates			Z	p
	x	y	z		
A. Somatosensory stimulation					
Normal controls					
Contralateral SI	−42	−26	58	5.53	<0.001
Contralateral SII	−42	−22	16	6.95	<0.001
Patients					
Contralateral SI	−36	−35	57	3.31	0.001
B. Auditory stimulation					
Normal controls					
Contralateral AI (area 41)	−40	−28	12	5.72	<0.001
Ipsilateral AI (area 41)	40	−30	11	4.97	<0.001
Contralateral AII (area 42/22)	−62	−36	18	4.66	<0.001
Patients					
Contralateral AI (area 41)	−31	−30	14	3.26	0.001

Note. Coordinates are defined in stereotaxic space of Talairach and Tournoux (1988). Z scores are for the peak voxel in each region. Contralateral is contralateral to the side of stimulation.

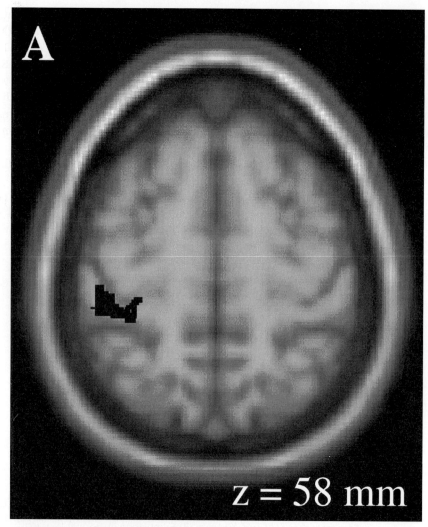

FIGURE 1. Regions of regional cerebral blood flow increase (in black) during somatosensory stimulation (A) and during auditory stimulation (B) in both vegetative patients, common to controls, projected on a transverse section of a normalized brain MRI. For display, results were thresholded at $p < 0.01$; z values represent the distance relative to the bicommisural line.

proportional scaling. We first identified the main effects of somatosensory and auditory stimulation in controls. Then, a conjunction analysis searched for activations common to each patient and the control population. A conjunction analysis of these contrasts identified the areas that showed activations common to both patients. Hereafter, we looked for the subject (individual patient versus controls) by condition (stimulation versus rest) interaction, looking for brain areas that would be less activated during stimulation in each patient than in controls. A conjunction of these interactions identified areas that were significantly less activated in common to both patients than in controls. The resulting set of voxel values for each contrast, constituting an SPM of the t statistic (SPM{t}) was transformed to the unit normal distribution (SPM{Z}) and thresholded at $p = 0.001$.

For controls, we used the voxel level corrected $p \leq 0.05$ threshold of significance. Having demonstrated significantly activated brain regions during somatosensory and auditory stimulation in controls we then restricted our analysis in patients to this set of areas and reported results at uncorrected $p \leq 0.001$. Due to the flipping of the data, results should not be interpreted as left- or right-sided, but as contralateral or ipsilateral to the side of stimulation.

III. RESULTS

A. Resting Metabolism

In both patients, CMRGlu in overall gray matter and in primary sensory (SI) and primary auditory (AI) cortices

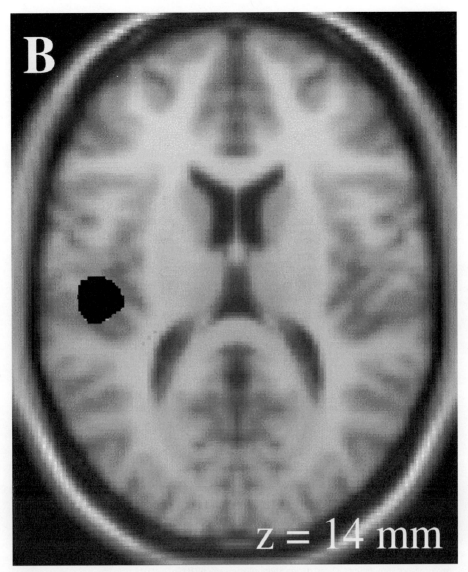

FIGURE 1.   (Continued).

(the peak voxel activating during auditory or somatosensory stimulation in controls) were more than 70% below normal values (Table 1).

B. Somatosensory Stimulation

In controls, comparison of median nerve stimulation to the resting state resulted in rCBF increase in contralateral thalamus, postcentral gyrus (SI), and lateral sulcus, encompassing the ventral bank of the parietal operculum (SII) (Table 2).

In both patients, somatosensory stimulation activated contralateral SI, as seen in controls (Table 2; Fig. 1A). Regions that activated less in patients than in controls were identified in contralateral area SII (coordinates –46 –22 18; $p \leq 0.001$; Fig. 2A).

C. Auditory Stimulation

In controls, auditory stimulation compared with rest resulted in increase of rCBF in bilateral transverse temporal gyrus (AI; Brodmann's area (BA) 41) and in contralateral superior temporal gyrus on its superior (BA 42) and lateral (BA 22) surface (Table 2).

In patients, comparison of activation during auditory stimulation, relative to the control group showed significant rCBF increase in contralateral AI (BA 41) (Table 2; Fig. 1B). Regions that showed a significant lack of activation in both patients relative to controls, were localized in the temporoparietal junction of the superior temporal sulcus contralateral to the side of click presentation (AII; coordinates –42 –47 20; $p = 0.001$; Fig. 2B).

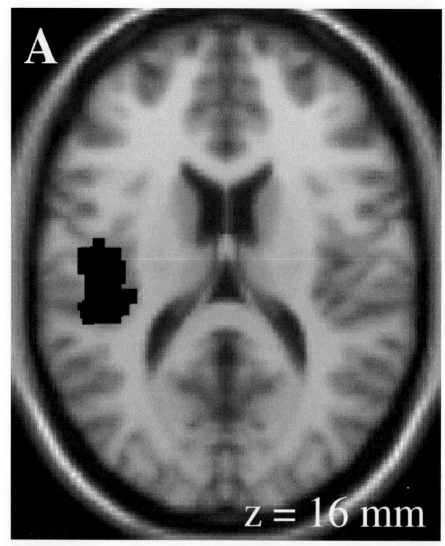

FIGURE 2. Regions showing significant less activation in patients compared to controls during somatosensory stimulation (A) and during auditory stimulation (B); $p < 0.01$.

IV. DISCUSSION

A. Methodological Considerations

Both patients were scanned during awake periods as demonstrated by polygraphic monitoring. Our activation paradigm compared external stimulation with the resting state (no stimulation). The used stimuli were passive, quantifiable, and reproducible and permitted simultaneous and simple acquisition of evoked potentials (SEP and AEP) during acquisition of CBF-PET data. Hence, essential information on the integrity of the peripheral and central somatosensory and auditory pathways was obtained using an independent method. These simple paradigms offered a robust activation of well-known somatosensory (Burton *et al.*, 1993; Fox *et al.*, 1987; Ibanez *et al.*, 1995; Seitz and Roland, 1992;

Tempel and Perlmutter, 1992) and auditory (Binder *et al.*, 1994; Engelien *et al.*, 1995; Hirano *et al.*, 1997; Strainer *et al.*, 1997; Zatorre *et al.*, 1992) neural networks in normal subjects. Another valuable experimental approach, which we choose not to follow in this study, would have been to use complex stimuli (de Jong *et al.*, 1997; Menon *et al.*, 1998; Owen *et al.*, 1999). We felt that, in a first step, simple stimulation had to be preferred in a "bottom-up" strategy of exploration of VS patients.

Functional imaging studies in VS patients are methodologically complex. A major problem in PET "activation" studies, and much more so in functional MRI, are movement artifacts. The large majority of patients in VS move their head, trunk, or limbs in meaningless ways. Our experimental paradigm demanded head immobility for nearly 4 h. Repeating scans sessions, in order to obtain a single accept-

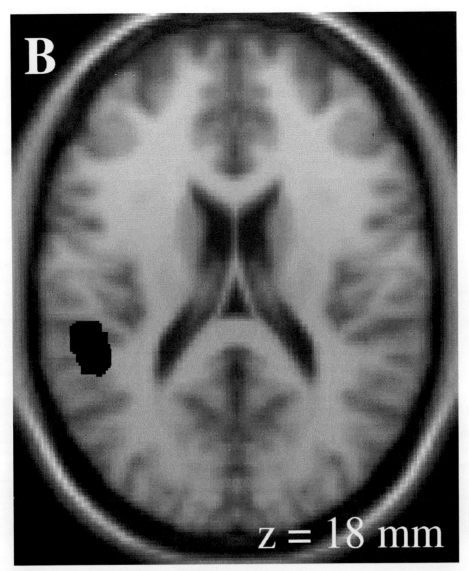

FIGURE 2.   (Continued).

able result, is limited by considerations of logistics, and most importantly, radiation burden. After consent of the family of the patient and following the recommendations of the Ethics Committee of the Faculty of Medicine of our University, both patients received neuromuscular blockers.

PET "activation" studies necessitate hemodynamic stability throughout the scanning session. External stimuli (e.g., sudden loud noise) often induce an arousal response and autonomic reactions (e.g., changes in pulse rate and blood pressure; The Multi-Society Task Force on PVS, 1994). The occurence of such hemodynamic changes between the *rest* and *activation* scans would compromise subsequent statistical analyses. Patient 1 needed light sedation because of autonomic instability (excessive transpiration, tachycardia, arterial hypertension). The use of these drugs is not likely to significantly modify the estimation of the regional distribution

of activity during stimulation. Indeed, if midazolam, at doses higher than those used in our patient, is known to decrease global CBF and metabolism by approximately 10% (Veselis *et al.*, 1997), results concerning regional changes are more contradictory, and none describe changes in somatosensory or auditory cortices. Some studies observed no regional differences (de Wit *et al.*, 1991; Forster *et al.*, 1982; Mathew and Wilson, 1991; Roy-Byrne *et al.*, 1993); others observed decreases in thalamus (Veselis *et al.*, 1997; Volkow *et al.*, 1995) and prefrontal cortex (Veselis *et al.*, 1997). The effect of midazolam on rCBF activation during somatosensory or auditory stimulation is not known. Previous PET studies in healthy subjects have shown sufentanyl-induced rCBF decreases in some frontal areas and cerebellum, while rCBF increased in cingulate, orbitofrontal, and medial prefrontal cortices and caudate nuclei (Firestone *et al.*, 1996). Thermoalgic stimulation during fentanyl administration does not induce stimulation-related rCBF alterations in somatosensory structures (Adler *et al.*, 1997).

B. Somatosensory Stimulation in Vegetative State

Notwithstanding the substantial metabolic impairment, peripheral sensory input still activated cortical area SI (encompassing BA 3b, 1 and 2) in both VS patients. Area 3b is a conduit for all cutaneous sensibility and is thought to be the first cortical step in the analysis of somatosensory inputs, areas 1 (texture perception) and 2 (stereognosis) being a second cortical step (Carlson, 1980; Kaas, 1993). Significantly less activation during somatosensory stimulation in VS than in controls, was found in SII. This area is considered as a higher order stage in a hierarchy of somatosensory areas. For instance, its destruction leaves a monkey unable to discriminate objects on the basis of their size and texture (Carlson and Burton, 1988; Murray and Mishkin, 1984) and causes deficits in tactile object recognition tasks in humans (Caselli, 1993). It is seen as the obligatory route taken by sensory inputs mediating memory and tactual learning (Friedman *et al.*, 1986).

C. Auditory Stimulation in Vegetative State

In patients as in controls, auditory stimulation activated contralateral AI. In alert monkeys, the AI core area encodes for the spectral and temporal features of sound (Heffner, 1997). Direct cortical stimulation of AI in humans resulted in (most often contralateral) perception of elementary sounds (e.g., ringing, humming, clicking, rushing, chirping, buzzing, knocking, rumbling) but never of words or voices (Penfield and Jasper, 1954). Regions that showed a significant lack of activation in patients compared to controls were localized in contralateral temporoparietal AII. Destruction of this area results in contralateral hemi-inattention, extinction, or frank neglect of auditory stimuli in both monkeys (Heilman *et al.*, 1971) and humans (Heilman and Valenstein, 1972).

D. Functional Significance of the Findings

Our results show that, despite a substantial decrease in cerebral metabolism, peripheral sensory input can still activate cortical primary somatosensory (SI) or primary auditory (AI) cortex in VS patients. In contrast, higher order stages in somatosensory (SII) or auditory (AII) processing, which are relays for the incoming sensory information to the limbic system, failed to activate. The present study does not permit to know to what extent vegetative patients can consciously experience external stimuli. Indeed, although it is unlikely, it cannot be excluded that the observed activation of SI or AI, may carry some conscious somatic perception. Further

research is needed to evaluate to what extent VS patients may be conscious of their external world.

Acknowledgements

P.M. is a Senior Research Associate at the Fonds National de la Recherche Scientifique de Belgique (FNRS). This research was supported by FNRS Grant 3.4536.99, by the Queen Elisabeth Medical Foundation (Belgium) and by Research Grants from the University of Liège.

References

Adler, L. J., Gyulai, F. E., Diehl, D. J., Mintun, M. A., Winter, P. M., and Firestone, L. L. (1997). Regional brain activity changes associated with fentanyl analgesia elucidated by positron emission tomography. *Anesth. Analg.* **84**: 120–126.

Ashburner, J., Andersson, J. L., and Friston, K. J. (1999). High-dimensional image registration using symmetric priors. *Neuroimage* **9**: 619–628.

Binder, J. R., Rao, S. M., Hammeke, T. A., Yetkin, F. Z., Jesmanowicz, A., Bandettini, P. A., Wong, E. C., Estkowski, L. D., Goldstein, M. D., Haughton, V. M., *et al.* (1994). Functional magnetic resonance imaging of human auditory cortex. *Ann. Neurol.* **35**: 662–672.

Burton, H., Videen, T. O., and Raichle, M. E. (1993). Tactile-vibration-activated foci in insular and parietal-opercular cortex studied with positron emission tomography: Mapping the second somatosensory area in humans. *Somatosens. Mot. Res.* **10**: 297–308.

Carlson, M. (1980). Characteristics of sensory deficits following lesions of Brodmann's areas 1 and 2 in the postcentral gyrus of Macaca mulatta. *Brain Res.* **204**: 424–430.

Carlson, M., and Burton, H. (1988). Recovery of tactile function after damage to primary or secondary somatic sensory cortex in infant Macaca mulatta. *J. Neurosci.* **8**: 833–859.

Caselli, R. J. (1993). Ventrolateral and dorsomedial somatosensory association cortex damage produces distinct somesthetic syndromes in humans. *Neurology* **43**: 762–771.

de Jong, B., Willemsen, A.T., and Paans, A.M. (1997). Regional cerebral blood flow changes related to affective speech presentation in persistent vegetative state. *Clin. Neurol. Neurosurg.* **99**: 213–216.

de Wit, H., Metz, J., Wagner, N., and Cooper, M. (1991). Effects of diazepam on cerebral metabolism and mood in normal volunteers. *Neuropsychopharmacology* **5**: 33–41.

Engelien, A., Silbersweig, D., Stern, E., Huber, W., Doring, W., Frith, C., and Frackowiak, R. S. (1995). The functional anatomy of recovery from auditory agnosia. A PET study of sound categorization in a neurological patient and normal controls. *Brain* **118**: 1395–1409.

Firestone, L. L., Gyulai, F., Mintun, M., Adler, L. J., Urso, K., and Winter, P. M. (1996). Human brain activity response to fentanyl imaged by positron emission tomography. *Anesth. Analg.* **82**: 1247–1251.

Forster, A., Juge, O., and Morel, D. (1982). Effects of midazolam on cerebral blood flow in human volunteers. *Anesthesiology* **56**: 453–455.

Fox, P. T., Burton, H., and Raichle, M. E. (1987). Mapping human somatosensory cortex with positron emission tomography. *J. Neurosurg.* **67**: 34–43.

Friedman, D. P., Murray, E. A., O'Neill, J. B., and Mishkin, M. (1986). Cortical connections of the somatosensory fields of the lateral sulcus of macaques: Evidence for a corticolimbic pathway for touch. *J. Comp. Neurol.* **252**: 323–347.

Friston, K., Ashburner, J., Frith, C., Poline, J.B., Heather, J., and Frackowiak, R. S. J. (1995a). Spatial realignment and normalization of images. *Hum. Brain Mapping* **2**: 165–189.

Friston, K. J. (1997). Analysing brain images: principles and overview. *In* "*Human Brain Function*" (R. S. J. Frackowiak, K. J. Friston, C. D. Frith,

R. J. Dolan, and J. C. Mazziotta, Eds.), pp. 25–41. Academic Press, San Diego.

Friston, K. J., Ashburner, J., Frith, C. D., Poline, J. B., Heather, J.D., and Frackowiak, R. S. J. (1995b). Spatial registration and normalization. *Hum. Brain Mapping* **3**: 1–25.

Friston, K. J., Holmes, A. P., Worseley, K. J., Poline, J. B., Frith, C. D., and Frackowiak, R. S. J. (1995c). Statistical parametric maps in functional imaging: A general linear approach. *Hum. Brain Mapping* **2**: 189–210.

Heffner, H. E. (1997). The role of macaque auditory cortex in sound localization. *Acta Otolaryngol.* **532** (Suppl): 22–27.

Heilman, K. M., Pandya, D. N., Karol, E. A., and Geschwind, N. (1971). Auditory inattention. *Arch. Neurol.* **24**: 323–325.

Heilman, K. M., and Valenstein, E. (1972). Auditory neglect in man. *Arch. Neurol.* **26**: 32–35.

Hirano, S., Naito, Y., Okazawa, H., Kojima, H., Honjo, I., Ishizu, K., Yenokura, Y., Nagahama, Y., Fukuyama, H., and Konishi, J. (1997). Cortical activation by monaural speech sound stimulation demonstrated by positron emission tomography. *Exp. Brain Res.* **113**: 75–80.

Ibanez, V., Deiber, M. P., Sadato, N., Toro, C., Grissom, J., Woods, R. P., Mazziotta, J. C., and Hallett, M. (1995). Effects of stimulus rate on regional cerebral blood flow after median nerve stimulation. *Brain* **118**: 1339–1351.

Kaas, J.H. (1993). The functional organization of somatosensory cortex in primates. *Anat. Anz.* **175**: 509–518.

Laureys, S., Faymonville, M. E., and Lamy, M. (2000a). Cerebral function in vegetative state studied by positron emission tomography. *In "2000 Yearbook of Intensive Care and Emergency Medicine"* (J. L. Vincent, Ed.) pp. 587–597. Springer-Verlag, Heidelberg.

Laureys, S., Faymonville, M. E., Luxen, A., Lamy, M., Franck, G., and Maquet, P. (2000b). Restoration of thalamo-cortical connectivity after recovery from persistent vegetative state. *Lancet* **355**: 1790–1791.

Laureys, S., Goldman, S., Phillips, C., Van Bogaert, P., Aerts, J., Luxen, A., Franck, G., and Maquet, P. (1999b). Impaired effective cortical connectivity in vegetative state: Preliminary investigation using PET. *Neuroimage* **9**: 377–382.

Laureys, S., Lemaire, C., Maquet, P., Phillips, C., and Franck, G. (1999c). Cerebral metabolism during vegetative state and after recovery to consciousness. *J. Neurol. Neurosurg. Psychiatry* **67**: 121.

Mathew, R. J., and Wilson, W. H. (1991). Evaluation of the effects of diazepam and an experimental anti-anxiety drug on regional cerebral blood flow. *Psychiatry Res.* **40**: 125–134.

Menon, D. K., Owen, A. M., Williams, E. J., Minhas, P. S., Allen, C. M., Boniface, S. J., and Pickard, J. D. (1998). Cortical processing in persistent vegetative state. *Lancet* **352**: 200.

Murray, E. A., and Mishkin, M. (1984). Relative contributions of SII and area 5 to tactile discrimination in monkeys. *Behav. Brain Res.* **11**: 67–83.

Owen, A. M., Menon, D. K., Williams, E. J., Minhas, P. S., Johnsrude, I. S., Scott, S. K., Allen, C. M., Boniface, S. J., Kendall, I. V., Downey, S. P. M. J., Antoun, N., Clarck, J. C., and Pickard, J. D. (1999). Functional imaging in persistent vegetative state. *Neuroimage* **9**(Suppl): 581. [Abstract]

Penfield, W., and Jasper, H. (1954). *"Epilepsy and the Functional Anatomy of the Human Brain"* pp. 113–116. Little, Brown, Boston.

Phelps, M. E., Huang, S. C., Hoffman, E. J., Selin, C., Sokoloff, L., and Kuhl, D. E. (1979). Tomographic measurement of local cerebral glucose metabolic rate in humans with (F-18)-2-fluoro-2deoxy-D-glucose: Validation of method. *Ann. Neurol.* 371–388.

Ribary, U., Schiff, N., Kronberg, E., Plum, F., and Llinas, R. (1998). Fractured brain function in unconscious humans: Functional brain imaging using MEG. *Neuroimage* **7**(Suppl): 106. [Abstract]

Roy-Byrne, P., Fleishaker, J., Arnett, C., Dubach, M., Stewart, J., Radant, A., Veith, R., and Graham, M. (1993). Effects of acute and chronic alprazolam treatment on cerebral blood flow, memory, sedation, and plasma catecholamines. *Neuropsychopharmacology* **8**: 161–169.

Seitz, R. J., and Roland, P. E. (1992). Vibratory stimulation increases and decreases the regional cerebral blood flow and oxidative metabolism: A positron emission tomography (PET) study. *Acta Neurol. Scand.* **86**: 60–67.

Strainer, J. C., Ulmer, J. L., Yetkin, F. Z., Haughton, V. M., Daniels, D. L., and Millen, S. J. (1997). Functional MR of the primary auditory cortex: An analysis of pure tone activation and tone discrimination. *Am. J. Neuroradiol.* **18**: 601–610.

Talairach, J., and Tournoux, P. (1988). *"Co-planar Stereotaxic Atlas of the Human Brain."* Thieme-Verlag, Stuttgart, 1988.

Tempel, L. W., and Perlmutter, J. S. (1992). Vibration-induced regional cerebral blood flow responses in normal aging. *J. Cereb. Blood Flow Metab.* **12**: 554–561.

The Multi-Society Task Force on PVS (1994). Medical aspects of the persistent vegetative state (1). *N. Engl. J. Med.* **330**: 1499–1508.

Veselis, R. A., Reinsel, R. A., Beattie, B. J., Mawlawi, O. R., Feshchenko, V. A., DiResta, G. R., Larson, S. M., and Blasberg, R. G. (1997). Midazolam changes cerebral blood flow in discrete brain regions: An H2(15)O positron emission tomography study. *Anesthesiology* **87**: 1106–1117.

Volkow, N. D., Wang, G. J., Hitzemann, R., Fowler, J. S., Pappas, N., Lowrimore, P., Burr, G., Pascani, K., Overall, J., and Wolf, A. P. (1995). Depression of thalamic metabolism by lorazepam is associated with sleepiness. *Neuropsychopharmacology* **12**: 123–132.

Zatorre, R. J., Evans, A. C., Meyer, E., and Gjedde, A. (1992). Lateralization of phonetic and pitch discrimination in speech processing. *Science* **256**: 846–849.

48

Impaired Cerebral Connectivity in Vegetative State

S. LAUREYS,[*,†] **M. E. FAYMONVILLE,**[‡] **S. GOLDMAN,**[§] **C. DEGUELDRE,**[†] **C. PHILLIPS,**[†,#]
B. LAMBERMONT,[‖] **J. AERTS,**[†] **M. LAMY,**[‡] **A. LUXEN,**[†] **G. FRANCK,**[*,†] **and P. MAQUET**[*,†]

[*]Department of Neurology, [†]Department of Anesthesiology and Intensive Care Medicine and [‖]Department of Internal Medicine,
CHU Sart Tilman, Liège, Belgium
[†]Cyclotron Research Center, University of Liège, Belgium
[#]The Wellcome Department of Cognitive Neurology, Institute of Neurology, Queen Square, London, United Kingdom
[§]PET/Biomedical Cyclotron Unit and Department of Neurology, ULB Erasme, Brussels, Belgium

Vegetative state (VS) is a condition of abolished awareness with persistence of arousal. Awareness is thought to represent an emergent property of cerebral neural networks. Our hypothesis was that part of the neural correlate underlying VS is an altered cortico-cortical and thalamo-cortico-thalamic connectivity. In a retrospective study on four patients in VS of different etiology we demonstrated impairment in effective connectivity between the left frontal cortices and the posterior cingulate cortex using [^{18}F]fluorodeoxyglucose-positron emission tomography. In the present prospective study we employed the same methods to assess regional cerebral glucose metabolism (rCMRGlu) and functional thalamo-cortico-thalamic connectivity in five patients in anoxic–ischemic VS. As expected, mean gray matter metabolism was decreased in VS patients compared to that seen in 53 controls (2.8 ± 0.6 versus 7.1 ± 1.3 mg/100 g/min). Using statistical parametric mapping (SPM) we identified cerebral regions where metabolism was relatively most impaired or, conversely, most spared. Our data showed a common pattern of impaired rCMRGlu in the prefrontal, premotor, and parieto-temporal association areas and the retrosplenial region (encompassing precuneus and posterior cingulate cortex) in VS. Relatively least impaired regions were observed in the brainstem (including the mesopontine reticular formation), hypothalamus, and basal forebrain. In a next step, we demonstrated that in VS patients both thalami have in common that they are less tightly connected with bilateral precuneus than in controls. These results suggest that impairment of human awareness seems related not only to a global impairment of cerebral function, but also to an impairment of cortico-cortical and thalamo-cortico-thalamic connectivity.

I. INTRODUCTION

Consciousness is thought to represent an emergent property of cortical and subcortical neural networks and their reciprocal projections. Its multifaceted aspects can be seen as expressions of various specialized areas of the cortex that are responsible for processing external and internal stimuli, short- and long-term storage, language comprehension and production, information integration and problem solving, and attention (Posner, 1994). One way to approach the study of human consciousness is to explore lesional cases where impairment of consciousness is the most prominent clinical sign. Vegetative state (VS) is such a condition wherein awareness is abolished while arousal persists (ANA Committee on Ethical Affairs, 1993; The Multi-Society Task Force on PVS, 1994).

Neuropathological studies of patients in persistent VS of nontraumatic origin show widespread cortical damage with involvement of the association cortices in conjunction with the primary and secondary sensory cortices (Kinney and Samuels, 1994). Previous positron emission tomography (PET) studies, in patients in VS of diverse origin and duration, showed a reduction in metabolism of overall cortex (De Volder *et al.*, 1990; Levy *et al.*, 1987; Momose *et al.*, 1989; Rudolf *et al.*, 1999; Tommasino *et al.*, 1995). The most profound reduction in metabolic activity has been described in the parieto-occipital and mesiofrontal cortices in persistent VS of anoxic origin (De Volder *et al.*, 1990), while infratentorial structures showed a less distinct hypometabolism (Rudolf *et al.*, 1999). We have previously shown an impairment in cortico-cortical connectivity between left frontal areas and posterior cingulate cortex in four patients in VS of different etiology and duration in a retrospective study (Laureys *et al.*, 1999a). The aim of the present prospective study was to assess thalamo-cortico-thalamic connectivity in VS employing the same methods. Regional cerebral glucose metabolism distribution (rCMR-Glu) was assessed in five cases of VS of anoxic/ischemic origin of different duration (5 to 38 days) by means of statistical parametric mapping (SPM) and [^{18}F]fluorodeoxyglucose ([^{18}F]FDG)-PET. We studied functional connectivity (Friston *et al.*, 1997b) in the bilateral thalamus of VS patients as compared to normal subjects. Given the small number of observations, this paper should be viewed as a preliminary record of an ongoing research.

II. MATERIALS AND METHODS

The study was approved by the Ethics Committee of the University of Liège. Informed consent was obtained by the family of all patients and for all control subjects. The control population consisted of drug-free, healthy volunteers of both sexes ($n = 53$; mean age, 42 ± 21 years).

Patient's characteristics are summarized in Table 1 ($n = 5$; mean age, 42 ± 13 years). At time of scanning, all patients fulfilled the international criteria for VS: (1) spontaneous eye opening; (2) spontaneous respiratory function, preserved blood pressure control, cardiac function, and thermoregulation; (3) no evidence of awareness of the environment; (4) no evidence of reproducible voluntary behavioral responses to any stimuli; (5) no evidence of language comprehension or expression; (6) intermittent wakefulness and behaviorally assessed sleep–wake cycles (ANA Committee on Ethical Affairs, 1993; The Multi-Society Task Force on PVS, 1994). All patients had preserved pupillary, corneal, and vestibulo-ocular reflexes. Auditory evoked potentials (AEP) and somatosensory evoked potential of the median nerve (SEP) were performed the day of scanning. AEP demonstrated preserved midbrain function in all patients. SEP showed the presence of P14 (medial lemniscus) and N20 (primary somatosensory cortex) potentials in all patients—except in patient 2 where only P14 was preserved. Patient 3 was scanned

in the acute and in the persistent phase of VS, only the data of the latter were included for further statistical analysis. In patient 1 no arterial line could be installed, hence no absolute quantification of PET data could be performed.

FDG-PET data were obtained on a Siemens CTI 951 scanner approximately 15–20 mm above the canthomeatal line and in two-dimensional mode. A transmission scan was performed to allow a measured attenuation correction. Five to 10 mCi of FDG was injected intravenously. Arterial blood samples were drawn during the whole procedure and cerebral metabolic glucose rates (CMRGlu) were calculated (Phelps *et al.*, 1979). In patients, EEG and vital parameters (temperature, blood pressure, O_2 saturation, and PCO_2) were monitored throughout the whole procedure. PET data were analyzed using the statistical parametric mapping software (SPM96 version; Welcome Department of Cognitive Neurology, Institute of Neurology, London, UK) implemented in MATLAB (Mathworks Inc., Sherborn, MA). Data from each patient were coregistered to individual T1-weighted MRI of the brain. All data were normalized to a standard stereotactic space (Talairach and Tournoux, 1988) and then smoothed with a 16-mm full-width half-maximum (FWHM) isotropic kernel (Friston, 1997a).

The design matrix included the 5 patients' scans and the 53 control subjects' scans. Global CMRGlu normalization was performed by proportional scaling. The first analysis identified brain regions where glucose metabolism was relatively most impaired in VS patients compared to that seen in controls. The second analysis identified brain regions where glucose metabolism was relatively least impaired in VS compared to that seen in controls. The resulting set of voxel values for each contrast, constituting a map of the t statistics [SPM{t}], was transformed to the unit normal distribution SPM{Z} and thresholded at $p < 0.001$ ($Z = 3.09$). The resulting foci were characterized in terms of peak height over the entire volume analyzed at a threshold of voxel-level corrected $p < 0.05$ (Friston, 1997a).

In a second step, we used a psychophysiological interaction analysis to test the hypothesis on altered functional thalamic connectivity in VS as compared to controls (Friston *et al.*, 1997b). The design matrix included the same scans as described above and took into account group differences in mean levels of glucose consumption. Now the analysis looked for brain regions that experienced a significant difference in reciprocal modulation with/from a selected voxel in left and right mid-thalamus (coordinates: $-12\ -16\ 6$ mm; $12\ -16\ 6$ mm). It assessed the difference in cerebral modulation with/from both of these thalamic voxels depending on the condition of being patient or normal control. Results were significant at voxel level corrected $p < 0.05$ ($Z \geq 4.19$).

TABLE 1. Patient's Demographic and PET Data

Patient					MRI	CMRGlu in mg/100 g/min	
No.	Gender	Age	Etiology of VS	EEG	(insult–MRI interval)	(insult–PET interval)	Outcome
1	F	39	Cardiorespiratory arrest (asthma)	Nonreactive 10 Hz	Basal ganglia hemorrhagic ischemic (day 14)	— (day 22)	Died after 3.5 months
2	M	37	Cardiorespiratory arrest (heroin overdose)	Nonreactive 8 Hz	Normal (day 9)	3.66 (day 5)	Remained in VS (3 months followup)
3	F	42	Respiratory arrest (hanging)	Nonreactive 8 Hz	Normal (day 21)	1.96 (day 5) 2.08 (day 38)	Remained in VS (3 months followup)
4	M	28	Cardiorespiratory arrest (methadone intake)	Nonreactive 5 Hz	Basal ganglia hemorrhagic ishemic (day 16)	3.13 (day 13)	Recovered (day 46)
5	F	63	Ischemic encephalopathy (iatrogenic hypothesion)	Reactive 6 Hz	Basal ganglia hemorrhagic ischemic, Leucoencephalopathy (day 17)	2.33 (day 36)	Recovered (day 51)

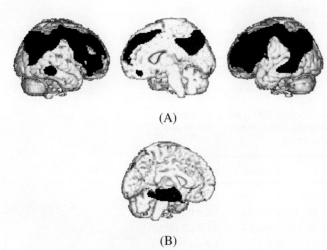

(A)

(B)

FIGURE 1. The common pattern of most impaired (A) and relatively least impaired (B) cerebral metabolism characterizing vegetative state patients. SPM{Z} thresholded at voxel and cluster level corrected $p < 0.05$ ($Z > 4.01$), normalized to the stereotaxic space of Talairach (Talairach and Tournoux, 1988), and projected on a normalized MRI template.

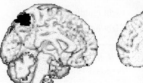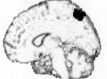

FIGURE 2. Localization of voxels that showed impaired connectivity with both thalami in patients in vegetative state compared to controls. SPM{Z} thresholded at voxel and cluster level $p < 0.05$ ($Z > 4.19$) projected on a normalized MRI template.

tified glucose metabolism in some selected cerebral areas are shown in Table 3.

We observed a significant difference in functional connectivity between bilateral thalamus and bilateral precuneus (BA 7; coordinates: –18 –56 58; $Z = 4.62$; p corrected $= 0.009$ and coordinates: 18 –70 62; $Z = 4.22$; p corrected $= 0.041$) in patients in VS as compared to controls (Fig. 2).

III. RESULTS

Mean gray matter glucose metabolism in each VS patient is shown in Table 1. A mean reduction of 60% was observed compared to controls (2.8 ± 0.6 versus 7.1 ± 1.3 mg/100 g/min, mean $\pm$ SD).

A common pattern of relatively most impaired rCMRGlu in the five VS patients was observed in bilateral frontal and parietal association cortices, the left postcentral, middle temporal and superior occipital gyrus, and the posterior cingulate cortex/precuneus (Fig. 1A and Table 2). Regions where metabolism was relatively least impaired were confined to the brainstem (encompassing the mesopontine reticular formation), hypothalamus, and basal forebrain (Fig. 1B). Quan-

IV. DISCUSSION

At the time of PET scanning all patients were in a VS according to the international criteria (ANA Committee on Ethical Affairs, 1993; The Multi-Society Task Force on PVS, 1994). It is important to distinguish between VS and *persistent* VS, the latter being defined as a VS that has continued or endured for at least 1 month. In the present study, we were interested in the impairments of functional thalamo-cortico-thalamic connectivity in VS independent of its duration or outcome. Patient 1 died, patients 2 and 3 remained in VS, and patients 4 and 5 recovered.

As expected, a substantial reduction in global cerebral metabolic rate was observed in our VS patients, ranging from 48 to 72%. On a regional basis, the presented data

TABLE 2.   Statistical Results and Localization of the Most Significant Voxels where rCMRGlu Is Decreased in Patients in Vegetative State Compared to Controls (Voxel-Level-Corrected $p < 0.05$)

Area (Brodmann's area)		X	Y	Z	Z score	p (corrected)
Precuneus/posterior cingular	(7/31)	2	−44	38	6.59	<0.001
Left middle frontal gyrus	(6)	−48	−2	56	6.38	<0.001
Left middle frontal gyrus	(8)	−42	28	46	5.76	<0.001
Left middle frontal gyrus	(9)	−42	12	38	6.19	<0.001
Left middle frontal gyrus	(10)	−30	58	12	5.48	<0.001
Left inferior frontal gyrus	(45/46)	−52	24	18	5.17	0.001
Left inferior frontal gyrus	(47)	−40	30	−4	4.93	0.002
Left inferior parietal lobule	(40)	−56	−56	38	6.18	<0.001
Left postcentral gyrus	(2)	−54	−20	52	4.93	0.004
Left middle temporal gyrus	(21)	−66	−42	−8	4.56	0.012
Left superior occipital gyrus	(19)	−22	−76	42	4.33	0.029
Right superior frontal gyrus	(6)	26	12	66	5.29	<0.001
Right middle frontal gyrus	(10)	34	54	20	5.12	0.001
Right middle frontal gyrus	(46/9)	46	28	26	4.81	0.001
Right inferior parietal lobule	(40)	50	−60	44	5.42	<0.001

Note. Coordinates are defined in the stereotaxic space of Talairach (Talairach and Tournoux, 1988).

TABLE 3.   Regional Glucose Metabolism (in mg/100 g.min) in Selected Voxels in Controls and Patients in Vegetative State (mean ± SD, % are Compared to Controls)

Area	x	y	z	Controls	Patients
Precuneus	−12	−56	44	8.3 ± 1.6	2.7 ± 0.9 (67%)
Left middle frontal gyrus	−48	−2	56	7.0 ± 1.4	2.2 ± 0.7 (68%)
Left inferior parietal lobule	−56	−56	38	6.3 ± 1.4	2.1 ± 0.7 (66%)
Left thalamus	12	−16	6	6.2 ± 1.3	2.9 ± 1.0 (53%)
Mesopontine region	−2	−30	−14	4.1 ± 1.0	2.5 ± 0.9 (39%)

Note. Coordinates are defined in the stereotaxic space of Talairach (Talairach and Tournoux, 1988).

confirm a significant dysfunction of the prefrontal, premotor, and parieto-temporal association areas (with a left sided predominance) and posterior cingulate cortex/precuneus as the common neural correlate in VS patients (Laureys *et al,* 1999a). Metabolism in the association cortices seemed most impaired (±70% lower than controls) while the brainstem (presumed to encompass the mesopontine reticular formation) was least impaired (±40% lower than controls); the thalamus showed an intermediate decrease in rCMRGlu. This pattern of metabolic impairment is in agreement with postmortem findings in persistent VS where involvement of the association cortices is reported as critical neuroanatomic substrate (Kinney and Samuels, 1994). On the contrary, regions that seemed relatively spared in our VS patients (brainstem, hypothalamus and basal forebrain), are known to be

relatively spared in persistent VS of anoxic–ischemic origin (Dougherty *et al.*, 1981; Ingvar *et al.*, 1978).

We observed that the functional relationships between both thalami and bilateral precuneus are different in VS patients than in controls. The thalamus contains both specific thalamo-cortical relay nuclei and so-called nonspecific intralaminar nuclei. The former are the necessary relay for all sensory afferent stimuli (except some olfactive information) (Steriade *et al.*, 1997). The latter have been implicated in the maintenance of thalamo-cortico-thalamic synchronous oscillations. Among these activities, 40 Hz oscillations seem to be deeply, although not exclusively, involved in conscious experience (Llinas *et al.*, 1998). Thus thalamic nuclei seem critical for the maintenance of human awareness. However, given the limited spatial resolution (16 mm FWHM smoothing of the data), a precise localization of thalamic nuclei cannot be attained in the present study.

Numerous neuropsychological studies in nonhuman primates suggest the role of the retrosplenial region (encompassing precuneus and posterior cingulate cortex) in spatial orientation and memory (Vogt *et al.*, 1992). In addition, clinical studies have implicated this region in anterograde and retrograde amnesia in humans (Rudge and Warrington, 1993; Valenstein *et al.*, 1987). Functional imaging studies have shown activation of the precuneus in episodic verbal memory retrieval (Shallice *et al.*, 1994), modulation of visual perception by attention (Dolan *et al.*, 1997) and mental imagery (Fletcher *et al.*, 1995). Interestingly, the precuneus and posterior cingulate cortex are one of the most active cerebral regions in conscious waking (Andreasen *et al.*, 1995) and are systematically one of the least active regions in unconscious or minimally conscious states such as halothane anesthesia (Alkire *et al.*, 1999), slow wave sleep (Maquet *et al.*, 1997), rapid eye movement sleep (Maquet *et al.*, 1996), and hypnotic state (Maquet *et al.*, 1999). Retrosplenial cortex is the site of earliest reductions in glucose metabolism (Minoshima *et al.*, 1997) in Alzheimer's disease. As has been presented at this meeting, it also seems to be the most metabolically impaired region in Wernicke–Korsakoff and postanoxic amnestic patients (Aupée *et al.*, 1999). These arguments suggest that the retrosplenial area, encompassing precuneus and posterior cingulate cortex, might represent part of the neural network subserving conscious experience.

As our statistical model included only influences from thalamic regions, the observed inferences about the psychophysiological interactions pertain to *functional* connectivity as defined by Friston *et al.* (1997b). It seems that the sum of activity in bilateral mid-thalamic regions differentially predicts activity in bilateral precuneus between the five examined patients in VS and the control population.

It remains controversial whether the observed metabolic impairment in VS reflects functional and potentially reversible damage or irreversible structural neuronal loss. Rudolf and coworkers argue for the latter, using [^{11}C]flumazenil as a marker of neuronal integrity in evaluating acute anoxic VS patients (see Chapter 49 in this volume). We plead for a capacity of partial neuronal functional recovery. In our five patients, of whom two recovered, structural imaging (MRI) showed no cortical lesions and yet functional imaging (FDG-PET) showed a predominantly cortical metabolic impairment, proving a structural–functional dissociation. Moreover, we recently published an intriguing case of VS after CO intoxication where MRI was normal (day 14) and global cerebral glucose utilization remained essentially the same during VS (4.5 mg/100 g/min) and after recovery to consciousness (4.7 mg/100 g/min). Hence, in this patient, recovery of awareness seemed to be related to a modification of the regional distribution of brain function

rather than to the global resumption of cerebral metabolism. The main decreases in metabolism seen during VS but not after recovery were found in parietal areas including retrosplenial cortex (Laureys *et al.*, 1999b). We hypothesize that impairment in cortico-cortical and thalamo-cortico-thalamic connectivity can explain part of the permanent, or, in some fortunate cases, transient, functional cortical impairment in VS.

This report provides the first and preliminary argument suggesting a disturbed thalamo-cortico-thalamic connectivity in VS patients. Impairment of human awareness seems related not only to a global impairment of cerebral function, but also to an impairment of cortico-cortical and thalamo-cortico-thalamic connectivity.

Acknowledgments

This research was supported by Fonds National de la Recherche Scientifique (FNRS) Grant 3.4571.97 and by the Queen Elisabeth Medical Foundation. P.M. is Research Associate at the Belgian FNRS.

References

Alkire, M. T., Pomfrett, C. J., Haier, R. J., Gianzero, M. V., Chan, C. M., Jacobsen, B. P., and Fallon, J. H. (1999). Functional brain imaging during anesthesia in humans: Effects of halothane on global and regional cerebral glucose metabolism. *Anesthesiology* **90**: 701–709.

ANA Committee on Ethical Affairs (1993). Persistent vegetative state: Report of the American Neurological Association Committee on Ethical Affairs. *Ann.Neurol.* **33**: 386–390.

Andreasen, N. C., O'Leary, D. S., Cizadlo, T., Arndt, S., Rezai, K., Watkins, G. L., Ponto, L. L., and Hichwa, R. D. (1995). Remembering the past: Two facets of episodic memory explored with positron emission tomography. *Am. J. Psychiatry* **152**: 1576–1585.

Aupée, A.-M., Desgranges, B., Eustache, F., Lalevée, C., de la Sayette, V., Viader, F., and Baron, J.-C. (1999). Mapping the neural network involved in the amnestic syndrome with FDG-PET and SPM. *J. Cereb. Blood Flow Metab.* **19**(Suppl 1): 777.

De Volder, A. G., Goffinet, A. M., Bol, A., Michel, C., de, B. T., and Laterre, C. (1990). Brain glucose metabolism in postanoxic syndrome. Positron emission tomographic study. *Arch. Neurol.* **47**: 197–204.

Dolan, R. J., Fink, G. R., Rolls, E., Booth, M., Holmes, A., Frackowiak, R. S., and Friston, K. J. (1997). How the brain learns to see objects and faces in an impoverished context. *Nature* **389**: 596–599.

Dougherty, J. H., Jr, Rawlinson, D. G., Levy, D. E., and Plum, F. (1981). Hypoxic–ischemic brain injury and the vegetative state: Clinical and neuropathologic correlation. *Neurology* **31**: 991–997.

Fletcher, P. C., Frith, C. D., Baker, S. C., Shallice, T., Frackowiak, R. S., and Dolan, R. J. (1995). The mind's eye-precuneus activation in memory-related imagery. *Neuroimage* **2**: 195–200.

Friston, K. J. (1997a). Analysing brain images: Principles and overview. *In* "Human Brain Function" (R. S. J. Frackowiak, K. J. Friston, C. D. Frith, R. J. Dolan, and J. C. Mazziotta, Eds.), pp. 25–41. Academic Press, San Diego.

Friston, K. J., Buechel, C., Fink, G. R., Morris, J., Rolls, E., and Dolan, R. J. (1997b). Psychophysiological and modulatory interactions in neuroimaging. *Neuroimage* **6**: 218–229.

Ingvar, D. H., Brun, A., Johansson, L., and Samuelsson, S.M. (1978). Survival after severe cerebral anoxia with destruction of the cerebral cortex: The apallic syndrome. *Ann. N. Y. Acad. Sci.* **315**: 184–214.

Kinney, H. C., and Samuels, M. A. (1994). Neuropathology of the persistent vegetative state: A review. *J. Neuropahtol. Exp. Neurol.* **53**: 548–558.

Laureys, S., Goldman, S., Phillips, C., Van Bogaert, P., Aerts, J., Luxen, A., Franck, G., and Maquet, P. (1999a). Impaired effective cortical connectivity in vegetative state: Preliminary investigation using PET. *Neuroimage* **9**: 377–382.

Laureys, S., Phillips, C., Lemaire, C., Luxen, A., Franck, G., and Maquet, P. (1999b). Cerebral metabolism during vegetative state and after recovery to consciousness. *J. Neurol. Neurosurg. Psychiatry* **67**: 121–122.

Levy, D. E., Sidtis, J. J., Rottenberg, D. A., Jarden, J. O., Strother, S. C., Dhawan, V., Ginos, J. Z., Tramo, M. J., Evans, A. C., and Plum, F. (1987). Differences in cerebral blood flow and glucose utilization in vegetative versus locked-in patients. *Ann. Neurol.* **22**: 673–682.

Llinas, R., Ribary, U., Contreras, D., and Pedroarena, C. (1998). The neuronal basis for consciousness. *Philos. Trans. R. Soc. London B* **353**: 1841–1849.

Maquet, P., Degueldre, C., Delfiore, G., Aerts, J., Peters, J. M., Luxen, A., and Franck, G. (1997). Functional neuroanatomy of human slow wave sleep. *J. Neurosci.* **17**: 2807–2812.

Maquet, P., Faymonville, M. E., Degueldre, C., Delfiore, G., Franck, G., Luxen, A., and Lamy, M. (1999). Functional neuroanatomy of hypnotic state. *Biol. Psychiatry* **45**: 327–333.

Maquet, P., Peters, J., Aerts, J., Delfiore, G., Degueldre, C., Luxen, A. and Franck, G. (1996). Functional neuroanatomy of human rapid-eye-movement sleep and dreaming. *Nature* **383**: 163–166.

Minoshima, S., Giordani, B., Berent, S., Frey, K. A., Foster, N. L., and Kuhl, D. E. (1997). Metabolic reduction in the posterior cingulate cortex in very early Alzheimer's disease. *Ann. Neurol.* **42**: 85–94.

Momose, T., Matsui, T., and Kosaka, N. (1989). Effect of cervical spinal cord stimulation (cSCS) on cerebral glucose metabolism and blood flow in a vegetative patient assessed by positron emission tomography (PET)

and single photon emission computed tomography (SPECT). *Radiat. Med.* **7**: 243–246.

Phelps, M. E., Huang, S. C., Hoffman, E. J., Selin, C., Sokoloff, L., and Kuhl, D. E. (1979). Tomographic measurement of local cerebral glucose metabolic rate in humans with (F-18)-2-fluoro-2deoxy-D-glucose: Validation of method. *Ann. Neurol.* **6**: 371–388.

Posner, M. I. (1994). Attention: The mechanisms of consciousness. *Proc. Natl. Acad. Sci. USA* **91**: 7398–7403.

Rudge, P., and Warrington, F. (1993). Selective impairment of memory and visual perception in splenial tumors. *Brain* **114**: 349–360.

Rudolf, J., Ghaemi, M., Haupt, W. F., Szelies, B., and Heiss, W. D. (1999). Cerebral glucose metabolism in acute and persistent vegetative state. *J. Neurosurg. Anesthesiol.* **11**: 17–24.

Shallice, T., Fletcher, P. C., Frith, C. D., Grasby, P., Frackowiak, R. S. J., and Dolan, R. J. (1994). Brain regions associated with acquisition and retrieval of verbal episodic memory. *Nature* **368**: 633–635.

Steriade, M., Jones, E. G., and McCormick, D. (1997). *"Thalamus."* Elsevier, Amsterdam, New York.

Talairach, J., and Tournoux, P. (1988). *"Co-planar Stereotaxic Atlas of the Human Brain."* Thieme-Verlag Stuttgart.

The Multi-Society Task Force on PVS (1994). Medical aspects of the persistent vegetative state. *N. Engl. J. Med.* **330**: 1499–1579.

Tommasino, C., Grana, C., Lucignani, G., Torri, G., and Fazio, F. (1995). Regional cerebral metabolism of glucose in comatose and vegetative state patients. *J. Neurosurg. Anesthesiol.* **7**: 109–116.

Valenstein, E., Bowers, D., Verfaillie, M., Heilman, K. M., Day, A., and Watson, R. T. (1987). Retrosplenial amnesia. *Brain* **110**: 1631–1646.

Vogt, B. A., Finch, D. M., and Olson, C. R. (1992). Functional heterogeneity in cingulate cortex: The anterior executive and posterior evaluative regions. *Cereb. Cortex.* **2**: 435–443.

49

Alterations of Cerebral Glucose Metabolism and Benzodiazepine Receptor Density in Acute and Persistent Vegetative State

J. RUDOLF,*,† Me. GHAEMI,† Mo. GHAEMI,* J. SOBESKY,† W. F. HAUPT,* B. SZELIES,*
M. GROND,* and W.-D. HEISS*,†

*Klinik für Neurologie der Universität zu Köln
†Max-Planck-Institut für Neurologische Forschung, Köln

The probability of recovery from severe anoxic injury is an important issue in critical care medicine. Clinical signs and laboratory or functional tests permit only an indirect estimation of the extent of structural brain damage in prolonged postanoxic states. With the intention of identifying positron emission tomography (PET) parameters with prognostic value, regional cerebral glucose metabolism was investigated with [^{18}F]-2-fluoro-2-deoxy-D-glucose ([^{18}F]FDG) and PET in 24 patients with acute (AVS, duration <1 month, n = 11) or persistent (PVS, duration >1 month, n = 13) vegetative state (VS). In an additional 9 patients with AVS (duration <1 month), regional cerebral glucose metabolism and benzodiazepine receptor distribution were assessed using [^{11}C]flumazenil ([^{11}C]FMZ)- and [^{18}F]FDG-PET. Overall glucose utilization was significantly reduced in VS in comparison to age-matched controls. Infratentorial stuctures showed a less distinct hypometabolism. Cortical metabolic rates in patients with PVS were significantly reduced when compared to patients studied in AVS. [^{11}C] FMZ-PET in AVS showed a considerable reduction of benzodiazepine receptor binding sites in all cortical regions that grossly corresponded to the extent of reduction of cerebral glucose metabolism assessed with [^{18}F]FDG-PET, while the cerebellum was spared from neuronal loss. The results of [^{11}C]FMZ-PET in AVS represent the first report on alterations of benzodiazepine receptor density following severe anoxic brain injury. They document the irreversible loss of supratentorial benzodiazepine receptors after global brain hypoxia. The comparison of [^{18}F]FDG and [^{11}C]FMZ-PET findings in AVS demonstrates that alterations of cerebral glucose consumption do not represent mere functional inactivation, but irreversible structural brain damage.

I. INTRODUCTION

The probability of recovery from severe anoxic injury is an important issue in critical care medicine. A reliable assessment of prognosis in vegetative state (VS) following cerebral anoxia is mandatory for all decisions concerning initiation or prolongation of extended intensive care procedures. Clinical signs and laboratory or functional tests (e.g. electroencephalography, evoked potentials) permit only an indirect estimation of the extent of structural brain damage in postanoxic states (Bates, 1991; Multi Society Task Force, 1994). The estimation of global and regional cerebral glucose consumption using positron emission tomography (PET) and [^{18}F]-2-fluoro-2-deoxy-D-glucose ([^{18}F]FDG) and the assessment of benzodiazepine receptor density with the benzodiazepine receptor ligand [^{11}C]flumazenil ([^{11}C]FMZ) permit insights into the extent of impairment of cerebral functions after severe global anoxia. [^{18}F]FDG-PET findings in VS are scarce, and their prognostic significance is not sufficiently established (de Volder *et al.*, 1990; Tommasino *et al.*, 1995; Rudolf *et al.*, 1999). So far, [^{11}C]FMZ-PET studies in VS have not been reported.

II. METHODS

Regional cerebral glucose metabolism ($rCMR_{glc}$) was investigated with [^{18}F]-2-fluoro-2-deoxy-D-glucose ([^{18}F]FDG) and positron emission tomography (PET) in 24 patients (group 1) with acute (AVS, duration <1 month, $n = 11$) or persistent (PVS, duration >1 month, $n = 13$) vegetative state (VS) following prolonged anoxia due to cardiorespiratory arrest. Upon examination, all patients were free of sedative drugs for more than 48 h and presented in a vegetative state: They were awake, but nonsentient, not obeying commands; verbal contact was impossible, and there was no meaningful response to sensory stimuli. Most of them showed one or more of the following phenomena: spontaneous eye opening, preservation of sleep–wake rhythm and intact primitive responses, e.g., swallowing reflexes. In accordance with current definitions (Quality Standards Subcommittee 1995), in patients with symptom onset within 1 month prior to the PET study, the syndrome was classified as acute (AVS), otherwise as persistent vegetative state (PVS).

The control group consisted of 18 healthy volunteers (14 male, 4 female), ages 28 to 73 years (mean age 46 years).

Regional cerebral glucose metabolism ($rCMR_{glc}$) was assessed on a positron scanner with 24 detector rings (ECAT EXACT, Siemens CTI, for technical data see Wienhard *et al.*, 1992) using the [^{18}F]FDG method (Reivich *et al.*, 1979). Scans between the 30 and 60 min of data acquisition and multiple arterialized venous blood samples were used to calculate CMR_{glc} based on the Sokoloff model, with adaptation of K_1 to measured activity (Heiss *et al.*, 1984, Wienhard *et al.*, 1985).

Group two consisted of nine additional patients with AVS (six men, three women, mean age 55 years [range 44–74 years], syndrome persistence <1 month) and nine healthy control subjects (six men, three women, mean age 44 years). Regional cerebral glucose metabolism and benzodiazepine receptor density were assessed using [^{11}C]FMZ- and [^{18}F]FDG-PET and the positron scanner ECAT EXACT HR (Siemens CTI, for technical data see Wienhard *et al.* (1994)). [^{18}F]FDG-PET studies were done as described above. [^{11}C]FMZ-PET was performed following intravenous bolus injection of approximately 740 mBq [^{11}C]FMZ, and the distribution and accumulation of the tracer was followed for 60 min by serial scanning. BZR density was estimated from the flumazenil binding 30 to 60 min after bolus injection. All PET images were coregistered to the individual CT along the anterior–posterior commissural line. Subsequently, cortical volumes of interest were defined on an anatomical basis on the individual CT scan. Since a quantification of receptor density was not generally feasible, relative values of [^{11}C]FMZ binding in comparison to averaged white matter activity were used for further analysis. As flumazenil binding can be reliably assessed only in the cortex, only cortical areas were used for comparative analysis (Heiss *et al.*, 1998).

III. RESULTS AND DISCUSSION

A. [^{18}F]FDG-PET in Acute and Persistent Vegetative State

In the first group of VS patients, overall glucose utilization (CMR_{glc}) was significantly reduced in comparison to age matched controls. The cerebral hypometabolism was most prominent in the supratentorial cortical regions, while basal ganglia and infratentorial stuctures were less affected. There were no relevant asymmetries in cortical glucose consumption in VS patients, a finding that corresponded with the absence of focal lesions on the CT scans of the VS patients. Table 1 summarizes the regional rates of glucose metabolism in patients in acute or persistent vegetative state as well as in controls. A striking difference in regional metabolic rates was found between patients studied in the acute and the persistent VS: Cortical metabolic rates in patients with PVS were significantly reduced when compared to patients studied in AVS ($p < 0.05$ for all cortical regions of interest except the frontal lobe). The absolute values for CMR_{glc} in AVS and PVS are given in Table 1, and the significant between-group differences are indicated by an asterisk.

The ongoing reduction of cerebral glucose metabolism with the transition from acute to persistent vegetative state reflects the progressive loss of residual cortical function following anoxic brain injury. These results of [^{18}F]FDG-PET in VS form the functional equivalent to the neuropathological findings of progressive Wallerian and transsynaptic degeneration after anoxic brain injury (Kinney and Samuels, 1994).

TABLE 1. Regional Metabolic Rates for Glucose ($rCMR_{Glc}$, mean $\pm$ SD in μmol/100 g/min) in Patients in an Acute (AVS, $n = 11$) or Persistent in μmol/100 g/min Vegetative State (PVS, $n = 13$), and in Healthy Control Subjects ($n = 24$)

Region	Controls	Acute VS	Persisting VS
Cortical regions			
Total	33 ± 5	27 ± 7	21 ± 6*
Frontal	34 ± 5	27 ± 7	21 ± 6
Parietal	33 ± 5	26 ± 7	21 ± 5*
Temporal	31 ± 4	27 ± 7	21 ± 6*
Occipital	33 ± 5	29 ± 8	22 ± 6*
Thalamus	33 ± 5	27 ± 7	21 ± 5*
Cerebellum	30 ± 4	28 ± 7	25 ± 6
Brainstem	24 ± 3	24 ± 5	23 ± 5

Note. Significant differences of $rCMR_{Glc}$ between AVS and PVS are indicated by an asterisk (*).

Left Frontal Cortical FMZ Binding

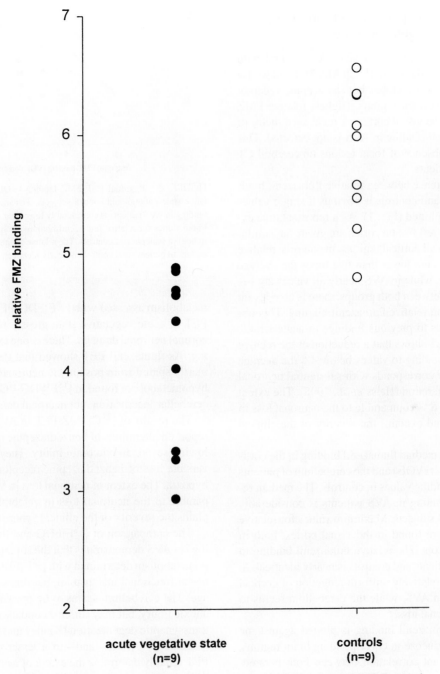

FIGURE 1. Relative benzodiazepine receptor density in the left frontal cortical VOI. In controls, relative flumazenil binding is not lower than 5, i.e., five times higher than the average white matter activity, while in AVS nearly all values are below this threshold.

However, these findings of [18F]FDG-PET in VS do not solve the question if alterations of cerebral glucose consumption determined with [18F]FDG-PET represent mere functional inactivation or irreversible structural brain damage. There is an evident need for PET studies using radiotracers of markers of neuronal integrity.

B. [18F]FDG- and [11C]FMZ-PET in Acute Vegetative State

In the group of patients in an acute VS studied with [18F]FDG and [11C]FMZ-PET, [11C]FMZ-PET showed a considerable reduction of global benzodiazepine receptor binding sites in all cortical regions. Highest relative FMZ binding is found in the visual cortex. A focal asymmetry of the cortical flumazenil binding in AVS is not detected. This corresponds to the absence of focal lesions on cerebral CT scans in the AVS patients.

The distinct difference between relative flumazenil binding in AVS patients and controls is obvious if single values for both groups are plotted (Fig. 1). As a representative example, data from the left frontal cortex are given, but similar results were found in all cortical regions. In controls, relative flumazenil binding is not lower than five times the average white matter activity, while in AVS nearly all values are below this threshold. Between both groups, there is no relevant overlap of the data on relative flumazenil binding. This observation corresponds to previous findings in acute stroke, where flumazenil-PET shows that a reduction of the relative cortical flumazenil binding to values below 4.5 the average white matter activity corresponds with substantial neuronal loss and manifest infarction (Heiss *et al.*, 1998). The extent of neuronal loss in VS is comparable to the neuronal loss in cerebral infarction and explains the severity of the clinical syndrome.

Figure 2 plots the median flumazenil binding in the cortical volumes of interest (VOIs) and the cerebellum of patients against the corresponding values in controls. The median regional flumazenil binding in AVS patients is considerably lower than in normal subjects. Maximum values for relative flumazenil binding are found in the visual cortex, both in patients and in controls. The relative flumazenil binding in the cerebellum in patients and controls is nearly identical. In summary, there is a relatively uniform reduction of cortical flumazenil binding in AVS, while the cerebellum seems to be spared from neuronal loss.

If the relative flumazenil binding is plotted against the metabolic rates for glucose in corresponding brain regions, a loose, but significant correlation between both parameters is found. As a representative example, the left central region is selected for presentation, but similar findings were observed in all cortical regions (Fig. 3). The reduction of relative cortical flumazenil binding corresponds approximately to the extent of reduction of cerebral glucose

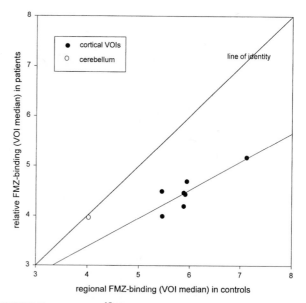

FIGURE 2. Regional [18F]FMZ binding (VOI median) in acute vegetative state patients and control subjects. The median regional flumazenil binding in AVS patients is considerably lower than in normal subjects. Maximum values for relative flumazenil binding are found in the visual cortex, in both in patients and controls. The relative flumazenil binding in the cerebellum in patients and controls is nearly identical.

metabolism assessed with [18F]FDG-PET. Thus, [18F]FDG-PET in acute vegetative state grossly reflects the extent of cortical neuronal damage. There is one exception to this finding: As flumazenil PET showed that the cerebellum is primarily spared from postanoxic neuronal loss, the cerebellar hypometabolism found in [18F]FDG-PET reflects functional cerebellar inactivation, not neuronal damage.

The results of [11C]FMZ-PET in AVS represent the first report on alterations of benzodiazepine receptor binding following severe anoxic brain injury. They document the irreversible loss of benzodiazepine receptors after global brain hypoxia. The extent of neuronal loss in VS seems to be comparable to the neuronal loss in cerebral infarction and explains the severity of the clinical syndrome.

The comparison of [18F]FDG and flumazenil-PET findings in AVS demonstrates that the supratentorial glucose hypometabolism determined with [18F]FDG-PET does not represent functional inactivation, but this structural brain damage. The cerebellum seems to be primarily spared from the anoxic injury, but may suffer secondary neuronal loss due to transsynaptic degeneration (Kinney and Samuels, 1994).

Thus, [11C]FMZ- and—to a lesser degree—[18F]FDG-PET reliably determine the extent of neuronal damage in the AVS, i.e., at an early stage following severe anoxic brain injury. Furthermore, flumazenil-PET may be helpful in establishing the prognosis as to possible recovery of consciousness and function in VS. However, larger series will be necessary to test this hypothesis.

Correlation Between Relative FMZ Binding and CMR$_{Glc}$ in the Left Central Region

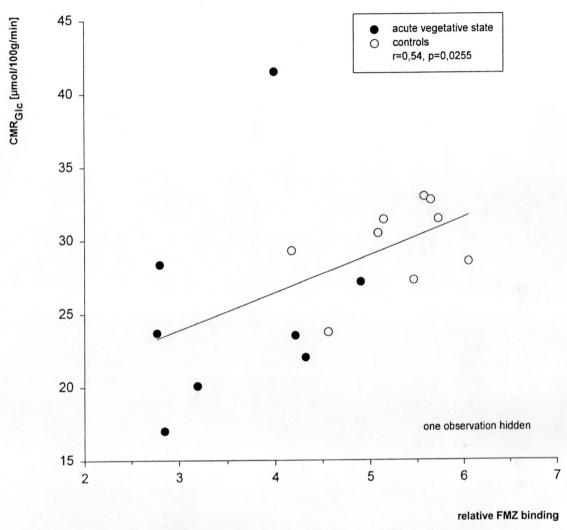

FIGURE 3. Correlation between relative benzodiazepine receptor density and glucose metabolism in the left central region. The reduction of relative cortical flumazenil binding corresponds approximately to the extent of reduction of cerebral glucose metabolism asessed with [^{18}F]FDG-PET.

References

Bates, D. (1991). Defining prognosis in medical coma. *J. Neurol. Neurosurg. Psychiatry* **54**: 569–571.

de Volder, A. G., Goffinet, A. M., Bol, A., Michel, C., de Barsy, T., and Laterre, C. (1990). Brain glucose metabolism in postanoxic syndrome. *Arch. Neurol.* **47**: 197–204.

Heiss, W. D., Grond, M., Thiel, A., Ghaemi, M., Sobesky, J., Rudolf, J., Bauer, B., and Wienhard, K. (1998). Permanent cortical damage detected by flumazenil positron emission tomography in acute stroke. *Stroke* **29**: 454–461.

Heiss, W. D., Pawlik, G., Herholz, K., Wagner, R., Göldner, H., and Wienhard, K. (1984). Regional kinetic constants and cerebral metabolic rate for glucose in normal human volunteers determined by dynamic positron emission tomography of (^{18}F)-2-fluoro-2-deoxy-D-glucose. *J. Cereb. Blood Flow Metab.* **4**: 212–223.

Kinney, H.C., and Samuels, M.A. (1994). Neuropathology of the persistent vegetative state. A review. *J. Neuropathol. Exp. Neurol.* **53**, 548–558.

Multi-Society Task force on PVS (1994). Medical aspects of the persistent vegetative state. First of two parts. *N. Engl. J. Med.* **330**: 1499–1508.

Quality Standards Subcommittee of the American Academy of Neurology: Practice Parameters (1995). Assessment and management of patients in the persistent vegetative state. *Neurology* **45**: 1015–1018.

Reivich, M., Kuhl, D., Wolf, A., Greenberg, J., Phelps, M., Ito, T., Casella, V., Fowler, J., Hoffmann, E., Alavi, A., Som, P., and Sokoloff, L. (1979). The (^{18}F)fluorodeoxyglucose method for the measurement of local cerebral glucose utilization in man. *Circ. Res.* **44**: 127–137.

Rudolf, J., Ghaemi, M., Ghaemi, M., Haupt, W. F., Szelies, B, and Heiss, W. D. (1999). Cerebral glucose metabolism in acute and persistent vegetative state. *J. Neurosurg. Anesthesiol.* **11**: 17–24.

Tommasino, C., Grana, C., Lucignani, G., Torri, G., and Fazio, F. (1995). Regional cerebral metabolism of glucose in comatose and vegetative state patients. *J. Neurosurg. Anesthesiol.* **7**: 109–116.

Wienhard, K., Dahlbohm, M., Eriksson, L., Michel, C., Bruckbauer, T., Pietrzyk, U., and Heiss, W. D. (1994). The ECAT EXACT HR: Performance of a new high-resolution positron scanner. *J. Comput. Assist. Tomogr.* **18**: 110–118.

Wienhard, K., Eriksson, L., Grootoonk, S., Casey, M., Pietzyk, U., and Heiss, W. D. (1992). Performance evaluation of the positron scanner ECAT EXACT. *J. Comput. Assist. Tomogr.* **16**: 804–813.

Wienhard, K., Pawlik, G., Herholz, K., Wagner, R., and Heiss, W. D. (1985). Estimation of local cerebral glucose utilization by positron emission tomography of ^{18}F-2-fluoro-2-deoxy-D-glucose: a critical appraisal of optimization procedures. *J. Cereb. Blood Flow Metab.* **5**: 115–125.

The Importance of Left Temporal Areas
for Efficient Recovery from Poststroke Aphasia

W.-D. HEISS, J. KESSLER, A. THIEL, M. GHAEMI, and H. KARBE

Max-Planck-Institut für neurologische Forschung and Neurologische Universitätsklinik Köln, Germany

The functional basis of recovery from poststroke aphasia is still controversial and the contribution of activated contralateral networks or preserved ipsilateral speech relevant to the improvement of impaired language function is still unclear (Cappa and Vallar, 1992). In order to relate recovered language function to distinct parts of the complex bilateral network $H_2^{15}O$-PET activation were performed during a word repetition in 23 right-handed aphasia patients 2 and 8 weeks after stroke. Patients were classified according to the site of lesion (frontal n = 7; subcortical n = 9; temporal n = 7) and their performance was compared with 11 control subjects. Fourteen brain regions representing eloquent and contralateral homotopic areas were selected and defined on coregistered magnetic resonance imaging scans for calculating blood flow. At baseline, differences in test performance were only found between the subcortical and temporal group. The extent of recovery, however, differed and was reflected in the activation. The frontal and subcortical group which showed a good outcome could reintegrate the left superior temporal gyrus into a functional language network, which was not possible for the temporal group. These results stress the importance of preserved speech relevant regions of the dominant hemisphere for an effective recovery of language function after ischemic stroke. These differential activation patterns suggest a hierarchy within the language-related network regarding effectiveness for improvement of aphasia; i.e., right hemispheric areas contribute, if left hemispheric regions are destroyed. Efficient restoration of language is usually only achieved if left temporal areas are preserved and can be reintegrated into the functional network.

I. INTRODUCTION

The involvement of areas within the subdominant, usually right hemisphere in language functions is still controversial (Zaidel, 1985; Gazzaniga *et al.*, 1984) and, due to differences in the hemispheric representation of language, may vary widely from one individual to another. As a consequence, the role of undamaged regions in the left hemisphere and of homotopic contralateral regions regarding recovery of aphasia after stroke is still a matter of debate. Since complete recovery of all components of language function in the majority of patients relies on undamaged eloquent regions of the left hemisphere (Vallar, 1990), a widely accepted concept (Kertesz, 1988; Gainotti, 1993) supposes that speech-related areas in the left hemisphere spared by the infarction are more efficient for the recovery of language function than homologous structures in the right hemisphere, which usually only contribute to incomplete improvement. Since previous functional neuroimaging studies were inconclusive due to restriction to small samples (Knopman *et al.*, 1984; Heiss *et al.*, 1997) and selected aphasia types (Weiller *et al.*, 1995), the effect of the site of the lesion on outcome was explored in patients with aphasia of different types and severities.

II. METHODS

Language performance and other cognitive abilities were assessed in 23 right-handed patients with aphasia 2 and 8 weeks after left hemispheric stroke (Table 1) and related to $H_2^{15}O$ PET activation patterns obtained by repeating words.

TABLE 1.   Patient Data with Aphasia Classification

Patient ID	Sex	Age (years)	Infarct volume (ccm)	Aphasia type
Subcortical				
S1	M	59	27.2	Anomic aphasia
S2	M	46	113.7	Wernicke aphasia
S3	M	63	29.3	Residual aphasia
S4	M	31	9.2	Broca aphasia
S5	F	66	7.8	Broca aphasia
S6	M	71	4.3	Residual aphasia
S7	F	70	30.3	Residual aphasia
S8	M	56	17.7	Broca aphasia
S9	M	44	18.2	Broca aphasia
Mean (SD)		50.62 (20.35)	28.63 (31.40)	
Frontal				
F1	F	77	65.3	Transcortical sensoric aphasia
F2	F	59	4.4	Wernicke aphasia
F3	M	65	72.5	Transcortical sensoric aphasia
F4	M	34	60.1	Residual aphasia
F5	M	49	62.2	Broca aphasia
F6	M	56	39.8	Residual aphasia
F7	M	65	15.1	Anomic aphasia
Mean (SD)		57.85 (12.65)	45.62 (24.66)	
Temporal				
T1	M	54	54.3	Wernicke aphasia
T2	F	71	18.9	Anomic aphasia
T3	F	72	19.5	Wernicke aphasia
T4	F	61	131.2	Wernicke aphasia
T5	M	54	9.9	Conduction aphasia
T6	F	40	22.8	Anomic aphasia
T7	M	44	154.3	Wernicke aphasia
Mean (SD)		56.57 (11.41)	58.7 (55.04)	

TABLE 2.   Significantly Activated Regions at $p < 0.0036$ by Repeating Words 2 (Initial) and 8 (Followup) Weeks after the Stroke

Region	Side	Subcortical Initial	Subcortical Followup	Frontal Initial	Frontal Followup	Temporal Initial	Temporal Follow-up	Control group ($n = 11$)
F3o	ri	3.97 (2.44)	—	3.75 (1.42)	—	—	—	—
	le	—	—	—	—	5.92 (2.97)	4.93 (2.74)	4.10 (2.68)
PT	ri	8.78 (6.57)	11.02 (4.42)	9.43 (4.27)	8.93 (2.87)	—	—	7.46 (3.83)
	le	7.07 (4.74)	12.21 (8.23)	—	11.72 (4.42)	—	—	9.51 (2.22)
T1p	ri	12.14 (8.61)	15.47 (1.98)	13.01 (4.41)	12.64 (3.07)	—	11.31 (3.59)	10.18 (3.60)
	le	—	11.95 (4.45)	—	10.61 (4.67)	—	—	10.55 (3.13)
H1	ri	—	9.65 (2.42)	—	10.95 (2.33)	—	—	7.94 (3.82)
	le	—	—	—	—	—	—	9.12 (3.80)
PRG1	ri	—	—	—	—	—	4.46 (1.68)	—
	le	—	—	—	—	—	6.78 (2.32)	5.60 (3.70)
SMC	ri	—	—	—	—	7.01 (3.32)	—	—
	le	—	—	—	—	—	6.86 (2.39)	—

Note. Mean and SD of regional relative differences.

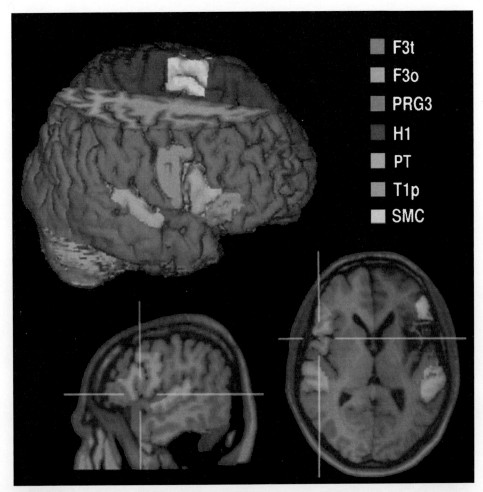

FIGURE 1. Set of regions of interest in a normal subject. Frontal ROIs and SMC were drawn on the surface; temporal ROIs were drawn on sagittal slices.

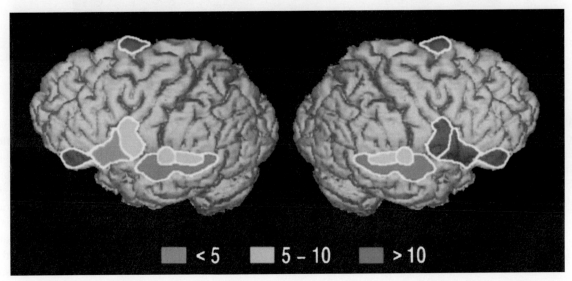

FIGURE 2. Significantly activated regions in 11 normal subjects. Mean values of regional relative differences in CBF between rest and activation conditions.

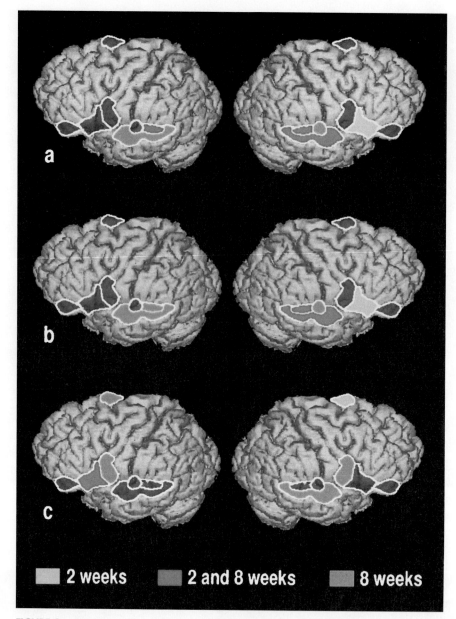

FIGURE 3. Significantly activated ROIs in aphasic patients at initial (2 weeks) and followup (8 weeks) PET. a) Subcortical, b) frontal, and c) temporal group.

CBF changes from rest to activation in the 23 patients and in 11 control subjects were calculated in 14 volumes of interest representing eloquent and contralateral homotopic areas defined on coregistered T1-weighted MRI (Fig. 1) and tested for significance using t-test and Bonferroni correction ($\alpha = 0.0036$).

III. RESULTS

The patients were classified according to infarct location into a frontal ($n = 7$), subcortical ($n = 9$), and tempo-

ral ($n = 7$) group. Infarct volumes showed large variability within the three groups (Table 1). At baseline differences were only found for test performance between the subcortical and temporal group. The extent of recovery, however, differed and was reflected in the activation pattern at baseline and at followup: The subcortical and frontal groups improved substantially in several verbal, the subcortical group also in nonverbal tests. All groups improved their error score in the Token test; however, the finally reached performance differed, with close to normal values in the subcortical group, a low error score in the frontal and a severe error score in the temporal group.

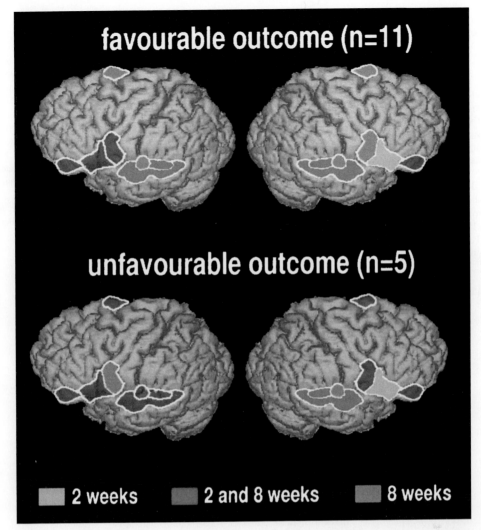

FIGURE 4. Significantly activated regions in aphasic patients with favorable outcome (relative improvement in Token test >50%) and unfavourable outcome (relative improvement in Token test <50%) at initial (2 weeks) and followup (8 weeks) PET.

Activation patterns were different from normal controls (Fig. 2) and among the groups and also changed differently in the course: The subcortical group always activated the right temporal areas; early after the stroke they used the right inferior frontal gyrus, an area not activated in the normal controls. With the improvement they reactivated the left temporal gyrus (Fig. 3, Table 2). This shift from right frontal to left temporal activation was also observed in the frontal group. In contrast, the group with temporal lesions—and with the most severe permanent deficits—activated the left inferior frontal gyrus (Broca area) all the time; with some improvement in language performance these patients activated the right temporal areas, the left precentral gyrus and even the supplementary motor cortex, but did not regain activation in the left temporal lobe (Fig. 3; Table 2). If patients with frontal and subcortical lesions were classified according to the degree of improvement, the difference in the activation of left temporal areas could also be demonstrated (Fig. 4).

IV. CONCLUSIONS

These lesion-dependent changes in activation patterns suggest a hierarchy of the various components of the language related network in effectiveness for improvement from aphasia: Right hemispheric areas contribute to some extent, if the more important left hemispheric regions are destroyed, but the extent of the right hemisphere's contribution varies widely depending on the individual's bilateral representation of language (Gainotti, 1993). Satisfactory outcome with restoration of language function, however, can be achieved only if the temporal eloquent cortical areas of the dominant hemisphere are morphologically preserved and can be reintegrated within the functional network.

References

Cappa, S. F., and Vallar, G. (1992). The role of the left and right hemispheres in recovery from aphasia. *Aphasiology* **6**: 359–372.

Gainotti, G. (1993). The riddle of the right hemisphere's contribution to the recovery of language. *Eur. J. Disorders. Communication.* **28**: 227.

Gazzaniga, M. S., Smylie, C. S., and Baynes, K. (1984). Profiles of right hemisphere language and speech following brain bisection. *Brain Language* **22**: 206–220.

Heiss, W.-D., Kessler, J., Karbe, H., Fink, G.R., and Pawlik, G. (1993) Cerebral glucose metabolism as a predictor of recovery from aphasia in ischemic stroke. *Arch. Neurol.* **50**: 958–964.

Heiss, W.-D., Karbe, H., Weber-Luxenburger, G., Herholz, K., Kessler, J., Pietrzyk, U., and Pawlik, G. (1997). Speech-induced cerebral metabolic activation reflects recovery from aphasia. *J. Neurol. Sci.* **145**: 213–217.

Kertesz, A. (1988). What do we learn from recovery from aphasia? *In* "*Functional Recovery in Neurological Disease*" (S. G. Waxman, Ed.), pp. 277–292. Raven Press, New York.

Knopman, D. S., Rubens, A.B., Selnes, O. A., Klassen, A. C., and Meyer, M. W. (1984). Mechanisms of recovery from aphasia: Evidence from serial Xenon 133 cerebral blood flow studies. *Ann. Neurol.* **15**: 530–535.

Vallar, G. (1990). Hemispheric control of articulatory speech output in aphasia. *In* "*Cerebral Control of Speech*" (G. R. Hammond, Eds.), pp. 387–416. North-Holland, Amsterdam.

Weiller, C., Isensee, C., Rijntjes, M., Huber, W., Müller, S., Bier, D., Dutschka, K., Woods, R. P., Noth, J., and Diener, C. (1995). Recovery from Wernicke's aphasia: A positron emission tomographic study. *Ann. Neurol.* **37**: 723–732.

Zaidel, E. (1985). Language in the right hemisphere. *In* "*The Dual Brain: Hemispheric Specialization in Humans*" (D. F. Benson, and E. Zaidel, Eds.), pp. 205–231. Guilford Press, New York.

DP-b99: A Novel Membrane-Targeted Compound Active against Global and Focal Cerebral Ischemia

M. KRAKOVSKY, M. POLYAK, I. ANGEL, and A. KOZAK

D-Pharm Ltd., Kiryat Weizmann Science Park, Rehovot, Israel

The neuroprotective properties of DP-b99 a novel membrane-targeted compound were assessed in a global forebrain ischemia model in gerbils, and focal ischemia model in rats. Administration of DP-b99 (5 or 10 µg/kg, i.p.) immediately postischemia, or 3 h after onset of reperfusion in the global forebrain ischemia model in gerbils resulted in a significant increase in the number of surviving hippocampal neurons and a significant decrease in the serum level of neuron specific enolase (NSE). In the rat MCAO model DP-b99, a single dose (5 µg/kg) significantly reduced the infarct volume even when administered up to 3 h post–MCAO. It was shown that the size of the infarct in all treated groups was reduced between 50–90% according to the brain region (subcortical or cortical, respectively), indicating a very large area of protection by DP-b99. The serum level of NSE correlates well with total infarct volume. These findings demonstrate that DP-b99 is able to prevent ischemic damage as assessed by histological and biochemical analysis. DP-b99 is a promising candidate for the treatment of stroke.

I. INTRODUCTION

Perturbed balance of divalent metal ions, such as calcium, copper, iron, or zinc, is thought to be a contributory factor to cell death following ischemic events, such as cerebrovascular stroke and cardiac infarction. A potential way of addressing this phenomenon is by using ion chelation in order to restore ionic balance. The therapeutic use of divalent metal ion chelators is, however, problematic due either to the nonselective nature of these compounds or to the perturbation of normal ionic balance that they may cause, leading to a high incidence of cardio- and hepatotoxic effects. This drawback may be overcome by restricting the action of such agents to the cell and intracellular membranes, which is the primary site of cellular damage. DP-b99 belongs to a novel family of neuroprotective drugs that are membrane-activated metal ion chelators. This drug candidate is designed such that metal ion chelation is limited to the vicinity of the biological membrane. It is hypothesized that DP-b99 may have the capacity to restore calcium homeostasis and thus moderate the cascade of events leading to cell death as a result of ischemic damage. In the present study, the effects of DP-b99 on ischemia are examined, using different routes and/or times of administration, in animal models of global and focal cerebral ischemia.

II. MATERIALS AND METHODS

Animals were housed and cared for in accordance with the National Institutes of Health guidelines for the use of experimental animals and the protocols were approved by the company chief veterinary surgeon.

Mongolian gerbils, weighing 70–80 g, (Harlan Laboratories, Israel) were housed in groups of two at $21 \pm 1°C$, with a 12-h light/dark cycle and with food and water provided *ad libitum*. Gerbils were initially anesthetized with halothane 4% and anesthesia was continued with halothane 0.8% and a 70% nitrous oxide, 30% oxygen mixture, using a face mask without intubations. Animals were food deprived prior to surgery. Both common carotid arteries (CCA) were occluded using arterial clips for 10 min. After removal of arterial clips to allow reperfusion, the animals were left to recover. Rectal temperature was monitored and maintained at 37°C during the operation using a Gaymar T/Pump and warming platform. The animals were sacrificed 7 days later by Pental (veterinary) overdose; their brains were perfused with 4% formaldehyde solution in PBS buffer, removed, and stored in the same solution for assessment of hippocampal neurons.

Five groups of gerbils were studied as follows: group 1 ($n = 6$) was sham operated animals without treatment; group 2 ($n = 10$) was injected with saline; groups 3 ($n = 8$) and 4 ($n = 9$) received a single dose of, respectively, 5 or 10 µg/kg of DP-b99 ip, immediately after onset of reperfusion; group 5 ($n = 8$) received a single dose of 10 µg/kg DP-b99 3h after onset of reperfusion.

Male rats (Sprague–Dawley, 300–340 g, Harlan Laboratories, Israel) were housed in groups of three at $21 \pm 1°C$, with a 12-h light/dark cycle, and food and water were provided *ad libitum*. Animals were not fasted prior to surgery. They were anaesthetized initially with halothane 4% and subsequently with halothane 1% and a 70% nitrous oxide, 30% oxygen mixture, using a face mask without intubations. The right middle cerebral artery (MCA) was occluded using the intraluminal thread technique (Zea Longa *et al.*, 1989; Belayev *et al.*, 1996) with a 3–O nylon monofilament coated with 1% poly-L-lysine solution. After MCA occlusion for 2 h, the filament was withdrawn to allow reperfusion, and the animals left to recover. Rectal temperature was monitored and maintained at 37°C during the operation (MCAO, 2 h) and during recovery from anesthesia using a Gaymar T/Pump and warming platform. After 72 h the animals were sacrificed by Pental (veterinary) overdose; their brains were perfused with 4% formaldehyde solution in PBS buffer, removed, and stored in the same solution for assessment of infarct volume.

DP-b99 was dissolved in saline and administered to rats at a dose of 10 µg/kg, through the ip or iv routes. There were four treatment groups as follows: group 1 was injected with saline ($n = 10$); group 2 ($n = 8$) received DP-b99 ip immediately after onset of reperfusion; group 3 ($n = 9$) received DP-b99 iv immediately after onset of reperfusion; group 4 ($n = 8$) received DP-b99 ip 3 h after the onset of reperfusion.

Brains were processed for analysis of global and of focal ischemic damage as follows: for global ischemic damage in gerbils paraffin sections 5 µm thick were taken from the dorsal hippocampus and stained with thionin for microscopic evaluation of the ischemic area. The CA-11, CA12, CA-2, and CA-3 subfields of the hippocampus were evaluated from each section. For the evaluation of focal ischemic damage in rats, each brain was frozen at $-80°C$, coronally sectioned with a cryomicrotome at $-20°C$ and slices (thickness 20 µm) taken every millimeter; slices were stained with thionin for 5 min and differentiated in methanol 95% for 30 s. At each coronal level, the infarct area at the contralat-

eral hemisphere, and the ipsilateral spared hemisphere, and the cortical and subcortical sites were delineated under microscopical examination of the tissue. Surface areas were determined by planimetry with ImageQuaNT image analysis software (Molecular Dynamics Inc., Bondoufle Cedex, France) of the histological sections and volumes were calculated by integrating lesion surfaces with the distance between slices (1 mm). Total, cortical and subcortical infarct volume (mm^3) and total infarct volume (expressed as percentage of the hemispheric volume) were calculated. The analysis was performed in a double-blind fashion.

Neuron-specific enolase (NSE) concentration was measured in gerbil and rat sera obtained 24 h and 72 h after occlusion, using an enzyme-linked immunoassay kit (Pharmatop and Pharmacia, Sweden). Values are expressed as mean $\pm$ SEM. Infarct volume comparisons between groups were performed by one-way ANOVA (group factor), followed when appropriate, by Dunnet post hoc tests with Bonferroni correction to control the familywise error rate. Statistical differences were considered significant at $p < 0.05$.

III. RESULTS

In the gerbil global forebrain ischemic model, DP-b99 (10 µg/kg, ip) significantly reduced the levels of NSE in the serum. This indicated that less nerve damage had occurred in the DP-b99-treated rats compared with the untreated animals ($p < 0.05$). DP-b99 exerted its neuroprotective effect to the same extent whether it was administered immediately after the ischemic event or 3 h later (Fig. 1). Histologically, neuronal damage was quantified by counting intact neurons in the hippocampus. It can be seen in Fig. 2 that DP-b99 increased the numbers of surviving cells in CA1, CA12, CA2, and CA3 regions 7 days postischemia. The most significant effects were observed in the CA2 and CA3 regions ($p < 0.001$).

A similar neuroprotective effect of DP-b99 was obtained in the rat MCA occlusion model. Serum levels of NSE, which are elevated in the event of ischemic insult (Barone *et al.*, 1993), increased approximately 2-fold compared to the shamoperated controls 24 h after the onset of reperfusion and 2.5-fold after 72 h. In contrast, in animals treated with DP-b99 (10 µg/kg) there were significantly lower levels of NSE in their serum compared to the vehicle-treated controls (Table 1). The effects observed were independent of the mode or time of drug administration and the reduction in NSE levels was maintained for at least 72 h, even when DP-b99 was given 3 h after the reperfusion.

As expected, MCAO produced a large infarct in the cortex and striatum of the rat brain. Microscopic examination showed a clear delineation between infarcted tissue (neuronal depletion and presence of dead neurons) and normal

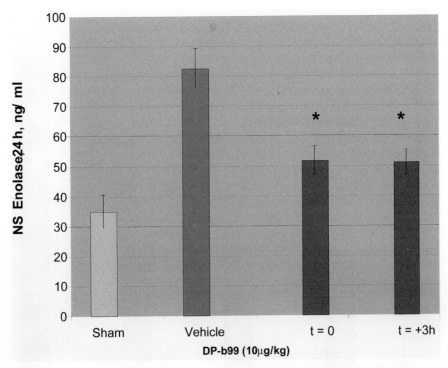

FIGURE 1. Neuron-specific enolase (NSE) levels 24 h postischemia in the presence or absence of DP-b99. Following 10 min of global forebrain ischemia in Mongolian gerbils, a single dose of DP-b99 was administered (10 μg/kg, ip), immediately ($t = 0$) or 3 h ($t = 3$) after the beginning of reperfusion. $^*p < 0.05$.

TABLE 1. Plasma NSE Level Following Transient Focal Cerebral Ischemia in Rat after DP-b99 Treatment

Treatment	Time after reperfusion	NSE (ng/mL) 24 h	p	NSE (ng/mL) 72 h	p
None—sham-operated	NA	7.9 ± 1.5		3.2 ± 0.5	
Vehicle	Immediately	17.0 ± 3.8	<0.001	20.1 ± 5.9	<0.001
DP-B99 10 μg/kg ip	Immediately	9.0 ± 2.9	<0.002	8.7 ± 3.1	<0.001
DP-B99 10 μg/kg iv	Immediately	8.8 ± 3.9	<0.001	12.7 ± 3.1	<0.01
DP-B99 10 μg/kg ip	3 h	6.8 ± 2.1	<0.001	10.0 ± 4.2	<0.001

tissue. The neuroprotective effect of DP-b99 is clearly visible in the rat brain sections shown in Fig. 3. In the absence of the drug, there is extensive cell damage as indicated by the white areas. When administered immediately following reperfusion, DP-b99 treatment prevented the neuronal damage almost completely. However, when administered 3 h after reperfusion, DP-b99 prevented neuronal damage from spreading further, but signs of the focal ischemic damage were still detectable.

The mean cortical and subcortical infarct volumes ($\pm$ SEM) are shown in Figs. 4a and 4b, respectively. In

all groups treated with DP-b99 the cortical infarct size was reduced by 70–90% compared to that seen in the vehicle-treated control group (Fig. 4a) and subcortical infarct size is reduced by about 50% (Fig. 4b). Regardless of the method of calculating infarct volume or the infarct location (cortical or subcortical), lesioned volumes are significantly higher in the vehicle-treated group than in the DP-b99-treated groups. There were no significant differences between groups 2 and 4. The body temperatures did not differ among the groups (data not shown).

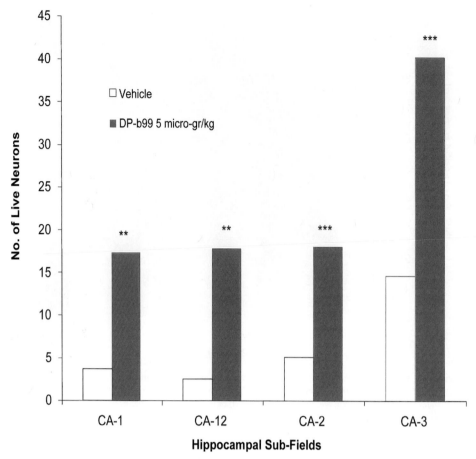

FIGURE 2. Vehicle-treated animals ($n = 10$) and DP-b99 (5 μg/kg, ip, single dose)-treated animals ($n = 7$). ** $p < 0.01$; *** $p < 0.001$.

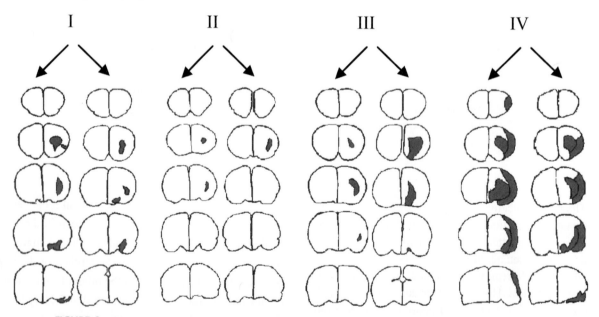

FIGURE 3. Representative distributions of lesioned areas in two animals of each group at five coronal levels are depicted. Each column represents data from one animal. Infarcted regions, shown in gray were quantified by histopathology. Group I, DP-b99 10 μg/kg administered ip immediately following reperfusion; Group II, DP-b99 10 μg/kg administered ip 3 h following reperfusion; Group III, DP-b99 10 μg/kg administered iv immediately following reperfusion, and Group IV, DP-b99 vehicle control.

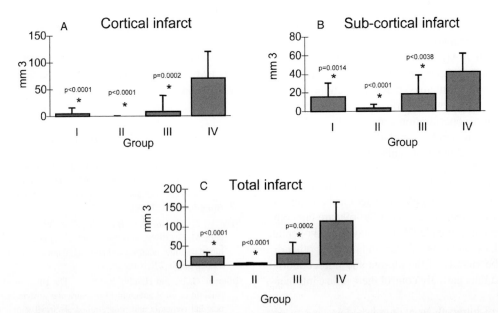

FIGURE 4. Infarct volumes (a) cortical, (b) subcortical, (c) total. Group I, DP-b99 10 μg/kg ip immediately following reperfusion. Group II, DP-b99 10 μg/kg ip 3 h following reperfusion. Group III, DP-b99 10 μg/kg iv immediately following reperfusion. Group IV, DP-b99 vehicle control.

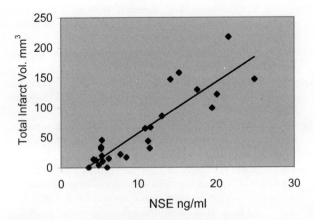

FIGURE 5. The correlation between NSE levels in serum (measured after 72 h) and infarct volume. The serum levels of NSE are found to correlate well with total infarct volume ($r = 0.79$).

The correlation between NSE levels in serum (measured after 72 h) and infarct volume in individual animals is depicted in Fig. 5. The serum levels of NSE are found to correlate well with total infarct volume ($r = 0.79$, $p < 0.001$).

IV. DISCUSSION

Disorders such as stroke and cardiac infarction are associated with ischemic cell injury and death. These may occur via excitotoxic necrosis or mechanisms of programmed cell death (apoptosis; Iadecola, 1999). A growing body of evidence indicates that in both degenerative processes, distur-

bance of Ca^{2+} homeostasis is an important factor in triggering cell damage. In particular, calcium influx into neurons and release from intracellular pools are thought to be major contributors to the neuronal cell death that occurs during brain ischemia (Silver and Erecinska, 1990). Subsequent to the rise of free intracellular calcium, generation of oxygen radicals is observed, which together with concurring mitochondrial dysfunction apparently underlie the degeneration process. Prevention of intracellular calcium accumulation has been attempted by blocking various calcium channels (Sadoshima et al., 1992). However, the lack of selectivity for neuronal tissue has resulted in variable efficacy in the treatment of cerebral ischemia and/or stroke and with unacceptable side-effects. Introduction of metal chelators into injured cells could maintain metal ion homeostasis and prevent cell damage, though some chelators were not always found to be effective (Xie et al., 1992) and some have adverse effects on the cardiac system (Billman, 1993, 1996; Blumenthal et al., 1999). Differently from other chelators, DP-b99 exerts its chelating capabilities in lipid-containing environments preferably. This characteristics limits its activity to the vicinity of the membrane, where cellular Ca^{2+} homeostasis could be finely regulated.

DP-b99 has demonstrated strong neuroprotection both in global and focal models of cerebral ischemia in two different models. In the MCAO rat model, its efficacy was large as it covered both cortical and subcortical brain regions. The results show that the size of the infarct in all the DP-b99 treatment groups is reduced by 70–90%, indicating a very large area of coverage and protection. In addition, we also show that blood serum NSE level measured 3 days after cerebral infarction proved to be a good predictor of infarct volume.

Two hours of MCAO (in rats) has been shown to lead to a profound perturbation of cell calcium metabolism (Kristian and Siesjo, 1996). In focal areas, massive calcium overload persists for 4–6 h following reperfusion, and in penumbral tissues the calcium concentration remains elevated (Kristian *et al.*, 1998). Zinc ion has similarly been implicated in neuronal death in focal brain ischemia (Kristian and Siesjo, 1997). *In vitro*, DP-b99 has the capacity to transfer different divalent cations (Ca^{2+}, Fe^{2+}, and Zn^{2+}) from water to octanol (data not shown). Although the exact mechanisms underlying the neuroprotective actions of DP-b99 are not completely elucidated, in principle DP-b99 may indeed operate, as hypothesized, by chelating calcium and possibly other divalent ions in the vicinity of the plasma and mitochondrial membranes, and thus partially control their influx into neurons.

Disturbance of intracellular alcium homeostasis can occur *via* a number of routes and mechanisms including voltage-operated calcium channels and excitatory amino acid channels, both NMDA and non-NMDA gated. Here, we have demonstrated the utility of a lipophilic Ca^{2+} chelating agent in preventing ischemic damage. This novel approach with its associated broad specificity, may ultimately prove more successful than selective calcium antagonist agents which block specific channels.

Stroke is a medical emergency that requires immediate intensive care as a heart attack. Although many compounds have been successfully tested in animal studies, in human trials they have not succeeded (De Keyser *et al.*, 1999). One reason for this may be failure to administer the drug within a specific time frame. A promising feature of DP-b99 is that it appears to have a relatively wide therapeutic window, since infarct volume is markedly reduced even when administration is delayed for up to 5 h following the start of occlusion.

The reduction in infarct size observed in this study is most impressive, apparently not dependent upon the route of administration, and the beneficial effects are still present at least 72 h after administration (as judged by attenuated serum NSE levels and infarct volume). It remains to be shown, however, that DP-b99 actually ameliorates rather than merely delays tissue damage.

V. CONCLUSION

DP-b99 demonstrates significant neuroprotective activity in animal model of global and focal cerebral ischemia, with a minimum of 70% protection even when administered 5 h after the onset of ischemia. This dramatic reduction in brain damage indicates that DP-b99 is a promising candidate for the treatment of stroke.

Acknowledgments

The technical and expert assistance of S. Roussel and the scientific input of E. T. MacKenzie from Cyceron, Caen, France, are greatly acknowledged. We also thank the technical assistance of M. Tolmasov and Z. Krakovsky.

References

Barone, F. C., Clark, R. K., Price, W. J., White, R. F., Feuerstein, G. Z., Storer, B. L., and Ohlstein, E. H. (1993). Neuron-specific enolase increases in cerebral and systemic circulation following focal ischemia. *Brain Res.* **623**: 77–82.

Belayev, L., Alonso, O. F., Busto, R., Zhao, W., Clemens, J. A., and Ginsberg, M. D. (1996). Middle cerebral artery occlusion in the rat by intraluminal suture: neurological and pathological evaluation of an improved model. *Stroke* **27**: 1616–1623.

Billman, G. E., and Hamlin, R. (1996). The effect of Mibefradil, a novel calcium channel antagonist on ventricular arrhythmias induced by myocardial ischemia and programmed electrical stimulation. *Pharmacol. Exper. Ther.* **277**: 1517–1526.

Billman, G. E. (1993). Intracellular calcium che3lator, BAPTA-AM, prevents cocaine-induced ventricular fibrillation. *Am. J. Physiol.* **265** (Heart Circ. Physiol. **34**): H1529–H1535.

Blumenthal, S. R., Williamss, T. C., Barbee, R. W., Watts, J. A., and Gordon, B. E. (1999). Effects of citrated whole blood infusion in response to hemorrhage. *Lab. Anim. Sci.* **49**: 411–417.

Choi, D. W. (1990). Cerebral hypoxia: Some new approaches and unanswered questions. *J. Neurosci.* **10**: 2493–2501.

Iadecola, C. (1999). Mechanism of cerebral ischemic damage. In "Cerebral Ischemia: Molecular and Cellular Pathophysiology," pp. 3–32. Humana Press, Clifton, NJ.

De Keyser, J., Sulter, G., and Luiten, P. G. (1998). Clinical trials with neuroprotective drugs in acute ischemic stroke: Are we doing the right thing? *Trends Neurosci.* **22**: 535–540.

Kristian, T., Gido, G., Schutz, A., and Siesjo, B. K. (1998). Calcium metabolism of focal and penumbral tissues in rats subjected to transient middle cerebral artery occlusion. *Exp. Brain Res.* **120**: 503–509.

Kristian, T., and Siesjo, B. K. (1996). Calcium-related damage in ischemia. *Life Sci.* **59**: 357–367.

Kristian, T., and Siesjo, B. K. (1997). Changes in ionoc fluxes during cerebral ischemia. *Int. Rev. Neurobiol.* **40**: 27–45.

Kristian, T., and Siesjo, B. K. (1998). Calcium in ischemic cell death. *Stroke* **29**: 705–718.

Sadoshima, S., Ibayashi, S., Nakane, H., Okada, Y., Ooboshi, H., and Fujishima, M. (1992). Attenuation of ischemic and postischemic damage to brain metabolism and circulation by a novel Ca channel antagonist, NC-1100, in spontaneously hypertensive rats. *Eur. J. Pharmacol.* **224**: 109–115.

Silver, I. A., and Erecinska, M. (1990). Intracellular and extracellular changes of [Ca2+] in hypoxia and ischemia in rat brain in vivo. *J. Gen. Physiol.* **95**: 837–866.

Xie, Y., Seo, K., Ishumaru, K., and Hossmann, K. A. (1992). Effect of calcium antagonists on postischemic protein biosyntheses in gerbil brain. *Stroke*, **23**: 87–92.

Zea Longa, E. Z., Weinstein, P. R., Carlson, S., and Cummins, R. (1989). Reversible middle cerebral artery occlusion without craniectomy in rats. *Stroke* **20**: 84–91.

PET Study in Patients with Neuropsychological Impairments in the Chronic State after Traumatic Diffuse Brain Injury

MITSUHITO MASE,* KAZUO YAMADA,* TAKASHI MATSUMOTO,* AKIHIKO IIDA,[†]
HIDEHIRO KABASAWA,[‡] TETSUO OGAWA,[‡] JUNKO ABE,[‡] and YURI NAGANO[‡]

*Department of Neurosurgery, Nagoya City University Medical School, Nagoya, Japan
[†]Departments of Radiology, Nagoya City Rehabilitation Center, Nagoya, Japan
[‡]Department Rehabilitation, Nagoya City Rehabilitation Center, Nagoya, Japan

The aim of this study is to clarify the relationship between cerebral blood flow and metabolism and neuropsychological impairments (disturbance of attention, memory, or information processing) in the chronic state after diffuse traumatic brain injury (TBI). Eight patients with these deficits and without abnormal findings on magnetic resonance imaging (MRI) were included in this study. Regional cerebral blood flow (rCBF), regional oxygen extraction fraction (rOEF), and regional metabolic rate of oxygen (rCMRO₂) were measured using positron emission tomography (PET) in the chronic state (mean, 9 months) after TBI. PET study showed mild decrease of rCBF and rCMRO₂ in all patients, which were divided into two types. Four patients showed the frontal type (relative decrease of rCBF and rCMRO₂ in bilateral frontal cortices), and the other showed the cerebellar type (relative decrease of rCBF and rCMRO₂ in bilateral occipital cortices and cerebellum). rOEF was normal in all patients. rCBF and rCMRO₂ were basically coupled. However, rCMRO₂ was more sensitive for detecting lesions than rCBF. Paced auditory serial addition task (PASAT), evaluating attention, of the frontal type tended to be disturbed more severely than that of the cerebellar type, whereas there was no significant difference. It is concluded that evaluation of cerebral blood flow and oxygen metabolism using PET can become an objective assessment of neuropsychological sequelae after traumatic diffuse brain injury.

I. INTRODUCTION

Diffuse traumatic brain injury, including diffuse axonal injury, can cause neuropsychological deficits such as problems with attention, concentration, memory, and information processing. These sequelae are sometimes very difficult understand for others. There are several neuropsychological testings for assessing the deficits, however, few objective examinations like neuroimaging. Cortical localization of these higher brain functions are still controversial. Conventional neuroimaging techniques such as MRI (magnetic resonance imaging) and CT (computed tomography) can not show lesions causing these deficits. Some authors reported changes of cerebral blood flow (CBF) and metabolism after brain injury in the acute (Abu-Judeh *et al.*, 1998; Yamaki *et al.*, 1996) and chronic states (Shiina *et al.*, 1998; Yamaki *et al.*, 1996). However, there was few reports to evaluate the relationship between CBF and metabolism and neuropsychological impairments after diffuse brain injury, except for CBF studies using single photon emission computed tomography (SPECT) (Laatsch *et al.*, 1997; Umile *et al.*, 1998). The aim of this study is to clarify the relationship using positron emission tomography (PET) for an objective assessment of these sequelae after diffuse brain injury.

II. MATERIALS AND METHODS

Eight patients (17–51, mean, 29 years old) with neuropsychological impairments clinically and without aphasia, agnosia, or motor weakness of extremities in the chronic state (mean, 9 months) after traumatic brain injury, who also had no abnormal lesion on MRI, were included in this study. PET scans (PCT 3600W, Hitachi, Japan) were obtained using ^{15}O-labeled gas steady-state technique to evaluate regional cerebral blood flow (rCBF), oxygen extraction fraction (rOEF), and regional metabolic rate of oxygen (rCMRO$_2$). Neuropsychological functions (IQ, attention, and memory) were evaluated using Japanese Wechsler Adult Intelligence Scale–Revised (WAIS-R), paced auditory serial addition task (PASAT), and Rey–Osterrieth complex figure test. The relationship between PET study and scores of these tests was analyzed.

III. RESULTS

Glasgow Coma Scale (GCS) of the patients on the initial admission were from 3 to 8 (mean 5.5), and periods of their consciousness disturbance were from 3 to 60 days (mean 18 days). The causes of traumatic brain injury were traffic accidents in all cases. Their initial diagnoses on admission to emergency hospitals were traumatic subarachnoid hemorrhage ($n = 3$), diffuse axonal injury ($n = 4$), and acute epidural hematoma ($n = 1$) which was surgically removed.

In the chronic state (mean: 9 months), all patients had an independent ability of daily life and no focal neurological deficits except for neuropsychological impairments. PET studies of all patients showed abnormal findings, which were divided into two types: frontal type (relative decrease of rCBF and rCMRO$_2$ in bilateral frontal cortices) (Fig. 1a) and cerebellar type (relative decrease of rCBF and rCMRO$_2$ in bilateral occipital cortices and cerebellum) (Fig. 2a), whereas their MRI were normal (Figs. 1b and 2b). Four patients showed the frontal type, and the other four patients showed the cerebellar type. The rCBF changes were coupled with rCMRO$_2$ changes (Fig. 3). However, the rCMRO$_2$ was more sensitive to detect lesions than rCBF in some patients (Fig. 4). rOEF of all patients were normal.

All patients had mild or severe disturbance of all neuropsychological tests. Results of PASAT (evaluating attention) of the frontal type tended to be disturbed more severely than those of the cerebellar type, whereas there was no significant difference (Table 1).

TABLE 1. Summary of Neuropsychological Tests of Each Group

Type	No. cases	WAIS-R FIQ	PASAT
Normal		90–110	48<
Frontal	4	72.6	22
Cerebellar	4	63.5	41

Note. WAIS-R, Japanese Wechsler Adult Intelligence Scale–Revised; PASAT, paced auditory serial addition task.

IV. DISCUSSION AND CONCLUSIONS

Diffuse and mild traumatic brain injury can produce a variety of neuropsychological deficits (Umile *et al.*, 1998). These deficits, such as disturbance of attention, concentration, memory, and information processing, are sometimes very difficult to understand for others, whereas neurological deficits like hemiparesis or aphasia are easily recognized and understood. This ambiguity is a great disadvantage when the patients return to work and also when deciding on reasonable compensation of accident insurance and social welfare. In order to confirm the presence or accurate severity of the neuropsychological deficits, simple and widely accepted objective examinations are needed.

In the present study, we clearly showed that the patients with neuropsychological impairments had abnormal changes of CBF and oxygen metabolism using PET. This can be an objective evidence of a presence of neuropsychological deficits in these patients. We showed that there were two types of PET findings and the frontal type (relative decrease of rCBF and rCMRO$_2$ in bilateral frontal cortices) had lower scores of PASAT. This suggested that the frontal lobe had an important role for attention. Kashima reported that frontal lobe lesions produced lower scores in the attention tests (Kashima *et al.*, 1986).

We chose the timing of the evaluation in the chronic state. The reasons are that intracranial conditions of the patients are not stable in the acute state after brain injury and that it is also very difficult to perform neuropsychological tests at that time. The results of the tests in the acute state are also not reliable, because they would change. The neuropsychological impairments after mild TBI largely resolve within 3 months (Levin *et al.*, 1987); however, some patients had these sequelae for longer periods, more than 6 months (Alves *et al.*, 1993). Therefore, about 9 months after TBI, we evaluated patients whose symptoms were stable.

We showed that rOEF was normal in all patients and rCBF and rCMRO$_2$ were basically coupled. Therefore, it might be enough to measure only CBF of the patients by SPECT to detect abnormalies. However, we also showed that changes of rCMRO$_2$ were more sensitive in detecting

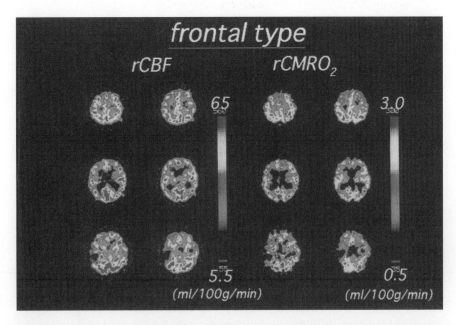

(a)

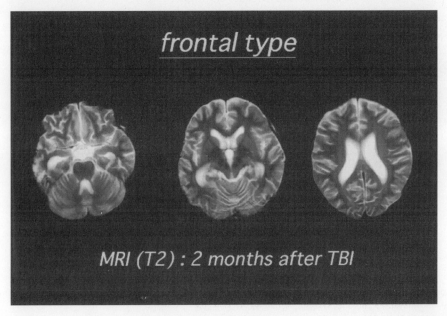

(b)

FIGURE 1.   PET (a) and MR images (b) of a 26-year-old female 2 months after traumatic brain injury are shown. Mild decrease of regional cerebral blood flow (rCBF) and regional metabolic rate of oxygen (rCMRO$_2$) are seen in bilateral frontal cortices, classified into the frontal type.

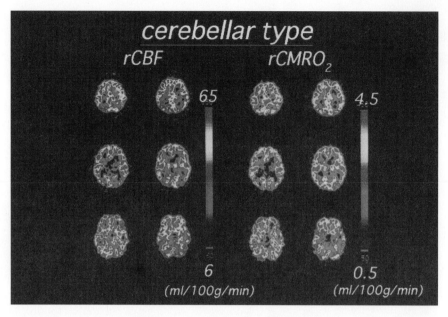

(a)

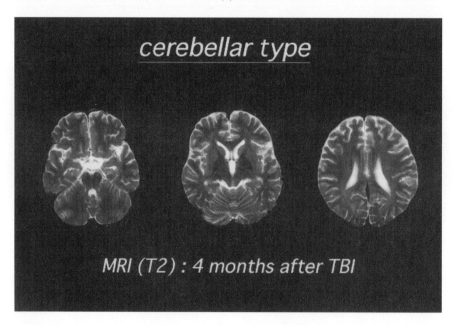

(b)

FIGURE 2.   PET (a) and MR images (b) of a 20-year-old man 4 months after traumatic brain injury are shown. Mild decrease of rCBF and rCMRO$_2$ are seen in bilateral occipital cortices and cerebellum, classified into the cerebellar type.

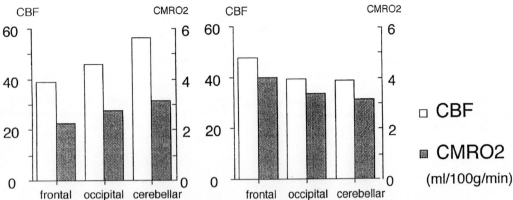

FIGURE 3. Mean rCBF and rCMRO₂ in each brain region of the frontal and cerebellar types are shown. Both parameters were coupled.

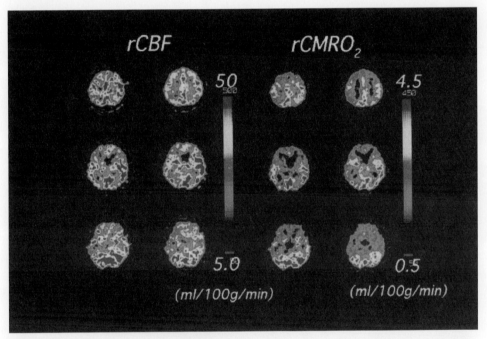

FIGURE 4. PET images of a 62-year-old man 3 months after traumatic brain injury are shown. Decrease of rCMRO₂ in bilatieral frontal cortices was prominent, whereas the rCBF change was mild.

the lesion then that seen with rCBF. Previous studies could not conclude that SPECT is able to predict neuropsychological test performance (Umile *et al.*, 1998). We believe that rCMRO₂ is a better index of the presence of the sequelae than only rCBF.

In conclusion, cerebral blood flow and metabolic evaluation using PET can become an objective examination of these sequelae. Further study will be needed to clarify the relationship between cerebral hemodynamic and metabolic changes and each higher brain function.

Acknowledgment

This work was supported by a Grant for Traffic Accident Medicine, the Marine and Fire Insurance Association of Japan, INC.

References

Abu-Judeh, H. H., Singh, M., Masdeu, J. C., and Abdel-Dayem, H. M. (1998). Discordance between FDG uptake and Technetium-99m-HMPAO brain perfusion in acute traumatic brain injury. *J. Nucl. Med.* **39**: 1357–1359.

Alves, W. M., Macciochi, S. N., and Barth, J. T. (1993). Postconcussive symptoms after uncomplicated mild head injury. *J. Head Trauma Rehabilitation* **8**: 48–59.

Kashima, H., Handa, T., Katoh, M., Honda, T., Sakuma, K., Muramatsu, T., Yoshino, A., Saitoh, H., and Ooe, Y. (1986). Disorders of attention due to frontal lobe lesion. *Shinkei-Shinpo (Jpn)* **30**: 847–858.

Levin, H. S., Mattis, S., Ruff R. M., Eisenberg, H. M., Marshall, L. F., Tabaddor, K., High, W. M., and Frankowski, R. F. (1987). Neurobehavioral outcome following minor head injury: A three-center study. *J. Neurosurg.* **66**: 234–243.

Laatsch, L., Jobe, T., Sychra, J., Lin, Q., and Blend, M. (1997). Impact of cognitive rehabilitation therapy on neuropsychological impairments as measured by brain perfusion SPECT: A longitudinal study. *Brain Injury* **11**: 851–863.

Shiina, G., Onuma, T., Kameyama, M., Shimosegawa, Y., Ishii, K., Shirane, R., and Yoshimoto, T. (1998). Sequential assessment of cerebral blood flow in diffuse brain injury by ^{123}I-Iodoamphetamine single-photon emission CT. *AJNR Am. J. Neuroradiol.* **19**: 297–302.

Umile, E. M., Plotkin, R. C., and Sandel, M. E. (1998). Functional assessment of mild traumatic brain injury using SPECT and neuropsychological testing. *Brain Injury* **12**: 577–594.

Yamaki, T., Imahori, Y., Ohmori, Y., Yoshino, E., Hohri, T., Ebisu, T., and Ueda, S. (1996). Cerebral hemodynamics and metabolism of severe diffuse brain injury measured by PET. *J. Nucl. Med.* **37**: 1166–1170.

Yamaki, T., Yoshino, E., Fujimoto, M., Ohmori, Y., Imahori, Y., and Ueda, S. (1996). Chronological positron emission tomographic study of severe diffuse brain injury in the chronic stage. *J. Trauma: Inj. Infect. Crit. Care* **40**: 50–56.

PATHOPHYSIOLOGY OF IDIOPATHIC BRAIN DISORDERS

53

Imaging Activated Microglia in the Aging Human Brain

ANNACHIARA CAGNIN, RALPH MYERS, ROGER N. GUNN, FEDERICO E. TURKHEIMER,
VIN J. CUNNINGHAM, DAVID J. BROOKS, TERRY JONES, RICHARD B. BANATI

MRC Cyclotron Unit, Imperial College School of Medicine, Hammersmith Hospital, Du Cane Road, London W12 ONN, UK

Microglia are the brain's resident tissue macrophages and constitute approximately 20% of all nonneuronal cells. They are normally in a quiescent resting state. However, they rapidly respond to most types of brain injury by expressing numerous immune system-related molecules and assuming functions typically seen in macrophages. This type of reaction to neuronal injury occurs not only at primary lesion sites but also in remote projection areas. In this study, we have used positron emission tomography (PET) and the tracer $[^{11}C](R)$-PK11195 to assess whether there is an age-dependent increase of activated microglia in the healthy normal brain as a response to the subtle pathological changes observed in aging. For quantification of the specific binding, we used a simplified reference tissue model. A reference input function was generated by two different methods: (a) the extraction of a concentration time–activity curve from a region of interest located over the cerebellum and (b) the segmentation of the dynamic scan data using cluster analysis. Regional mean binding potential (BP) values and a "lesion load," i.e., the fraction of voxels with significantly increased binding, were calculated by volume of interest (VOI) analysis on parametric binding potential maps. Using cluster analysis, a significant (r = 0.72) age-related increase of $[^{11}C](R)$-PK11195 binding was found in the thalamus, but not in the cerebral cortex. This age-related effect, however, was reduced to a trend (r = 0.43) when the data were analyzed using the reference input function generated from the cerebellar reference tissue. We hypothesize that the age-related increase of the $[^{11}C](R)$-PK11195 binding in the thalamus is due to an "amplification" of the thalamic signal by virtue of the high density of corticothalamic projections received from the entire cortex, while the underlying subtle and diffuse cortical pathology could not be detected by our technique.

I. INTRODUCTION

$[^{11}C](R)$-PK11195 (1-(2-chlorophenyl)-N-methyl-N-(1-methylpropyl)-3 isoquinoline carboxamide) is a highly specific ligand of the "peripheral benzodiazepine binding site" (PBBS) (Benavides *et al.*, 1988). The PBBS is also found to coprecipitate with the outer membrane of mitochondria (Anholt *et al.*, 1986) and is, therefore, also referred to as the "mitochondrial benzodiazepine receptor." However, there is also disproportionately high binding to nonmitochondrial fractions of brain extract and mitochondria-free cells (Hertz, 1993; Olson *et al.*, 1988). In the brain, PBBS are exclusively found on nonneuronal cells, primarily activated microglia or, in the case of a disrupted blood–brain barrier, on invading cells of mononuclear-phagocyte lineage (Conway *et al.*, 1998; Myers *et al.*, 1991; Benavides *et al.*, 1988). Although astrocytes in cell culture conditions may express a significant number of PBBS (Itzhak *et al.*, 1993), they have not been found to do so *in vivo* (Conway *et al.*, 1998; Banati *et al.*, 1997; Myers *et al.*, 1991). This observation has been confirmed by the absence of any significant binding of $[^{11}C](R)$-PK11195 in patients (clinically stable and seizure-free) with hippocampal sclerosis, a neuropathology dominated by reactive astrocytes (Banati *et al.*, 1999).

Microglia are a population of small, highly ramified, and normally dormant tissue macrophages that in the normal brain comprise approximately 20% of all nonneuronal cells (Banati *et al.*, 1994). While their resting function is still unclear, they have been shown to respond to a wide variety of pathological stimuli by the expression of surface proteins typically found on peripheral macrophages, proliferation, secretion of cytokines and also the upregulation of PBBS (Kreutzberg, 1996; Banati *et al.*, 1993; Banati and Graeber, 1994). The transformation from a resting state to an activated state often occurs well before any other overt sign of tissue damage or cell death can be seen. This highly sensitive process has, therefore, been used to detect active brain disease and can serve as a surrogate marker of subtle neuronal injury in experimental lesion models as well as human diseases (Kreutzberg, 1996). The expression of PBBS closely follows the temporal and spatial distribution pattern of activated microglia/brain macrophages. Hence, imaging activated microglia with $[^{11}C](R)$-PK11195 PET may potentially be used as a generic tool to localize and monitor ongoing subtle brain disease.

In a normal brain, without active pathology, the presence of activated microglia is minimal. However, "normal" aged brain shows typical pathological changes, such as subclinical neuronal loss and accompanying glial cell reactions. This study focused on the question whether and where $[^{11}C](R)$-PK11195 PET detects any age-dependent regional increases in activated microglia in a group of healthy normal subjects over a wide age range.

II. METHODS

A. Subjects

We studied 14 (8 male/6 female) healthy subjects between the ages of 32 and 80 years (mean age 57.3 years). Age-related systemic diseases (heart failure, high blood pressure, cerebrovascular diseases, diabetes mellitus) and neurological impairment were ruled out by general medical screening and a formal neurological examination. The presence of a relevant cognitive decline was excluded by performing a general neuropsychological assessment (Mini Mental State Examination). Each subject underwent a volumetric T1-weighted MRI brain scan obtained from a 1.0-T Picker MRI scanner for the purpose of coregistration with the PET image and exclusion of incidental pathology in any of the subjects.

B. Data Acquisition

The PET study was performed on a CTI/Siemens ECAT 953B PET scanner used in 3D acquisition mode. The enantiomer $[^{11}C](R)$-PK11195 was injected as a bolus 30 s after the acquisition scan started. The mean tracer dose was 362 ± 32 MBq with a specific activity of 37.3 ± 1.2 GBq/mmol.

Dynamic data were collected over 60 min as 18 temporal frames. Attenuation correction factors were determined using a 15-min transmission scan. Scatter correction was achieved using a dual energy window method (Grootoonk *et al.*, 1996). Data were reconstructed with a ramp filter producing an image resolution of 5.8 mm (full-width at half-maximum) at the center of the field of view.

C. Kinetic Modelling and Analysis

The quantification of the $[^{11}C](R)$-PK11195 binding at the voxel level was obtained using a simplified reference tissue model (Lammertma and Hume, 1996; Gunn *et al.*, 1997) and cluster analysis (Ashburner *et al.*, 1996; Acton *et al.*, 1997; Gunn *et al.*, 1998; Myers *et al.*, 1999). Parametric maps of $[^{11}C](R)$-PK11195 binding potential were generated using a basis function implementation of a simplified reference tissue model (Gunn *et al.*, 1997).

The selection of a reference region may introduce a bias since an *a priori* assumption that the ROI is devoid of specific binding has to be made. This is particularly relevant under conditions where global changes in binding may occur. Cluster analysis offers an alternative "data-led" approach for the extraction of the normal ligand kinetic that can serve as the reference input function. Cluster analysis allows the automatic segmentation of the dynamic raw data into a number of clusters (in our case 10) of concentration time–activity curves (TACs) on the basis of the shape of the TAC. The process then associates each voxel of the image with one of the cluster curves according to the likelihood with which the TAC of the voxel belongs to one of the clusters. In normal brain, the majority (around 90%) of the voxels are segregated into two clusters, one representing the TAC mainly from the skull and scalp and one representing the TAC of voxels mostly located in the cerebral cortex (Fig. 1). The latter was identified as the kinetic of the ligand in normal brain tissue. A normalized mean TAC (population input kinetic) was calculated from the normal ligand kinetics (from each of the 14 subjects) as previously identified by cluster analysis. The suitability of the TAC extracted by cluster analysis as the individual normal ligand kinetic was confirmed by testing for dissimilarity with the normal population input kinetic (χ^2 test <0.95; Fig. 2). In this statistical comparison, the tested normal ligand kinetic was not contained in the group of TACs that constituted the population input kinetic. The results obtained by the cluster analysis approach were compared with those using the reference input kinetic generated from a cerebellar VOI. In the latter approach, each individual's T1-weighted MRI scan was coregistered with the subject's dynamic $[^{11}C](R)$-PK11195 image, which aided the definition of a bilateral cerebellar reference region of 16 cm^3 encompassing three slices and located in the middle of each of the cerebellum hemispheres.

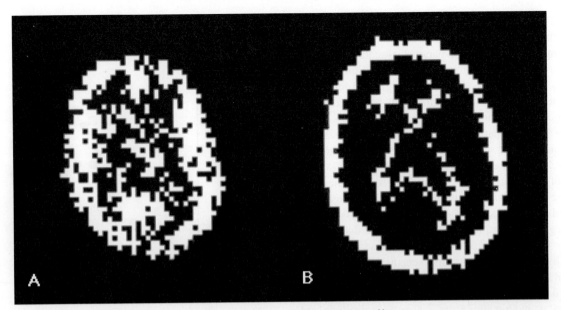

FIGURE 1. Binary mask images derived from cluster analysis of a dynamic $[^{11}C](R)$-PK11195 scan of a normal subject. The two images show the distribution of voxels characterized by the two main clusters. (A) The mask image of a cluster which encompasses predominantly the cerebral cortex. (B) Mask image showing the cluster which corresponds to scalp and white matter.

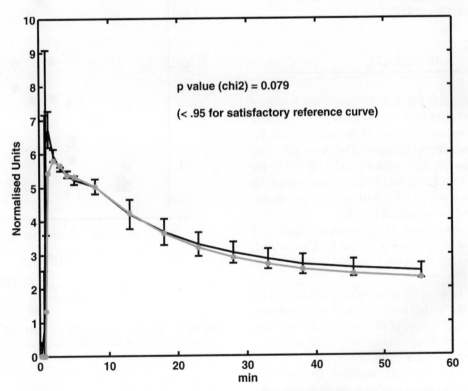

FIGURE 2. The TAC extracted by cluster analysis in a single subject as a suitable reference kinetic (gray line) and the normal population kinetic (black line) were tested for dissimilarity. Here, the individual TAC is not significantly dissimilar from the normal population kinetic (χ^2 test < 0.95) and can, therefore, serve as a reference input function.

The BP maps, obtained with both cluster analysis and a cerebellar input function, were transformed into MRI space using the transformation parameters acquired from the coregistration of the PET image to the MRI (Studholme *et al.*, 1997). VOIs corresponding to the right and left thalamus and temporal lobe were anatomically identified on the volumetric MRI and applied to the parametric maps of BP. From the VOI two parameters were derived: the mean BP value (a measure of specific binding of the tracer) and the "lesion load," i.e., the numbers of voxels of the VOI (as percentage of the volume) above a threshold fixed at the noise level. The individual noise level from the BP images was calculated over three slices of scalp-edited brain tissue taken at the level of cerebral hemispheres above the thalamus. A model for robust statistical estimation of noise which is not affected by the presence of sparse specific signal was used (Rey, 1983). Voxels above threshold (i.e., values above two standard deviations of the calculated background level) were considered as having significant, i.e., pathologically raised, $[^{11}C](R)$-PK11195 binding,

$$\text{lesion load} = \text{volume of tissue with BP} > \text{BP}_T, \quad (1)$$

where BP_T $(= 2\text{BP}_{SD})$ is the binding potential threshold, BP_{SD} $(= \text{MAD}/0.6745)$ is the standard deviation of the background level, and MAD is the maximum absolute deviation (0.6745 is the factor chosen for calibration with the normal distribution) (Rey, 1983).

III. RESULTS

Using $[^{11}C](R)$-PK11195 and the simplified reference tissue model, combined with a cluster analysis approach, we found an age-related increase in the mean $[^{11}C](R)$-PK11195 BP values in the thalamus. The mean BP value was 0.40 ± 0.08 for the right thalamus and 0.38 ± 0.08 for the left thalamus. The correlation coefficient with age for the right thalamus was $r = 0.71$ $(p < 0.01)$ and for the left thalamus was $r = 0.65$ $(p < 0.02)$. (Fig. 3A).

The statistical significance of the correlation increased when the binding was expressed as "lesion load" $(r = 0.81$, $p < 0.01$, for the right thalamus; $r = 0.84$, $p < 0.01$, for the left thalamus) (Fig. 3B).

In our group of subjects the mean value of BP_{SD} used for the calculation of the "lesion load" was 0.19 with a standard deviation of 0.02. The volume of the thalamus did not show a significant change with age (mean number of voxels for the bilateral thalamus, 8656 ± 859).

Using the subjects' individual cerebellar TAC as reference input function, a weak correlation between $[^{11}C](R)$-PK11195 binding and age was seen in the right thalamus, while statistical significance was not reached in the left thalamus (right thalamus, $r = 0.53$, $p = 0.05$; left thalamus,

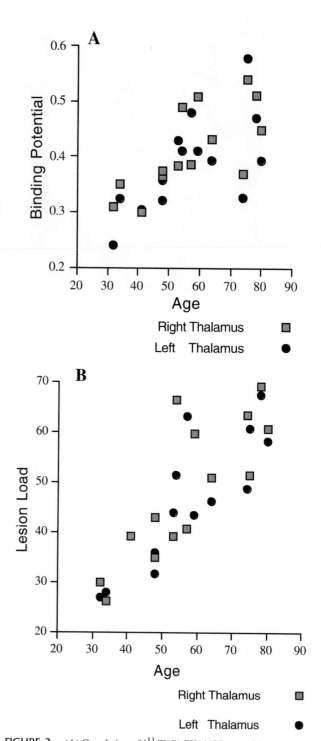

FIGURE 3. (A) Correlation of $[^{11}C](R)$-PK11195 mean BP values with age in right (□) and left (●) thalamus using cluster analysis and a simplified reference tissue model. Right thalamus, $r = 0.72$ $(p < 0.01)$; left thalamus, $r = 0.65$ $(p < 0.02)$. (B) Correlation of "lesion load" (expressed as percentage of VOI with a BP value above the noise level) with age in right and left thalamus using cluster analysis. Right thalamus, $r = 0.81$ $(p < 0.01)$; left thalamus, $r = 0.84$ $(p < 0.01)$.

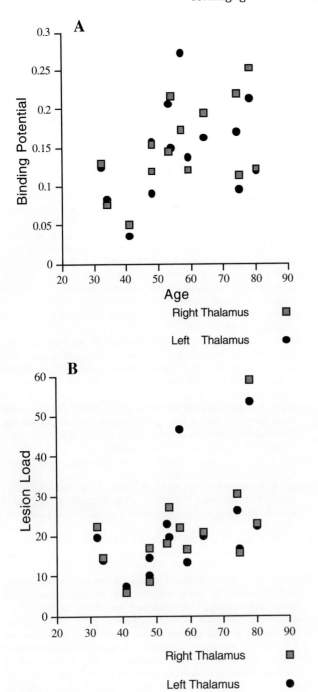

FIGURE 4. (A) Correlation of $[^{11}C](R)$-PK11195 mean BP values with age in right (□) and left (●) thalamus using a cerebellar TAC as reference input function. Right thalamus, $r = 0.53$ ($p = 0.05$); left thalamus, $r = 0.35$ ($p > 0.1$). (B) Correlation of "lesion load" (expressed as percentage of VOI with BP values above threshold) with age in right and left thalamus using a cerebellar TAC as reference input function. Right thalamus, $r = 0.53$ ($p = 0.05$); left thalamus, $r = 0.47$ ($p < 0.1$).

$r = 0.35$, Fig. 4A). The same findings were obtained with the "lesion load" (right thalamus, $r = 0.53$, $p = 0.05$; left thalamus, $r = 0.47$, $p < 0.1$) (Fig. 4B).

In the temporal lobe, the mean $[^{11}C](R)$-PK11195 BP value was 0.135 ± 0.2 using cluster analysis and approximately zero using the cerebellar input function. $[^{11}C](R)$-PK11195 binding in the temporal lobe, expressed either as regional mean BP or as "lesion load," did not show any correlation with age, regardless of whether a cerebellar input function or cluster analysis was used.

IV. DISCUSSION

In this study, we have found an age-dependent increase of $[^{11}C](R)$-PK11195 binding in the thalamus of a group of normal subjects ages 32 to 80 years. The thalamus was chosen as region of interest because of its dense direct anatomical connections with most cortical areas. Our working hypothesis was that $[^{11}C](R)$-PK11195 may be used as a measure of microglial activation in the thalamus, where it occurs as a secondary phenomenon due to the subtle, global cortical pathology found in normally aging brains. Experimental lesion studies have shown that microglial activation does not only occur at the primary site of neuronal injury but also in remote areas into which the axons of the lesioned neurons project (Banati *et al.*, 1997). This has also been demonstrated in experimental animal models of focal ischemia using autoradiography with $[^{3}H]$PK11195 (Dubois *et al.*, 1988; Myers *et al.*, 1991) and in immunohistochemical studies of microglial activation following neocortical lesions (Ancarin *et al.*, 1999; Sorensen *et al.*, 1996). In both cases, it was possible to localize a regional increase of activated microglia in the thalamus as a sign of a secondary response associated with the primary focal damage in the cortex.

Normal brain aging is associated with a variety of pathological changes affecting predominantly cortical regions: mild loss of neurons, dendritic and synaptic compensatory changes, presence of sparse senile plaques and glial activation. *Postmortem* immunohistochemical studies in monkeys and mice have provided evidence for an age-related, diffuse increase of activated microglia in the white matter, in the temporal lobe and around senile plaques (Sheng *et al.*, 1998; Ogura *et al.*, 1994). These neuropathological changes coincide with age-related functional alterations, such as the decrease in cerebral metabolism of glucose (CMRglc) or O_2 consumption (CMRO$_2$). The latter shows a rate of decrease of 6% per decade in the neocortex (Petit-Taboue *et al.*, 1998; Marshall *et al.*, 1992). Clinically, these changes may underlie the subtle cognitive decline typical of elderly people. This age-related decline is characterized by a nonspecific pattern of impairment of higher cognitive functions involving different cortical domains. The neuronal substrate of such cognitive impairment in the elderly appears to be a "reduced ability to focus neural activity" (disconnectivity theory), resulting in the failure to potentiate the functional activation of areas needed for the task and an inability to suppress areas not critical for the task performance (Esposito *et al.*, 1999). Following this theory, the functional impairment observed in the healthy elderly is due to a diffuse derangement of networks (system failure) rather than a defect in a defined anatomical or functional domain.

Compared to the neocortex, the thalamus seems relatively spared by aging. In fact, the thalamus does not show any significant structural (decrease of volume) or functional (decrease of CMRglc) alterations due to aging (Murphy *et al.*, 1992; Marshall *et al.*, 1992). This suggests that the activation of microglia found in the present study is not the result of aging of the thalamus *per se*. The absence of overt macroscopic and microscopic pathological changes or functional impairment that is attributable to primary pathology in the thalamus indicates that the age-related increase in $[^{11}C](R)$-PK11195 binding is a secondary effect of deafferentation due to subtle but widespread cortical pathology. The finding of only minimal binding in the cerebral cortex (temporal lobe) may be due to an insufficient sensitivity of $[^{11}C](R)$-PK11195 PET for the detection of very low-grade cortical pathology in "normal" aging brains. In contrast, the high density of direct cortico-thalamic projections converging into the thalamus may lead to a cumulative thalamic signal representing a regional "amplification" of the widespread, subthreshold cortical pathology.

Methodologically, this paper shows how quantification of the weak $[^{11}C](R)$-PK11195 signal can be achieved using 3D acquisition, a reference tissue model, and cluster analysis. The advantage of using the cluster analysis approach (over a cerebellar input function derived from a manually defined region of interest) is that no *a priori* decision with respect to an anatomically defined reference region has to be made. The cluster analysis may be particularly relevant in the analysis of $[^{11}C](R)$-PK11195 data from patients with a diffuse pathology where no region can be considered suitable as a reference region, as is often the case in neurodegenerative diseases. Previous autoradiographic studies showed that in normal brain $[^{11}C](R)$-PK11195 binding is very low in the cerebellar hemispheres (Doble *et al.*, 1987). Yet, using the cerebellar input function, a small amount of binding in the cerebellum was sufficient to reduce the calculated BP values in both thalamus and temporal lobe, which in turn weakened the correlation of $[^{11}C](R)$-PK11195 binding with age found by cluster analysis.

In summary, the detection of subtle changes in microglial activation requires particular care in the selection of a reference input function with minimal contamination with specific signal. Cluster analysis is suitable for the analysis of $[^{11}C](R)$-PK11195 image data in "normal" brain aging and in neurological conditions characterized by a global and/or very subtle brain pathology.

Acknowledgments

The project is supported by the Medical Research Council. A.C. is supported by "Marie Curie" fellowship from the European Community in the "Training and Mobility of Researchers Programme in Biomedicine." R.B.B. received funding from the Max-Planck-Institute for Neurobiology (Munich), the German Research Foundation (DFG), and the Multiple Sclerosis Society (UK).

References

Acton, P. D., Pilowsky, L. S., Costa, D.C., and Ell, P. J. (1997). Multivariate cluster analysis of dynamic iodine-123 iodobenzamide SPET dopamine D2 receptor images in schizophrenia. *Eur. J. Nucl. Med.* **24**: 111–118.

Anholt, R. R., Pedersen, P. L., DeSouza, E. B., and Snyder, S. H. (1986). The peripheral-type benzodiazepine receptor. Localisation to the mitochondrial outer membrane. *J. Biol. Chem.* **261**: 776–783.

Ancarin, L., Gonzalez, B., Castro, A. J., and Castellano, B. (1999). Primary cortical glial reaction versus secondary thalamic glial response in the excitotoxically injured young brain: Microglia/macrophage response and major histocompatibility complex class I and II expression. *Neuroscience* **89**: 549–565.

Ashburner, J., Haslam, J., Taylor, C., Cunningham, V., and Jones, T. (1996). A cluster analysis approach for the characterization of dynamic PET data. *In "Quantification of Brain Function Using PET"* (R. Myers, V. Cunningham, D. Bailey, and T. Jones, Eds.), pp. 301–306. Academic Press, San Diego.

Banati, R. B., Myers, R., and Kreutzberg, G. W. (1997). PK ('peripheral benzodiazepine')-binding sites in the CNS indicate early and discrete brain lesions: Microautoradiographic detection of [3H]PK11195 binding to activated microglia. *J. Neurocytol.* **26**: 77–82.

Banati, R. B., Gegrman, J., Schubert, P., and Kreutzberg, G. W. (1993). Cytotoxicity of microglia. *Glia* **7**: 111–118.

Banati, R. B., Goerres, G. W., Myers, R., Gunn, R. N., Turkeimer, F. E., Brooks, D. J., and Jones, T. (1999). $[^{11}C](R)$-PK11195 PET imaging of activated microglia in vivo in Rasmussen's encephalitis. *Neurology.* **49**: 1682–1688.

Banati, R. B., and Graeber, M. B., (1994). Surveillance, intervention and cytotoxicity: is there a protective role of microglia? *Dev. Neurosci.* **16**: 114–127.

Benavides, J., Cornu, P., Dennis, T., Dubois, A., Hauw, J. J., MacKenzie, E. T., Sazdovitch, V., and Scatton, B. (1988). Imaging of human brain lesions with an omega 3 site radioligand. *Ann. Neurol.* **24**: 708–712.

Conway, E. L., Gundlach, A. L., and Craven, J. A. (1998). Temporal changes in glial fibrillary acidic protein messenger RNA and [3H]PK11195 binding in relation to imidazoline-I2-receptor and alpha 2-adrenoreceptor binding in the hippocampus following transient global forebrain ischaemia in the rat. *Neuroscience* **82**: 805–817.

Doble, A., Malgouris, M., Daniel, M., Imbault, F., Basbaum, A., Uzan, A., Gueremy, C., and Le Fur, G. (1987). Labelling of peripheral-type benzodiazepine binding sites in human brain with [3H]PK 11195: anatomical and subcellular distribution. *Brain Res. Bull.* **18**: 49–61.

Dubois, A., Benavides, J., Peny, B., Duverger, D., Fage, D., Gotti, B., MacKenzie, E. T., and Scatton, B. (1988). Imaging of primary and remote ischaemic and excitotoxic brain lesions. An autoradiographic study of peripheral type benzodiazepine binding sites in the rat and cat. *Brain Res.* **445**: 77–90.

Esposito, G., Kirkby, B. S., Van Horn, J. D., Ellmore, T. M., and Berman, K. F. (1999). Context-dependent, neural system-specific neurophysiological concomitants of ageing: mapping PET correlates during cognitive activation. *Brain.* **122**: 963–979.

Gunn, R. N., Lammerstma, A. A., and Cunningham, V. J. (1998). Parametric imaging of ligand-receptor interactions using a reference tissue model and cluster analysis. *In "Quantitative Functional Brain Imaging with Positron Emission Tomography"* (R. Carson, M. Daule, P. Witherspoon, and Herscovitch, Eds.), pp. 401–406. Academic Press, San Diego.

Gunn, R. N., Lammertsma, A. A., Hume, S. P., and Cunningham, V. J. (1997). Parametric imaging of ligand-receptor binding in PET using a simplified reference region model. *Neuroimage* 6: 279–287.

Grootoonk, S., Spinks, T. J., Sashin, D., Spyrou, N. M., and Jones, T. (1996). Correction for scatter in 3D brain PET using a dual energy window method. *Phys. Med. Biol.* 41: 2757–2774.

Hertz, L. (1993). Binding characteristics of the receptor and coupling to transport proteins. *In "Peripheral Benzodiazepine Receptor"* (Gliessen Crouse Eds.), pp. 27–51. Academic Press, London.

Itzhak, Y., Baker, L., and Norenberg. (1993). Characterisation of the peripheral-type benzodiazepine receptors in cultured astrocytes labeled with [3H]PK 11195: evidence for multiplicity. *Glia* 9: 211–218.

Kreutzberg, G. W. (1996). Microglia: A sensor for pathological events in the CNS. *Trends Neurosci.* 19: 312–318.

Lammertsma, A. A., and Hume, S. P. (1996). Simplified reference tissue model for PET receptor studies. *Neuroimage* 4:153–158.

Marshall, G., Rioux, P., Petit Taboue,' M. C., Sette, G., Travere, J. M., Le Poec, C., Courtheoux, P., Derlon, J. M., and Baron, J. C. (1992). Regional cerebral oxygen consumption, blood flow, and blood volume in healthy human aging. *Arch. Neurol.* 49: 1013–1020.

Myers, R., Manjil, L. G., Frackowiak, R. S. J., and Cremer, J. E. (1991). [3H]PK 11195 and the localisation of secondary thalamic lesions following focal ischaemia in rat motor cortex. *Neurosci. Lett.* 339: 1054–1055.

Myers, R., Manjil, L. G., Cullen, B. M., Price, G. W., Frackowiak, R. S. J., and Cremer, J. E. (1991). Macrophage and astrocyte population in relation to [3H]PK 11195 binding in rat cerebral cortex following a local ischaemic lesion. *J. Cereb. Blood Flow Metab.* 11: 314–322.

Myers, R., Gunn, R. N., Cunningham, V. J., Banati, R. B., and Jones, T. (1999). Cluster analysis and the reference tissue model in the analysis of clinical [^{11}C](R)-PK11195 PET. *J. Cereb. Blood Flow Metab.* 19: S789.

Murphy, D., DeCarli, C., Shapiro, M., Rapoport, S., and Horwitz, B. (1992). Age-related differences in volumes of subcortical nuclei, brain matter, and cerebrospinal fluid in healthy men as measured with magnetic resonance imaging. *Arch. Neurol.* 49: 839–844.

Ogura, K., Ogawa, M., and Yoshida, M. (1994). Effects of ageing on microglia in the normal rat brain: immunohistochemical observations. *Neuroreport* 10: 1224–1226.

Olson, L. M., Ciliax, B. J., Mancini, W. R., and Young, A. B. (1988). Presence of peripheral-type like binding sites on human erytrocyte membranes. *Eur. J. Pharmacol.* 152: 47–53.

Petit-Taboue, Landeau, B., Desson, J. F., Desgranges, B., and Baron, J. C. (1998). Effects of healthy aging on the regional cerebral metabolic rate of glucose assessed with statistical parametric mapping. *Neuroimage* 3: 176–184.

Ramsay, S. C., Weiller, C., Myers, R., Cremer, J. E., Luthra, S. K., Lammertsma, A. A., and Frackowiak, R. S. J. (1992). Monitoring by PET of macrophage accumulation in brain after ischaemic stroke. *Lancet* 339: 1054–1055.

Rey, J. J. (1983). *"Introduction to Robust and Quasi-Robust Statistical Methods"* pp. 124–130. Springer-Verlag, Berlin-Heidelberg.

Sheng, J. G., Mark, R. E., and Griffin, W. S. (1998). Enlarged and phagocytic, but not primed, interleukin-1 alpha-immunoreactive microglia increase with age in normal human brain. *Acta Neuropathol.* 3: 229–234.

Sorensen, J. C., Dalmau, I., Zimmer, J., and Finsen, B. (1996). Microglial reactions to retrograde degeneration of racer-identified thalamic neurons after frontal sensorimotor cortex lesions in adult rats. *Exp. Brain Res.* 11: 203–212.

Studholme, C., Hill, D. L. G., and Hawkes, D. J. (1997). Automated three-dimensional registration of magnetic resonance and positron emission tomography brain images by multiresolution optimization of voxel similarity measures. *Med. Phys.* 24: 25–35.

54

The Functional Anatomy of the Amnesic Syndrome Studied with Resting [18F]FDG-PET and SPM: Revealing Dysfunctional Networks in Specific Neuropsychological Syndromes

A. M. AUPÉE, B. DESGRANGES, F. EUSTACHE, C. LALEVÉE, V. DE LA SAYETTE, F. VIADER, and J. C. BARON

INSERM U320, Cyceron and Dept of Neurology, University of Caen, France

In this retrospective study, we compared the resting CMR-glc of five patients suffering from the amnesic syndrome to that of healthy age-matched subjects, using SPM. The aims of this study were (i) to assess a metabolic group pattern and (ii) to assess individual significant changes. The results from the voxel-by-voxel group analysis documented that the amnesic syndrome is associated with hypometabolism in the thalamus, the posterior cingulate cortex, and the mesial prefrontal cortex, with extension to some neocortical language areas of the left hemisphere. The individual analysis showed: (i) that this group pattern was found in essentially each patient, regardless of the cause of the amnesic syndrome; (ii) that hypometabolism in the hippocampal region and the anterior cingulate gyrus was present in some cases individually, suggesting dysfunction in an extended Papez's circuit; and (iii) that additional hypometabolic regions may be related to superimposed neuropsychological impairments. The SPM method was used here for the first time to reveal a common network of synaptic dysfunction in a neuropsychological syndrome regardless of etiology and to analyze each patient's metabolic pattern in relation to individual neuropsychological profiles. The result indicates that this should be a powerful method in functional neuropsychology, with widespread applications in different neuropsychological syndromes.

I. INTRODUCTION

The amnesic syndrome (AS) is an organic syndrome, characterized by an inability to learn new information (an-terograde amnesia) and the failure to recall facts or events that have occurred before the onset of the illness (retrograde amnesia). Retrograde amnesia may extend to several years or even decades but usually spares the very remote events (Victor, *et al.*, 1989), pointing to the existence of a temporal gradient (Albert *et al.*, 1979; Kopelman *et al.*, 1999). The AS is also characterized by sparing of the other cognitive functions, i.e., short-term memory, semantic memory, implicit memory and intelligence. Classically, two subtypes of AS are distinguished, the hippocampal and the diencephalic AS. The former is illustrated by the well-known patient, HM, who exhibited a severe amnesia after bilateral surgical removal of the medial temporal lobes (Scoville and Milner, 1957). This mechanism is assumed to underlie the AS which sometimes follows brain anoxia, as cardiocirculatory arrest and asphyxia are known to induce selective neuronal death, especially affect the hippocampus (Volpe *et al.*, 1984; Ouchi *et al.*, 1998). The prototype for diencephalic AS is the Wernicke–Korsakoff (W-K) syndrome which occurs after thiamine deficiency generally affecting alcoholic patients (Victor *et al.*, 1989). The pathology of the W-K syndrome consists of petaechial haemorrhage and neuronal death affecting diencephalic structures such as the medial thalamus and mammillary bodies, as well as brain-stem structures such as the locus coeruleus (Victor *et al.*, 1989). In the W-K syndrome, frontal signs can be observed in addition to global amnesia. Thus, some W-K patients have difficulties in performing tasks like the Wisconsin card sorting or verbal fluency tests (Hunkin *et al.*, 1994; Parkin and Leng, 1996). Confabulations constitute another hallmark of the W-K syndrome, but whether or not they occur as a way of compensating memory impairment is controversial (Kopelman, 1987). Amnesia in the W-K syndrome was initially related to alterations in the mammillary bodies, disrupting Papez's hippocampo–mamillo–thalamus circuit (Delay and Brion, 1954). However, Victor *et al.* (1989) and others (Markowitsch *et al.*, 1993; Squire *et al.*, 1979), attribute the memory impairment mainly to bilateral damage in the medio-dorsal thalamic nuclei.

Thus, the AS may result from damage to any part of Papez's circuit itself, and possibly also to the amygdala–mediodorsal thalamus–ventromedial prefrontal cortex circuit, i.e., an extended Papez's circuit or limbic–paralimbic system (Mishkin, 1993). However, neither in the W-K syndrome nor in the postanoxic syndrome does structural imaging such as CT or MRI reveal the real extent of the lesions that are mainly microscopic. Furthermore, structural imaging is unable to reveal functional changes along these circuits, and even more so functional impairments that can occur in structures remote from but connected to the area of primary damage (Feeney and Baron, 1986). Because positron emission tomography (PET) allows the study of cerebral parameters such as resting cerebral blood flow or glucose consumption, which are closely related to synaptic activity, PET makes it possible to study dysfunction in the neural networks involved in the AS in the absence of clear-cut structural lesions.

Although previous studies using PET have reported that the AS is associated with resting hypometabolism in Papez's circuit (Fazio *et al.*, 1992; Kuwert *et al.*, 1993; Joyce *et al.*, 1993), all were based on the ROI method which is hypothesis-driven and assesses only a fraction of the brain parenchyma. In addition, all of these studies concerned group analysis; i.e., the individual metabolic profile was not assessed.

The present study had two aims: (1) to map the hypometabolic network in a group of patients with typical AS and no focal lesion revealed by structural imaging; and (2) to assesss the individual metabolic pattern in relation to the neuropsychological profile. To these ends, we have mea-

sured the cerebral glucose consumption (CMRglc) with PET and $[^{18}F]$-fluoro-2-deoxy-D-glucose ($[^{18}F]FDG$) in a cohort of five amnesic patients.

II. MATERIALS AND METHODS

A. Patients

This retrospective study comprises five patients with AS. Three patients were alcoholic W-K syndrome, and two were postanoxia cases (demographic data are summarised in Table 1). Except one patient (VD) in whom memory was assessed at the bedside only, the patients were submitted to a neuropsychological examination of (i) episodic long-term memory, assessed by the Wechsler Memory Scale, 1965, and the California Verbal Learning Test (Delis *et al.*, 1987); (ii) short-term memory, assessed by the Brown–Peterson paradigm (1958); (iii) intelligence, evaluated by the Raven matrices PM 38 (1981) and the Wechsler Adult Intelligence Scale (1991); and (iv) frontal lobe functions, assessed by the Stroop test (1935) and verbal fluency (Cardebat *et al.*, 1990) tests. Standard MRI was performed in all subjects with the exception of one subject (VD) who underwent a CT scan. The findings with structural imaging are summarized in Table 1.

B. PET Method

The CMRglc was measured for each subject. The PET studies were performed in a resting state, with eyes closed, in dimmed light, with reduced ambient noise, using $[^{18}F]$fluoro-2-deoxy-D-glucose ($[^{18}F]FDG$) and a high-resolution seven-slice PET camera (LETI-TTV 03); the effective resolution for ^{18}F was $5.5 \times 5.5 \times 9$ mm x, y, z (Penniello *et al.*, 1995). The patient was positioned with reference to the glabella and the inion bony landmarks as determined on a lateral skull X-ray (Fox *et al.*, 1985); the latter was obtained while the patient was lying on the PET couch, with the head gently restrained in a Laitinen's stereotaxic frame. Following

TABLE 1. Demographic Data

Case	Age (years)	Sex	Aetiology	Structural imaging
1 (DG)	40	M	Anoxia	Mild cortical atrophy (MRI)
2 (HJC)	47	M	Anoxia	Normal (MRI)
3 (BT)	48	F	Wernicke–Korsakoff	Third-ventricle dilatation (MRI)
4 (PF)	39	M	Wernicke–Korsakoff	Mild cortical atrophy, third-ventricle dilatation (MRI)
5 (VD)	51	F	Wernicke–Korsakoff	Moderate cortico-subcortical atrophy, third-ventricle dilatation (CT)

[68]Ga transmission scans, 5–10 mCi of FDG were injected as a bolus, and blood samples were obtained *via* a radial artery catheter for determination of the time course of FDG in plasma and the average plasma glucose content (Phelps *et al.*, 1979). From 50 to 60 min postinjection, seven PET planes were acquired parallel to, and from −4 to +68 mm relative to, the GI plane (which corresponds approximately to −25 to +47 mm relative to the bicommissural line). During PET data acquisition, head motion was continuously monitored with, and whenever necessary corrected according to, laser beams projected onto ink marks drawn over the forehead skin. The seven [18F]FDG planes were transformed into parametric maps of CMRGlc according to the operational equation of Phelps *et al.* (1979), which uses the subject's [18F]FDG plasma time course and mean plasma glucose content, and the standard set of [18F]FDG transfer coefficients and the lumped constant published by these authors.

C. Data Analysis

The PET data were implemented on a SUN workstation for image processing. The SPM 95 (Statistical Parametric Mapping; Wellcome department of Cognitive Neurology, London, UK) and ANALYZE (Biodynamic Research Unit, Mayo Clinic, Rochester, MN) softwares were used to reorient each PET data sets onto Talairach's template (Talairach and Tournoux, 1988). Manipulations of the image matrix (stereotaxic normalization) and statistical calculations were carried out with MATLAB (Mathworks, Sherborn, MA). Following masking by isocontours, the seven PET slices were interpolated to create a three-dimensional space (Deiber *et al.*, 1993; Desgranges *et al.*, 1998). The M 95 then realigned the data set parallel to the intercommissural line, and reset and interpolated it in three dimensions to correspond with the 26 axial slices of the Talairach and Tournoux Atlas (1988). Because of the limited axial field of view of the PET camera used (i.e., 81 mm), the brain could not be studied in its entirety. Each voxel obtained from the transformed images represented a parallelepiped with sides of 2 mm and a thickness of 4 mm. The slices were therefore 4 mm apart, like the planes in the Talairach atlas. In order to minimize "edge effects," only those voxels with values >80% of the mean for the whole brain were selected for statistical analysis. PET images were filtered with a three-dimensional Gaussian filter of 14 mm in order to attenuate anatomical interindividual differences and to reduce noise in the data.

Statistical analyses were then processed with the SPM 96 software (Wellcome Department of Cognitive Neurology, London, UK). Both a group analysis and an individual analysis were carried out. Given the fact that the high degree of intersubject variability in global glucose utilization might blur differences in individual specific pattern and also in the group analysis, metabolic values were expressed in each analysis as relative rates of glucose utilization. To this end, each voxel value was normalized to the global metabolism value, using the Proportional Scaling routine. In the first step, a group analysis was performed in order to map the hypometabolic network in the group of patients (the reverse contrast, namely increases in the patients relative to controls, was not assessed). This analysis consisted in comparing statistically the group of the five AS subjects to an aged-matched control group ($n = 9$, mean age $= 47 \pm 5.6$ years). For this group comparison, we used the threshold of $p < 0.001$ (uncorrected, $Z = 3.09$). In the second step, individual analyses were performed, in order to assess the across-subject consistency of the pattern found in the group analysis, and to relate individual PET features to individual neuropsychological profiles. To this end, we statistically compared each patient to a control group consisting of 10 healthy subjects arranged so that the mean age matched that of each particular patient. As these individual analyses were performed on a hypothesis-driven basis after the group analysis, the $p < 0.01$ (uncorrected, $z = 2.33$) threshold was used.

III. RESULTS

A. Group Study (Table 2, Fig. 1)

The group analysis revealed significantly decreased adjusted CMRglc in the AS group as compared to the control group in the thalamus, posterior cingulate and mesial prefrontal cortex bilaterally, and the left supramarginal gyrus and middle temporal gyrus.

B. Individual Analysis

1. Neuropsychological Data (Table 3)

All four patients assessed with formal tests exhibited impairment in episodic memory as assessed by the CVLT (California Verbal Learning Test) or by the Wechsler Memory Scale. All patients exhibited normal performance at frontal lobe tests except HJC who had frontal signs. Moreover, all patients (except BT; QI = 72) had normal intelligence as assessed by Raven matrices or by Wechsler Intelligence Scale.

2. PET Data (Tables 4 and 5)

No clear differential hypometabolic pattern emerged between anoxic and W-K patients.

The pattern observed in the group analysis was found in essentially each of the five patients (see Table 4). Table 5 shows the additional areas with significant ($p < 0.01$) hypometabolism in the individual analysis. No clear-cut relationship between the individual metabolic and neuropsychological patterns emerged (see below).

IV. DISCUSSION

A. Group Study

The group analysis revealed decreased resting relative glucose consumption bilaterally in the medio-dorsal/

TABLE 2.   Results for the Group Analysis

Regions	Brodmann areas	x	y	z	Z(P)
Thalamus					
R anterior nuclei[a]		—	—	—	—
L anterior nuclei[a]		—	—	—	—
R medio-dorsal nuclei[a]		—	—	—	—
L medio-dorsal nuclei		−4	−12	4	4.66
R ventrolateral nuclei		6	−12	4	3.73
R posterior Cingulate/retrosplenial cortex	BA 30/23	2	−50	16	4.67
L posterior Cingulate/retrosplenial cortex[a]	BA 30/23	—	—	—	—
Left inferior supramarginal gyrus	BA 40	−58	−18	20	4.06
R mesial prefrontal cortex	BA 10	2	60	20	3.70
L mesial prefrontal cortex[a]	BA 10	—	—	—	—
L middle temporal gyrus[b]	BA 21	−60	−12	−8	3.21

[a] Regions that were clearly identified on the SPM printout, but for which no peak was listed.
[b] Region not visible on Fig. 1 (size = 2 voxels).
Note. R, right; L, left.

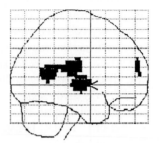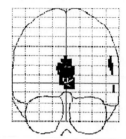
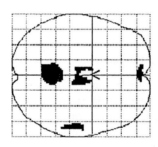

FIGURE 1.   SPM "glass brain" template showing the significantly hypometabolic regions found in the amnesic group, with $p < 0.001$ (uncorrected). The right hemisphere corresponds to the left side of the figure.

TABLE 3.   Neuropsychological Results of the Amnesic Patients

	Episodic memory	Executive functions	Intelligence
DG	Impaired	Normal	Normal
HJC	Impaired	Impaired	Normal
BT	Impaired	Normal	Impaired
PF	Impaired	Normal	Normal

TABLE 4.   Individual Analysis: Findings Relative to the Group Pattern

Hypometabolic regions observed in the group analysis	DG (Anoxia)	HJC (Anoxia)	BT (W-K)	PF (W-K)	VD (W-K)
Thalamus	$p < 0.001$	$p < 0.001$	$p = 0.004$	$p < 0.001$	$p < 0.001$
Posterior cingular/retrosplenial cortex (BA 23/30)	$p < 0.001$			$p < 0.001$	$p < 0.001$
Mesial frontal cortex (BA 10)		$p = 0.003$	$p = 0.005$	$p < 0.001$	$p < 0.001$
Left middle temporal gyrus (BA 21)		$p = 0.004$		$p < 0.001$	
Left supramarginal gyrus (BA 40)	$p < 0.001$	$p < 0.001$	$p < 0.001$	$p = 0.007$	$p < 0.001$

Note. W-K, Wernicke-Korsakoff's syndrome; BA, Brodmann's area.

TABLE 5. Additional Hypometabolic Areas Observed in the Individual Analyses ($p < 0.01$, Uncorrected)

Hypometabolic regions	DG	HJC	BT	PF	VD
Hippocampal region	R	R		L/R	
Anterior cingulate gyrus (BA 24/32)	R			L	L/R
Dorsolateral prefrontal cx (BA 46)		L			
Inferior temporal gyrus (BA 20)			R	R	
Superior temporal gyrus (BA 22/39)				R	L

Note. L, left; R, right; Cx, cortex; BA, Brodmann's area.

anterior/ventro-lateral thalamic nuclei, the posterior cingulate/retrosplenial cortex, the mesial prefrontal cortex bilaterally and the left middle temporal and inferior supramarginal gyri. The first three bilaterally affected regions are part of an extended Papez's circuit, while the latter two are part of the verbal semantic and phonologic system. Thus, the AS appears to be related to a dysfunction of extended Papez/limbic circuit, however, with neocortical language areas being unexpectedly involved.

Concerning the involvement of Papez's circuit in the AS, our results are consistent with previous PET studies which used the ROI approach of image analysis (Fazio *et al.*, 1992; Kuwert *et al.*, 1993; Joyce *et al.*, 1993). Although the finding of thalamic hypometabolism can be related to the well-known fact that Wernicke's encephalopathy affects the medio-dorsal, anterior, and centro-median thalamus and the mamillary bodies (Victor *et al.*, 1989), our finding of hypometabolism in the posterior cingulate and mesial prefrontal cortices is less straightforward. It might reflect either a neural loss, or a mechanism of disconnection (*via* the thalamic damage) within Papez's extended circuit. Earlier PET studies have clearly documented the frequent existence of neocortical hypometabolism affecting parts of subcortico-cortical networks after subcortical (especially thalamic) damage (Baron *et al.*, 1986; Okada *et al.*, 1991; Della Sala *et al.*, 1993). Moreover, Clarke *et al.* (1994) showed significant hypometabolism in the posterior cingulate cortex in an amnesic patient who had an isolated left polar thalamic infarct; to explain this finding, the authors implicated the existence of direct projections from the mediodorsal thalamus to the posterior cingulate cortex. In Alzheimer's disease, the degree of impairment in episodic memory is significantly related to the severity in glucose hypometabolism in the posterior cingulate cortex (Desgranges *et al.*, 1998).

Our observation of significant resting hypometabolism in the left supramarginal and middle temporal gyri has not been reported so far in PET studies of AS patients. However, this finding would be consistent with Paller *et al.* (1997) who reported glucose hypometabolism in the left inferior parietal cortex in W-K patients, but the CMRGlc was measured during a recognition memory task. The authors attributed

this finding to thalamo-cortical disconnection (Baron *et al.*, 1986), which could in turn explain long-term memory storage impairment. Kim *et al.* (1994) reported a SPECT study of an amnesic patient with left-sided thalamic infarct and in whom diaschisis in the left fronto-temporo-parietal cortex was observed. One prevalent hypothesis of memory consolidation is that after some years, the memories are deposited in neocortical areas (Squire and Zola, 1997). Activation studies in healthy subjects have revealed that the left supramarginal gyrus is preferentially involved in verbal working memory, notably in the articulatory loop (Paulesu *et al.*, 1993). Moreover, Penniello *et al.* (1995) reported a relation between glucose hypometabolism in the left supramarginal gyrus and phonological system impairment in Alzheimer's disease patients. However, activation studies in healthy subjects have shown that this structure is also involved in long-term verbal episodic memory retrieval (Andreasen *et al.*, 1995; Schacter *et al.*, 1996), which could be related to our findings.

B. Individual Analysis

Interestingly, the pattern observed in the group study was present individually in essentially each of the five patients, whether of nutritional or anoxic etiology (thalamus, 5/5; posterior cingulate/retrosplenial cortex, 4/5; mesial frontal lobe, 4/5; left middle temporal gyrus, 2/5; left supramarginal cortex, 5/5), suggesting that the pattern of functional disruption of extended Papez's circuit (with presumably secondary effects upon the left neocortex) is associated with the AS regardless of its cause, and hence of the exact lesion topography within this circuit. It should be reminded here that alcoholic subjects without cognitive impairment show only a mild mesial frontal cortex hypometabolism (Samson *et al.*, 1986).

The individual analysis also revealed interesting findings concerning other brain areas. Hippocampal region hypometabolism was observed in three subjects, i.e., both postanoxic cases (in agreement with known neuropathology) and one W-K patient. This finding is consistent with both the Ouchi *et al.* (1998) report of significant bilateral hypometabolism of the hippocampus head (containing the CA1 field) in 10 AS cases, mainly of posthypoxia etiology, and the Kuwert *et al.* (1993) report of a significant mesial temporal hypometabolism in a group of seven posthypoxia cases; both studies used the ROI method. However, Joyce *et al.* (1993) did not find significant hippocampal region hypometabolism in W-K patients, which might explain our negative group finding. The neuropathology of W-K does not classically mention hippocampal damage, although one study reported neuronal changes at morphometry in two cases (Mayes *et al.*, 1988).

Anterior cingulate hypometabolism was found in three patients, two of them with W-K, but not in the group analysis. Earlier ROI-based studies reported significant anterior cingulate hypometabolism in the AS (Fazio *et al.*, 1992; Joyce *et al.*, 1994). The anterior cingulate receives direct input from the anterior thalamus.

This lack of significant hypometabolism in the hippocampal and anterior cingulate in our group analysis may be due to (1) too small samples, exposing to type II errors (i.e., false negative); (2) the use of a too-stringent Z threshold in the group analysis; and (3) the use of an 80% threshold for the SPM analysis (see Materials and Methods), which might have excluded important brain areas. To address this issue, we performed *post hoc* group analyses, which revealed that, even with a less stringent statistical threshold ($p = 0.01$), significant hypometabolism was present neither in the anterior cingulate gyrus nor in the hippocampal region. However, when the group analysis was performed with a 40% threshold, there was significant hypometabolism at the $p < 0.01$ (uncorrected $z = 2.33$) level, in the right perirhinal, peri-amygdalian, and enthorhinal cortices. Thus, it appears that the AS is associated with dysfunction in the extended Papez's circuit including the hippocampal area, though this can be shown only with less conservative thresholds probably because of the small sample used here.

We also attempted to relate these individual metabolic patterns with neuropsychological profiles. There was at least one interesting concordance. Thus, HJC exhibited frontal signs, his anterograde amnesia was more severe than DG's and accordingly he also expressed dorsolateral prefrontal cortex hypometabolism (BA 46), a region implicated in both executive functions and episodic memory (Garnier *et al.*, 1998; Tulving *et al.*, 1994). However, we were unable to discern further correlations of interest in this small sample, but further studies will be necessary.

In conclusion, our voxel-by-voxel group analysis documented that the AS is associated with hypometabolism in an extended Papez's circuit, but also in some neocortical areas of the left hemisphere possibly involved in verbal episodic memory. Moreover, the individual analysis indicated that (1) this group pattern was found in essentially each patient, regardless of etiology, suggesting that dysfunction in this network is specifically related to the neuropsychological syndrome of the AS; (2) hippocampal anterior cingulate hypometabolism was also observed individually; and (3) other hypometabolic regions may be in accordance with additional neuropsychological impairments. The SPM method was used here for the first time to reveal a common network of synaptic dysfunction in a neuropsychological syndrome regardless of etiology and to analyze each patient's metabolic pattern in relation to individual neuropsychological profiles. The results indicate that this should be a powerful method in functional neuropsychology, with widespread applications in different neuropsychological syndromes.

References

Albert, M. S., Butters, N., and Levin, J. (1979). Temporal gradient in the retrograde amnesia of patients with alcoholic Korsakoff's disease. *Arch. Neurol.* **36**: 211–216.

Andreasen, N. C., O'Leary, D. S., Arnt, S., Cizalo, T., Hurtig, R., Rezai, K., Watkins, G. L., Ponto, L. B., and Hichwa, R. D. (1995). Short-term and long-term verbal memory: a positron emission tomography study. *Proc. Natl. Acad. Sci. USA* **92**: 5111–5115.

Baron, J. C., D'Antona, R., Pantano, P., Serdaru, M., Samson, Y., and Bousser, M. G. (1986). Effects of thalamic stroke on energy metabolism of the cerebral cortex. A positron tomography study in man. *Brain* **6**: 1243–1259.

Brown, J. (1958). Some tests of the decay theory of immediate memory. *Quart. J. Exp. Psychol.* **10**: 12–21.

Cardebat, D., Doyon, B., Puel, M., Goulet, P., and Joanette, Y. (1990). Evocation lexicale formelle et sémantique chez des sujets normaux: Performances et dynamique de production en fonction du sexe, de l'âge et du niveau d'étude. *Acta Neurol. Belg.* **90**: 207–217.

Clarke, S., Assal, G., Bogousslavsky, J., Regli F., Townsend, D. W., Leenders, K. L., and Blecic, S. (1994). Pure amnesia after unilateral left polar thalamic infarct: Tomographic and sequential neuropsychological and metabolic (PET) correlations. *J. Neurol. Neurosurg. Psychiatry* **57**: 27–34.

Deiber, M. P., Pollak, P., Passingham, R., Landais, P., Gervason, C., Cinotti, L., *et al.* (1993). Thalamic stimulation and suppression of parkinsonian tremor. *Brain* **116**: 267–279.

Delay, J., and Brion, S. (1954). Syndrome de Korsakoff et corps mamillaires. *Encéphale* **43**: 193–200.

Delis, D. C., Kramer, J. H., Kaplan E., and Ober, B. A. (1987). "The California Verbal Learning Test." Research edition. Psychological cooperation, New York.

Della Sala, S., Laicona, M., Spinnler, H., and Trivelli, C. (1993). Autobiographical recollection and frontal damage. *Neuropsychologia* **31**: 823–839.

Desgranges, B., Baron, J. C., De La Sayette, V., Petit-Taboué, M. C., Benali, K., Landeau, B., Lechevalier, B., and Eustache, F. (1998). The neural substrates of memory systems impairment in Alzheimer's disease: A PET study of resting brain glucose utilization. *Brain* **121**: 611–631.

Fazio, F., Perani, D., Gilardi, M. C., Colombo, F., Cappa, S. F., Vallar, G., Bettinardi, V., Paulesu, L., Alberoni, M., Bressi, S., Franceschi, M., and Lenzi, G. L. (1992). Metabolic impairment in human amnesia: A PET study of memory networks. *J.Cerebr. Blood Flow Metabol.* **12**: 253–258.

Feeney, D., and Baron, J. C. (1986). Diaschisis. *Stroke* **17**: 817–830.

Fox, P. T., Perlmutter, J. S., and Raichle, M. E. (1985). A stereostatic method of anatomical localization for positron emission tomography. *J. Comput. Assist. Tomogr.* **9**: 41–153.

Garnier, C., Enot-Joyeux, F., Jokic, C., Le Thiec, F., Desgranges, B., and Eustache, F. (1998). Une évaluation des fonctions exécutives chez les traumatisés crâniens: l'adaptation du test des six éléments. *Rev. Neuropsychol.* **8**: 1–29.

Hunkin, N. M., Parkin, A. J., and Longmore, B. E. (1994). Aetiological variation in the amnesic syndrome: Comparisons using the list discrimination task. *Neuropsychologia* **32**: 819–825.

Joyce, E. M., Rio, D. E., Ruttimann, U. E., Rohrbaugh, J. W., Martin, P. R., Rawlings, R. R., and Eckardt, M. J. (1993). Decreased cingulate and precuneate glucose utilization in alcoholic Korsakoff's syndrome. *Psychiatr. Res.* **54**: 225–239.

Kim, M. H., Hong, S. B., and Roh, J. K. (1994). Amnesia syndrome following left anterior thalamic infarction: With intrahemispheric and crossed cerebro-cerebellar diaschisis on brain SPECT. *J. Korean Med. Sci.* **9**: 427–431.

Kopelman, M. D. (1987). Two types of confabulation. *J. Neurol. Neurosurg. Psychiatry* **50**: 1482–1487.

Kopelman, M. D., Stanhope, N., and Kingsley, D. (1999). Retrograde amnesia in patients with diencephalic, temporal lobe or frontal lesions. *Neuropsychologia* **8**: 939–958.

Kuwert, T., Hömberg V., Steinmetz, H., Unverhau S., Langen, K.-J., Herzog H., and Feinendegen, L. E. (1993). Posthypoxic amnesia: Regional cerebral glucose consumption measured by positron emission tomography. *J. Neurol. Sci.* **118**: 10–16.

Markowitsch, H. J., von Cramon Y., and Schuri, U. (1993). Mnestic performance profile of a bilateral diencephalic infarct patient with preserved intelligence and severe amnesic disturbances. *J. Clin. Exp. Neuropsychol.* **15**: 627–652.

Mayes, A. R., Meudell, P. R., Mann, D., and Pickering, A. (1988). Location of lesions in Korsakoff's syndrome: Neuropsychological and neuropathological data on two patients. *Cortex* **24**: 367–388.

Mishkin, M. (1993). Cerebral memory circuits. *In "Exploring Brain Functions: Models in Neuroscience"* (T. A. Poggio, and D. A. Glaser, Eds.), pp. 113–125. Wiley, New York.

Okada, Y., Sadoshima, S., Fujii, K., Kuwabara, Y., Ichiya, Y., and Fujishima, M. (1991). Cerebral blood flow and metabolism in an amnestic patient with left thalamic infarction. A positron emission tomographic study. *Jpn. J. Med.* **30**: 367–372.

Ouchi, Y., Nobezawa, S., Okada, H., Yoshikawa, E., Futatsubashi, M., and Kaneko, M. (1998). Altered glucose metabolism in the hippocampal head in memory impairment. *Neurology* **51**: 136–142.

Paller, K. A., Acharya, A., Richardson, B. C., Plaisant, O., Shimamura, A. P., Reed, B. R., and Jagust, W. J. (1997). Functional neuroimaging of cortical dysfunction in alcoholic Korsakoff's syndrome. *J. Cogn. Neurosci.* **9**: 227–293.

Parkin, A. J., and Leng, N. R C. (1996). *"L'amnésie en question: Neuropsychologie du syndrome amnésique."* Presses Universitaires de Grenoble.

Paulesu, E., Frith, C. D., and Frackowiak, R. S. J. (1993). The neural correlates of the verbal component of working memory. *Nature* **25**: 342–345.

Penniello, M. J., Lambert, J., Eustache, F., Petit-Taboue, M. C., Barre, L., Viader, F., Morin,P., Lechevalier, B., and Baron, J. C. (1995). A PET study of the functional neuroanatomy of writing impairment in Alzheimer's disease. The role of the left supramarginal and left angular gyri. *Brain* **118**: 697–706.

Phelps, M. E., Huang, S. C., Hoffman, E. J., Selin, C., Skoloff, L., and Kuhl, D. E. (1979). Tomographic measurement of local cerebral glucose metabolic rate in humans with (F-18) 2-fluoro-2-deoxy-D-glucose: Validation of method. *Ann. Neurol.* **6**: 371–388.

Raven, J. (1981). *"Progressive Matrices."* Editions scientifiques et psychologiques, Issy-Les Moulineaux.

Samson, Y., Baron, J. C., Feline, A., Bories, J., and Crouzel, C. (1986). Local cerebral glucose utilisation in chronic alcoholics: A positron tomographic study. *J. Neurol. Neurosurg. Psychiatry* **49**: 1165–1170.

Scoville, W. B. and Milner, B. (1957). Loss of recent memory after bilateral hippocampal lesions. *J. Neurol. Neurosurg. Psychiatry* **20**: 11–21.

Schacter, D. L., Reiman, E., Curran, T., Sheng Yun, L., Bandy, D., Kathleen, B., and Roediger, H. L. (1996). Neuroanatomical correlates of veridical and illusory recognition memory: Evidence from positron emission tomography. *Neuron.* **17**: 267–274.

Squire, L. R., and Zola, S. M. (1997). Amnesia, memory and brain systems. *Philos. Trans. R. Soc. London B* **352**: 1663–1673.

Stroop, J. R. (1935). Studies of interferences in serial verbal reactions. *J. Exp. Neurol.* **18**: 643–662.

Talairach, J., and Tournoux, P.(1988). *"Co-planar Sterotaxic Atlas of the Human Brain. 3-Dimensional Proportional System: an Approach to Cerebral Imaging"* Thieme Medical, Stuttgart, New York.

Tulving, E., Kapur, S., Craik, F. I. M., Moscovitch, M., and Houle, S. (1994). Hemispheric encoding/retrieval aymmetry in episodic memory: Positron emission tomography findings. *Proc. Natl. Acad. Sci. USA* **91**: 2016–2020.

Victor, M., Adams, R. D., and Collins, G. H. (1989). *"The Wernicke–Korsakoff Syndrome and Related Disorders Due to Alcoholism and Malnutrition,"* 2nd ed. Davis, Philadelphia.

Volpe, B. T., Herscovitsch, P., and Raichle, M. E. (1984). PET evaluation of patients with amnesia after cardiac arrest. *Stroke* **15**: 196.

Wechsler, D. (1968). *"Echelle d'Intelligence Pour Adultes."* Centre de Psychologie Appliquée, Paris. [éd. française]

Wechsler, D. (1969). *"Echelle Clinique de Mémoire (Forme 1)."* Centre de Psychologie Appliquée, Paris. [éd; française)]

[^{11}C]β-CFT Binding in Parkinson's Disease with and without Dementia Measured by Positron Emission Tomography: Statistical Parametric Mapping Analysis

ETSUJI YOSHIKAWA,[*] **YASUOMI OUCHI,**[†] **TOSHIHIKO KANNO,**[†] **HIROYUKI OKADA,**[*]
MASAMI FUTATSUBASHI,[*] **SHUJI NOBEZAWA,**[†] **TATSUO TORIZUKA,**[†]
and MASANOBU SAKAMOTO[‡]

[*]*Central Research Laboratory, Hamamatsu Photonics K.K., Hamakita, Japan*
[†]*Positron Medical Center and*
[‡]*Department of Neurology, Hamamatsu Medical Center, Hamamatsu, Japan*

Dopamine transporter concentration in the striatum was reported to be reduced in patients with Parkinson's disease (PD) as well as in Alzheimer's disease patients with extrapyramidal symptoms. The purpose of the present study was to investigate changes in striatal dopamine uptake sites in PD with regard to the presence of dementia. The striatal dopamine transporter levels were measured using positron emission tomography with a dopamine transporter probe [^{11}C]2-β-cartomethoxy-3β-(4-fluorophenyl) tropane ([^{11}C]β-CFT) in drugnaive PD patients without dementia (Hoehn and Yahr I–II) and with dementia (Hoehn and Yahr II). The volumes of the striatum in both groups were calculated using data from magnetic resonance imaging in order to evaluate structural atrophy. Striatal β-CFT uptake was compared pixel by pixel using statistical parametric mapping (SPM96). Our volumetric study showed that there was no significant difference in striatal volume between the two groups. The SPM showed that β-CFT binding was significantly reduced in the putamen in nondemented PD patients, while a further reduction was observed in the caudate and the putamen (more anteriorly) in demented PD patients. This caudate reduction was correlated with intelligence scores. Our results confirmed that the dopamine transporter in the putamen is vulnerable in PD patients and suggested that presynaptic dopaminergic dysfunction involving the caudate may be important for the pathophysiology of PD with dementia.

I. INTRODUCTION

Previous *in vivo* dopamine transporter imaging studies based on region of interest (ROI) methods showed that transporter binding in the putamen was severely affected, while the binding in the caudate was substantially spared in early PD. A recent positron emission tomography (PET) study with ROI-based analysis showed that the striatal uptake of [^{11}C]2-β-carbomethoxy-3β-(4-fluorophenyl) tropane (β-CFT), dopamine transporter marker, in patients with Alzheimer's disease was correlated with the severity of extrapyramidal symptoms (Rinne *et al.*, 1998). The manual definition of ROIs requires *a priori* information on the target regions, which may bias PET results. In contrast, statistical parametric mapping (SPM) on receptor binding studies can measure intensity differences on voxel-by-voxel basis without subjectivity (Koepp *et al.*, 1998). In the present study, we compared the striatal β-CFT uptake in patients with Parkinson's disease (PD) with that in patients with PD plus dementia using SPM96 to investigate whether dopamine transporter binding densities would differ in the striatum of PD with dementia.

II. SUBJECTS AND METHODS

Six drug-naive patients with PD without dementia (PDnd) (Hoehn and Yahr I–II) (mean age ± SD, 66.8 ± 6.7 years), six drug-naive PD patients with dementia (PDd) (Hoehn and Yahr II) (68.4 ± 8.2 years) and six normal subjects (53.8 ± 12.8 years) participated in this study. Disease durations in PD patients were 1.6 ± 3.8 and 1.9 ± 0.7 years, respectively. Patients who reported cognitive impairment before onset of the motor symptoms of PD were excluded. In neuropsychological tests, mini mental state examination (MMSE) scores were 27.2 in PDnd and 17.8 in PDd, and total IQ scores of WAIS-R were 101.0 in PDnd and 69.0 in PDd (except for two unexamined patients). The Unified Parkinson's Disease Rating Scale (UPDRS) disclosed no differences in severity of motor functions between the two groups. Neither patients nor normal subjects had medical treatment before the present PET study. L-Dopa treatment after PET study was effective for parkinsonian symptoms in all PD patients.

To determine the orientation of PET scanning parallel to the anterior–posterior (AC–PC) commisure line, magnetic resonance imaging (MRI) was performed just before PET examination using a static magnet (0.3 T MRP7000AD, Hitachi, Japan) with the three-dimensional acquisition mode (Ouchi *et al.*, 1999).

PET was performed using a high-resolution head scanner (SHR2400, Hamamatsu Photonics K.K., Hamamatsu, Japan) with five detector rings yielding nine-slice images simultaneously (Yamashita *et al.*, 1990). The gantry of this tomograph could be tilted from −20° to +90°. The in-plane spatial resolution was 2.7 mm and the resolution in the axial plane was 5.5 mm full-width at half-maximum (FWHM) with an 80-mm axial field of view. The gantry was set parallel to the intercommissural line determined by MRI (Ouchi *et al.*, 1999). After a 20-min transmission scan, dynamic scanning began at the start of slow bolus intravenous injection (taking 1 min) of 450 MBq of β-CFT (time frames: 4×30, 20×60, 14×300 s). Two summation images constructed for each scan using data collected from 0 to 20 min (perfusion image) and from 60 to 90 min (distribution image) after tracer injection were used for analyses with SPM96 (Wellcome Department of Cognitive Neurology, UK).

Each volume of the striatum was obtained by calculating the product of the semicircular area drawn on the MRI and the slice thickness. One-way analysis of varience (ANOVA) was used to compare the volumes among groups. β-CFT images for SPM analysis were reconstructed from data of summed tracer concentration recorded from 60 to 90 min and normalized by the cerebellar count [(striatum-cerebellum)/cerebellum]. All distribution images were realigned to the AC–PC line and transformed into a standard stereotactic anatomical space (Talairach and Tournoux, 1988) with parameters obtained in the course of perfusion image normalization. A global normalization process was omitted because the tracer activity in the specific binding region was previously normalized. Each image was smoothed

with an isotropic Gaussian kernel of 8 mm. The resultant set of voxel values for the between-group contrasts constituted t-statistic maps (SPM{t}), which were then transformed to the unit normal distribution (SPM{Z}). The threshold was set at $p < 0.01$ corrected for multiple comparison (Friston *et al.*, 1995).

III. RESULTS

Volumetric analysis showed that the volume of the whole striatum was not different among the three groups (control, PDnd, PDd; 4.39 ± 0.16, 4.25 ± 0.34, 4.19 ± 0.25 cm^3, respectively).

Compared with normal controls, β-CFT uptake was significantly reduced in the bilateral putamen with dominant reduction posteriorly in the right putamen in PDnd (Table 1; Fig. 1A), while in PDd β-CFT uptake was significantly decreased in the right caudate and the bilateral putamen more anteriorly (Table 1; Fig. 1B). No significant deferences in brain foci were observed between PDnd and PDd groups even after the threshold was reduced to $p < 0.05$ uncorrected.

SPM with correlation analysis showed that in the demented PD group β-CFT uptake in the bilateral caudate and the left anterior putamen were significantly reduced as a function of the decrease in MMSE scores (Fig. 2).

IV. DISCUSSION

The present SPM results showed that putaminal β-CFT binding was more affected in early PD patients without dementia, whereas β-CFT binding in the putamen as well as the caudate was lower in demented PD patients than that in normal controls. A previous [^{18}F]fluorodopa (FDOPA) PET study reported that progression of PD resulted in dopamine storage reductions in the posterodorsal putamen at an early

TABLE 1. Significant Binding Reduction Foci in Comparison with Normal Controls

Group	Brain region	Coordinates (*xyz*)	*Z*
PDnd	R putamen (posterior)	30 −10 2	6.55
	L putamen (posterior)	−28 −2 2	5.99
PDd	R putamen (posterior)	30 −12 2	5.73
	R putamen (anterior)	24 0 6	5.26
	R caudate	16 8 12	5.11
	L putamen (anterior)	−24 2 −2	4.88

Note. R, right; L, left. "Anterior" or "posterior" in the putamen denotes the side anterior or posterior to the anterior commissure line on the Talairach atlas.

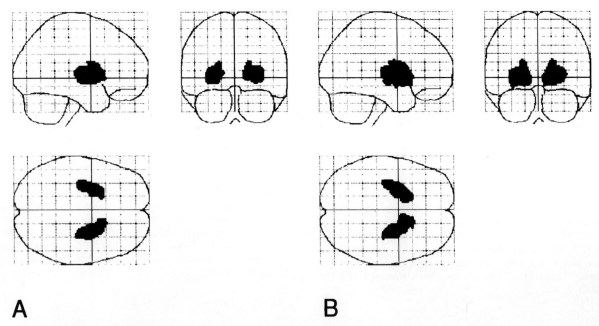

FIGURE 1. SPM projections of reductions in dopamine transporter binding sites in PDnd (A) and PDd (B) compared with normal controls ($p < 0.01$, corrected). Details of coordinates and Z values are given in Table 1.

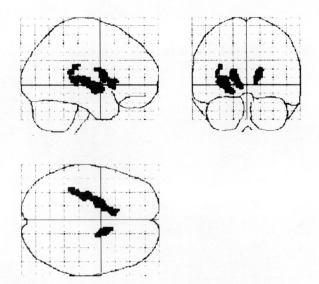

FIGURE 2. Brian regions showing a significant correlation between decreased β-CFT uptake and scores of MMSE. The correlation was more marked in the bilateral caudate and the left anterior putamen.

phase and in the anteroventral putamen at an advanced stage of disease, while disease progression showed less of an effect on the caudate FDOPA uptake (Brooks *et al.*, 1990). Consistent with this finding, our SPM study demonstrated that more posterior regions of the putamen were affected in patients with PD without dementia (Fig. 1A). In contrast, β-CFT uptake in patients with PD with dementia was decreased in the putamen more anteriorly and the caudate (Fig. 1B). A histopathological study showed that, in PD pa-

tients with dementia, there was neuronal degeneration in the medial substantia nigra, neurons which project preferentially to the caudate nucleus. Thus, alterations in caudate β-CFT binding might be a key phenomenon associated with PD with dementia.

β-CFT reductions in the bilateral caudate and the left anterior putamen were correlated with a decrease in MMSE scores (Fig. 2). This result further supported the idea that the occurrence of dementia implicates the striatum more anteriorly. The primate caudate nucleus receives projections from the frontal association cortex or the cingulate cortex, while the putamen exclusively receives afferents from the sensorimotor cortex, suggesting that the caudate function is related to complex and associative types of behavior. Therefore, the anterior extension of dopaminergic dysfunction observed in the striatum in our demented PD patients supports this speculation. In the present study, we excluded patients with cognitive impairment or hallucinations which developed earlier than the onset of Parkinsonism in order to avoid heterogeneity of the demented population. Since Alzheimer patients with Parkinsonian signs exhibited reduction in β-CFT uptake equally in the caudate and the putamen in contrast to PD cases showing predominantly putaminal reduction (Rinne *et al.*, 1998), the caudate dysfunction might be a prerequisite factor related to the occurrence of dementia in patients with Parkinsonism.

In conclusion, SPM for dopamine transporter binding site densities confirmed that the presynaptic dopaminergic function is affected predominantly in the posterior segment of the putamen in early PD and suggested that alterations in the caudate dopaminergic system may reflect the pathophysiology in PD with dementia.

Acknowledgment

We thank Dr. Norihiro Sadato (Okazaki National Institute) for data analysis.

References

Brooks, D. J., Ibanez, V., Sawle, G. V., Quinn, N., Lees, A. J., Mathias, C. J., Bannister, R., Marsden, C. D., and Frackowiak, R. S. (1990). Differing patterns of striatal ^{18}F-dopa uptake in Parkinson's disease, multiple system atrophy, and progressive supranuclear palsy. *Ann. Neurol.* **28**: 547–555.

Friston, K. J., Holmes, A. P., Worsley, K. J., Poline, J. P., Frith, C. D., and Frackowiak, R. S. J. (1995). Statistical parametric mapping in functional imaging: A general linear approach. *Hum. Brain. Mapping* **2**: 189–210.

Koepp, M. J., Gunn, R. N., Lawrence, A. D., Cunningham, V. J., Dagher, A., Jones, T., Brooks, D. J., Bench, C. J., and Grasby, P. M. (1998). Evidence for striatal dopamine release during a video game. *Nature* **393**: 266–268.

Ouchi, Y., Yoshikawa, E., Okada, H., Futatsubashi, M., Sekine, Y., Iyo, M., and Sakamoto, M. (1999). Alterations in binding site density of dopamine transporter in the striatum, orbitofrontal cortex, and amygdala in early Parkinson's disease: Compartment analysis for beta-CFT binding with positron emission tomography. *Ann. Neurol.* **45**: 601–610.

Rinne, J. O., Sahlberg, N., Ruottinen, H., Nagren, K., and Lehikoinen, P. (1998). Striatal uptake of the dopamine reuptake ligand [^{11}C]beta-CFT is reduced in Alzheimer's disease assessed by positron emission tomography. *Neurology* **50**: 152–156.

Talairach, J., and Tournoux, P. (1988). *"Co-planer Stereotaxic Atlas of the Human Brain: 3-Dimensional Proportional System: An Approach to Cerebral Imaging."* Georg Thieme, Stuttgart.

Yamashita, T., Uchida, H., Okada, H., Kurono, T., Takemori, T., Watanabe, M., Shimizu, K., Yoshikawa, E., Ohmura, T., Satoh, N., and Tanaka, E. (1990). Development of a high resolusion PET. *IEEE Trans. Nucl. Sci.* **37**: 594–599.

56

Assessment of the Performance of FDG and FMZ PET Imaging against the Gold Standard of Invasive EEG Monitoring for the Detection of Extratemporal Lobe Epileptic Foci in Children

OTTO MUZIK,* DIANE C. CHUGANI,*,† CHENGGANG SHEN,* CSABA JUHASZ,*
HANS-MARTIN VON STOCKHAUSEN,‡ and HARRY T. CHUGANI*,†,§

*Departments of *Radiology, †Pediatrics, and §Neurology*
Wayne State University, Children's Hospital of Michigan, Detroit, Michigan
‡Max-Planck Institute for Neurological Research, Cologne, Germany

Although positron emission tomography (PET) imaging with fluorodeoxyglucose (FDG) and flumazenil (FMZ) is being used clinically to localize epileptogenic regions in patients with refractory epilepsy, the electrophysiological significance of the identified PET abnormalities remains poorly understood. We studied five children (three boys, two girls, ages 4–15 years, mean age 10 years) suffering from extratemporal lobe epilepsy, who underwent FDG and FMZ-PET imaging together with intracranial electroencephalographic (EEG) monitoring. EEG electrode positions relative to the brain surface were identified using a new approach which allows accurate coregistration of a planar X-ray image with the MRI image volume. Cortical areas of abnormal glucose metabolism or FMZ binding were determined objectively based on asymmetry measures derived from homotopic cortical areas. A receiver operating characteristics analysis was performed to determine the specificity and sensitivity of PET-defined abnormalities against the gold standard of intracranial EEG data. FMZ-PET detected at least part of the seizure onset zone in all subjects, whereas FDG-PET failed to detect the seizure onset region in one patient. Both FMZ and FDG-PET showed poor performance for detection of rapid seizure spread. Our results indicate that FMZ-PET is significantly more sensitive than FDG-PET for the detection of cortical regions of seizure onset and frequent spiking in patients with extratemporal lobe epilepsy, whereas both FDG and FMZ-PET show low sensitivity in the detection of cortical areas of rapid seizure spread. The application of FMZ-PET in guiding subdural electrode placement in re-

fractory extratemporal lobe epilepsy will enhance coverage of the epileptogenic zone.

I. INTRODUCTION

The success of epilepsy surgery relies upon accurate identification of the extent of cortex responsible for generating epileptic seizures, coupled with the demonstration that the epileptogenic zone can be removed without causing unacceptable neurological sequelae. However, delineation of the epileptogenic zone in extratemporal lobe epilepsy is not always successful as scalp electroencephalographic (EEG) recordings often are not adequate for localizing the seizure onset. As a result, patients with extratemporal lobe epilepsy often require highly invasive intracranial electrodes to be placed for further localization of the seizure onset.

Advances in neuroimaging techniques have largely contributed to improved results of epilepsy surgery for extratemporal lobe epilepsy. In order to decrease the number of patients requiring invasive EEG monitoring or to guide the placement of intracranial electrodes, PET tracers showing the neurofunctional status of cortical tissue have been developed, including 2-deoxy-2-[^{18}F]fluoro-D-glucose (FDG) and [^{11}C]flumazenil (FMZ). In patients with frontal lobe epilepsy, it was reported that the cortical focus of decreased FMZ binding corresponded well with the location of seizure onset determined with subdural electrodes, whereas the hypometabolic zones shown on FDG-PET appeared larger than the FMZ focus (Savic *et al.*, 1995). In fact, these large zones of glucose hypometabolism visualized by FDG-PET are difficult to interpret and possibly include areas of seizure propagation, secondary foci, and cortical–cortical diaschisis, in addition to the primary seizure focus (DaSilva *et al.*, 1997). As both PET and EEG are believed to provide complementing information relevant for surgical intervention, a better understanding of the correlation between functional PET imaging and electrophysiological EEG recordings is necessary.

The present study was undertaken to determine, using a new method which allows coregistration of intracranial electrodes with PET and MRI image volumes (vonStockhausen *et al.*, 1998), the relationship between FDG and FMZ-PET abnormalities to three well-established electrophysiological parameters (seizure onset, rapid seizure spread, and frequent interictal spiking) in patients with extratemporal lobe epilepsy.

II. MATERIALS AND METHODS

A. Electrophysiological Procedure

Five children (three boys, two girls, ages 4–15 years, mean age 10 years) with clinically intractable extratemporal lobe epilepsy based on seizure semiology and scalp ictal EEG were studied. Patients underwent long-term scalp video–EEG monitoring in order to identify the location of seizure onset, to observe seizure spread, and to characterize interictal abnormalities. In addition, all patients underwent chronic video–EEG monitoring with subdural electrodes in order to record ictal events and interictal activity. Subdural electrode placement was guided by the seizure semiology, seizure onset area as determined by scalp ictal EEG, and FMZ and FDG cortical abnormalities. Identification of electrodes involved in the seizure onset of habitual clinical seizures (defined as a localized, sustained, rhythmic /semirhythmic or spiking EEG pattern with frequency >2 Hz) as well as electrodes involved in the propagation of ictal activity (seizure spread areas, defined as areas involved in seizure activity within 10 s of seizure onset) during ictal episodes was determined by chronic subdural EEG monitoring. In addition, electrodes with frequent interictal spiking (defined as mean occurrence of >10 spikes/min) were identified.

B. PET Scanning Procedure

PET studies were performed using the CTI/Siemens EXACT/HR positron tomograph in 2D mode located at Children's Hospital of Michigan. This scanner has a 15-cm field of view and generates 47 image planes with a slice thickness of 3.125 mm. The reconstructed image in-plane resolution obtained is 5.5 ± 0.35 mm at full-width half-maximum

(FWHM) and 6.0 ± 0.49 mm in the axial direction (reconstruction parameters: Shepp–Logan filter with 0.3 cycles/pixel cutoff frequency). FDG and FMZ were synthesized using a Siemens RDS-11 cyclotron.

Patients were fasted for 4 h prior to the PET procedure. Scalp EEG was monitored during the FDG uptake period (first 30 min after FDG administration), and during the entire FMZ-PET scan in order to document ictal activity. Children who were unable to cooperate during the scanning phase were sedated with either nembutal or chloral hydrate. Both FDG and FMZ-PET studies were performed on two separate occasions in the interictal state. Measured attenuation correction was applied to all images using a 15-min attenuation scan. For FDG, a venous line was established for injection of FDG (5.3 MBq/kg), followed by a static 20-min emission scan of the brain 30 min after tracer administration. For FMZ, a static 10-min scan was acquired 10 min following intravenous administration of the tracer (15 MBq/kg). Activity images for this time frame show a good agreement with the parametric images of FMZ volume of distribution (Frey *et al.*, 1991; Niimura *et al.*, 1999).

C. MRI and X-ray Procedure

All patients underwent a volumetric spoiled gradient echo (SPGR) MRI scan (256×256 image matrix, 124 planes with 1.5 mm thickness, no skip). In addition, a planar X-ray image was acquired with the subdural electrode grid in place. This was done at the patient's bedside using a portable X-ray unit. Three differently shaped metallic fiducial markers were placed at anatomically well-defined locations on the patients' head (left/right ear lobe and lateral eye corner on the side of the grid) in order to allow reconstruction of the acquired planar view at later time points. The X-ray film cassette was positioned adjacent to the electrode grid and exposed from the contralateral side perpendicular to the film plane at a distance of at least 100 cm in order to minimize geometric distortions. Finally, in order to obtain the most identifiable position, a view was chosen projecting the three fiducials onto the film as a triangle with maximal area (Fig. 1). In the case that all three fiducials were imaged close to a line the acquisition was repeated with an altered view.

D. PET Image Analysis

FDG, FMZ, and MRI SPGR image volumes were coregistered using MPI-Tool which is a software package employing a multipurpose three-dimensional registration technique (Pietrzyk *et al.*, 1994). The coregistration method is highly interactive and is based on the simultaneous alignment of image contours which are exchanged in three orthogonal cuts through the brain.

The extent of regional cortical abnormalities of brain glucose metabolism and FMZ binding was identified using an

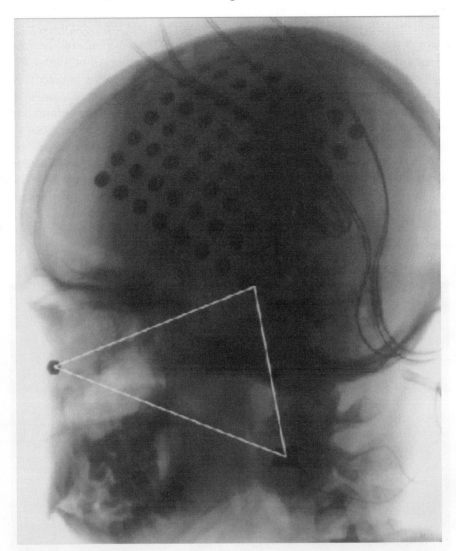

FIGURE 1.   X-ray image obtained from a representative patient with subdural grid electrodes implanted. Three differently shaped metallic fiducial markers were placed on the patients head prior to X-ray imaging. The three fiducial markers create a coordinate triplet (depicted in white).

objective method based on a semiautomated software package applied to all supratentorial planes of the PET studies (Muzik *et al.*, 1998). This procedure allows the definition of abnormal cortical areas of glucose metabolism or FMZ binding based on asymmetry measures derived from homotopic cortical areas according to a predefined cutoff threshold. Those cortical areas which exceeded a given asymmetry threshold were marked at the side which was decreased. Focal increases on one side were distinguished from potential relative decreases on the contralateral side by visual inspection of the regions of asymmetry compared to the remaining cortex. A "marked file" containing the cortical regions with abnormalities was created for each PET study and further processed using the 3D-Tool software package (von Stockhausen *et al.*, 1998). This software package combines meth-

ods for segmentation, visualization, and quantitative analysis of coregistered multimodality volume data. The brain was automatically segmented from MRI data using morphological operations and three-dimensional surface views were created. Functional data obtained from the "marked" PET image volumes were projected onto the brain surface using reverse gradient fusion (Stokking *et al.*, 1994).

E. Determination of Electrode Position on Brain Surface

In order to reconstruct surface views corresponding to the planar X-ray image showing the subdural grid, three virtual markers were defined in the SPGR MRI image volume at positions corresponding to those defined in the X-ray image. In order to obtain the necessary anatomical information, the skin was segmented in the MRI image volume displaying the ears and the face of the patient. Using these landmarks, virtual markers were defined at the left/right ear lobe and at the temporal eye corner, thus creating a coordinate triplet derived from the MRI image volume (Fig. 2). The advantage of this method is that no physical fiducial markers need to be used during the MRI scan, which significantly increases the feasibility in clinical routine. Subsequently the X-ray film was digitized and converted to a JPEG image. This image was then submitted into the 3D-Tool software package and the three fiducial markers were identified, representing a coordinate triplet derived from the X-ray image. Starting from a lateral view of the brain surface, an iterative algorithm minimized the differences between the previously

defined MRI and X-ray coordinate triplets (von Stockhausen *et al.*, 1997). In short, the two coordinate triplets were first centered to their center of gravity followed by adjustment of the three euler angles in order to maximize the similarity between the triangles. As can be seen in Fig. 3 the similarity between the two coordinate triplets is in direct proportion to the sum of the distances $S = d_1 + d_2 + d_3$. To adjust the euler angles, an iterative optimization algorithm was used employing discrete parameter increments. In order to achieve excellent accuracy in connection with a clinically feasible computation time, the iterative algorithm was applied repeatedly with succesively finer step sizes. The best surface view estimate obtained at a certain step size was the starting point for the next iteration run. Finally, the zoom

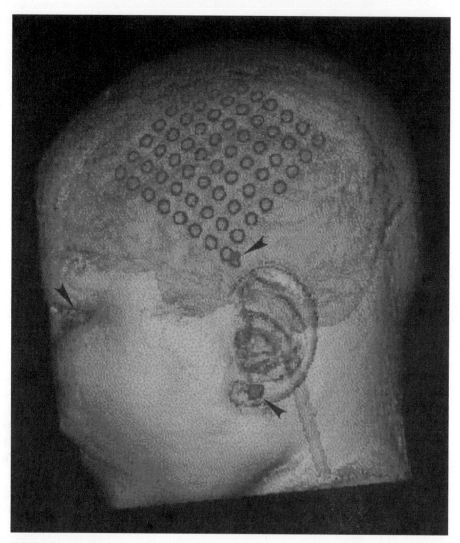

FIGURE 2. The figure shows the outline of the subdural grid electrodes and the three fiducial markers obtained from the planar X-ray image. Anatomical information obtained from surface rendering of the patient's head was used to adjust the three-dimensional surface view to correspond to the same view as the planar image. Note the agreement between the three virtual fiducial markers defined in the MRI image volume (black arrows) and the fiducials obtained from the X-ray image.

was determined as the ratio between corresponding sides of the two triangles (L_1/l_1 or L_2/l_2 or L_3/l_3 in Fig. 3). As a result, a surface view was created which corresponded to the planar X-ray image and where the location of electrodes directly projected onto the brain surface (Fig. 2). The accuracy of this procedure was reported previously as 1.24 ± 0.66 mm with a maximal misregistration of 2.7 mm (von Stockhausen *et al.*, 1997). Following coregistration of the subdural EEG grid with the surface view, virtual spheres with 5 mm diameter were created at those locations on the brain surface, which represented a projection of the electrodes onto the sur-

face. Each sphere was color coded to allow easy recognition of seizure onset, seizure spread, frequent interictal spiking, and normal electrodes (Fig. 4). These spheres marked permanently the position of the electrodes on the cortical surface and allowed assessment of the electrode grid location relative to the "marked" cortical areas from varying view angles.

F. Receiver Operating Characteristics Analysis

In order to assess the concordance between PET imaging and EEG data, we performed a receiver operating characteristics (ROC) analysis (Metz, 1978). Marked abnormalities obtained from three different asymmetry cutoff thresholds (10, 12, and 15%) were compared with the gold standard of subdural EEG data. In the context of the ROC analysis, a true positive event was recorded when an electrophysiologically positive electrode at least partially overlapped the abnormal zone as defined by PET. Conversely, a false negative event was recorded when the electrophysiologically positive electrode failed to overlap an abnormal zone by PET. Furthermore, the number of false positive (normal EEG electrode over marked region) and true negative (normal EEG electrode over unmarked region) decisions was determined for all implanted electrodes. The analysis was performed for FDG and FMZ PET marked abnormalities defined at all three cutoff thresholds independently for seizure onset, seizure spread, and frequent interictal spiking electrodes.

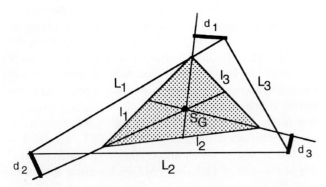

FIGURE 3. The figure shows the two coordinate triplets which are initially centered to their center of gravity (S_G). The stippled triangle represents the coordinate triplet derived from the virtual fiducials defined in the MRI image volume. The three-dimensional surface view is varied in order to minimize the sum of the distances d_1, d_2, and d_3. Once a global minimum is found, the zoom is calculated as the ratio between corresponding sides of the two triangles (L_1/l_1 or L_2/l_2 or L_3/l_3).

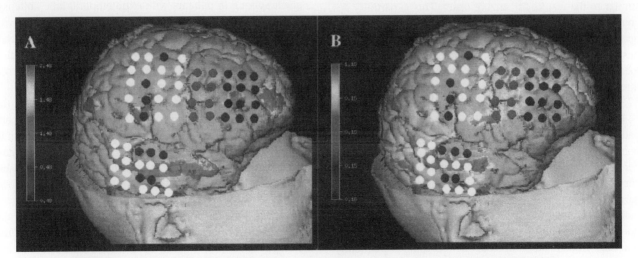

FIGURE 4. Surface rendered MRI overlaid with functional PET data and subdural grid electrode locations. Regions with greater than 10% asymmetry were marked with a value equal to two times the maximum value found in the PET image volume. Thus, dark regions on surface represent marked cortical regions with greater than 10% asymmetry. Three 4×5 subdural grid electrode arrays were implanted for chronic EEG monitoring. Electrodes are color coded: black, seizure onset; light grey, rapid seizure spread; dark grey, frequent interictal spiking only; white, normal. (A) FMZ PET correctly detected the region of seizure onset and some electrodes corresponding to seizure spread. (B) FDG PET failed to detect the region of seizure onset and covered only some electrodes corresponding to seizure spread and one electrode corresponding to frequent spiking.

III. RESULTS AND DISCUSSION

Cortical regions with decreased FDG and FMZ at the 10% threshold together with grid electrode data are shown for one patient (Fig. 4). During intracranial EEG recording three seizures were recorded, which originated at one electrode (black colored electrode in Figs. 4A and 4B) located on the electrode array in the frontal lobe. Whereas FMZ-PET (Fig. 4A) correctly identified the area of seizure onset, FDG-PET (Fig. 4B) failed to detect the seizure onset area.

More detailed analysis revealed that the sensitivity for detecting areas of seizure onset was best for FMZ at the 10% threshold ($81 \pm 9\%$ of the electrodes overlying the seizure onset) and decreased to 57 ± 10 and $37 \pm 12\%$ for the 12 and 15% thresholds, respectively. Sensitivity for detecting the seizure onset zone with FDG was lower than that for FMZ at all thresholds. The sensitivity for FDG at the 12% threshold ($32 \pm 12\%$, $P = 0.03$) was significantly lower than for FMZ and was also lower at the 10% ($53 \pm 13\%$, $P = 0.06$) and 15% thresholds ($16 \pm 6\%$, $P = 0.08$), respectively. There was no significant difference with regard to specificity between FMZ and FDG at any threshold (specificity in the range 68–86%).

We found a very low sensitivity for both FMZ and FDG to detect cortical areas of seizure spread. At the 10% threshold, sensitivity was $31 \pm 9\%$ for FMZ and $34 \pm 9\%$ for FDG ($P = 0.51$), respectively. These values declined to $19 \pm 8\%$ for FMZ and $10 \pm 6\%$ for FDG at the 15% threshold ($P = 0.28$).

The sensitivity for detecting areas of frequent interictal spiking was always higher for FMZ than for FDG and was similar to the results for seizure onset. This is partially accounted for by the fact that the set of electrodes with frequent interictal spiking almost always included the seizure onset electrodes. Furthermore, in one patient FDG failed to localize the area of frequent interictal spiking. At the 10% threshold, sensitivity was $80 \pm 7\%$ for FMZ and was only $57 \pm 13\%$ for FDG ($P = 0.15$).

One of the main objectives of presurgical investigation in patients with medically refractory epilepsy is to define the boundaries of the epileptogenic region to be resected in order to achieve complete seizure control. Toward this goal, chronic intracranial EEG evaluation remains the gold standard. However, this method also has serious limitations for defining the epileptogenic region. The accuracy of localization using intracranial electrodes highly depends on selection bias. In fact, subdural electrode recording is frequently unreliable to localize seizure onset in the absence of other localizing information (Uematsu et al., 1990). Regional or diffuse seizure onset or onset at the edge of an electrode array likely represents spread from unsampled areas (Jayakar et al., 1991). Successful localization is poor without imaging clues and a high-risk of sampling error exists especially in frontal lobe epilepsy where a widespread distribution of the epileptogenic brain tissue responsible for the patient's habitual seizures can also contribute to a false or missed localization (Quesney et al., 1992). Moreover, physiological effects of large subdural grids may impact on quality or character of recorded signals as well. Because intracranial EEG recording carries a risk of morbidity and possible mortality, it is appropriate only when reliable conclusions cannot be obtained by less invasive methods.

Unlike the limited sampling available with intracranial monitoring, PET imaging allows assessment of the entire brain in order to define the location and extent of all functionally abnormal regions. It is well known that gross or microscopic abnormalities of the cortical architecture and/or functional alterations related to interictal and ictal epileptiform discharges may produce functional abnormalities detected by PET (Robitaille et al., 1992). In order to study the correlation between these functional abnormalities and electrophysiological recordings, we have applied recent advances in image processing which allow precise coregistration of the MRI image volume (showing the positions of the grid electrodes) with FDG and FMZ-PET data sets. Using this approach, which allows highly accurate comparison of the sites of FDG and FMZ-PET cortical abnormalities, we now demonstrate that the location of FMZ-PET abnormalities included the cortical region identified as the seizure onset zone by subdural EEG in all patients studied, while FDG-PET was able to detect the seizure onset region in only four of five cases. The ROC analysis demonstrated that FMZ-PET is significantly more sensitive than FDG-PET for the detection of cortical regions of seizure onset and frequent spiking in patients with extratemporal lobe epilepsy. However, both FDG and FMZ-PET showed low sensitivity in the detection of cortical areas of rapid seizure spread. Thus, functional changes in cortex resulting in facilitated seizure spread do not appear to result in changes of glucose metabolism or benzodiazepine receptor binding.

Interestingly, a region of decreased FMZ binding in the parietal cortex which has no electrographic correlate was observed in several patients. Decreased parietal and thalamic glucose metabolism has been reported in patients with temporal lobe epilepsy (Henry et al., 1990), and in these cases, the parietal region rarely evolves into a seizure focus following temporal lobectomy. Furthermore, it has been reported that FMZ binding alterations in cortical areas outside the epileptic focus normalize following surgical removal of the epileptic focus (Savic et al., 1998). Regardless of the mechanism responsible for these additional regions of cortical FMZ binding abnormalities, their presence precludes the use of FMZ-PET alone to define the boundaries of the area that needs to be removed for complete seizure control. However, FMZ-PET should be an effective tool for guiding subdural electrode placement to ensure coverage of the epileptogenic zone in patients with extratemporal lobe epilepsy.

Acknowledgments

We thank the staff of the Children's Hospital of Michigan PET center for their help in performing all studies. These studies were supported in part by funding from NIH Grant NS-34488.

References

DaSilva, E. A., Chugani, D. C., Muzik, O., and Chugani, H. T. (1997). Identification of frontal lobe epileptic foci in children using positron emission tomography. *Epilepsia* **38**: 1198–208.

Frey, K., Holthoff, V., Koeppe, R., Jewett, D., Kilbourne, M., and Kuhl, D. (1991). Parametric in vivo imaging of benzodiazepine receptor distribution in human brain. *Ann. Neurol.* **30**: 663–72.

Henry, T. R., Mazziotta, J. C., Engel, J., Jr., Christenson, P. D., Zhang, J. X., and Phelps, M. E. (1990). Quantifying interictal metabolic activity in human temporal lobe epilepsy. *J. Cereb. Blood Flow Metab.* **10**: 748–757.

Jayakar, P., Duchowny, M., Resnick, T. J., and Alvarez, L. A. (1991). Localization of seizure foci: pitfalls and caveats. *J. Clin. Neurophysiol.* **8**: 414–431.

Metz, C. E. (1978). Basic principles of ROC analysis. *Semin. Nucl. Med.* **8**: 283–298.

Muzik, O., Chugani, D. C., Shen, C., da Silva, E. A., Shah, J., and Shah, A. (1998). Objective method for localization of cortical asymmetries using positron emission tomography to aid surgical resection of epileptic foci. *Comput. Aided Surg.* **3**: 74–82.

Niimura, K., Muzik, O., Chugani, D. C., Shen, C., and Chugani, H. T. (1999). [^{11}C]Flumazenil PET: activity images versus parametric images for the detection of neocortical epileptic foci. *J. Nucl. Med.* **40**: 1985–1991.

Pietrzyk, U., Herholz, K., and Fink, G. (1994). An interactive technique for three dimensional image registration: validation for PET, SPECT, MRI and CT brain studies. *J. Nucl. Med.* **35**: 2011–2018.

Quesney, L. F., Constain, M., Rasmussen, T., Olivier, A., and Palmini, A. (1992). Presurgical EEG investigation in frontal lobe epilepsy. *Epilepsy Res. Suppl.* **5**: 55–69.

Robitaille, Y., Rasmussen, T., Dubeau, F., Tampieri, D., and Kemball, K. (1992). Histopathology of nonneoplastic lesions in frontal lobe epilepsy. Review of 180 cases with recent MRI and PET correlations. *Adv. Neurol.* **57**: 499–513.

Savic, I., Blomqvist, G., Halldin, C., Litton, J. E., and Gulyas, B. (1998). Regional increased in [^{11}C]flumazenil binding after epilepsy surgery. *Acta Neurol. Scand.* **97**: 279–286.

Savic, I., Thorell, J. O., and Roland, P. (1995). [^{11}C]Flumazenil positron emission tomography visualizes frontal epileptogenic regions. *Epilepsia* **36**: 1225–1232.

Stokking, R., Zuiderveld, H., Hulshoff-Pol, H., and Viergever, M. (1994). Integrated visualization of SPECT and MR images for frontal lobe damaged regions. *In "Visualization in Biomedical Computing"* (E. Robb, Ed.), pp. 282–290. SPIE Press, Bellingham, WA.

Uematsu, S., Lesser, R., and Fisher, R. (1990). Resection of the epileptogenic area in critical cortex with the aid of a subdural electrode grid. *Stereotact. Funct. Neurosurg.* **54**: 34–45.

von Stockhausen, H. M., Pietrzyk, U., and Herholz, K. (1998). "3D-Tool"— A software system for visualization and analysis of coregistered multi-modality volume datasets of individual subjects. *NeuroImage* **7**: S799. [abstract]

von Stockhausen, H. M., Thiel, A., Herholz, K., and Pietrzyk, U. (1997). A convenient method for topographical localization of intracranial electrodes with MRI and a conventional radiograph. *NeuroImage* **5**: S514. [abstract]

Benzodiazepine Receptor Binding in Japanese Subtype of Hereditary Spastic Paraplegia with a Thin Corpus Callosum

MASAYUKI UEDA, KIYOSHI IWABUCHI,* MASAHIRO MISHINA, TATSUSHI KAMIYA, HIDEKI NAGATOMO, MICHIO SENDA,† AKIRO TERASHI, and YASUO KATAYAMA

Division of Neurology, Second Department of Internal Medicine, Nippon Medical School
*Department of Neurology and Psychiatry, Kanagawa Rehabilitation Center
†Positron Medical Center, Tokyo Metropolitan Institute of Gerontology

We performed [¹¹C]flumazenil positron emission tomography study in four patients suffering from autosomal recessive hereditary spastic paraplegia with a thin corpus callosum, and in seven normal volunteers. Benzodiazepine receptor (BZR) binding was reduced in the thalamus in all patients, and in the medial frontal cortex in two patients. In contrast, the BZR binding was increased in the cerebellar hemisphere in two patients and also in the putamen in two patients. Cortical BZR binding was relatively preserved except for the medial frontal cortex. The findings indicate a central GABAergic involvement in the disorder and suggest possibility of pharmacological interventions focused on the central BZRs as a treatment for the disorder.

I. INTRODUCTION

Autosomal recessive hereditary spastic paraplegia (HSP) with a thin corpus callosum is a rare neurodegenerative disorder characterized by slowly progressive spastic paraparesis and mental impairment developing in the second decade in addition to an extremely thin corpus callosum (Iwabuchi *et al.*, 1990, 1994; Ueda *et al.*, 1998). The disorder is relatively accumulated in the Japanese population, and may be essentially Japanese subtype of HSP. We have previously reported two siblings with the disorder showing severely decreased thalamic metabolism on positron emission tomography (PET) using [¹⁸F]fluoro-2-deoxy-D-glucose ([¹⁸F]FDG) (Ueda *et al.*, 1998). We have also suggested that the thalamic

involvement may be a characteristic finding in addition to the thin corpus callosum, and that the involvement is a possible cause of dementia, in the disorder (Ueda *et al.*, 1998). Recent studies have demonstrated abnormal central benzodiazepine receptor (BZR) binding in various neurodegenerative disorders with [¹¹C]flumazenil ([¹¹C]FMZ) PET and have provided important information on the degenerative process in such disorders (Holthoff *et al.*, 1993). However, the receptor binding has not been investigated in the rare Japanese HSP. The purpose of the present study was to determine central BZR binding using [¹¹C]FMZ-PET in the HSP.

II. SUBJECTS AND METHODS

A. Patients

Patients were four men ages 26, 28, 26, and 31 years (Patients 1, 2, 3, and 4, respectively) from two separate families, who shared the following clinical features (Ueda *et al.*, 1998); (1) normal motor development, (2) slowly progressive spastic paraplegia and dementia developing from the late first or early second decade, (3) probable autosomal recessive inheritance, (4) extremely thin corpus callosum on MRI (Fig. 1), and (5) thalamic hypometabolism on [¹⁸F]FDG-PET (Fig. 2). Precisely, it has not been established whether the disorder is genetically homogeneous and whether all patients with the disorder have thalamic involvement. Thus, we included only HSP patients showing

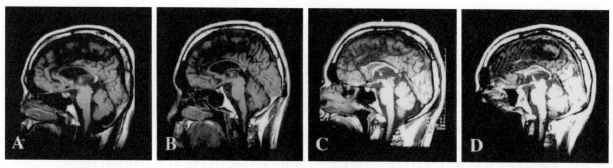

FIGURE 1. Sagittal T1-weighted magnetic resonance images (MRI) in four patients with hereditary spastic paraplegia with a thin corpus callosum. MRI showed extremely thin corpus callosum in all patients. (A) Patient 1, (B) patient 2, (C) patient 3, (D) patient 4.

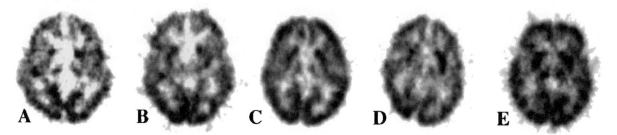

FIGURE 2. Positron emission tomographic (PET) images using [^{18}F]fluoro-2-deoxy-D-glucose in four patients with hereditary spastic paraplegia with a thin corpus callosum and in a normal volunteer. PET images revealed thalamic hypometabolism in all patients compared with that seen normal volunteer. (A) Patient 1, (B) patient 2, (C) patient 3, (D) patient 4, (E) normal volunteer.

both a thin corpus callosum and a thalamic hypometabolism in the present study, to avoid possible heterogeneity of the subjects. Two siblings (Patients 1 and 2) were included in our previous [^{18}F]FDG-PET study (Ueda *et al.*, 1998), and the other two siblings (Patients 3 and 4) in the previous communication (Iwabuchi *et al.*, 1994) were subsequently evaluated with [^{18}F]FDG-PET to confirm the thalamic hypometabolism and are now included in the present study. Table 1 shows their clinical characteristics. Age at symptom onset of Patients 1, 2, 3, and 4 were 12, 12, 9, and 15 years, respectively. Neither cerebellar signs nor sensory disturbances were remarkable in all patients. They were negative for anti-human lymphotropic virus type-1 (HTLV-1) antibody, and their leukocyte lysozomal enzyme activities and plasma very long chain fatty acid levels were within normal limits. Beside the extremely thin corpus callosum, their MRI revealed bilateral medial frontal atrophy. No abnormal MRI findings were observed in them in either the cerebellum or the brainstem.

TABLE 1. Clinical Characteristics in Four Patients with Hereditary Spastic Paraplegia with a Thin Corpus Callosum

	Patient 1	Patient 2	Patient 3	Patient 4
Age at exam (years)	26	28	26	31
Sex	Male	Male	Male	Male
Family	I	I	II	II
Consanguinity	None	None	+	+
Abnormality in parents	None	None	None	None
Age at onset (years)	12	12	9	15
Duration (years)	14	16	17	16
Spastic paraplegia	+	+	+	+
Mental impairment	+	+	+	+
Cerebellar sign	None	None	None	None
Extrapyramidal sign	None	None	None	None
Sensory impairment	None	None	None	None
Corpus callosum (MRI)	Thin	Thin	Thin	Thin
Thalamus ([^{18}F]FDG-PET)	Hypo	Hypo	Hypo	Hypo

B. Methods

The subjects received an intravenous bolus injection of [^{11}C]FMZ (175–537 MBq; sp. act., 22.9 ± 11.6 GBq/μmol) following an 8-min transmission scan. A static emission scan was subsequently performed with HEADTOME IV tomograph (Shimadzu, Co., Kyoto, Japan) for 20 min from 20 min after injection. This system simultaneously acquired 14 parallel slices with a center-to-center interslice distance

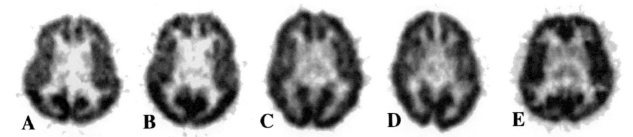

FIGURE 3. Positron emission tomographic images using [^{11}C]flumazenil in four patients with hereditary spastic paraplegia with a thin corpus callosum and in a normal volunteer. thalamic ligand uptake appeared to be reduced in all patients compared with normal volunteer. (A) Patient 1, (B) patient 2, (C) patient 3, (D) patient 4, (E) normal volunteer.

of 6.5 mm, and the images were reconstructed with in-plane resolution of 7.5 mm full-width at half-maximum (FWHM) and Z-axis resolution of 9.5 mm FWHM. The retention of [^{11}C]FMZ at 20–40 min postinjection is reported to be strongly correlated with the distribution volume (DV = K_1/k_2) of [^{11}C]FMZ obtained from the kinetic analysis in a dynamic scan and therefore represents relative BZR binding capacity (Toyama *et. al.*, 1997). Seven normal healthy volunteers with similar age (mean age, 21.7 years) were also evaluated, and data from them were used for normal control values. All the patients and the control subjects gave written informed consent to the PET scan. At the PET study, none of the patients or control subjects were taking medications which are known to influence BZR binding. MRI was performed using MAGNEX 1.5 Tesla machine (Shimadzu Co., Kyoto, Japan) in Patients 1 and 2, Magnetom Vision Plus 1.5 Tesla machine (Siemens AG., München, Germany) in Patients 3 and 4, and SIGNA 1.5 Tesla machine (General Electric, Inc., WI) in healthy volunteers.

Image manipulations were carried out on an Indy workstation (Silicon Graphics Inc., CA), using a medical image processing software Dr. View (Asahi Kasei Joho System Co. Ltd., Tokyo, Japan) and Automatic Medical Image Registration (AMIR) software version 3.1 (Research Institute for Brain and Blood Vessels, Akita, Japan). The PET images were three-dimensionally registered to MRI of each subject to provide anatomical information for placing regions of interest (ROIs). Circular ROIs of 5 mm diameter were manually drawn over 10 regions in each hemisphere on the MRI in the same way as our previous PET study (Ueda *et al.*, 1998) and applied automatically to the registered PET images. Values from structures identified in both hemispheres were averaged for each subject. The [^{11}C]FMZ uptake in each ROI was normalized by the value in the occipital cortex, which is involved neither pathologically nor metabolically in the disorder, according to the previous [^{18}F]FDG-PET study (Ueda *et al.*, 1998). The value was considered abnormal if more than two standard deviations above or below the control mean.

III. RESULTS AND DISCUSSION

In visual inspection of the PET images, thalamic and medial frontal [^{11}C]FMZ uptake appeared to be decreased in all patients compared with that seen in normal volunteers (Fig. 3). Table 2 lists the [^{11}C]FMZ uptake, reflecting the regional BZR binding, relative to the occipital cortex in these patients and in controls. The uptake values were decreased in both the anterior and the posterior thalamus in all patients, in the medial frontal cortex in two patients, and in the parietal cortex in one patient. In addition, the uptake values were increased in the cerebellar hemisphere in two patients and also in the putamen in two patients. In contrast, the temporal, lateral frontal cortices and caudate demonstrated normal levels.

Gamma-aminobutyric acid (GABA) is the most widely distributed inhibitory neurotransmitter in the central nervous system (Matsumoto, 1989). GABA receptors contain several binding sites for pharmacologically specific agents such as BZR (Matsumoto, 1989). Since these sites are coupled to one another (Matsumoto, 1989), *in vivo* measurement of BZR binding may yield important information on synaptic function, neuronal integrity, and distinction of drug-action sites in the brain. [^{11}C]FMZ, a BZR ligand which can be used with PET, makes it possible to visualize central BZR/GABA receptors and to measure the density of them *in vivo* since [^{11}C]FMZ DV is linearly related to the density of available BZR sites (Frey *et al.*, 1991). Alterations in central BZR binding on PET may result from a change in neuronal density per volume of cortex, or in BZR density per neuron.

Quantitative [^{11}C]FMZ-PET is usually carried out using a kinetic model with a dynamic scan, and the [^{11}C]FMZ DV is estimated in the model. However a dynamic scan requires subjects to keep stable over 1.5 h with their head fixed in a PET machine as well as to cannulate their radial artery for arterial blood sampling. Because of mental impairment in the present patients, it might be difficult to perform a dynamic scan of them. Consequently, we analyzed them with a static scan and estimated regional abnormalities in central BZR binding using relative [^{11}C]FMZ uptake to the occipital

TABLE 2.   Regional Benzodiazepine Receptor Binding Assessed with Relative Uptake of [^{11}C]Flumazenil (% of Occipital Cortex) in Four Patients of Hereditary Spastic Paraplegia with a Thin Corpus Callosum, and in Seven Normal Controls

| | Family-I | | Family-II | | |
| | Patient 1 | Patient 2 | Patient 3 | Patient 4 | Controls ($N = 7$) (mean $\pm$ SD) |
Age (years):	26	28	26	31	21.7 $\pm$ 0.8
Cerebeller hemisphere	80.4†	90.1†	69.9	73.7	66.0 $\pm$ 4.5
Caudate	55.0	55.1	42.4	59.8	50.0 $\pm$ 5.3
Putamen	60.7	72.6†	52.1	66.9†	53.6 $\pm$ 4.2
Anterior thalamus	39.3*	31.9*	42.2*	40.6*	54.3 $\pm$ 5.0
Posterior thalamus	39.5*	43.9*	36.7*	38.4*	51.9 $\pm$ 3.1
Temporal cortex	96.7	103.9	94.7	95.3	98.6 $\pm$ 2.8
Medial frontal cortex	86.4*	93.5	90.2	79.9*	99.4 $\pm$ 5.6
Lateral frontal cortex	88.5	98.9	91.4	91.3	96.4 $\pm$ 7.2
Parietal cortex	90.3	94.5	88.3*	103.0	97.3 $\pm$ 3.9

*Below mean -2 SD of control mean.
†Above mean $+2$ SD of control mean.

cortex. Since the occipital cortex is pathologically (Iwabuchi *et al.*, 1990; Ferrer *et al.*, 1995) and metabolically (Ueda *et al.*, 1998) normal in the disorder, the relative uptake values may be acceptable for semiquantification.

The present study indicated a decrease in thalamic BZR binding in all patients, an increase in cerebellar and putaminal BZR binding in two patients, and almost normal cortical BZR binding except for the medial frontal cortex. Previous postmortem studies have shown severe thalamic degeneration in the disorder (Iwabuchi *et al.*, 1990; Ferrer *et al.*, 1995). In addition, an immunohistochemical study showed decreased calbindin D28K immunoreactive neurons in the disorder, suggesting probable involvement in GABAergic system (Ferrer *et al.*, 1995). Based on these neuropathologic findings, the decreases in thalamic BZR binding probably result from degeneration. In addition, our previous PET study have demonstrated medial frontal hypometabolism of glucose in the disorder (Ueda *et al.*, 1998). The present data, combined with our previous report, suggest that the medial frontal hypometabolism is caused by not only remote effects of thalamic involvement but also by medial frontal neuronal loss. Furthermore, a pathologic study has revealed occasional neuronal migration disorder such as ectopic neurons in the disorder, indicating maldevelopment of the central nervous system (Iwabuchi *et al.*, 1990). Neuronal migration disorders reportedly demonstrate various abnormal BZR binding on [^{11}C]FMZ-PET, consisting of both increases and decreases, in the affected cortex (Richardson *et al.*, 1996), although they do not necessarily demonstrate morphologic abnormalities on MRI (Chugani *et al.*, 1990). It remains uncertain precisely why the cerebellar and putaminal BZR binding was increased in the present patients, although one possible explanation of the increases is that the alteration may reflect disorders in neuronal migration or receptor upregula-

tion. However it is not possible to distinguish this possibility using this PET method, nothing but postmortem quantitative analysis of neuronal and BZR densities in the disorder can resolve this question.

These findings indicate central GABAergic involvement in the disorder. Furthermore, the finding of relatively preserved cortical BZR binding except for the medial frontal cortex suggests the possibility of pharmacological interventions focused on the central BZRs as a treatment for the disorder.

References

Chugani, H. T., Shields, W. D., Shewmon, D. A., Olson, D. M., Phelpes, M. E., and Peacock, W. J. (1990). Infantile spasms. I. PET identifies focal cortical dysgenesis in cryptogenic cases for surgical treatment. *Ann. Neurol.* **27**: 406–413.

Ferrer, I., Olivé, M., Rivera, R., Narberhaus, P. B., and Ugarte, A. (1995). Hereditary spastic paraparesis with dementia, amyotrophy and peripheral neuropathy. A neuropathological study. *Neuropathol. Appl. Neurobiol.* **21**: 255–261.

Frey, K. A., Holthoff, V. A., Koeppe, R. A., Jewett, D. M., Kilbourn, M. R., and Kuhl, D. E. (1991). Parametric in vivo imaging of benzodiazepine receptor distribution in human brain. *Ann. Neurol.* **30**: 663–672.

Holthoff, V. A., Koeppe, R. A., Frey, K. A., Penney, J. B., Markel, D. S., Kuhl, D. E., and Young, A. B. (1993). Positron emission tomography measures of benzodiazepine receptors in Huntington's disease. *Ann. Neurol.* **34**: 76–81.

Iwabuchi, K., Kubota, Y., Hanihara, T., and Nagatomo, H. (1994). Three patients of complicated form of autosomal recessive hereditary spastic paraplegia associated with hypoplasia of the corpus callosum. *No to Shinkei* **46**: 941–947.

Iwabuchi, K., Yagishita, S., Amano, N., Yokoi, S., Honda, H., Tanabe, T., Kinoshita, J., and Kosaka, K. (1990). An autopsy case of complicated form of spastic paraplegia with amyotrophy, mental deficiency, sensory impairment, and parkinsonism. *No to Shinkei* **42**: 1075–1083.

Matsumoto, R. R. (1989). GABA receptors: are cellular differences reflected in function? *Brain. Res. Rev.* **14**: 203–225.

Richardson, M. P., Koeppe, M. J., Brooks, D. J., Fish, D. R., and Duncan, J. S. (1996). Benzodiazepine receptors in focal epilepsy with cortical dysgenesis: An ^{11}C-flumazenil PET study. *Ann. Neurol.* **40**: 188–198.

Toyama, H., Yamada, T., Uemura, K., Kimura, Y., Ishii, K., Ohyama, M., Mishina, M., and Senda, M. (1997). Quantitative mapping of benzodiazepine receptor concentration in a static PET study. *NeuroImage* **5** (Suppl. 3): A34.

Ueda, M., Katayama, Y., Kamiya, T., Mishina, M., Igarashi, H., Okubo, S., Senda, M., Iwabuchi, K., and Terashi, A. (1998). Hereditary spastic paraplegia with a thin corpus callosum and thalamic involvement in Japan. *Neurology* **51**: 1751–1754.

PHARMACOLOGY AND EXPERIMENTAL TOMOGRAPHY: THE FUTURE

58

A Panel Discussion on the Future of Pharmacology and Experimental Tomography[1]

GITTE MOOS KNUDSEN, Copenhagen, Denmark: This slide (Fig. 1) shows the development in the number of drugs annually marketed in Denmark during the past 20 years. An increasing number of new drugs are being marketed every year, at the moment about 20 new drugs yearly, but these represent only a minor fraction of all drugs being developed. Thousands of compounds are under consideration before they eventually reach the market and this necessitates a strategy for selection of drugs. The strategy for selection of drugs is an approach that may be useful also for tracer ligands.

How come there are so many new drugs being developed every year? This is partially be cause of the development of new methodologies such as

- combinatorial chemistry
- high throughput screening assays
- cloned human receptor lines
- identification of novel drug targets through genomics research.

We may adapt the strategies employed for the selection of drugs. Conversely, PET can also assist in drug developement by, for example,

- demonstrating penetration into effect compartment
- quantitating therapeutic action (occupancy, K_d)
- defining kinetics in effect compartment
- identifying pharmacodynamic effects
- defining the dose ranges to be used in clinical efficacy trials,

thereby both speeding up the time to market and reducing costs considerably.

In order to address many of the questions raised by the pharmaceutical industry, it is clear that more suitable and specific tracers for more systems are required. So, the question is, how can we facilitate the development of new tracers? This could be done by

- interaction at an early stage between users, modelers, and radiochemists
- new strategies for improvement of BBB (blood–brain barrier) passage
- new strategies to prevent metabolism
- improved exchange of information

---

[1]Chairman: GITTE MOOS KNUDSEN; panelists: Antony Gee, Adriaan Lammertsma, Robert Koeppe, Michio Senda.

- establishment of better criteria for ligand selection
- experimental support for selection of ligands
- harmonization of legislation regarding approval of radiolabeled pharmaceuticals.

Better criteria for ligand selection instead of a—more or less—trial-and-error procedure every time a new ligand is tested would be extremely useful.

Finally, harmonization of legislation regarding the approval of radiolabeled ligands would be appropriate. As it is now, every single PET center has to go through the efforts of collecting information with regard to toxicology, dosimetry, etc. This is an enormously time-consuming procedure with a little impact. Maybe it should be considered to centralize, and even commercialize, this part of the work. Within individual countries, you have to get radiolabeled ligands approved for each single tracer, which is inefficient and time consuming. A harmonization of this aspect would be welcomed.

TONY GEE, Cambridge, UK: I think that was a good presentation, highlighting some important aspects of PET ligand and drug development. Actually, collaboration between PET centers and the pharmaceutical industry has been evident for quite a number of years, although the extent of the collaboration may be debated. For example, the PET community relies to a certain extent on industry making available novel, selective and potent receptor ligands and their labeling precursors, in addition to the toxicology data required for tracer administration to man. Where would the PET community be without radioligands like Raclopride, WAY 100,635, MDL 100, 907, SCH 23390, etc.

However, often, the compounds that the PET community get to know about from pharmaceutical companies aren't always the most useful for producing PET ligands. You've got to remember that industry is producing hundreds of compounds for each particular molecular target, and maybe the ones that the PET community gets access to in the end are only the few that have been proposed as potential therapeutics. As you know, very few of the PET ligands that we use for *in vivo* imaging have succeeded as drugs, and vice versa. It seems likely that the physiochemical properties of therapeutic drugs may not always have the desirable properties we require for a successful PET imaging. That is, the physiochemical properties of a

PET ligand and the physiochemical properties of a therapeutic drug are not necessarily the same.

If we could open up the channels between the PET community and pharmaceutical companies, it might make available the libraries of compounds that no one outside the company ever sees. This would enable the screening of large numbers of compounds, in a manner similar to that which was presented earlier, to select compounds that do have physiochemical properties suitable for PET ligands. Achieving this access to compound libraries, however, may be a difficult task, as this would require the company to release a wealth of potentially sensitive information to the external centers.

I want to address another point now, which is that a lot of drug discovery is done *in vitro*. One of the big strengths of the PET and SPECT techniques is that you actually have the opportunity to look at how novel compounds work in living man. I am sure that most pharmaceutical industries should take this as a golden opportunity, as we are all aware that many *in vivo* effects and phenomena don't necessarily occur *in vitro*. And it's not really surprising: *In vitro* assays utilizing a slice of tissue incubated at 25°C are not necessarily influenced by, for example, second messenger systems, positive cooperativity, or interneuronal communication. However, all of these effects have the potential to influence the action of a drug *in vivo*. By understanding how the brain works *in vivo* we have to start developing new models of, for example, our understanding of not only receptor binding, but also how receptor binding influences metabolism, enzyme kinetics, second

messengers, and interneuronal communication and how they work in concert. We should maybe stop taking the reductionist point of view of looking at one receptor in isolation and possibly out of context. Take the example of clozapine. Now why is that such a good drug? It is certainly very "dirty," acting at many receptors, but it is a superb drug. So, this understanding of balance and how things work as a whole, I believe, can be effectively addressed only *in vivo*. The pharmaceutical industry cannot possibly get all the information it requires for drug development using *in vitro* screening methods. And *in vivo* studies in nonhuman species are often misleading. *In vivo* PET in humans is a golden opportunity for the drug industry.

ADRIAAN LAMMERTSMA, Amsterdam, Holland: Basically, I will raise a few points similar to those mentioned by Gitte, but from a different perspective. Instead of concentrating on how we can improve what we are doing, I will discuss what we presently can do for drug development.

In my opinion, there are three different types of studies that can be performed. One is to label a drug and measure its uptake to see if this is sufficient and to check whether it gets in the right compartment or at the targeted site of action. This is simple from a theoretical point of view, but in practice it is the most difficult type of study because of the radiochemistry involved, the presence of *in vivo* metabolism, modeling issues, etc. Consequently, as an alternative, in the past there have been studies of another type, looking at a function, such as flow or metabolism, before and after administration of a

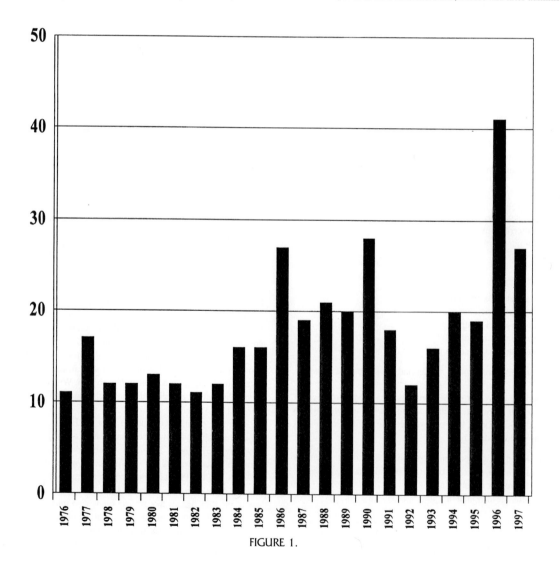

FIGURE 1.

drug. Probably the most exciting type of study is to look at receptor occupancy or enzyme activity for drugs that are directly active at that particular site. In the Symposium last Tuesday, I showed two examples of measuring dose–response curves. In this way, only a few subjects are needed to give a good idea of the optimal dose. In addition, the biological half-life can be measured. That would help in deciding how often a day the drug should be given. If this procedure doesn't work and you arrive at a dose that appears to be completely incorrect in clinical trials, then actually even more information is obtained. In that case, the theory behind the development of the drug is incorrect, and you have to think again.

The remaining points are basically some specific issues that I would like to bring up as discussion points. I think that an important advantage of PET for animal studies is a more efficient use of animal models of disease, especially if the animal model is expensive, and the same animal needs to be used again. It also leads to less variability, because there is no interanimal variability. There is the gene imaging aspect, which is coming up as an exciting new development, and perhaps Ron Blasberg can comment on that later on. Tony already mentioned the potential for misinterpretation in the extrapolation from *in vitro* to *in vivo* studies. At this meeting we have seen a couple of examples with unexpected findings in the *in vivo* situation. With PET we have a much more complex system with interaction of receptors systems, and if you look *in vitro* at one separate receptor you can arrive at the wrong conclusion. The same is actually true for *in vivo* animal models, because of possible species differences. The strength of clinical PET is that we can go directly to human *in vivo* studies and monitor what is going on. Also, it gives a possibility to discriminate between differential effects at a regional level. For example we saw an example from the Hammersmith group, where therapy seemed to have a different effect on WAY100635 binding to pre- and postsynaptic receptors. In the long run, we have a chance to provide individualized therapy, i.e., to test whether a compound will work in an individual patient and, if it doesn't, to change

to something else. This approach would also be useful in clinical trials, because you could create more homogenous patient groups. Often the responders versus nonresponders problem obscures the results from trials.

ROBERT KOEPPE, Ann Arbor, MI, USA: One of the things that we have heard several times is that we ought to be able to get more information out of our data. One way that we are just starting to do this, is by looking at two tracers almost at the same time. If we want to look at more than one neuroreceptor system at a time or want to look at interactions between two different systems, two tracers can be injected 10–20 min apart and time–activity curves for both tracers can be imaged in a single scanning session. This overhead (Fig. 2) shows a two-tracer study with [^{11}C]PMP, which is an acetylcholinesterase enzyme marker, injected first and then [^{11}C]flumazenil (FMZ) injected 20 min later. Of course, in this type of approach, as opposed to equilibrium studies, you have to do a lot more metabolite analysis. This next figure (Fig. 3) is an example of blood curves obtained for a dual-tracer study when FMZ is injected first, where the open circles are what is actually measured. The peaks of these curves are much higher of course, for the total [^{11}C] activity in plasma than the authentic [^{11}C] activity. By performing a couple of different separations, you can distinguish metabolites from both authentic tracers as are shown by the filled circles. Then by analyzing at least several points throughout the later time period in order to separate the two authentic tracers, you can get total PMP, total flumazenil, and both authentic curves, which are shown as the solid curves. We can then take both of these input functions, perform a single kinetic analysis, looking at both tracers with parallel compartment models, and, in one scan, estimate parameters from two different ligands at once. Thus, in a single scan session, when you normally would be doing one scan for each of two tracers, you could perform both a baseline and an intervention study, measuring each parameter for the two tracers twice.

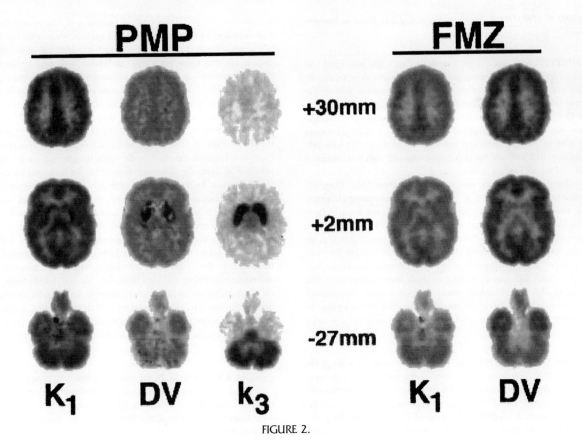

FIGURE 2.

FMZ:PMP Arterial Plasma Input Functions

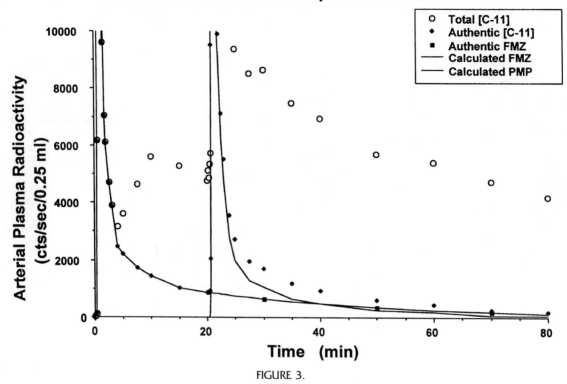

FIGURE 3.

STEPHEN STROTHER, Minneapolis, MN, USA: It seems like you are estimating an awful lot of parameters.

R. KOEPPE: Well, that's a good question, but when you inject two tracers separated somewhat in time, you have substantially more kinetic information to estimate from. It is not that you are estimating four or five parameters from a single time–activity curve in which you perturb a system and then watch it reach equilibrium. You have a second perturbation, so people may be pleasantly surprised that there really is quite a bit more "kinetic content" to these studies, if you want to call it that. We have performed studies with different tracer combinations, and with both orders of injection, and we are looking at these data now.

This slide (Fig. 2) shows one such study with PMP and FMZ. Three different brain levels are shown. K_1, K_1/k_2, and k_3 parameters are calculated for PMP, while K_1 and total distribution volume parameters are calculated for FMZ. These are pixel-by-pixel estimates using two input functions, a single kinetic model estimating all five parameters plus blood volume at the same time, and the quality is very good! These studies yielded the same relative transport rates (PMP about twice FMZ), the same distribution volume for flumazenil and same k_3 for PMP as single-tracer studies show. The image quality also is very similar, almost as good as doing single-tracer studies. So in conclusion, this is one possible approach in which we may be able to look at interactions among different neurotransmitter systems much more easily.

MICHIO SENDA, Tokyo, Japan: I am not developing new drugs but I am collaborating with basic neurosciencts and radiochemists to apply the PET technology to pharmacology. We are doing simultaneous measurements of microdialysis and the PET imaging in the rat. With the microdialysis technique, a probe is inserted into the brain of a living animal. The probe is perfused, and we collect what we believe is extracellular fluid, in which

the concentration of a compound can be measured. The microdialysis technique was applied within the animal PET camera. We have studied the effect of cholecystokinin (CCK). CCK was first demonstrated in the alimentary tract, as you may know, but it also functions as brain neurotransmitter with its receptors closely related to the dopaminergic system. In this experiment we measured the concentration of dopamine in the dialysate. When CCK was added to the perfusate, we found an increase in dopamine representing the endogenous release of dopamine. When a simultaneous PET measurement was done, a decrease in the binding of [^{11}C]raclopride was found on the perfused side. By combining these two independent techniques, each of them would validate the result by the other, since we are not yet completely sure how accurate the dopamine release can be measured by changes in [^{11}C]raclopride binding, as there have been heated discussion about this during this meeting. On the other hand, one may not be sure whether the microdialysate represents the extracellular space fluid. So, combining these techniques will strengthen the results.

For the application of PET technology to pharmacology, I have here summarized what challenges we are facing. First of all, we need good labeled ligands. The ligand has to have a high brain uptake, it must have rapid and reversible binding and high selectivity. Second, I don't know why, but the pharmacologists and basic neuroscientists love rats, so we need a PET camera that has a resolution good enough to image small animals. In the third place, we also have to tune the data analysis to apply to small animals, with appropriate partial volume correction and quantification without blood sampling.

G. M. KNUDSEN: Well, a number of issues have been raised here.

Antero, you presented earlier on today data on some criteria that may be suitable for selection of ligands. How can we do a better job in selecting these ligands? Do you think that the methodology that you are employing could be extended and used?

A. ABRUNHOSA: I think what I showed today were some preliminary results where we can correlate tracer *in vivo* behavior with some basic properties. I think if it increases our understanding of which specific physicochemical properties are important for each one of the *in vivo* parameters, that would be quite good. But first you need to define what you are looking for. So you probably have to ask the modelers which specific parameters will be crucial for the *in vivo* behavior of the molecules.

G. M. KNUDSEN: Modelers: which is the gold standard for a good ligand? I would like to hear, what is the gold standard if the target-to-background isn't good enough as criterium, how do we select it then?

A. ABRUNHOSA: I am afraid I don't have the answer. I tried to address what actually is affecting specific to nonspecific binding because that seems to be the problem in many specific target systems, but obviously there are more complex issues than that.

TERRY JONES: We have seen some examples where PET can be used for drug development and there are others. However, it is clear to me that we are skirting around the real issue. This is the need to forge long-term partnerships with drug developers and the pharmaceutical industry in particular. Unless this happens, I am absolutely convinced that the field of PET will not realize its full potential for clinical research and for drug development. At this forum, we are in effect talking to ourselves since there are present relatively few pharmacologists and representatives from industry.

We need scientific partnerships with the pharmaceutical industries to the extent that when they initiate a drug discovery program, with a focus on a specific therapeutic target, they also initiate, in parallel, a radioligand or radiotracer discovery programme for imaging. This they will do once they accept the need to use imaging to provide proofs of principle to show that a therapeutic target is differentially expressed in diseased tissue, that it can be modulated with the drug being developed, and that the drug has the desired pharmacodynamic effect. A commitment to developing imaging probes to a level where medicinal chemists and drug discoverers search for and develop ligands and tracers, using the criteria we PET scientists are able to define, would really advance what our field has to offer for drug development. In effect, we need the drug industry to share the risk in radioligand and tracer discovery alongside the contributions from the PET centers in radiochemistry, scanner technology, biomathematical modeling, and clinical applications. Unless such partnerships happen, this field is going to wither. As Antero Abrunhosa showed, the risk in discovering suitable ligands is comparable to that for finding new drugs and yet his own IT work illustrates where we could begin to link with industry and mine its libraries of molecules for potential imaging probes. Another advantage gained from involvement with medicinal chemists is that precursor compounds would be more readily available for radiolabelling, thereby speeding up the radiochemistry development of a new tracer.

I also believe that small animal PET scanning would be a positive means for gaining commitment from the drug developers. The main attraction here would not, in the first instance, be to save on the numbers of animals sacrificed but the means to study, over time, pharmacokinetic and pharmacodynamic changes in a nonstandard animal disease model. It would also help them to develop scanning paradigms which eventually will evolve for use in human studies. It would be a way of exposing drug discoverers and developers to *in vivo* imaging and the restrictions associated with the interpretation of image data. Here they do not appreciate the restrictions in our field and there is a culture gap which needs to be crossed. Industrial partnerships will also bring with them the pace of industrial development which is often lacking in academia due to nonmobility of funds at a relatively short notice. Finally, industrial involvement would add pharmacological expertise to the field of PET, thereby helping to explore new experimental paradigms and in the interpretation of data produced for both normal and diseased tissue.

T. GEE: Yes, I think that you are right in a way, Terry, but we shouldn't really develop the worlds best drugs for rats. Okay, what works in rats doesn't always work in humans, so of course, if you can evaluate drugs in humans, then that's better. You optimize your chances of getting a good drug. Where I think I disagree with you is when you say, that the PET community understands what the signal is doing. Do we really know what our tracers are telling us?

T. JONES: We have to get to the point of pharmacological understanding of the effect of nonspecific binding and hence of appreciating the problems of interpreting signals from *in vivo* recordings.

T. GEE: Yes, but the PET community still doesn't fully understand why two tracers targeting the the same binding site behave so differently. For, example, the debate about raclopride and NMSP has been going on for years, and we still don't fully understand the differences. Now, a part of this whole idea of industry—PET collaboration is not only to develop drugs but also to develop our understanding of what the tracers are actually telling us.

A. LAMMERTSMA: Terry, you made a long plea for partnership from a PET angle. Perhaps we can have a comment from the pharmaceutical industry. Don, do you want to comment on these views from Terry?

G. M. KNUDSEN: How many in here are from the pharmaceutical industry? Can you raise your hands?
—About five people.

DONALD BURNS, Merck: I think it is difficult to start off with a major partnership with a drug company. I think one of the problems you have is the thought that you don't understand everything leaves the impression that you don't understand anything. You actually do understand quite a bit, and what you do understand can be useful to drug industry. So what you need to do is match up where you can answer in nice clear valid way with questions that are important to drug companies, and I think that you can actually do that. But you do need to get that kind of message to people in drug companies, who for the large part aren't here. And I think that if they were here, they would be scared away by all the debate about what it is that you don't understand and would decide to come back in about 10 years when you understand more.

M. SENDA: May I ask if there are any pharmaceutical companies that have invested a big sum of money on building a PET center or maintaining PET centers in countries other than Japan?

D. BURNS: I am not aware of any that have invested in their own PET center. I have an understanding that there are some close collaborations between some academic centers and drug companies, but I don't really know the details of it. We have invested a lot of money in PET in the past few years.

M. SENDA: In Japan there are two PET centers on which the pharmaceutical companies invested a large sum. They built the centers and maintain the centers.

From the audience: I would like to add a comment. A graduate student friend of mine who now works for Schering-Plough has told me that Schering-Plough has invested with a number of other pharmaceutical companies in Northern New Jersey to develop a PET center with whole-body imaging of animals so that they instead of doing ^{14}C labeling of drug distribution in kinetics, they will do PET labeling in order to view the whole-body drug distribution, so that they can do the whole body drug distribution in larger animals other than just rats hopefully, the dog and the pig being more physiologically similar to humans. But they are not emphasizing brain, they are looking at whole body and tissue distribution of these compounds for toxicology purposes.

T. GEE: We should think with a bit more vision about how we develop drugs. I mean, the PET community has put a lot of effort in recent years into probing interneuronal communication, how can we benefit from this? Is anyone here familiar with Steve Dewey and the Brookhaven group's work on γ vinyl GABA? The drug indirectly modulates dopaminergic tone and [^{11}C]raclopride binding via the GABAergic system. What are your thoughts about using that technique for designing drugs that act indirectly at a target neuron? There must be more creative ways of looking at targeting drugs. Now, can you design drugs that indirectly act at your target site? Conceivably, if there is interneuronal communication in the brain, you could perturb your GABAergic system, for example, and get a change in dopaminergic function. Couldn't we possibly use this to indirectly tune the action of a drug in the body?

T. JONES: We could put more emphasis on pharmacodynamic studies, matching the receptor study. Is that what you are saying?

T. GEE: No. Lets say you have got a compound X binding at the receptor Y. Now, you might be able to increase it's efficacy by combining with another drug Z, that indirectly modulates the action of drug X at receptor Y. This is something that you can't possibly predict *in vitro*. This is a challenging field, and it's going to take an enormous amount of intellectual weight and partnerships to develop this field.

G. M. KNUDSEN: Maybe we could get back to the question regarding development of tracers. I don't think the pharmaceutical industry really will get something out of working with PET until we have better ligands and for more systems. So maybe I can ask the modelers: it has been stated during the past days that we need some more interaction between the modelers and the radiochemists, so what do you require from a good ligand? Could you give some hints to that? And if you want to use a database approach for analyzing how to create a good tracer, what should be the criteria for a successful ligand?

A. LAMMERTSMA: I think, there has been quite a bit of interaction today, because modelers gave most of the radiochemistry talks. Maybe Bob has some feelings about this?

R. KOEPPE: I think, the point is a good one that what makes a good tracer for a pharmaceutical company, is probably not a good tracer for kinetic modeling from an estimation point of view and vice versa. Certainly, for PET tracers labeled with ^{11}C, and some with ^{18}F, and even some iodine-labeled tracers, the concept has gotten lost that if you inject something into the brain and none of it comes back out because it's bound so highly, you can derive absolutely no information about what is going on inside. The only way you can learn about the different stages or steps of the process you are interested in, without doing blocking studies, is to have a substantial portion of the tracer come back out of the brain. That's the whole reason for wanting high-uptake and even high-clearance rate drugs that can then allow a whole distribution of binding levels to be determined. You still have to couple this with challenges or manipulations of the system to be able to prove sensitivity and specificity to changes in the parameter you are trying to quantify. I think we need to get away from the tracers that go in and have say 99%, for example, of what gets into the brain ending up bound to the receptors. In this scenario, the radiolabel is all bound; it is all trapped; it has all been metabolized or hydrolized but **NOT in proportion to anything that we are interested in quantifying.**

G. M. KNUDSEN: Can we say that there is an optimal k_{off} and k_{on} to give a good ligand? Do we know that?

R. KOEPPE: Well, that depends on B_{max}, because it is really the combination of the number of binding sites, the number of enzyme molecules, etc., multiplied by either the on rate or the hydrolysis rate, so that is why it is so difficult to make the prediction of what is going to work. It also depends on the half-life of the tracer. You can afford to have a higher ratio of k_3 over k_4, generally meaning a higher affinity ligand, when you are using a iodine-labeled tracer with SPECT, than you can afford with a ^{11}C-labeled tracer. I think from a kinetic point of view, at least for ^{11}C-labeled tracers, the ratio of the k_3 parameter to the k_2 parameter for irreversible ligands, should be somewhere around 0.5, maybe a bit less for some tracers, maybe a little higher for others, and certainly not higher than 1.0. Above this, you will start to run into flow-limited uptake. For rapidly reversible tracers such as raclopride or flumazenil, I think the optimal range of the k_3 over k_4 ratio depends on the level of nonspecific binding. You would like nonspecific binding to be as low as possible. However, if you are starting with a nonspecific binding level of around 20, like the MCN5652 tracer, then things are different. In general, we would like ratios of maybe around 2 or 3 at the low end, to, at most, about 20 at the high end. When you get in the range of total distribution volumes of around 60 to 100, you will not have good sensitivity to changes in binding. You may have all the tracer specifically bound, yet you won't have any sensitivity to biologically meaningful changes in binding.

RONALD BLASBERG: To follow up on Antero's discussion, we have not talked about designing ligands within families, with subtle alterations in their structure, developing design, the type of ligand to fit the modeling criteria that Bob just illustrated. And I'd like to hear, what Mike's comments are, and the radiochemists, why that really hasn't been done. I mean, we have been in this business for 15 years now, and I have not been aware of a systematic effort of trying to modulate a basic family of ligands to meet the kinetic requirements of PET imaging.

MICHAEL KILBOURN: In a lot of cases there haven't been enough available structural alterations of a parent compound that made enough sense that one could predict changes in kinetics, which is what we are talking about. Bob and I have had this discussion so often so that it gets tiring. We can probably speak for each other. All he wants me to do is change the kinetics of everything that I first make, and with a lot of the drugs that we get, we can't change the molecules enough without losing the affinity for some other reason. There are some examples where people did, for example, with the spiperone molecule. The PET group at St. Louis and number of other people made a lot of derivatives of spiperone, but to tell you the truth, they were all analyzed by striatum–cerebellum ratios. We didn't know better then. I'd like to go back and do all that work over again, trying to establish the kinetics of all those parameters. And maybe then we would have found one that was better than spiperone itself or *N*-methyl-spiperone.

A lot of the radiochemists in the field unfortunately still don't understand the simple concept of flow limitation and I have no excuse for it. And that is where the modelers have to work a little better than their education.

RICHARD CARSON: We have worked with a lot of analogs of the WAY100635 looking for both a high-affinity and a low-affinity ligand, one perhaps for structural and receptor information and one for trying to modulate the serotonin release. And certainly we find that when we look at the rat, we look at the ratios, we look at the kinetics, we look at the blocking; we get hints on what might be good and we take it into primates and humans and we lose a little. And that is the problem, that all we need to lose is sometimes a factor of two. We get a little less lipophilic, a little more protein binding, a little more nonspecific. The nice target-to-background ratio of 3 and 4 is going to 1.5, and there is just not enough signal there. We need a way for us to come up with some generalizations, if that is possible. And when we start with the rat, what are we gonna need in the rat so that we get it to humans.

R. KOEPPE: I wonder if the ratios, as things change within species, will be more reliable. So maybe even with rat/human species differences, if you have done at least one in human study, and you have done an entire series of studies in rats, you can extrapolate the change using ratios better than the change in absolute numbers.

T. GEE: There are some examples of rational design of PET ligands, it is not a totally hopeless situation. One that springs to mind just now is the tropane analogs, for example, where you tend to get irreversible binding, in other words, "high k_3 with low k_4"-like behavior.

M. KILBOURN: I kind of disagree. If you look at the history on the tropane parameters, how many of them made, I mean there is a whole list. And everybody at the different laboratories is making a specific flavor of it, and most of it without really any good rationale that they ever wrote down. Or maybe they just didn't tell the rest of us why they had to have a fluorinated one that had their characteristics. Why did we make 30 or 40 of them?

T. GEE: I think that there are rationale that have been used to do that. If you are looking, for example, at the N-methyl group on CIT or CFT, the idea is to reduce the nitrogen electron density, which reduces the affinity of the compound with the result that you start to see a k_4 component in the binding kinetics. So, there is some rationale behind that, and it is not totally hopeless. But this approach has only been used in designing maybe 3 out of 30 tropane derivatives.

R. BLASBERG: I would disagree a little bit with Mike, at least with the tone of the implication for the future. The pharmaceutical industry is now heavily committed, to structural analysis of enzyme systems that we are trying to target. The crystallographers have defined a three-dimensional structure of many enzyme active sites, and that information is being used in various computer models which suddenly change the design of specific drugs. Some pharmaceutical industry groups are more further advanced in this than others, but this is the type of partnership that Terry is talking about, that this group needs to get into to basically convince the pharmaceutical industry to tap into this resource that they have with the end product, as Terry indicated, a useful ligand for their end purposes as well as for the finding of a specific enzyme receptor system.

So, I really think that we need to get out of ourselves and to tap new resources, new areas of expertise that we haven't discussed in the past 15 years, and it is out there. It just needs to be sought, identified, and set up.

M. KILBOURN: A comment on that, Ron. If you read the pharmacology literature and you read that sort of editorial thing, it is actually kind of schizophrenic. You have a lot of people trying to do things at a genuine level or at a crystallography level and thinking that they are going to do rational drug design. And then you look and see what the drug industry is doing now. Now they are generating combinatorial libraries, and they are getting essentially the equivalent of screening things in a masses and they are fishing, and they have found an efficient way of finding drugs, it is not rational any longer. It's just like saying: let's make everything we can, test it against everything we can, find something that looks interesting and then give it to the medicinal chemist who is back doing what he did before. Let's change this, let's change that and get the properties that we need for a drug. So, it is really kind of interesting. There was a very nice editorial, that the idea of rational drug design did never really pan out to the industry.

T. GEE: There is still a lot of rational design going on in identifying lead compounds. It is the elaboration of the prosthetic groups on a rationally designed lead structure that is often performed on a large scale and seemingly on a "nonrational" basis.

G. M. KNUDSEN: I think we have to close. Thank all of you for your contribution to this panel discussion.

Index

5